An Elementary Approach to
Design and Analysis
of Algorithms

Primers in Electronics and Computer Science

Print ISSN: 2516-6239
Online ISSN: 2516-6247

(*formerly known as ICP Primers in Electronics and Computer Science* — ISSN: 2054-4537)

Series Editor: Mark S. Nixon *(University of Southampton, UK)*

This series fills a gap in the market for concise student-friendly guides to the essentials of electronics and computer science. Each book will address a core elements of first year BEng and MEng courses and will stand out amongst the existing lengthy, dense and expensive US textbooks. The texts will cover the main elements of each topic, with clear illustrations and up-to-date references for further reading.

Published

Vol. 4   *An Elementary Approach to Design and Analysis of Algorithms*
by Lekh Raj Vermani and Shalini Vermani

Vol. 3   *Image Processing and Analysis: A Primer*
by Georgy Gimel'farb and Patrice Delmas

Vol. 2   *Programming: A Primer – Coding for Beginners*
by Tom Bell

Vol. 1   *Digital Electronics: A Primer – Introductory Logic Circuit Design*
by Mark S. Nixon

Primers in Electronics and Computer Science ○—⟨—○ Vol. 4

An Elementary Approach to
Design and Analysis
of Algorithms

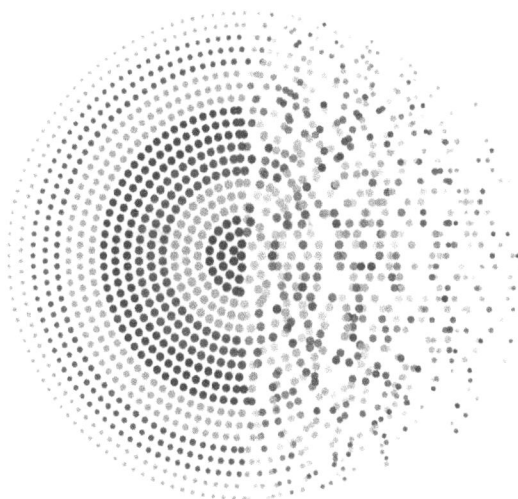

Lekh Raj Vermani
Kurukshetra University, India

Shalini Vermani
Apeejay School of Management, India

Ｗ❺ World Scientific

NEW JERSEY · LONDON · SINGAPORE · BEIJING · SHANGHAI · HONG KONG · TAIPEI · CHENNAI · TOKYO

Published by

World Scientific Publishing Europe Ltd.

57 Shelton Street, Covent Garden, London WC2H 9HE

Head office: 5 Toh Tuck Link, Singapore 596224

USA office: 27 Warren Street, Suite 401-402, Hackensack, NJ 07601

Library of Congress Cataloging-in-Publication Data

Names: Vermani, L. R. (Lekh R.), author. | Vermani, Shalini, author.
Title: An elementary approach to design and analysis of algorithms / by Lekh Raj Vermani
    (Kurukshetra University, India) and Shalini Vermani (Apeejay School of Management, India).
Description: New Jersey : World Scientific, 2019. | Series: Primers in electronics and
    computer science ; volume 4 | Includes bibliographical references and index.
Identifiers: LCCN 2018056450 | ISBN 9781786346759 (hc : alk. paper)
Subjects: LCSH: Algorithms--Study and teaching.
Classification: LCC QA9.58 .V47 2019 | DDC 518/.1--dc23
LC record available at https://lccn.loc.gov/2018056450

British Library Cataloguing-in-Publication Data
A catalogue record for this book is available from the British Library.

For any available supplementary material, please visit
https://www.worldscientific.com/worldscibooks/10.1142/Q0201#t=suppl

Desk Editors: Aanand Jayaraman/Jennifer Brough/Shi Ying Koe

Typeset by Stallion Press
Email: enquiries@stallionpress.com

To the memory of my Parents

Lekh Raj Vermani

Preface

Informally speaking, an algorithm is a well-defined set of instructions or rules for solving a given problem. An algorithm accepts a value or a set of values as input and as the instructions are followed in a logical fashion, there results a value or a set of values as the final solution. The value or the set of values finally arrived at is called the output. Another important feature of an algorithm is that it must terminate in a finite number of steps. Euclid's division algorithm is a most elementary algorithm which leads, besides many other things, to the greatest common divisor of two integers not both zero. Algorithms form an integral part of mathematics for solving problems in combinatorial mathematics. With the advent of computers, algorithms have become an integral and a very important part of theoretical computer science. For solving any problem simple or complex on computers, a programmer has to be extremely well versed in algorithms as he is the one who has to convert an algorithm, which may normally be written in simple English language, into computer program, i.e., he has to convert the algorithm into a language understood by the machine on which the problem is to be solved. In this text book, we are not concerned with "programs" written into machine language but will be content only with various algorithms. The number of recursive steps involved in the algorithm gives an estimate of time taken on the machine for solving a problem and is called the time complexity of the algorithm. How much memory of the machine is going to be involved is called the space complexity of the algorithm. However, in this book, the main concern is with time complexity and space complexity is not considered.

The design and analysis of algorithms today forms one of the core areas within computer science. This is clear from the fact that one to two one semester courses form a compulsory part of the study leading to a degree in computer science, computer engineering and information technology of almost every university.

The text begins with elementary sorting algorithms such as bubble sort, insertion sort, selection sort and quick sort, asymptotic notations are then introduced and are discussed at length. We then discuss the coloring problem and the n-queen problem by comprehensive method known as backtracking. As part of the divide and conquer approach the max-min problem and Strassen's matrix multiplication algorithm are discussed. Job sequencing problem and Huffman code are studied as part of greedy algorithms. Matrix chain multiplication, multistage graphs and optimal binary search trees are discussed as part of dynamic programming. Among graph algorithms discussed are breadth first search, depth first search, Prim's algorithm, Kruskal's algorithm, Dijkstra's algorithm and Bellman–Ford algorithm for single source shortest paths and Floyd–Warshall algorithm and Johnson's algorithm for all pairs shortest paths. Then flow networks, polynomial multiplication, FFT and DFT are discussed. The Rabin–Karp algorithm and Knuth–Morris–Pratt algorithm are discussed for string matching. Some other sorting networks are then discussed. Most problems we come across can, in general, be solved in polynomial time but not all. There may be problems which may need exponential time for their completion. Problems are classified as P, NP and NP-complete. The last chapter is devoted to study NP-completeness.

In the choice of topics discussed in the book we were guided by the course contents of the relevant paper(s) in computer science, computer engineering and information technology of some universities in the region. The relevance of the topics chosen is also evident from the fact that eight of these are among the 12 topics covered (may be at a much advanced level) in the 4 week courses (4–8 hours per week) as part of Algorithms Specialization Course of the Stanford University.

Our effort has been to present the material at an elementary level and in an easy to read and understand manner. In fact this has been our chief concern throughout. For the purpose, several solved examples have been given some as motivation for a topic and others as applications. Some exercises for the student to solve in order to further test the understanding of the subject are given. Maturity in logical mathematical reasoning including the principle of mathematical induction and knowledge of elementary algebra (not groups, rings and fields), some graph theory is required.

While preparing the text, the following books: Aho, Hopcroft and Ullman [1, 2]; Cormen *et al.* [5]; Horowitz, Sahni and Rajasekaran [14] and Jungnickel [15], together with some downloaded material were freely used. These books, we consider, are at a much higher level and are more exhaustive whereas the present book may be considered only as an introductory course (to be used as a text book) in the subject. All efforts have been made to avoid any verbatim inclusion of any material from these books and the other literature. Similarity, if any, of the material presented in the book with that in any of the sources mentioned is most sincerely acknowledged. Indeed, there is no claim to originality except that the material has been presented in an extremely simple and easy to understand fashion.

The authors express their most sincere thanks and appreciation to the authors of these books and the authors of other literature consulted and used.

The authors would like to thank Mr. Harish Chugh for being the first to request them to write such a text after our book on Discrete Mathematics was completed and to Professor R.P. Sharma regarding his advice in connection with the conversion of the manuscript from a word file to Latex. The second author would like to thank the authorities of Apeejay School of Management, Dwarka, New Delhi for providing the necessary facilities during the preparation of the text. The authors would also like to place on record their appreciation to Ms. Jennifer and her team for giving the text the present shape.

L.R. Vermani
Shalini Vermani

About the Authors

Lekh Raj Vermani is a retired Professor of Mathematics from Kurukshetra University, Kurukshetra. After his retirement, he was Visiting Professor/Principal/Director at Jai Prakash Mukand Lal Institute of Technology, Radaur (Yamuna Nagar) and Panipat Institute of Engineering and Technology, Smalkha (Panipat). He held the position of Commonwealth Post-doctoral Fellowship at the University of Exeter, England and was also Visiting Fellow/Scientist/Professor at Tata Institute of Fundamental Research, Bombay; University of Manitoba, Winnipeg, Canada; Techlov Institute of Mathematics, Moscow; Panjab University, Chandigarh; Harish-Chandra Research Institute, Allahabad; Indian Statistical Institute, Delhi; and Himachal Pradesh University, Shimla. He is a Fellow of the National Academy of Sciences, India. He is the author of *Lectures in Cohomology of Groups, Elements of Algebraic Coding Theory, An Elementary Approach to Homological Algebra* and *A Course in Discrete Mathematical Structures* — the last one in collaboration with Dr. Shalini Vermani. He obtained his Masters degree in Mathematics and his Ph.D. from Kurukshetra University.

Dr. Shalini Vermani is Associate Professor in Apeejay School of Management, New Delhi in the Department of Information Technology and Operations Management. She has more than 17 years of teaching experience. Her area of interest is Information Security and Cryptography. She has published various papers in national and international journals and has also written a book, *A Course on Discrete Mathematical Structures*, published by Imperial College Press, London. Dr. Vermani has also reviewed various papers for national and international journals/conferences.

Contents

Chapter 10.　All Pairs Shortest Paths　321

Chapter 11.　Flow Networks　359

Chapter 12.　Polynomial Multiplication, FFT and DFT　387

Chapter 13.　String Matching　399

Chapter 1

Algorithms

In the subject of computer science the word algorithm, in general, refers to a method that can be used by a computer for the solution of a problem. As such the word algorithm does not simply mean a process, technique or method. Strictly speaking, it is a finite collection of well-defined computational techniques each of which is clear and unambiguous, capable of being performed with finite effort and terminates in a finite amount of time. Before an algorithm comes to play, a problem to be solved on computer is first converted into a mathematical problem called mathematical model. An algorithm is different from a program (computer program). Although algorithm is an easily understood procedure giving a set of logical instructions for solving the problem at hand, a program is written in a computer language giving the instructions that the computer understands.

1.1. Some Sorting Algorithms

We begin our study of algorithms by considering some sorting algorithms. Here we are given a sequence of non-negative integers $a_1, a_2, \ldots, a_n$, not necessarily all distinct, the aim is to obtain a permutation of the a's to get the numbers in a non-decreasing (or non-increasing) order i.e., to obtain a sequence $a_{1'}, a_{2'}, \ldots, a_{n'}$ with $\{a_{1'}, a_{2'}, \ldots, a_{n'}\} = \{a_1, a_2, \ldots, a_n\}$ and $a_{1'} \leq a_{2'} \leq \cdots \leq a_{n'}$. We restrict ourselves to obtaining a non-decreasing sequence.

1.1.1. Bubble sort

Among the sorting algorithms, this is most perhaps the easiest. In this algorithm a_n is first compared with a_{n-1} and if $a_n < a_{n-1}$, a_n and a_{n-1} are interchanged. Then a_n is compared with a_{n-2} and if $a_n < a_{n-2}$, a_n and a_{n-2} are interchanged and the process stops when we come to an a_i such that $a_i < a_n$. The process is then followed with a_{n-1}, then a_{n-2} etc.

Algorithm 1.1. Bubble Sort Algorithm

---
1. For $i = 1$ to $n - 1$, do
2. for $j = n$ to $i + 1$
3. if $a_j < a_{j-1}$, then
4. interchange (swap) $a_j \leftrightarrow a_{j-1}$

We will better explain this algorithm through two examples: one a sequence with all terms distinct and the other when all the terms are not necessarily distinct. We may not restrict ourselves to sequences of numbers alone but may consider sequences over any ordered set.

Example 1.1.   Using bubble sort, rearrange the sequence

Patiala, Ernakulam, Karnal, Amritsar, Venus, Shimla

in alphabetical order.

Solution. This is a sequence of six terms all distinct. So i takes the values from 1 to 5. We express the sequence as a column denoting each word by the first letter of the word. We shall fully describe the internal loops for every i (refer to Table 1.1).

Observe that in the case $i = 1$, there is no change for $j = 5$, in the case of $i = 2$, there is no change for $j = 6, 5, 4$ and for $i = 3$, there is no change for $j = 6, 5$. At this stage the sequence is already sorted and, so, for $i = 4$ and 5, no changes are required. Hence the final sorted sequence is A, E, K, P, S, V or

Amritsar, Ernekulam, Karnal, Patiala, Shimla, Venus

Example 1.2. Using bubble sort, rearrange the sequence 2, 3, 6, 1, 0, 4, 8, 0 in non-decreasing order.

Table 1.1

For i = 1				For i = 2		For i = 3
P	P	P	P	A	A	A
E	E	E	A	P	E	E
K	K	A	E	E	P	K
A	A	K	K	K	K	P
V	S	S	S	S	S	S
S	V	V	V	V	V	V
$j = 6$	$j = 4$	$j = 3$	$j = 2$	$j = 3$	$j = 4$	

Table 1.2(a)

For i = 1						For i = 2			
2	2	2	2	2	2	0	0	0	0
3	3	3	3	3	0	2	2	2	2
6	6	6	6	0	3	3	3	3	0
1	1	1	0	6	6	6	6	0	3
0	0	0	1	1	1	1	0	6	6
4	4	0	0	0	0	0	1	1	1
8	0	4	4	4	4	4	4	4	4
0	8	8	8	8	8	8	8	8	8
$j = 8$	$j = 7$	$j = 5$	$j = 4$	$j = 3$	$j = 2$	$j = 6$	$j = 5$	$j = 4$	$j = 3$

Table 1.2(b)

For i = 3			For i = 4	
0	0	0	0	0
0	0	0	0	0
2	2	2	1	1
3	3	1	2	2
6	1	3	3	3
1	6	6	6	4
4	4	4	4	6
8	8	8	8	8
$j = 6$	$j = 5$	$j = 4$	$j = 7$	

Solution. This is a sequence of eight terms. So i takes values from 1 to 7. We express the sequence as a column and fully describe the internal loops for every i (as given in Tables 1.2(a) and 1.2(b)).

Observe that the array obtained in the last column of Table 1.2(b) is already sorted and, so, no changes are needed for $i = 5, 6, 7$. Also on the way, no changes were needed for $i = 1$ and $j = 6$, for $i = 2$ and $j = 8, 7$, for $i = 3$, $j = 8, 7$ and for $i = 4$, $j = 8, 6, 5$. Hence the sorted sequence is $0, 0, 1, 2, 3, 4, 6, 8$.

1.1.2. Insertion sort

In this sorting method, we start with the index $i = 2$ and compare the keys $a[1], a[2]$. If $a[2] < a[1]$, we interchange (swap) $a[1] \leftrightarrow a[2]$. Otherwise, leave these as they are. Here we are writing $a[i]$ for a_i. Thus a_1, a_2 are in sorted order. Suppose that $a_1, a_2, \ldots, a_i$ are in sorted order. Then, we consider a_{i+1}. If $a_{i+1} < a_i$, swap $a[i] \leftrightarrow a[i + 1]$. Then compare the present $a[i]$ with $a[i - 1]$ and continue the above process of swapping till we arrive at a position j such that the key value $a[j - 1] < a[j]$. Thus the terms $a_1, a_2 \ldots, a_{i+1}$ are in sorted order. The process ends when we finally arrive at $i = n$ and that we have put a_n in its proper place. Thus the algorithm may be stated as follows.

Algorithm 1.2. Insertion Sort Algorithm

---

1. For $i = 2$ to n do
2. move a[i] forward to the position $j \leq 2$ in a step by step manner such that
3. $a[i] < a[k]$ for $j \leq k < i$, and
4. either $a[j - 1] \leq a[i]$ or $j = 1$

More fully, it may be stated as

Algorithm 1.2′.

---

1. For $i = 2$ to n do begin
2. for $j = i$ to 2
3. while $a[j] < a[j - 1]$ (provided $j \geq 2$) do begin
4. interchange $a[j] \leftrightarrow a[j - 1]$
5. end
6. end

We explain the procedure through the two examples considered for bubble sort.

Example 1.3. Sort the sequence of Example 1.1 using insertion sort.

Solution. The given sequence is P, E, K, A, V, S. We have

$$P, E, K, A, V, S \overset{i=2}{\to} E, P, K, A, V, S \overset{i=3}{\to} E, K, P, A, V, S$$
$$\overset{i=4}{\to} (E, K, A, P, V, S \to E, A, K, P, V, S \to A, E, K, P, V, S)$$
$$\overset{i=5}{\to} A, E, K, P, V, S \overset{i=6}{\to} A, E, K, P, S, V.$$

In the step (or pass) $i = 4$, we have to move A to its proper place and it first moves past P, then past K and finally past E. These steps for the pass $i = 4$ have been placed in parenthesis.

Example 1.4. Sort the sequence of Example 1.2 using insertion sort.

Solution. The given sequence is sorted as follows (observe that no change is needed for the passes $1, 2, 3$):

$$2, 3, 6, 1, 0, 4, 8, 0 \overset{i=4}{\to} (2, 3, 1, 6, 0, 4, 8, 0 \to 2, 1, 3, 6, 0, 4, 8, 0$$
$$\to 1, 2, 3, 6, 0, 4, 8, 0) \overset{i=5}{\to} (1, 2, 3, 0, 6, 4, 8, 0 \to 1, 2, 0, 3, 6, 4, 8, 0$$
$$\to 1, 0, 2, 3, 6, 4, 8, 0 \to 0, 1, 2, 3, 6, 4, 8, 0) \overset{i=6}{\to} 0, 1, 2, 3, 4, 6, 8, 0$$
$$\overset{i=8}{\to} (0, 1, 2, 3, 4, 6, 0, 8 \to 0, 1, 2, 3, 4, 0, 6, 8 \to 0, 1, 2, 3, 0, 4, 6, 8$$
$$\to 0, 1, 2, 0, 3, 4, 6, 8 \to 0, 1, 0, 2, 3, 4, 6, 8 \to 0, 0, 1, 2, 3, 4, 6, 8).$$

Observe that no changes are needed for the passes $i = 2, 3, 7$. The finally sorted sequence is $0, 0, 1, 2, 3, 4, 6, 8$ as before.

1.1.3. Selection sort

In the selection sort algorithm, we look for the smallest element (key) in $a[1], a[2], \ldots, a[n]$, say it is $a[i]$. We interchange $a[1] \leftrightarrow a[i]$. Then in the revised sequence, we look for the smallest element, which we again call $a[i]$ and this time we interchange (swap) $a[2] \leftrightarrow a[i]$. In general, if after $j - 1$ steps or passes we come to the sequence $a[1], \ldots, a[j], \ldots, a[n]$ in which the sequence $a[1 \ldots j] (= (a[1], \ldots, a[j]))$ is already sorted, then we look for the smallest element $a[i]$ in $a[j \ldots n]$ and swap that $a[i]$ with $a[j]$. We continue doing so until we arrive at sorted sequence $a[1 \ldots n-1]$. Then we look for the smaller of the two elements $a[n-1]$, $a[n]$ and swap that with $a[n-1]$ finally getting the sorted sequence $a[1], a[2], \ldots, a[n]$. In a formal manner, the algorithm may be stated as follows.

Algorithm 1.3. Selection Sort Algorithm

1. For $i = 1$ to $n-1$ do
2. choose the smallest element $a[j]$ in $a[i \ldots n]$
3. swap $a[i] \leftrightarrow a[j]$
4. call low key $a[i]$
5. for $j = i + 1$ to $n - 1$
6. repeat steps 2, 3, 4
7. end

We now apply this sorting algorithm to sort the strings of Examples 1.1 and 1.2.

Example 1.5. Using selection sort algorithm, sort the sting of Example 1.1.

Solution. The given string is P, E, K, A, V, S which we may call $a[1 \ldots 6]$. In the first step, we look for the smallest element (key) for the whole sequence and find that it is $a_4 = A$. Swap a_4 with a_1 to get the sequence A, E, K, P, V, S. Now we look for the smallest element (low key) in $a[2 \ldots 6]$ which is $a[2] = E$. We need to swap a_2 with a_2 and, so, no change is needed and the sequence after two passes (steps) is again A, E, K, P, V, S. Looking for the low key in the sequence $a[3 \ldots 6]$, we find that a_3 itself is the low key and swapping does not lead to a new sequence. Now looking for low key in the sequence $a[4 \ldots 6]$, we find that a_4 itself is the low key and swapping does not change this sequence. Finally, observe that $a_6 = S < a_5 = V$, so swap $a_5 \leftrightarrow a_6$ and we get the sorted sequence A, E, K, P, S, V.

Symbolically, we may represent the above steps as in Table 1.3 and the final sorted sequence appears as the last column in the table.

Table 1.3

P	A	A	A	A	A
E	E	E	E	E	E
K	K	K	K	K	K
A	P	P	P	P	P
V	V	V	V	V	S
S	S	S	S	S	V
$\overrightarrow{i=1}$	$\overrightarrow{i=2}$	$\overrightarrow{i=3}$	$\overrightarrow{i=4}$	$\overrightarrow{i=5}$	

Example 1.6. Sort the sequence of Example 1.2 using the selection sort.

Solution. We indicate the steps symbolically as follows.

$$2, 3, 6, 1, 0, 4, 8, 0 \xrightarrow{i=1} 0, 3, 6, 1, 2, 4, 8, 0 \xrightarrow{i=2} 0, 0, 6, 1, 2, 4, 8, 3$$
$$\xrightarrow{i=3} 0, 0, 1, 6, 2, 4, 8, 3 \xrightarrow{i=4} 0, 0, 1, 2, 6, 4, 8, 3 \xrightarrow{i=5} 0, 0, 1, 2, 3, 4, 8, 6,$$
$$\xrightarrow{i=6} 0, 0, 1, 2, 3, 4, 8, 6 \xrightarrow{i=7} 0, 0, 1, 2, 3, 4, 6, 8.$$

Hence, the final sorted sequence is 0, 0, 1, 2, 3, 4, 6, 8.

1.1.4. Quick sort

Algorithm 1.4. Quick Sort Algorithm

---
1. given a sequence $a[1 \ldots n]$
2. choose a pivot (= some kind of a central element) $v = a[m]$
3. partition $a[1 \ldots n]$ into two subsequences $a[1 \ldots p]$ and $a[p+1 \ldots n]$ such that $a[i] \leq v$ for every i, $1 \leq i \leq p$ and $a[j] > v$ for every j, $j = p+1, \ldots, n$
4. quick sort the subsequences $a[1 \ldots p]$, $a[p+1 \ldots n]$ in place
5. end

Thus in quick sort there are two major steps. One is the choice of a pivot and, then, is the partitioning of the sequence around the pivot.

One of the earliest method of choosing a pivot is to look for the leftmost consecutive pair of unequal entries and to choose the larger of these two entries as a pivot. Suppose that the pivot chosen in this manner is $a[m]$. (There are two possibilities for the consecutive pair of entries

a_{m-1}, a_m being distinct. Either $a_{m-1} < a_m$ in which case a_m itself is the pivot or $a_m < a_{m-1}$ in which case we swap a_{m-1} and a_m and rename a_{m-1} as a_m and a_m as a_{m-1} so that a_m again is the pivot.) Then all the entries of the sequence $a[i]$ with $i < m$ are either equal or $a[1] = \cdots = a[m-2]$ and $a[m-1] < a[m-2]$. Sorting of the sequence $a[1 \ldots m-1]$ is easy. Either $a[1 \ldots m-1]$ is already sorted or $a[m-1], a[1], \ldots, a[m-2]$ is sorted. Thus the problem reduces to sorting the sequence $a[m \ldots n]$. Hoar (cf. [5]) has given a simple procedure for partitioning a sequence $a[m \ldots n]$ around $a[m]$. It works as under:

Starting left, look for the first i such that $a[i] \geq v$ and starting from right look for the first j such that $a[j] < v$. Interchange $a[i] \leftrightarrow a[j]$. Continue this process till we find a sequence in which there is a k such that $a[i] \leq v$ with $m + 1 \leq i \leq k$ and $a[j] > v$ for all $j \geq k + 1$. Finally interchange $a[m] \leftrightarrow a[k]$, if necessary. Thus

$$\text{for } i = m + 1, j = n$$
$$\text{if } a[i] \leq v, a[j] > v, \text{ set } i = m + 2, j = n - 1$$
$$\text{if } a[i] \leq v, a[j] \leq v, \text{ set } i = m + 2, j = n$$
$$\text{if } a[i] \geq v, a[j] > v, \text{ set } i = m + 1, j = n - 1$$

Algorithm 1.5. Procedure Partition 1

Partition $a[m \ldots p]$ around $a[m] = v$
{partition $a[m], \ldots, a[p]$ into two sequences $a(m + 1, \ldots, k - 1), a(k + 1, \ldots, p)$ with $a[i] \leq v$ for $i = m + 1, \ldots, k - 1, a[j] > v$ for $j = k + 1, \ldots, p, a[k] = v$}
begin
$i = m$
$j = p$
$a[m] = v$

    1. repeat $i = i + 1$ until $a[i] > v$
    2. repeat $j = j - 1$ until $a[j] \leq v$
    3. interchange $a[i] \leftrightarrow a[j]$ if $i < j$
    4. repeat 1, 2, 3 until $i > j$
    5. return i interchange $a[m] \leftrightarrow a[i]$

end

Algorithm 1.6. Procedure Partition 2

1. Partition $a[m \ldots p]$ so keys $a[i] \leq$ pivot are on the left and keys $a[j] \geq$ pivot are on the right
2. $k =$ pivot index, pivot key $a[k] = v$
3. begin
4. $i = m, j = p$
5. repeat
6. $i = i + 1$ until $a[i] > v$
7. $j = j - 1$ until $a[j] \leq v$
8. interchange $a[i] \leftrightarrow a[j]$
9. return i
10. end (partition)

The two partition procedures are similar, the only difference being that in procedure 2, the pivot index is already decided and occupies its right position while in procedure 1, the partitioning key $a[m]$ is shifted to its right position after all the keys $\leq a[m]$ have been collected together on the left side and all the keys $> a[m]$ are collected on the right side.

There is yet another partition procedure which is also variation of the above two procedures — it is more like procedure 1.

Let the pivot key $= v$ be determined without occupying its rightful position in the sequence. This key is supposed to sit outside the sequence: either on the left or on the right. For a change, suppose that it sits on the right. Start scanning the sequence from left to right. Keep all the beginning consecutive keys that are $\leq v$ unchanged. Suppose we come to keys $a[i], a[i + 1]$ such that $a[i] > v, a[i + 1] \leq v$. Then interchange $a[i] \leftrightarrow a[i+1]$. Change $i = i+1$. Compare $a[i+1]$ and $a[i+2], \ldots$. Keep all the keys $a[i+1], \ldots$ that are $> v$ intact till we arrive at a situation where $a[j] > v, a[j + 1] \leq v$. Then interchange $a[j + 1] \leftrightarrow a[i + 1]$. Continue the process till all keys $\leq v$ have been collected together on the left side without changing their relative positions and all keys $> v$ on the right side.

For the sake of simplicity, we take array $a[1 \ldots n]$ to be sorted and pivot x. In case the pivot is not occurring either on the extreme right or on the extreme left but occurs somewhere in the array, the above partitioning procedure may be stated as follows.

Choose i such that $a[j] < x$ for $1 \le j < i$ but $a[i] > x$. Choose j such that $a[k] > x$ for $i \le k < j$ but $a[j] < x$. Group together $a[i], \ldots, a[j-1]$ and consider it one block (we may think of it as $a[i]$). Interchange $a[i], a[j]$. Continue the process till we arrive at an array which looks like

(i) $a[k] < x$, for $1 \le k < i, a[i] = x, a[k] < x$, for $i < k < j$ and $a[k] > x$ for $j \le k \le n$ or

(ii) $a[k] < x$, for $1 \le k < i, a[k] > x$, for $i \le k < j, a[j] = x$ and $a[k] > x$ for $j < k \le n$.

In case (i), regarding $a[i+1], \ldots, a[j-1]$ as one block, interchange $a[i], a[i+1]$ so that the pivot x appears in the position $a[j]$. In case (ii), regarding $a[i], \ldots, a[j-1]$ as one block, interchange $a[i], a[j]$ so that the pivot x appears in the position $a[i]$.

We illustrate the above partition procedures through a couple of examples.

Example 1.7. Partition the array $65, 90, 75, 70, 45, 60, 85, 50, 80, 55$ with pivot $x = 70$.

Solution. We use procedure partition 1 and the above modified partition leaving the details.

(i)	The array	swap
	$65, 90, 75, 70, 45, 60, 85, 50, 80, 55$	$2 \leftrightarrow 10$
	$65, 55, 75, 70, 45, 60, 85, 50, 80, 90$	$3 \leftrightarrow 8$
	$65, 55, 50, 70, 45, 60, 85, 75, 80, 90$	$4 \leftrightarrow 6$
	$65, 55, 50, 60, 45, 70, 85, 75, 80, 90$	

(ii)	The array	swap(s)
	$65, [90], 75, (70), 45, 60, 85,50,80,55$	
	$65, [90, 75], (70), 45, 60, 85, 50, 80, 55$	
	$65, 45, (70), [90, 75], 60, 85, 50, 80, 55$	$2 \leftrightarrow 5$
	$65, 45, (70), 60, [90, 75], 85, 50, 80, 55$	$4 \leftrightarrow 6$
	$65, 45, (70), 60, [90, 75, 85], 50, 80, 55$	
	$65, 45, (70), 60, 50, [90, 75, 85], 80, 55$	$5 \leftrightarrow 8$
	$65, 45, (70), 60, 50, [90, 75, 85, 80], 55$	
	$65, 45, (70), 60, 50, 55, [90, 75, 85, 80]$	$8 \leftrightarrow 10$
	$65, 45, 60, 50, 55, (70), 90, 75, 85, 80$	$3 \leftrightarrow 4, 4 \leftrightarrow 5, 5 \leftrightarrow 6$

We have shown the pivot within parentheses and the elements grouped together have been shown within the brackets [ ]. Observe that the total number of swaps involved is 5 (counting the last three swaps as just 1).

Example 1.8. Partition the array $65, 90, 75, 70, 45, 60, 85, 50, 80, 55$ with pivot $x = 75$.

Solution. We apply procedure partition 1 and the modified partition procedure leaving the details.

(i) The array   swap
  $65, 90, 75, 70, 45, 60, 85, 50, 80, 55$
  $65, 55, 75, 70, 45, 60, 85, 50, 80, 90$ $2 \leftrightarrow 10$
  $65, 55, 50, 70, 45, 60, 85, 75, 80, 90$ $3 \leftrightarrow 8$
  $65, 55, 50, 70, 45, 60, 75, 85, 80, 90$ $7 \leftrightarrow 8$
  Observe that the number of swaps used is 3.

(ii) The array   swap(s)
  $65, 90, (75), 70, 45, 60, 85, 50, 80, 55$
  $65, [90], (75), 70, 45, 60, 85, 50, 80, 55$
  $65, 70, (75), [90], 45, 60, 85, 50, 80, 55$ $2 \leftrightarrow 4$
  $65, 70, (75), 45, [90], 60, 85, 50, 80, 55$ $4 \leftrightarrow 5$
  $65, 70, (75), 45, 60, [90], 85, 50, 80, 55$ $5 \leftrightarrow 6$
  $65, 70, (75), 45, 60, [90, 85], 50, 80, 55$
  $65, 70, (75), 45, 60, 50, [90, 85], 80, 55$ $6 \leftrightarrow 8$
  $65, 70, (75), 45, 60, 50, [90, 85, 80], 55$
  $65, 70, (75), 45, 60, 50, 55, [90, 85, 80]$ $7 \leftrightarrow 10$
  $65, 70, 45, 60, 50, 55, (75), 90, 85, 80$ $3 \leftrightarrow 4, 4 \leftrightarrow 5, 5 \leftrightarrow 6$

The total number of swaps is 6 on counting the last three swaps as just one.

We now consider a couple of examples for Quicksort algorithm.

Example 1.9. Sort the sequence $2, 3, 6, 1, 0, 5, 4, 9, 8, 0$ using quick sort.

Solution. Scanning the sequence from left to right we find that $a[1] = 2, a[2] = 3$ are distinct and $a[2] > a[1]$. So, the pivot is $a[2] = 3$.

Algorithm 1.7. Procedure Quicksort $a[m \ldots p]$

{sort the elements $a[m], \ldots, a[p]$ in non-decreasing order}

1. find pivot
2. pivot index = k, pivot key = $a[k]$
3. partition $a[m \ldots p]$ with pivot index k and pivot key = $a[k]$
4. quick sort

$$a[m \ldots k]$$

$$a[k+1 \ldots p]$$

5. end
6. end; {quick sort}

We partition the sequence using procedure partition 2 with $a[2]$ as its pivot.

$$2, 3, 6, 1, 0, 5, 4, 9, 8, 0 \rightarrow 2, 0, 6, 1, 0, 5, 4, 9, 8, 3$$

$$\rightarrow 2, 0, 0, 1, 6, 5, 4, 9, 8, 3$$

Now partition $2, 0, 0, 1$ with pivot $v = 2$ and $6, 5, 4, 9, 8, 3$ with pivot $v = 6$. We then get

$$2, 0, 0, 1 \rightarrow 1, 0, 0; 2 \quad 6, 5, 4, 9, 8, 3 \rightarrow 3, 5, 4 \quad 9, 8, 6$$

Now partition $1, 0, 0$ with pivot $v = 1$; $3, 5, 4$, with pivot $v = 5$ and $9, 8, 6$ with pivot $v = 9$. We then get the subsequences in order as $0, 0, 1; 2; 3, 4, 5; 6, 8, 9$. Concatenating these subsequences we get the sorted form of the given sequence $0, 0, 1; 2; 3, 4, 5; 6, 8, 9$.

Example 1.10. Sort the sequence of Example 1.9 using quick sort with procedure partition 1.

Solution. The given sequence is $2, 3, 6, 1, 0, 5, 4, 9, 8, 0$. Partition it using Procedure 1 around 2. Then

$$2, 3, 6, 1, 0, 5, 4, 9, 8, 0 \rightarrow 2, 0, 6, 1, 0, 5, 4, 9, 8, 3$$

$$\rightarrow 2, 0, 0, 1, 6, 5, 4, 9, 8, 3 \rightarrow 1, 0, 0, 2, 6, 5, 4, 9, 8, 3$$

$\rightarrow 0, 0, 1, 2, 6, 5, 4, 3, 8, 9 \rightarrow 0, 0, 1, 2, 3, 5, 4, 6, 8, 9$

$\rightarrow 0, 0, 1, 2, 3, 4, 5, 6, 8, 9 \rightarrow 0, 0, 1, 2, 3, 4, 5, 6, 8, 9$

$\rightarrow 0, 0, 1, 2, 3, 4, 5, 6, 8, 9$

which is the final sorted form of the sequence.

Observe that on the way we have used procedure 1 to partition the sequences $1, 0, 0$ around 1; $6, 5, 4, 9, 8, 3$ around 6 and $3, 5, 4$ around 5.

Example 1.11. Sort the sequence of Example 1.9 using quick sort with procedure partition 3.

Solution. The given sequence is $2, 3, 6, 1, 0, 5, 4, 9, 8, 0$. Let us consider 4 as pivot. The given sequence may then be rewritten as $4|2, 3, 6, 1, 0, 5, 9, 8, 0$.

The elements $2, 3$ are less than 4 and 6 is greater than 4. So we get the initial blocks $4|2, 3|6, 1, 0, 5, 9, 8, 0$.

Then swap block of $1, 0$ with 6 to get $4|2, 3, 1, 0|6, 5, 9, 8, 0$. Now swap 6 and 0 to get $4|2, 3, 1, 0, 0|5, 9, 8, 6$. Now shifting the pivot to its rightful position, we get $2, 3, 1, 0, 0|4, 5, 9, 8, 6$.

Use 2 as pivot for the first subsequence and 6 as pivot for the second subsequence, this sequence takes the form

$$2|3, 1, 0, 0||6|4, 5, 9, 8 \rightarrow 2|1, 3, 0, 0||6|4, 5, 9, 8$$

$$\rightarrow 2|1, 0, 3, 0||4, 5, 6||9, 8 \rightarrow 2|1, 0, 0, 3||4, 5, 6||8, 9$$

(using 9 as pivot for the last subsequence)

$$1, 0, 0||2, 3||4, 5, 6||8, 9 \rightarrow 1|0, 0||2, 3||4, 5, 6||||8, 9$$

$$\rightarrow 0, 0, 1||2, 3||4, 5, 6||8, 9$$

Hence the sorted sequence is $0, 0, 1, 2, 3, 4, 5, 6, 8, 9$.

1.2. Algorithm Performance/Algorithm Complexity

There are two types of complexity of an algorithm. One of these is the time complexity meaning the running time of an algorithm depending on the size of the data while the other is space complexity meaning the memory space an algorithm needs. It is important to study the complexity of an algorithm. More than one algorithms are sometimes (in fact, in general) available for solving a given problem. For example, we have seen four different algorithms for sorting a given sequence of numbers (or elements of an ordered set). Which of the available algorithms should be preferred depends on their performance — the running time required by the algorithm and the memory space required. An algorithm which takes less time on machine and also requires less memory space is, in general, preferred over an algorithm that has more running time and or more memory space for a data of the same size. However, sometimes a simpler written algorithm may be preferred over an algorithm which is very complicatedly written. For all this, we have to be quite precise about what an algorithm is and we need to make clear how the input data and the time and the memory space needed are measured.

Time complexity of an algorithm A is a function f, where $f(n)$ is the maximal number of steps or computations that the algorithm A needs to solve an instance of a problem having an input data of size n. The word "steps" may mean the number of arithmetic operations, comparisons of numbers, etc. It may not, sometimes, be possible to compute the complexity $f(n)$ of an algorithm exactly. In such cases, we have to be content with how the function $f(n)$ grows.

While talking of complexity of an algorithm, we mean time complexity unless stated otherwise.

As an illustration, we compute the complexity of the sorting algorithms considered in the earlier section.

(a) The inner loop of the bubble sort repeats $(n - i)$ times and since i varies from 1 to $n - 1$, the number of computations involved is $\sum_{i=1}^{n-1}(n - i) = \sum_{i=1}^{n-1} i = \frac{n(n-1)}{2}$. Each such computation involved consists of comparison of two numbers followed by swapping of the two numbers, if necessary. Each such computation requires a constant number of time units. Thus, the complexity or running time of the bubble sort is $\leq \frac{cn(n-1)}{2} < cn^2$, where c is a constant.

Observe that in the inner loop j varies from n to $i + 1$, i.e., we start with a_n, compare it with a_{n-1} and if $a_n < a_{n-1}$, we swap a_n with a_{n-1}. Then a_n is compared with a_{n-2} and swap $a_n \leftrightarrow a_{n-2}$ if $a_n < a_{n-2}$ and we continue the process till we arrive at an i such that $a_i < a_n$ where the process stops so far as a_n is concerned. Then we consider the number a_{n-1} which may now occupy the nth place in the sequence. As the very name suggests, you may compare it with the movement of an air bubble stuck in water bottle.

(b) Insertion sort works precisely in the reverse order to the order in which the bubble sort works. The inner loop in this case moves j from i to 2 i.e., j takes $i-1$ values and then i takes values from 2 to n. In the internal loop, $i-1$ comparisons and swaps are required and each comparison followed by a swap operation, if necessary, requires a constant units of time. The total number of steps is $\sum_{i=2}^{n}(i-1) = \sum_{i=1}^{n-1} i = \frac{n(n-1)}{2}$. Therefore, the running time of the insertion sort is $T(n) \leq \frac{cn(n-1)}{2} < cn^2$, where c is a constant (which is the number of units of time needed for one move consisting of comparison of two numbers and a possible swap).

(c) Finding the smallest element in the sequence $a[1 \ldots n]$ needs $n-1$ comparisons. Once this is done, one swap is needed. Once the smallest element is placed in the first position from left, we are left with a sequence of length $n-1$. Finding the smallest element of this sequence needs $n-2$ comparisons followed by one swap. Thus, for this sorting procedure, a total number of comparisons followed by possible swaps $\leq (n-1) + (n-2) + \cdots + 1 = \frac{n(n-1)}{2}$. Hence the running time of the selection sort $\leq \frac{cn(n-1)}{2} \leq cn^2$.

We find that the complexity of each of the three sorting algorithms is $\leq cn^2$.

(d) Quick sort works quite differently. In this case, we first choose a pivot and then divide the sequence into two subsequences, say S_1 and S_2, which are such that if v is the pivot, then every entry of S_1 is $< v$ while every entry of S_2 is $\geq v$ with v sitting between S_1 and S_2. Once having done this, quick sort is applied to S_1 and S_2. The process continues till the sequence is completely sorted.

The given sequence of elements is $a[m \ldots p]$ with $p - m + 1 = n$. The pivot chosen $a[k]$ is, say, the ith smallest entry in the sequence. The index i could take any one of the n values from 1 to n and every i is equally likely. So, the probability of choosing i is $\frac{1}{n}$. Partitioning of the sequence requires $n-1$ comparisons and one swap, if necessary, for every comparison. Each step consisting of a comparison and a swap, if necessary, takes a constant amount of time. Therefore, partitioning of the sequence takes at most cn time units, where c is a constant. Therefore, the running time $T(n)$ for quick sort satisfies, for $n \geq 2$,

$$T(n) \leq cn + \frac{1}{n}\sum_{i=1}^{n}\{T(i-1) + T(n-i)\}, \qquad (1.1)$$

where $T(i-1)$, $T(n-i)$ are respectively, the running time on using quick sort for the subsequences S_1 and S_2. It is clear that $T(0) = T(1) = b - a$ constant. Observe that

$$\sum_{i=1}^{n} T(i-1) = T(0) + T(1) + \cdots + T(n-1)$$

$$= T(n-1) + T(n-2) + \cdots + T(1) + T(0)$$

$$= \sum_{i=1}^{n} T(n-i).$$

Therefore (1.1) takes the form

$$T(n) \leq cn + \frac{2}{n}\sum_{i=0}^{n-1} T(i) \qquad (1.2)$$

Using induction on n, we prove that $T(n) \leq kn \log n$, where $k = 2b + 2c$. For $n = 2$,

$$T(2) \leq 2c + T(0) + T(1) = 2(c + b) = k < k \log 4 = k(2\log 2).$$

Now, suppose that for any i, $2 \leq i < n$, $T(i) \leq ki \log i$. Then, by (1.2) and by induction hypothesis,

$$T(n) \leq cn + \frac{2}{n}\sum_{i=0}^{n-1} T(i)$$

$$= cn + \frac{4b}{n} + \frac{2}{n}\sum_{i=2}^{n-1} T(i) \leq cn + \frac{4b}{n} + \frac{2k}{n}\sum_{i=2}^{n-1} i \log i. \qquad (1.3)$$

Let $f(x) = x \log_e x$, where x is a positive real variable. If $x' > x$, then $\log x' \geq \log x$ and therefore, $x' \log x' > x \log x$. Thus $f(x)$ is a strictly

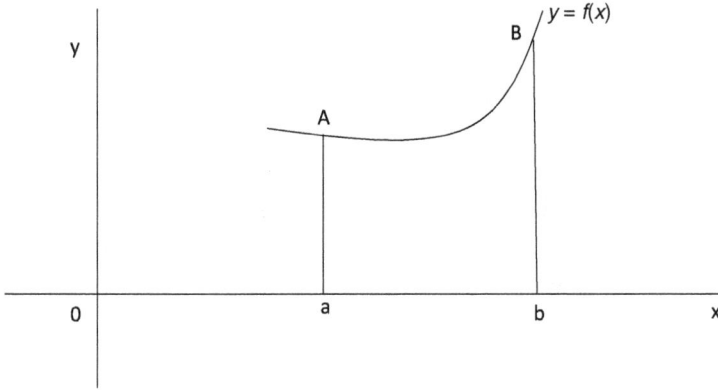

Fig. 1.1

increasing function of x. Therefore $f(x)$ is Reiman integrable. For any real numbers $a, b, 2 \le a < b$, $\int_a^b f(x)dx$ equals the area $AabB$ bounded by the lines aA, bB parallel to y-axis, the x-axis and the curve $y = f(x)$ (Fig. 1.1). Also $\sum_{i=2}^{n-1} f(i)$ is the sum of the line segments $(i, f(i))$ all parallel to the y-axis and lying within the area $AabB$ with $a = 2$ and $b = n - 1$. Therefore,

$$\sum_{i=2}^{n-1} i \log i \le \int_2^{n-1} x \log_e x \, dx \le \int_2^n x \log_e x \, dx$$

$$= \frac{x^2}{2} \log_e x \mid - \int_2^n \frac{x^2}{2} \cdot \frac{1}{x} dx$$

$$= \frac{n \log n}{2} - 2 \log_e 2 - \left(\frac{n^2}{4} - 1 \right).$$

$$(1.4)$$

Observe that $\frac{x^2}{2} \log_e x \mid$ stands for evaluating $\frac{x^2}{2} \log_e x$ for $x = n$ and $x = 2$ and taking their difference. From (1.3) and (1.4), we get

$$T(n) \le cn + \frac{4b}{n} + kn \log n - \frac{kn}{2} + (1 - 2 \log 2) \frac{2k}{n}.$$

As $k = 2b + 2c$ and $n \ge 2$, it follows that $cn + \frac{4b}{n} - \frac{kn}{2} + \frac{2k}{n} - \frac{4k \log 2}{n} \le 0$. Hence $T(n) \le kn \log_e n$.

1.3. Heap Sort

We finally discuss another sorting algorithm called heap sort.

Recall that a tree is called a binary tree if every node has at most two children. Suppose that an array $a[1 \ldots n]$ is represented as a **binary tree** T with $a[1]$ as its root and $a[i]$ going from left to right i.e., $a[1]$ has $a[2]$ as its left child and $a[3]$ as its right child, then $a[2]$ has $a[4]$ as its left child and $a[5]$ as its right child, $a[3]$ has $a[6]$ as its left child and $a[7]$ as its right child and so on. In general for any node $a[i]$ its left child is $a[2i]$, and its right child is $a[2i + 1]$ (if one exists).

The binary tree T representing an array $a[1 \ldots n]$ is called a **max heap** if for every $i \geq 1, a[i] \geq a[2i]$ and $a[i] \geq a[2i + 1]$. It is called a **min heap** if for every $i \geq 1, a[i] \leq a[2i]$ and $a[i] \leq a[2i + 1]$. Representing an array as a max heap or a min heap is called **build heap**. Also by a heap we mean a data represented as a binary max heap or a binary min heap.

Henceforth, by a heap we shall mean a max heap unless explicitly mentioned otherwise.

Given any array $a[1 \ldots n]$ we can always build a heap for it (or heapify it). Once this is done, the element at the root is the largest element of the array. We swap the root element with the last element of the heap and then consider the last element as detached from the data. The remaining $n - 1$ elements may not be a heap. We build a heap for the remaining $n - 1$ elements. Swap the element at the root with the last element of the heap formed and, again, considered as detached so that it comes just before the last element detached. Continue the process till we arrive at the data in sorted order.

Formally we have the following heap sort algorithm.

Algorithm 1.8. Procedure Heap Sort

---
1. given an array $a[1 \ldots n]$ of length n
2. build heap from $a[1 \ldots n]$ (again denoted by $a[1 \ldots n]$)
3. for $i = n$ to 2
4. swap $a[i] \leftrightarrow a[1]$
5. detach $a[i]$
6. repeat steps 2 to 5 till we arrive at two elements $a[1], a[2]$ (and the elements detached in proper sorted form)
7. end

Observe that build max heap $a[1 \ldots n]$ or heapify $a[1 \ldots n]$ has to be done $n - 1$ times. Therefore, the complexity of the heap sort is $\leq c$ $(n \times complexity\ of\ build\ heap\ a[1 \ldots n])$, where c is a constant. Thus, in order to apply heap sort, we need to give procedure for build max heap $a[1 \ldots n]$ and also find its complexity.

Given a data $a[1 \ldots 3]$ of length 3 arranged in the form of a binary tree with root $a[1]$. If $a[1] \geq \max\{a[2], a[3]\}$, the data is already in max heap form. Otherwise swap $a[1] \leftrightarrow \max\{a[2], a[3]\}$ and the data is in max heap form.

Observe that an array $a[1 \ldots n]$ arranged in the form of a binary tree, the node $a[i]$ has $a[2i]$, $a[2i + 1]$ as its children while the node $a[i]$ has $a\left[\left\lfloor \frac{i}{2} \right\rfloor\right]$ as its parent/ancestor (here $\left\lfloor \frac{i}{2} \right\rfloor$ denotes the largest integer $\leq \frac{i}{2}$). The height of the binary tree is $\log n$. To build a heap out of the given data, we have to compare every node $a[i]$ with its parent $a\left[\left\lfloor \frac{i}{2} \right\rfloor\right]$ and swap $a[i] \leftrightarrow a\left[\left\lfloor \frac{i}{2} \right\rfloor\right]$ if $a[i] > a\left[\left\lfloor \frac{i}{2} \right\rfloor\right]$.

Thus, to build a heap, every node needs at most $\log n$ comparisons and necessary swaps, if any. There being a total of n nodes, the running time of build heap procedure is $\leq c(n \log n)$. Also, for the process of build heap we start with a leaf of the data tree with largest entry. The process build heap can be performed on line (without drawing actual binary tree representing the data array). We explain this process through some examples.

Example 1.12. Build a heap for the array 6, 5, 3, 4, 1, 2, 9, 8, 7.

Solution.

Data									swap	
$a[1]$	$a[2]$	$a[3]$	$a[4]$	$a[5]$	$a[6]$	$a[7]$	$a[8]$	$a[9]$	$a[i] \leftrightarrow a\left[\left\lfloor \frac{i}{2} \right\rfloor\right]$	
6	5	3	4	1	2	9	8	7	4,	7
6	5	3	7	1	2	9	8	4	5,	7
6	7	3	5	1	2	9	8	4	6,	7
7	6	3	5	1	2	9	8	4	5,	8
7	6	3	8	1	2	9	5	4	6,	8
7	8	3	6	1	2	9	5	4	7,	8
8	7	3	6	1	2	9	5	4	3,	9
8	7	9	6	1	2	3	5	4	8,	9
9	7	8	6	1	2	3	5	4		

As $a\left[\left\lfloor\frac{i}{2}\right\rfloor\right] > a[i]$ for every i, no further swaps are needed and we have got a heap.

Example 1.13. Build a heap for the array 4, 7, 8, 6, 1, 2, 3, 5.

Solution.

Data								swap
$a[1]$	$a[2]$	$a[3]$	$a[4]$	$a[5]$	$a[6]$	$a[7]$	$a[8]$	$a[i] \leftrightarrow a\left[\left\lfloor\frac{i}{2}\right\rfloor\right]$
4	7	8	6	1	2	3	5	4, 8
8	7	4	6	1	2	3	5	

As $a\left[\left\lfloor\frac{i}{2}\right\rfloor\right] > a[i]$ for every i, no further swaps are needed and we get a heap: 8 7 4 6 1 2 3 5

Example 1.14. Build a heap for the array 5, 7, 4, 6, 1, 2, 3.

Solution.

Data							swap
$a[1]$	$a[2]$	$a[3]$	$a[4]$	$a[5]$	$a[6]$	$a[7]$	$a[i] \leftrightarrow a\left[\left\lfloor\frac{i}{2}\right\rfloor\right]$
5	7	4	6	1	2	3	5, 7
7	5	4	6	1	2	3	5, 6
7	6	4	5	1	2	3	

As $a\left[\left\lfloor\frac{i}{2}\right\rfloor\right] > a[i]$ for every i, no further swaps are needed and we get a heap:

$$7 \quad 6 \quad 4 \quad 5 \quad 1 \quad 2 \quad 3$$

Example 1.15. Build a heap for the array data 3, 6, 4, 5, 1, 2.

Solution.

Data						swap
$a[1]$	$a[2]$	$a[3]$	$a[4]$	$a[5]$	$a[6]$	$a[i] \leftrightarrow a\left[\left\lfloor\frac{i}{2}\right\rfloor\right]$
3	6	4	5	1	2	3, 4
4	6	3	5	1	2	4, 6
6	4	3	5	1	2	4, 5
6	5	3	4	1	2	

No further swaps are needed and, so, the heap is

$$6 \quad 5 \quad 3 \quad 4 \quad 1 \quad 2$$

Example 1.16. Build a heap for the array data 2, 5, 3, 4, 1.

Solution.

Data					swap	
$a[1]$	$a[2]$	$a[3]$	$a[4]$	$a[5]$	$a[i] \leftrightarrow a\left[\left\lfloor \frac{i}{2} \right\rfloor\right]$	
2	5	3	4	1	2,	3
3	5	2	4	1	3,	5
5	3	2	4	1	3,	4
5	4	2	3	1		

No further swaps are needed and, so, the heap is

$$5 \quad 4 \quad 2 \quad 3 \quad 1$$

Example 1.17. Build a heap for the array data 1, 4, 2, 3.

Solution.

Data				swap	
$a[1]$	$a[2]$	$a[3]$	$a[4]$	$a[i] \leftrightarrow a[\lfloor \frac{i}{2} \rfloor]$	
1	4	2	3	1,	2
2	4	1	3	2,	4
4	2	1	3	2,	3
4	3	1	2		

No further swaps are needed and, so, the heap is

$$4 \quad 3 \quad 1 \quad 2$$

Example 1.18. Build a heap for the array data 2, 3, 1.

Solution. For this we need only one swap 2 $\leftrightarrow$ 3 and the resulting heap is 3, 2, 1.

We now have available all the intermediate steps to heap sort the array of Example 1.12.

Example 1.19. Heap sort the sequence 6, 5, 3, 4, 1, 2, 9, 8, 7.

Solution. We first build a heap for the data which is already done in Example 1.12.

Heap		swap	detach	sorted array	
9	7 8 6 1 2 3 5 4				
4	7 8 6 1 2 3 5 9	4 ↔ 9	9		
4	7 8 6 1 2 3 5			9	build heap cf. Ex. 1.13
8	7 4 6 1 2 3 5				
5	7 4 6 1 2 3 8	5 ↔ 8	8		
5	7 4 6 1 2 3			8, 9	build heap cf. Ex. 1.14
7	6 4 5 1 2 3				
3	6 4 5 1 2 7	3 ↔ 7	7		
3	6 4 5 1 2			7, 8, 9	build heap cf. Ex. 1.15
6	5 3 4 1 2				
2	5 3 4 1 6	2 ↔ 6	6		
2	5 3 4 1			6, 7, 8, 9	build heap cf. Ex. 1.16
5	4 2 3 1				
1	4 2 3 5	1 ↔ 5	5		
1	4 2 3			5, 6, 7, 8, 9	build heap cf. Ex. 1.17
4	3 1 2				
2 3	1 4	2 ↔ 4	4		
2 3	1			4, 5, 6, 7, 8, 9	build heap cf. Ex. 1.18
3 2	1				
1 2	3	1 ↔ 3	3		
1 2				3, 4, 5, 6, 7, 8, 9	build heap
2 1					
1 2		1 ↔ 2	2		
1				2, 3, 4, 5, 6, 7, 8, 9	

and we get the final sorted sequence

$$1 \quad 2 \quad 3 \quad 4 \quad 5 \quad 6 \quad 7 \quad 8 \quad 9$$

Exercise 1.1.

1. Arrange the sequence of Example 1.18 in non-decreasing order using (a) bubble sort, (b) insertion sort, (c) selection sort and (d) quick sort.
2. Arrange the sequences of Examples 1.2, 1.7 and 1.9 using heap sort.

Chapter 2

Growth of Functions

Given a real-valued function f defined on the set of real numbers, by the growth of f we mean how rapidly or slowly it increases in comparison with another real-valued function when the real number x increases. So for the growth of functions, there need to be considered two functions. How rapidly or slowly the function f increases in comparison with another function g is expressed in terms of certain notations called asymptotic notations. In this chapter, we are concerned with this concept. Of course, asymptotic means that how the two functions behave only for large values of x.

2.1. Introduction

Let f be a function from the set R of real numbers to R. By the growth of the function f we mean to say how the function $f(x)$ grows when the variable x increases or becomes large. In the analysis of algorithms we are generally concerned with the number of algebraic computations involved as the running time of an algorithm is proportional to this number. The number of computations depends on the size n (which is a positive integer) of the problem instance. Thus we come across functions $f : N \rightarrow N$, where N denotes the set of all natural numbers (or non-negative integers). Also there may be more than one algorithm for solving a certain problem so that for a given problem we may get two functions, say $f(n)$ and $g(n)$, giving the performance of the two algorithms (for the same problem). Which one of the two algorithms may be preferred will normally depend on the comparative growth of the two functions $f(n)$

and $g(n)$. In this chapter, we study this problem of comparative study of functions. A function $f(n)$ which grows more slowly than the function $g(n)$ is called a **superior function** (to $g(n)$). In fact we are not concerned with the growth of two functions for all n but only for large n. We in such situations, say that we are concerned with asymptotic behavior of the two functions.

2.2. Asymptotic Notations

In this section, we introduce some asymptotic notations which are used to compare the running times of different algorithms for the same problem. Before we do that, we define two types of asymptotic functions.

Definition. A function f from N to R (or from reals to reals) is called **asymptotically non-negative** if there exists a positive constant n_0 such that $f(n) \geq 0$ for all $n \geq n_0$ and f is called **asymptotically positive** if there exists a positive constant n_0 such that $f(n) > 0$ for all $n \geq n_0$.

It is clear that every asymptotically positive function is asymptotically non-negative. However, the converse is not true in general. For example, the function $n + (-1)^n n, n \in Z$ (the set of all integers) is asymptotically non-negative but is not asymptotically positive. A simple induction on n shows that $(n - 1)! - n^2 > 0$ for all $n \geq 6$ and, so, it is an asymptotically positive function. The function $f(n) = 1 + \cos n, n \in N$, oscillates between 0 and 2 and, so, is asymptotically non-negative but is not asymptotically positive (it is indeed a non-negative function and not just asymptotically non-negative).

2.2.1. *O*-notation (big oh notation)

Let g be a function defined from N to R or (R to R). Define $O(g)$ $(= O(g(n))) = \{f \mid f$ is a function from N to R for which there exist positive constants c and n_0, n_0 an integer, such that $|f(n)| \leq c|g(n)|$ for all $n \geq n_0\}$.

If $f \in O(g)$, by abuse of notation, we also write $f = O(g)$ or even $f(n) = O(g(n))$ and say that $f(n)$ is **big oh of** $g(n)$. For positive integers n and k, by $O(n^k)$ we mean $O(g)$, where g is the function defined by $g(x) = x^k$.

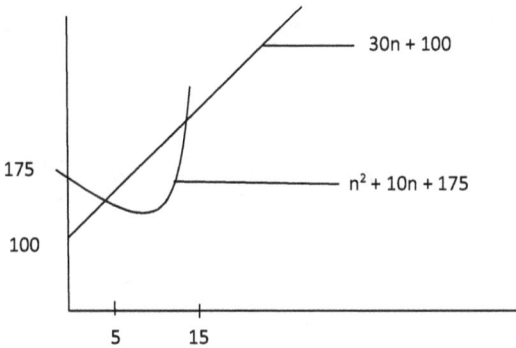

Fig. 2.1

Example 2.1. Consider $f(n) = 30n + 100$ and $g(n) = n^2 + 10n + 175$. The polynomial $f(n)$ being of degree 1 represents a straight line while $g(n)$ represents a parabola with vertex $(-5, 150)$. The straight line meets the parabola in two points $(5, 250)$ and $(15, 550)$. Beyond the point $(15, 550)$, the parabolic curve lies above the straight line. Therefore $30n + 100 = O(n^2 + 10n + 175)$. Observe that

$$g(n) - f(n) = n^2 - 20n + 75$$
$$= (n-5)(n-15) \geq 0 \text{ for all } n \geq 15$$

or that

$$|f(n)| \leq c|g(n)| \text{ for all } n \geq 15,$$

where $c = 1$. Therefore, $f(n) = O(g(n))$ (see Fig. 2.1).

Example 2.2. If $f(n) = an^3 + bn^2 + cn + d$, where a, b, c, d are constants and $a > 0$, then $f(n) = O(n^3)$.

Proof. Observe that

$$an^3 - |b|n^2 - |c|n - |d| \leq an^3 + bn^2 + cn + d$$
$$\leq an^3 + |b|n^2 + |c|n + |d|$$

for all $n \geq 1$. Let n_0 be a fixed positive integer. Then for all $n \geq n_0$, $\frac{1}{n_0^i} \geq \frac{1}{n^i}$ and $\frac{-1}{n_0^i} \leq \frac{-1}{n^i}$. Therefore, for all $\geq n_0$,

$$a - \frac{|b|}{n_0} - \frac{|c|}{n_0^2} - \frac{|d|}{n_0^3} \leq a + \frac{b}{n} + \frac{c}{n^2} + \frac{d}{n^3} \leq a + \frac{|b|}{n_0} + \frac{|c|}{n_0^2} + \frac{|d|}{n_0^3}.$$

Take n_0 the smallest positive integer $\geq 2 \max \left\{ \frac{|b|}{a}, \sqrt{\frac{|c|}{a}}, \sqrt[3]{\frac{|d|}{a}} \right\}$. Then $\frac{|b|}{n_0} \leq \frac{a}{2}, \frac{|c|}{n_0^2} = \frac{a}{4}, \frac{|d|}{n_0^3} \leq \frac{a}{8}$. Therefore,

$$\left(a - \frac{a}{2} - \frac{a}{4} - \frac{a}{8} \right) \leq a - \frac{|b|}{n_0} - \frac{|c|}{n_0^2} - \frac{|d|}{n_0^3}$$

$$\leq a - \frac{|b|}{n} - \frac{|c|}{n^2} - \frac{|d|}{n^3}$$

$$\leq a + \frac{|b|}{n} + \frac{|c|}{n^2} + \frac{|d|}{n^3}$$

$$\leq a + \frac{|b|}{n_0} + \frac{|c|}{n_0^2} + \frac{|d|}{n_0^3}$$

$$\leq a + \frac{a}{2} + \frac{a}{4} + \frac{a}{8}$$

or

$$\frac{a}{8} \leq a + \frac{b}{n} + \frac{c}{n^2} + \frac{d}{n^3} \leq \frac{15}{8}a$$

which implies that

$$\frac{a}{8}|n^3| \leq |f(n)| \leq \frac{15}{8}a|n^3| \text{ for all } n \geq n_0.$$

Taking $c_1 = \frac{8}{a}, c_2 = \frac{15}{8}a$, we have proved that $f(n) = O(n^3)$ and $n^3 = O(f(n))$.

Observe that taking n_0 the smallest positive integer $\geq \max \left\{ \frac{|b|}{a}, \sqrt{\frac{|c|}{a}}, \sqrt[3]{\frac{|d|}{a}} \right\}$ or $\frac{3}{2} \max \left\{ \frac{|b|}{a}, \sqrt{\frac{|c|}{a}}, \sqrt[3]{\frac{|d|}{a}} \right\}$ or even $\frac{5}{3} \max \left\{ \frac{|b|}{a}, \sqrt{\frac{|c|}{a}}, \sqrt[3]{\frac{|d|}{a}} \right\}$ does not help us in proving our claim that $n^3 = O(f(n))$. □

Using the argument as in Example 2.2, we can now prove the following general result.

Theorem 2.1. *If* $f(n) = a_0 n^k + a_1 n^{k-1} + \cdots + a_{k-1} n + a_k$, *where* $a_0, a_1, \ldots, a_k$ *are constants and* $a_0 > 0$, *is a polynomial of degree k, then* $f(n) = O(n^k)$ *and* $n^k = O(f(n))$.

Proof. For all $n \geq 1$, we have

$$a_0 - \frac{|a_1|}{n} - \frac{|a_2|}{n^2} - \cdots - \frac{|a_k|}{n^k} \leq a_0 + \frac{a_1}{n} + \frac{a_2}{n^2} + \cdots + \frac{a_k}{n^k}$$

$$\leq a_0 + \frac{|a_1|}{n} + \frac{|a_2|}{n^2} + \cdots + \frac{|a_k|}{n^k}$$

Let n_0 be a fixed positive integer. Then for all $n \geq n_0$, $\frac{1}{n_0^i} \geq \frac{1}{n^i}$, $i = 1, 2, \ldots, k$. Therefore, for all $n \geq n_0$, we have

$$a_0 - \frac{a_1}{n_0} - \frac{a_2}{n_0^2} - \cdots - \frac{a_k}{n_0^k}$$

$$\leq a_0 - \frac{a_1}{n} - \frac{a_2}{n^2} - \cdots - \frac{a_k}{n^k}$$

$$\leq \frac{1}{n^k} f(n) \leq a_0 + \frac{|a_1|}{n} + \frac{|a_2|}{n^2} + \cdots + \frac{|a_k|}{n^k}$$

$$\leq a_0 + \frac{a_1}{n_0} + \frac{a_2}{n_0^2} + \cdots + \frac{a_k}{n_0^k} \tag{2.1}$$

Take $n_0 = $ the least positive integer $\geq 2 \max \left\{ \frac{|a_1|}{a_0}, \sqrt{\frac{|a_2|}{a_0}}, \sqrt[3]{\frac{|a_3|}{a_0}}, \ldots, \sqrt[k]{\frac{|a_k|}{a_0}} \right\}$. Then relation (2.1) implies that for all $n \geq n_0$, $a_0 - \frac{a_0}{2} - \frac{a_0}{2^2} - \cdots - \frac{a_0}{2^k} \leq \frac{1}{n^k} f(n) \leq a_0 + \frac{a_0}{2} + \frac{a_0}{2^2} + \cdots + \frac{a_0}{2^k}$ or $\frac{1}{2^k} a_0 \leq \frac{1}{n^k} f(n) \leq (2 - \frac{1}{2^k}) a_0$ or $c_1 |n^k| \leq |f(n)| \leq c_2 |n^k|$, where $c_1 = \frac{a_0}{2^k}$, $c_2 = (2 - \frac{1}{2^k}) a_0$. Hence $f(n) = O(n^k)$ and $n^k = O(f(n))$. $\square$

Remark. The proof of $f(n) = O(n^k)$ is much simpler than the one given above. In fact, for all $n \geq 1$, we have $|f(n)| \leq (a_0 + |a_1| + \cdots + |a_k|) n^k$ and indeed, for this part we do not even need a_0 to be positive as we could replace a_0 in the above inequality by $|a_0|$. However, the harder proof above yields the existence of positive constants c and n_0 with $c|n^k| \leq |f(n)|$ for all $n \geq n_0$.

It immediately follows from Theorem 2.1 that for any polynomial $f(n)$ and constant a, $O(f(n)+a) = O(f(n))$. In particular, $O(n^k + a) = O(n^k)$ for any positive integers n and k. However, for any function $f(x)$ and constant a, do we have $f(x) + a = O(g(x))$ and $af(x) = O(g(x))$ if $f(x) = O(g(x))$? Suppose that $f(x) = O(g(x))$. It is fairly easy to see that $af(x) = O(g(x))$. Now suppose that $g(x)$ has an asymptotic lower bound, i.e., there exist positive constants c and n_0 such that $g(x) \geq c$ for all $x \geq n_0$. As $f(x) = O(g(x))$, there exist positive constants c_1 and n_1 such that $|f(x)| \leq c_1|g(x)|$ for all $x \geq n_1$. Let $n_2 = \max\{n_0, n_1\}$. Then, for all $x \geq n_2$,

$$|f(x) + a| \leq |f(x)| + |a| \leq c_1|g(x)| + \frac{|a|}{c}c$$

$$\leq c_1|g(x)| + \frac{|a|}{c}|g(x)| = \left(c_1 + \frac{|a|}{c}\right)|g(x)|$$

Therefore $f(x) + a = Og(x))$.

We next consider some simple but important properties that the asymptotic notation O satisfies.

Definition. Given functions $f, g : A \to B$, where A, B are subsets of real numbers, then $\max\{f, g\}$ is a function $h : A \to B$ such that for every $x \in A, h(x) = \max\{f(x), g(x)\}$.

Example 2.3. Let $f(x) = 3x + 5$, $g(x) = x^2$ and $h = \max\{f, g\}$. Then h is given by

x	$f(x)$	$g(x)$	$h(x)$
1	8	1	$8 = f(1)$
2	11	4	$11 = f(2)$
3	14	9	$14 = f(3)$
4	17	16	$17 = f(4)$
5	20	25	$25 = g(5)$

A simple induction on x shows that $h(x) = g(x)$ for all $x \geq 5$, while as above, $h(x) = f(x)$ for $x = 1, 2, 3, 4$.

Remark. Since in the analysis of running time of algorithms negative values usually do not appear, in the definitions of asymptotic notations we may ignore the modulus sign $|\ldots|$ and may take $f(x)$ in place of $|f(x)|$ etc. We invoke this observation whenever we find it more convenient.

Theorem 2.2. *If $f(n) \in O(g(n))$ and $g(n) \in O(h(n))$, then $f(n) \in O(h(n))$.*

Proof. Exercise. □

Theorem 2.3. *If $f_1 \in O(g_1)$ and $f_2 \in O(g_2)$, then $f_1 + f_2 \in O(\max\{|g_1|, |g_2|\})$.*

Proof. Suppose that $f_1 \in O(g_1)$ and $f_2 \in O(g_2)$. Then there exist positive constants c_1, c_2, n_1, n_2 such that

$$|f_1(n)| \leq c_1|g_1(n)| \quad \text{for all } n \geq n_1$$

and

$$|f_2(n)| \leq c_2|g_2(n)| \quad \text{for all } n \geq n_2.$$

Let $\max\{|g_1|, |g_2|\} = h$. Then for every n,

$$h(n) = \max\{|g_1(n)|, |g_2(n)|\}.$$

For $n \geq n_1 + n_2$, we have

$$
\begin{aligned}
|(f_1 + f_2)(n)| = |f_1(n) + f_2(n)| &\leq |f_1(n)| + |f_2(n)| \\
&\leq c_1|g_1(n)| + c_2|g_2(n)| \leq c_1 h(n) + c_2 h(n) \\
&= (c_1 + c_2)h(n).
\end{aligned}
$$

Therefore $f_1 + f_2 \in O(h)$. □

Corollary 2.4. *If $f(n), g(n) \in O(h(n))$, then $f(n) + g(n) \in O(h(n))$.*

Corollary 2.5. *Let $f_1 \in O(g_1)$, $f_2 \in O(g_2)$ and $g_1 \in O(g_2)$. Then $f_1 + f_2 \in O(g_2)$.*

Proof. Since $f_1 \in O(g_1)$ and $g_1 \in O(g_2)$, it follows from Theorem 2.2 that $f_1 \in O(g_2)$. The result then follows from Corollary 2.4. □

Remark. Replacing max by min we may ask: If $f_1(n) \in O(g_1(n))$ and $f_2(n) \in O(g_2(n))$, is it always true that $f_1 + f_2 \in O(\min\{|g_1|, |g_2|\})$? This is not true. For example $n \in O(n)$, $n^2 \in O(n^2)$, $\min\{n, n^2\} = n$ but

$n + n^2 \notin O(n)$. For, if $n + n^2 \in O(n)$, there exist positive constants c and n_0, n_0 an integer, such that for all $n \geq n_0$,

$$n + n^2 \leq cn \quad \text{or} \quad n \leq c - 1 \tag{2.2}$$

Taking $n_1 = \max\{n_0, c\}$, we have $n_1 \geq n_0$ and $n_1 \geq c > c - 1$ which contradicts (2.2).

Theorem 2.6. *Let $f(n) + g(n) \in O(h(n))$ and $f(n)$, $g(n)$ be asymptotically non-negative. Then $f(n) \in O(h(n))$.*

Proof. There exist non-negative constants c, n_1, n_2, n_3 such that

$$f(n) \geq 0 \text{ for all } n \geq n_1, g(n) \geq 0 \text{ for all } n \geq n_2 \quad \text{and}$$

$$|f(n) + g(n)| \leq c|h(n)| \text{ for all } n \geq n_3.$$

Let $n_0 = n_1 + n_2 + n_3$. Then for all $n \geq n_0$

$$0 \leq f(n) \leq f(n) + g(n) = |f(n) + g(n)| \leq c|h(n)|$$

and the result follows. □

Theorem 2.7. *If $f(n) \in O(g(n))$, $f'(n) \in O(g'(n))$, then $f(n)f'(n) \in O(g(n)g'(n))$.*

Proof. Exercise □

Theorem 2.8. *If $f(n)g(n) \in O(h(n))$ and $g(n)$ has a positive asymptotically lower bond, then $f(n) \in O(h(n))$.*

Proof. Let $f(n)$, $g(n)$ and $h(n)$ be functions such that $f(n)g(n) \in O(h(n))$ and $g(n)$ has a positive asymptotically lower bond. Then there exist positive constants a, b, n_1, n_2 such that

$$g(n) \geq a \text{ for all } n \geq n_1 \quad \text{and} \quad |f(n)g(n)| \leq b|h(n)| \text{ for all } n \geq n_2.$$

For all $n \geq n_1 + n_2$,

$$a|f(n)| \leq |f(n)|g(n) = |f(n)g(n)| \leq b|h(n)| \quad \text{or} \quad |f(n)| \leq \frac{b}{a}|h(n)|$$

and, therefore, $f(n) \in O(h(n))$. □

In case the condition on $g(n)$ being asymptotically positive is dropped, the result of Theorem 2.8 may not be true.

Example 2.4. Let

$$f(n) = \begin{cases} n & \text{if } n \text{ is a multiple of 3,} \\ 0 & \text{otherwise,} \end{cases}$$

$$g(n) = \begin{cases} 0 & \text{if } n \text{ is a multiple of 3,} \\ n & \text{otherwise,} \end{cases}$$

and $h(n) = 0$ for all n. Then $f(n)g(n) = 0$ for all n and, so, $f(n)g(n) \in O(h(n))$.

If $f(n) \in O(h(n))$, there exists a positive constant n_0 such that $f(n) = 0$ for all $n \geq n_0$. We can choose an $n \geq n_0$ which is a multiple of 3. For this n, $f(n) = n \neq 0$ which is a contradiction. Hence $f(n) \neq O(h(n))$.

The restriction on $f(n)$, $g(n)$ being asymptotically non-negative is essential in Theorem 2.6. For example, if $f(n) = n$, $g(n) = -n$ and $h(n) = 0$, then $f(n) + g(n) = O(h(n))$ whereas $f(n) \neq O(h(n))$. Observe that $g(n)$ is not asymptotically non-negative.

As applications of the general results, we consider some examples.

Example 2.5. (i) $3^n + n^3 = O(3^n)$ and (ii) $12n^2 3^n + n^3 = O(n^2 3^n)$.

Solution. A simple induction on n shows that $n^3 \leq 3^n$ for all $n \geq 3$ and then $n^3 \leq n^2 3^n$ for all $n \geq 3$. Therefore $n^3 = O(3^n)$ and $n^3 = O(n^2 3^n)$. Also $3^n = O(3^n)$ and $12n^2 3^n = O(n^2 3^n)$. Hence $3^n + n^3 = O(3^n)$ and $12n^2 3^n + n^3 = O(n^2 3^n)$.

Example 2.6. Suppose that there are two computer algorithms such that Algorithm 1 has complexity $n^2 - 2n + 1$ and Algorithm 2 has complexity $\frac{n^2}{2} + 2n + 3$. Then the two complexities are both in $O(n^2)$. The second algorithm complexity has smaller leading coefficient and, so, is expected to be smaller than complexity of first algorithm. A simple induction on n shows that

$$n^2 - 2n + 1 \geq \frac{n^2}{2} + 2n + 3 \quad \text{for all } n \geq 9.$$

Therefore, the second algorithm is indeed faster than the first one and, so, it needs to be preferred to Algorithm 1.

Example 2.7. $3^n = O(n!)$, where $n!$ stands for factorial n.

Solution. For $n \geq 1$,

$$3^n = 3.3\ldots.3(n \text{ terms}) \ < 9.3.4..n < 9.1.2.3.4\ldots n = 9.n!$$

which implies that $3^n = O(n!)$.

2.2.2. Ω-notation (big omega notation)

Definition. Given a function $g(n)$, we define $\Omega(g(n)) = \{f(n)|$ there exist positive constants c and n_0 such that $c|g(n)| \leq |f(n)|$ for all $n \geq n_0\}$.

If $f(n) \in \Omega(g(n))$, we also express it by writing $f(n) = \Omega(g(n))$ or simply saying that $f(n)$ is $\Omega(g(n))$.

It is clear from the definitions of O and Ω that $f(n) = \Omega(g(n))$ if and only if $g(n) = O(f(n))$.

In Theorem 2.1, we proved that if $f(n) = a_0 n^k + a_1 n^{k-1} + \cdots + a_{k-1}n + a_k$, a_i real numbers and $a_0 > 0$, is a polynomial of degree k, there exists a positive integer n_0 such that for all $n \geq n_0$, $c_1 |n^k| \leq |f(n)|$, where $c_1 = \frac{a_0}{2^k}$. This shows that $f(n) \in \Omega(n^k)$.

In view of the connection between the notations O and Ω, we can have some general properties satisfied by Ω corresponding to the properties satisfied by the O notation proved in Theorems 2.2–2.8. Precise statements of these results are left as exercises.

2.2.3. θ-notation (big theta notation)

Definition. Given a function $g(n)$, we define $\theta(g(n)) = \{f(n) \mid$ there exist positive constants c_1, c_2, n_0 such that $c_1 |g(n)| \leq |f(n)| \leq c_2 |g(n)|$ for all $n \geq n_0\}$.

If $f(n) \in \theta(g(n))$, we also express it by writing $f(n) = \theta(g(n))$. It is immediate from the definitions of asymptotic notations O, Ω and θ that

$$f(n) \in \theta(g(n)) \text{ if and only if } f(n) \in O(g(n)) \text{ and } f(n) \in \Omega(g(n)).$$

Example 2.8. (i) $f(n) = \frac{1}{3}n^2 - 2n = \theta(n^2)$; (ii) $3n^2 \neq \theta(n)$.

Solution. (i) For $n \geq 6$, $\frac{1}{3}n^2 - 2n \geq 0$, so we need to find positive constants c_1, c_2, n_0 (≥ 6) such that for all $n \geq n_0$,

$$c_1 n^2 \leq \frac{1}{3}n^2 - 2n \leq c_2 n^2 \quad \text{or}$$

$$c_1 \leq \frac{1}{3} - \frac{2}{n} \leq c_2 \tag{2.3}$$

Taking $c_2 = \frac{1}{3}$, the right-hand side in (2.3) holds for all $n > 0$ and, in particular, for $n \geq n_0$ (≥ 6). However, for $n = 6$, the left-hand inequality in (2.3) implies that $c_1 = 0$ whereas we are looking for $c_1 > 0$. But for $n \geq 7$, both the inequalities in (2.3) are possible. Taking $n_0 = 8$, we find that for $n \geq 8$,

$$\frac{1}{3} - \frac{2}{n} \geq \frac{1}{3} - \frac{1}{4} = \frac{1}{12}.$$

So, we take $n_0 = 8$, $c_1 = \frac{1}{12}$, $c_2 = \frac{1}{3}$. We find that, for all $n \geq 8$, $\frac{1}{12}n^2 \leq \frac{1}{3}n^2 - 2n \leq \frac{1}{3}n^2$ so that $\frac{1}{3}n^2 - 2n = \theta(n^2)$.

Observe that we could have taken $n_0 = 7$ but then we need to take $c_1 = \frac{1}{21}$ and $c_2 = \frac{1}{3}$.

(ii) Suppose that there exist positive constants c_1, c_2, n_0 such that for all $n \geq n_0$,

$$c_1 n \leq 3n^2 \leq c_2 n \quad \text{or} \quad c_1 \leq 3n \leq c_2.$$

Since $3n$ is a strictly increasing function, we may choose n_0 such that $3n_0 > c_2$ which will contradict $3n \leq c_2$ for all $n \geq n_0$. However, the left-hand inequality can be satisfied by taking $c_1 = n_0 = 1$. Therefore, although $3n^2 = \Omega(n)$, $3n^2 \neq \theta(n)$.

Example 2.9. $3n^2 - 19n = \theta(n^2)$.

Solution. Observe that for $n \geq 7$, $3n^2 - 19n \geq 0$ and, so, for such an n, $|3n^2 - 19n| = 3n^2 - 19n$. Also $|n^2| = n^2$ for every such n. So, we need to find positive constants c_1, c_2 and $n_0 \geq 7$ such that for all $n \geq n_0$,

$$c_1 n \leq 3n - 19 \leq c_2 n. \tag{2.4}$$

For $n \geq 10$, we find that $n \leq 3n - 19 \leq 3n$. Thus, by taking $c_1 = 1$, $c_2 = 3$ and $n_0 = 10$, (2.4) holds. Hence $3n^2 - 19n = \theta(n^2)$.

Remarks.

1. The proof of $3n^2 \neq \theta(n)$ also works for $3n^{k+1} \neq \theta(n^k)$ for every $k \geq 1$.
2. Examples 2.8(i) and 2.9 are particular cases of the result of Theorem 2.1.

In Theorem 2.1 we have proved that $(n + a)^k = \theta(n^k)$ for all real numbers a and positive integers k. We can in fact prove this result when k is any positive real number.

Example 2.10. For any real constants a, b, $b > 0$, $(n + a)^b = \theta(n^b)$.

Solution. Choose n_0 a positive integer such that $n_0 + a > 0$. We need to, then, find positive constants c_1, c_2 such that

$$c_1 n^b \leq (n + a)^b \leq c_2 n^b \quad \text{or} \quad c_1 \leq \left(1 + \frac{a}{n}\right)^b \leq c_2 \text{ for all } n \geq n_0.$$

Observe that

$$\left(1 - \frac{|a|}{n}\right)^b \leq \left(1 + \frac{a}{n}\right)^b \leq \left(1 + \frac{|a|}{n}\right)^b.$$

Rechoose n_0 such that $n_0 \geq 2|a|$ as well. Then, for all $n \geq n_0$,

$$\left(\frac{1}{2}\right)^b \leq \left(1 - \frac{|a|}{n_0}\right)^b \leq \left(1 - \frac{|a|}{n}\right)^b \leq \left(1 + \frac{a}{n}\right)^b$$

$$\leq \left(1 + \frac{|a|}{n}\right)^b \leq \left(1 + \frac{|a|}{n_0}\right)^b \leq \left(\frac{3}{2}\right)^b.$$

Take $c_1 = \frac{1}{2^b}$, $c_2 = (\frac{3}{2})^b$ both of which are positive and we have, for all $n \geq n_0$,

$$c_1 n^b \leq (n + a)^b \leq c_2 n^b.$$

Hence $(n + a)^b = \theta(n^b)$.

Theorem 2.9. *Let $|h_1(n)| \leq |f(n)| \leq |h_2(n)|$ for sufficiently large n and $h_1(n) = \Omega(g(n))$ but $h_2(n) = O(g(n))$. Then $f(n) = \theta(g(n))$.*

Proof. There exist positive constants c_1, c_2, n_1, n_2, n_3 such that

$$|h_1(n)| \leq |f(n)| \leq |h_2(n)| \quad \text{for all } n \geq n_3, \tag{2.5}$$

$$c_1|g(n)| \leq |h_1(n)| \quad \text{for all } n \geq n_1, \tag{2.6}$$

$$|h_2(n)| \leq c_2|g(n)| \quad \text{for all } n \geq n_2. \tag{2.7}$$

Let $n_0 = \max\{n_1, n_2, n_3\}$. Then the relations (2.5)–(2.7) hold for all $n \geq n_0$. Therefore, for all $n \geq n_0$, we have $c_1|g(n)| \leq |f(n)| \leq c_2|g(n)|$ and, so, $f(n) = \theta(g(n))$. $\square$

As an application of this result, we have the following result.

Example 2.11. For a fixed integer $k \geq 1$, $\sum_{i=1}^{n} i^k = \theta(n^{k+1})$.

Solution. $\sum_{i=1}^{n} i^k \leq \sum n^k = n^{k+1}$ which is in $O(n^{k+1})$. Let $\lceil \frac{n}{2} \rceil$ be the smallest integer $\geq \frac{n}{2}$. Then (we can suppose that $n > 1$)

$$\sum_{i=1}^{n} i^k \geq \sum_{i=\lceil \frac{n}{2} \rceil}^{n} i^k \geq \sum_{i=\lceil \frac{n}{2} \rceil}^{n} \left(\frac{n}{2}\right)^k$$

$$= \left(n - \left\lceil \frac{n}{2} \right\rceil + 1\right) \left(\frac{n}{2}\right)^k \geq \left(\frac{n}{2}\right)^{k+1} = \left(\frac{1}{2}\right)^{k+1} n^{k+1}$$

and $\left(\frac{1}{2}\right)^{k+1} n^{k+1} \in \Omega(n^{k+1})$. It then follows from Theorem 2.9 (or more directly from the definition of θ) that $\sum_{i=1}^{n} i^k = \theta(n^{k+1})$.

2.3. Some Asymptotic Notations Using Limits

There are some asymptotic notations that involve the concept of limit of a function. Recall that a function $f(x)$ approaches a finite limit a when x approaches ∞ (infinity), if given any $c > 0$, there exists an $x_0 > 0$ such that $|f(x) - a| < c$ for all $x \geq x_0$. We are particularly interested in the case when $a = 0$. We are also interested in the case $a = \infty$. We say that $f(x)$ approaches ∞ when x approaches ∞ if given any positive M, there exists an $x_0 > 0$ such that $|f(x)| > M$ for all $x \geq x_0$.

Definition. Given a function $g(n)$, we define $o(g(n)) = \{f(n)|$ for every $c > 0$, there exists a positive number n_0 such that $|f(n)| < c|g(n)|$ for all $n \geq n_0\}$.

If $f(n) \in o(g(n))$, we say that $f(n)$ is "**little oh**" of $g(n)$ and also indicate it by writing $f(n) = o(g(n))$. Observe that every function $f(n)$ in $o(g(n))$ is surely in $O(g(n))$. However, the converse is not true in general, as there do exist functions $f(n)$, $g(n)$ such that $f(n)$ is in $O(g(n))$ but is not in $o(g(n))$. For example, $3n^2 = O(n^2)$ but $3n^2 \neq o(n^2)$ because if we take $c = 2$, there is no n_0 such that $3n^2 < 2n^2$ for all $n \geq n_0$.

In view of the definition of limit, we find that $f(n)$ is in $o(g(n))$ if and only if $\lim_{n\to\infty} \frac{f(n)}{g(n)} = 0$; i.e., relative to $g(n)$, $f(n)$ becomes insignificant for sufficiently large n.

Definition. Given a function $g(n)$, we define $\omega(g(n)) = \{f(n)|$ given any $c > 0$, there exists an $n_0 > 0$ such that $c|g(n)| \leq |f(n)|$ for all $n \geq n_0\}$.

If $f(n)$ is in $\omega(g(n))$, we say that $f(n)$ is "**little omega**" of $g(n)$ and we express it by writing $f(n) = \omega(g(n))$. It is immediate from the definitions of Ω and ω that if $f(n)$ is in $\omega(g(n))$, then $f(n)$ is in $\Omega(g(n))$. However, the converse is not true in general. Before considering an example in support of this we observe (in view of the definition of limit) that $f(n)$ is in $\omega(g(n))$ if and only if $\lim_{n\to\infty} \frac{f(n)}{g(n)} = \infty$.

Given $c > 0$, there exists an $n_0 > 0$ such that $|f(n)| < c|g(n)|$ for all $n \geq n_0$ if and only if $|g(n)| > \frac{1}{c}|f(n)|$ for all $n \geq n_0$. Hence $f(n) = \omega(g(n))$ if and only if $g(n) = o(f(n))$.

Example 2.12. $\frac{n^3}{3} = \omega(n^2)$, $\frac{n^3}{3} \neq \omega(n^3)$ and $n^3 = \Omega(n^3)$.

Solution. Let $c > 0$ be given. Take $n_0 = 3c + 1$. Then for all $n \geq n_0$,

$$3cn^2 < (3c + 1)n^2 \leq n.n^2 = n^3 \quad \text{or} \quad cn^2 \leq \frac{n^3}{3}.$$

Therefore $\frac{n^3}{3} = \omega(n^2)$.

If $\frac{n^3}{3} = \omega(n^3)$, then given any $c > 0$, there exists an $n_0 > 0$ such that for all $n \geq n_0$, $cn^3 \leq \frac{n^3}{3}$ or $3c \leq 1$. We could have taken $c = 1$. Then there is no n_0 such that $3c \leq 1$ for all $n \geq n_0$. Hence $\frac{n^3}{3} \neq \omega(n^3)$.

That $\frac{n^3}{3} = \Omega(n^3)$ follows from Theorem 2.1.

Given real numbers a, b, we know that either $a < b$ or $a = b$ or $a > b$. We now prove that the corresponding result for O, Ω, o and ω does not hold for asymptotically non-negative functions.

Example 2.13. Consider functions $f(n) = n$, $g(n) = n^{1+\cos n}$. Suppose that there exist constants $c > 0$ and $n_0 > 0$, such that for all $n \geq n_0$,

$$n \leq cn^{1+\cos n} \text{ or } \frac{1}{c} \leq n^{\cos n}. \tag{2.8}$$

Choose n an odd multiple of π with $n > \max\{n_0, c\}$. Then $\cos n = -1$ and (2.8) becomes $n \leq c$ which contradicts the choice of n. Hence $f(n) \neq O(g(n))$.

Next suppose that there exist constants $c > 0$ and $n_0 > 0$, n_0 an integer, such that for all $n \geq n_0$,

$$0 \leq cn^{1+\cos n} \leq n \text{ or } n^{\cos n} \leq \frac{1}{c}. \tag{2.9}$$

Now, choose n an even multiple of π with $n > \max\{n_0, \frac{1}{c}\}$. Then $\cos n = 1$ so that (2.9) becomes $n \leq \frac{1}{c}$ which contradicts the choice of n. Hence $f(n) \neq \Omega(g(n))$.

Observe that $\lim_{n \to \infty} \frac{n^{1+\cos n}}{n} = \lim_{n \to \infty} n^{\cos n}$ does not exist because for large enough values of n, $\cos n$ can be 1 or -1 according as we take n an integral multiple of 2π or an odd integral multiple of π. Hence neither $f(n) = o(g(n))$ nor $f(n) = \omega(g(n))$.

For any positive function $h(n)$, $\frac{h(n)}{h(n)} = 1$ so that $\lim_{n \to \infty} \frac{h(n)}{h(n)} = 1$ which proves that neither $h(n) = o(h(n))$ nor $h(n) = \omega(g(n))$.

Remark. If $f(n) = o(g(n))$, then $f(n) \neq \Omega(g(n))$ while if $f(n) = \omega(g(n))$, then $f(n) \neq O(g(n))$.

Definition. Functions $f(n)$ and $g(n)$ are said to be **asymptotically comparable** if $f(n) = \lambda(g(n))$ where λ is any one of the asymptotic notations O, Ω, θ, o or ω. Otherwise $f(n)$ and $g(n)$ are said to be **asymptotically non-comparable**.

Example 2.13 gives two asymptotically non-comparable functions. Two other such functions are given in the following example.

Example 2.14. The functions $f(n) = \sqrt{n}$ and $g(n) = n^{\cos n}$ are asymptotically non-comparable.

Proof. Suppose that $\sqrt{n}$ is in $O(n^{\cos n})$. Then there exist positive constants c, n_0 such that for all $n \geq n_0$, $\sqrt{n} \leq cn^{\cos n}$. Let m be a number of the form $2k\pi + \frac{\pi}{2}, k$ a positive integer and $m > \max\{n_0, c^2\}$. Then

$$\sqrt{m} \leq cm^{\cos m} = c \quad \text{or} \quad m \leq c^2,$$

which contradicts the choice of m. Hence $\sqrt{n}$ is not in $O(n^{\cos n})$.

If $\sqrt{n}$ is in $\Omega(n^{\cos n})$, there exist positive constants c, n_0 such that for all $n \geq n_0$,

$$0 < cn^{\cos n} \leq \sqrt{n}.$$

Let m be a positive number which is a multiple of 2π and $m > \max\{n_0, \frac{1}{c^2}\}$. Then $cm^{\cos m} \leq \sqrt{m}$ or $m \leq \frac{1}{c^2}$ which contradicts the choice of m. Hence $\sqrt{n}$ is not in $\Omega(n^{\cos n})$. Then, also $\sqrt{n}$ is not in $\theta(n^{\cos n})$.

Since $f(n) \in o(g(n))$ implies that $f(n) \in O(g(n))$ and $f(n) \in \omega(g(n))$ implies that $f(n) \in \Omega(g(n))$, it follows from above that $\sqrt{n}$ is neither in $o(n^{\cos n})$ nor in $\omega(n^{\cos n})$.   □

Remark. To prove that functions $f(n)$ and $g(n)$ are non-comparable, it is enough to prove that $f(n)$ is neither in $O(g(n))$ nor in $\Omega(g(n))$.

Exercise 2.1. Decide if the following pairs of functions are asymptotically comparable or not.

(a) $\sqrt{n}$ and $n^{\sin n}$;
(b) n and $n^{1+\sin n}$;
(c) $\sqrt{n}$ and $n^{\tan n}$.

Lemma 2.10. *Let a, b be real numbers with $b > 0$. Then $|a| \leq b$ if and only if $-b \leq a \leq b$.*

Proof. Suppose that $|a| \leq b$.

If $a > 0$, then $a = |a| \leq b$ and, so, $-b \leq -a \leq a \leq b$, i.e., $-b \leq a \leq b$.

If $a < 0$, then $-a = |a| \leq b$ and, so, $-b \leq -(-a) = a$. Thus $-b \leq a \leq b$.

Therefore, in either case, $-b \leq a \leq b$.

Conversely, suppose that $-b \leq a \leq b$.

If $a < 0$, then $-b \leq a$ implies $-a \leq b$ or $|a| \leq b$. If $a > 0$, then $|a| = a \leq b$ (by hypothesis). Hence $-b \leq a \leq b$ implies that $|a| \leq b$.   □

Corollary 2.11. *If a, b are real numbers, then $|a| - |b| \leq |a + b| \leq |a| + |b|$.*

Proof. It follows from the lemma that $-|a| \leq a \leq |a|$ and $-|b| \leq b \leq |b|$. Therefore $-(|a| + |b|) \leq a + b \leq |a| + |b|$ which implies that $|a + b| \leq |a| + |b|$

Again $|a| = |a + b - b| = |a + b + (-b)| \leq |a + b| + |-b| = |a + b| + |b|$. Therefore $|a| - |b| \leq |a + b|$. $\qquad\square$

Theorem 2.12. *Let $f(n)$, $g(n)$ be asymptotically positive functions with $\lim_{n \to \infty} \frac{f(n)}{g(n)} = a$, $0 < a < \infty$. Then $f(n) = \theta(g(n))$.*

Proof. Since $\lim_{n \to \infty} \frac{f(n)}{g(n)} = a$, given any c, $0 < c < a$, there exists an $n_0 > 0$ such that for all $n \geq n_0$,

$$\left| \frac{f(n)}{g(n)} - a \right| < c \quad \text{or} \quad -c < \frac{f(n)}{g(n)} - a < c \quad \text{or} \quad a - c < \frac{f(n)}{g(n)} < a + c.$$

Since $a - c > 0$, $a + c > 0$ and $f(n)$, $g(n)$ are asymptotically positive, we have for all $n \geq n_0$ (the integer n_0 is suitably rechosen)

$$(a - c)|g(n)| < |f(n)| < (a + c)|g(n)| \text{ which implies that}$$

$$f(n) = \theta(g(n)).$$

$\qquad\square$

Remark. Observe that without the restriction $f(n)$, $g(n)$ asymptotically positive, we could not have $\lim_{n \to \infty} \frac{f(n)}{g(n)}$ positive. For example, if $f(n) = n \cos n$ and $g(n) = n$, then limit of the quotient $\frac{f(n)}{g(n)} = \cos n$ does not exist. Also, if $f(n) = -n$, $g(n) = n$, then $\frac{f(n)}{g(n)} = -1$ and its limit as $n \to \infty$ is -1.

2.4. Asymptotic Notations in Equations

We have used asymptotic notations in mathematical formulas such as $n = O(n^2)$, $n^3 = \Omega(n^2)$, $2n^3 - 3n^2 + 2n - 5 = \theta(n^3)$, etc. However, there may be equations in which one asymptotic notation may appear on the left and another appears on the right. We may need to give an interpretation to the notations in such cases. For example, what do we understand by

$$2n^3 - 3n^2 + O(n) = O(n^3), \tag{2.10}$$

$$2n^3 - 3n^2 + 2n + 1 = 2n^3 + \theta(n^2), \tag{2.11}$$

$$2n^3 + \theta(n^2) = \theta(n^3). \tag{2.12}$$

Such equations are interpreted by the rule:

In whatever way an unknown function on one side of the equation is chosen, there is a way to choose an unknown function on the other side of the equation so that the equation remains valid.

Thus, for Eq. (2.10), no matter how a function $f(n)$ in $O(n)$ is chosen, there is a way to choose a function $g(n)$ in $O(n^3)$ such that the equation $2n^3 - 3n^2 + f(n) = g(n)$ holds. If $f(n) \in O(n)$, there exist positive constants c, n_0 (n_0 an integer) such that $|f(n)| \leq cn$ for all $n \geq n_0$. Then $|2n^3 - 3n^2 + f(n)| \leq |2n^3 - 3n^2| + |f(n)| \leq |2n^3 - 3n^2| + cn$ for all $n \geq n_0$. But $2n^3 - 3n^2$ is in $O(n^3)$. Then there exist constants d, n_1, n_1 an integer such that $|2n^3 - 3n^2| \leq dn^3$ for all $n \geq n_1$. Thus $|2n^3 - 3n^2 + f(n)| \leq dn^3 + cn$ for all $n \geq n_0 + n_1$. Since $dn^3 + cn$ is in $O(n^3)$, it follows that $2n^3 - 3n^2 + f(n)$ is in $O(n^3)$.

Regarding (2.11), there exists a function $g(n)$ in $\theta(n^2)$ for which $2n^3 - 3n^2 + 2n + 1 = 2n^3 + g(n)$. Observe that $|-3n^2 + 2n + 1| \leq |-3n^2| + |2n + 1| = 3n^2 + 2n + 1 \leq 4n^2$ for all $n \geq 3$ and $|-3n^2 + 2n + 1| \geq |-3n^2| - |2n + 1| = 3n^2 - 2n - 1 \geq n^2$ for all $n \geq 2$.

Thus for all $n \geq 3$, $n^2 \leq |-3n^2 + 2n + 1| \leq 4n^2$ so that $g(n) = -3n^2 + 2n + 1 \in \theta(n^2)$. We thus find that $2n^3 - 3n^2 + 2n + 1 = 2n^3 + g(n)$ and $g(n) \in \theta(n^2)$.

In the case of (2.12), no matter how a function $g(n)$ in $\theta(n^2)$ is chosen, there exists a function $h(n)$ in $\theta(n^3)$ such that $2n^3 + g(n) = h(n)$. Let $g(n)$ be in $\theta(n^2)$. Then there exist positive constants c_1, c_2 and a positive integer n_0 such that for all $n \geq n_0$, $c_1 n^2 \leq |g(n)| \leq c_2 n^2$ so that $2n^3 + c_1 n^2 \leq 2n^3 + |g(n)| \leq 2n^3 + c_2 n^2 \leq (c_2 + 2)n^3$.

Then $|2n^3 + g(n)| \leq 2n^3 + |g(n)| \leq (c_2 + 2)n^3$ for all $n \geq n_0$. Also $|2n^3 + g(n)| \geq 2n^3 - |g(n)| \geq (2 - \frac{c_1}{n})n^3 \geq n^3$ for all $n \geq \max\{c_1, n_0\}$. It then follows that $2n^3 + g(n) \in \theta(n^3)$.

2.5. Comparison of Growth of Some Typical Functions

We observed in Theorem 2.2 that the asymptotic notation O has transitive property, i.e., if $f(n)$ is in $O(g(n))$ and $g(n)$ is in $O(h(n))$, then $f(n)$ is in $O(h(n))$. Observe that for all $n \geq 1$, $3^{n+1} \leq c3^n$ with $c = 3$ showing that $3^{n+1} = O(3^n)$. By the same token $3^{n+2} = O(3^{n+1})$ and, therefore, $3^{n+2} = O(3^n)$. Then $3^{n+3} = O(3^n)$. We can continue this process a finite number of times. We cannot continue this infinitely as the following example shows.

Example 2.15. $3^{2n} \neq O(3^n)$.

Solution. If $3^{2n} = O(3^n)$, there exist positive constants c, n_0 such that for all $n \geq n_0$, $0 \leq 3^{2n} \leq c3^n$ or $3^n \leq c$.

However, given $c > 0$, we can always find an n' such that $3^{n'} > c$. For $n = \max\{n', n_0\}$, we have $3^n \leq c$ and $3^n > c$ which cannot hold together. Hence $3^{2n} \neq O(3^n)$. In particular $3^{2n} \neq \theta(3^n)$.

Remarks.

1. Since $\frac{3^{n+1}}{3^n} = 3$ which is a constant and, so, cannot tend to 0 or ∞ as n tends to infinity, neither $3^{n+1} = o(3^n)$ nor $3^{n+1} = \omega(3^n)$. On the other hand $\frac{3^{2n}}{3^n} = 3^n$ which tends to ∞ as n tends to ∞. Therefore $3^{2n} = \omega(3^n)$.
2. Since $f(n) = O(f(n))$ for every function $f(n)$, the above example also proves that if $f(n) = O(g(n))$, then $f(n)$ need not be in $O(g(\frac{n}{2}))$.
3. Since $3^n = O(3^{2n})$, the above example also shows that if $f(n) = O(g(n))$, then $g(n)$ need not be in $O(f(n))$.

Example 2.16. $an^{k+1} \neq \theta(n^k)$ for any positive integers a and k.

Solution. Suppose that k, a are positive integers and that $an^{k+1} = \theta(n^k)$. Then there exist positive constants c_1, c_2, n_0, n_0 an integer, such that for all $n \geq n_0$,

$$c_1 n^k \leq an^{k+1} \leq c_2 n^k \text{ or } c_1 \leq an \leq c_2.$$

Choose $n > \max\{n_0, \frac{c_2}{a}\}$. For this n we cannot have $an \leq c_2$. Hence $an^{k+1} \neq \theta(n^k)$.

In this section, we consider growth of some typical functions. Among other things we prove that powers dominate logs and exponentials dominate powers.

Powers dominate logs

Although we have come across logarithms earlier, we consider these formally here and recall some simple properties.

Recall that if a, b are positive real numbers, $b > 1$, then $\log_b a$ is the real number x such that $a = b^x$. The real number b is called the base of the logarithm. We are generally concerned with logarithms with base 2 or the irrational number $e \left(= \sum_{n=0}^{\infty} \frac{1}{n!} \right)$ or 10. We generally write $\log n$ for $\log_2 n$ (**binary logarithm**) and $\ln n$ for $\log_e n$ (**natural logarithm**). Let a, b be real numbers both greater than 1. Let $\log_b a = u$ and $\log_a b = v$. Then $a = b^u$, $b = a^v$ so that $a = a^{uv}$ and $uv = \log_a a = 1$. Therefore $\log_b a = \frac{1}{\log_a b}$.

Let c be another real number greater than 1. Let $\log_b a = x$. Then $a = b^x$. Therefore $\log_c a = \log_c b^x = x \log_c b$. Therefore $\log_b a = x = \frac{\log_c a}{\log_c b}$.

Lemma 2.13. *If* $\lim_{x \to \infty} f(x) = a$, *then* $\lim_{x \to \infty} |f(x)| = |a|$.

Proof. Suppose that $\lim_{x \to \infty} f(x) = a$.

If $a = 0$, given $\varepsilon > 0$, there exists an $x_0 > 0$ such that for all $x \geq x_0$, $|f(x)| < \varepsilon$ which implies that $\lim_{x \to \infty} |f(x)| = 0$.

Suppose that $a \neq 0$. Let $\varepsilon > 0$ be given.

If $a > 0$, let k be the least positive integer > 1 such that $a - \frac{\varepsilon}{k} > 0$. Let $x_0 > 0$ such that for all $x \geq x_0$,

$$|f(x) - a| \leq \frac{\varepsilon}{k} \text{ or } a - \frac{\varepsilon}{k} \leq f(x) \leq a + \frac{\varepsilon}{k}$$

Since $a - \frac{\varepsilon}{k} > 0$, we have for all $x \geq x_0$,

$$|a| - \varepsilon = a - \varepsilon < a - \frac{\varepsilon}{k} \leq |f(x)| \leq a + \frac{\varepsilon}{k} < a + \varepsilon = |a| + \varepsilon.$$

Therefore $||f(x)| - |a|| < \varepsilon$.

If $a < 0$, choose k to be the smallest positive integer > 1 such that $a + \frac{\varepsilon}{k} < 0$. Let $x_0 > 0$ be such that for all $x \geq x_0$, $|f(x) - a| < \frac{\varepsilon}{k}$. Then for all $x \geq x_0$, $a - \frac{\varepsilon}{k} < f(x) < a + \frac{\varepsilon}{k}$ and, so, $|a| - \varepsilon \leq |a| - |\frac{\varepsilon}{k}| \leq |a + \frac{\varepsilon}{k}| < |f(x)| < |a - \frac{\varepsilon}{k}| \leq |a| + |\frac{\varepsilon}{k}| \leq |a| + \varepsilon.$

Therefore $||f(x)| - |a|| < \varepsilon$ for all $x \geq x_0$.

Hence, whether $a > 0$ or $a < 0$, there exists an $x_0 > 0$ such that for all $x \geq x_0$, $||f(x)| - |a|| < \varepsilon$. Therefore $\lim_{x \to \infty} |f(x)| = |a|$.

We need the following rule for finding limit of a quotient of two functions. □

L'Hospital's Rule. *If* $\lim_{x \to \infty} f(x) = \lim_{x \to \infty} g(x) = \infty$ *or* $\lim_{x \to \infty} f(x) = \lim_{x \to \infty} g(x) = 0$ *and the derivatives* $f'(x), g'(x)$ *both exist, then*

$$\lim_{x \to \infty} \frac{f(x)}{g(x)} = \lim_{x \to \infty} \frac{f'(x)}{g'(x)}.$$

Proposition 2.14. *Let* a, b *be real numbers both greater than* 1. *Then* $\log_a x = \theta(\log_b x)$.

Proof. $\lim_{x \to \infty} \frac{|\log_a x|}{|\log_b x|} = \lim_{x \to \infty} \left| \frac{\log_a x}{\log_b x} \right| = \lim_{x \to \infty} \left| \frac{\log_e x}{\log_e a} \frac{\log_e b}{\log_e x} \right| =$
$\lim_{x \to \infty} \left| \frac{\log_e b}{\log_e a} \right| = \alpha$ (say) which is a constant, $0 < \alpha < \infty$.
Hence $\log_a x = \theta(\log_b x)$. □

In view of the above result, for considering the growth of a logarithmic function it is immaterial as to which real number greater than 1 is taken as base of logarithm. We may then write lg for logarithm when the base is not indicated or it is 2. We write $\ln f(x)$ for the natural logarithm $\log_e f(x)$.

Let x be a real variable. Then any polynomial in the variable $\lg x$ is called a **logarithmic polynomial** or **polylogarithm**. For comparing the growth of polynomials and polylogarithms, it is enough to compare the growth of n^k and $(\log_b n)^l$ for k, l integers both ≥ 1. In this direction we have the following example.

Example 2.17. (Powers dominate logarithmic powers). Let a, b and k be positive real numbers with $k, b > 1$. Then

$$(\log_b n)^a = O(n^k) \text{ but } (\log_b n)^a \neq \theta(n^k).$$

Solution. In view of Proposition 2.14, we may take $b = 2$. Let $\lg n = x$. Then $n = 2^x$ and $n^k = 2^{kx}$. When $n \to \infty$, x also approaches ∞. Now

$$\frac{x^a}{2^{kx}} \lessgtr \frac{(x+1)^a}{2^{k(x+1)}} \text{ if and only if } 2^k \lessgtr \left(1 + \frac{1}{x}\right)^a.$$

When $x \to \infty$, $(1 + \frac{1}{x})^a \to 1$ which is less than 2^k. Therefore $\left\{ \frac{x^a}{2^{kx}} \right\}$ is a strictly decreasing monotonic sequence of positive terms. Therefore

$$\lim_{x \to \infty} \frac{(\lg n)^a}{n^k} = \lim_{x \to \infty} \frac{x^a}{2^{kx}} = 0.$$

Hence $(\lg n)^a = o(n^k)$. In particular, $(\lg n)^a = O(n^k)$

If $(\lg n)^a = \Omega(n^k)$, there exist positive constants c and n_0 such that $0 \leq cn^k \leq (\lg n)^a$ or $0 \leq c2^{kx} \leq x^a$ for all $x \geq \lg n_0$. Since the sequence $\frac{2^{kx}}{x^a}$ is a strictly increasing monotonic sequence of positive terms, it diverges to ∞. Therefore, we can choose x large enough so that $\frac{2^{kx}}{x^a} > \frac{1}{c}$ which contradicts $c2^{kx} \leq x^a$ for all $x \geq \lg n_0$. Hence $(\lg n)^a$ is not in $\Omega(n^k)$, and, therefore, $(\log_b n)^a \neq \theta(n^k)$.

Comparison of growth of $\lg n!$ and $n \lg n$

We have already considered the expression $n!$ (factorial n) for non-negative integer n which is formally defined as

$$n! = \begin{cases} 1 & \text{if } n = 0, \\ n(n-1)! & \text{if } n > 0. \end{cases}$$

We have proved earlier a simple observation that $n!$ is in $O(n^n)$. However a much better result is available. For proving this improvement and also comparing the growth of $\lg n!$ and $n \lg n$, we need Stirling's formula for $n!$, namely

$$n! = \sqrt{2\pi n} \frac{n^n}{e^{n-\alpha_n}},$$

where α_n is a real number satisfying $\frac{1}{12n+1} < \alpha_n < \frac{1}{12n}$ for all $n \geq 1$.

Example 2.18. (i) $n! = o(n^n)$, (ii) $\lg n! = \theta(n \lg n)$, (iii) $n! = \omega(k^n)$, $k > 1$ a real number.

Solution. (i) $\frac{(n+1)!}{(n+1)^{n+1}} = \frac{n!}{(n+1)^n} < \frac{n!}{n^n}$ for all $n \geq 1$. Therefore $\left\{ \frac{n!}{n^n} \right\}$ is a strictly decreasing monotonic sequence of positive terms. Therefore

$$\lim_{n \to \infty} \frac{|n!|}{|n^n|} = \lim_{n \to \infty} \frac{n!}{n^n} = 0.$$

Hence $n! = o(n^n)$. In particular, $n! = O(n^n)$.

(ii) By Stirling's formula,

$$n! = \sqrt{2\pi n}\,\frac{n^n}{e^{n-\alpha_n}},$$

where α_n is a real number satisfying $\frac{1}{12n+1} < \alpha_n < \frac{1}{12n}$ for all $n \geq 1$. Then

$$\frac{1}{(12n+1)n\lg n} < \frac{\alpha_n}{n\lg n} < \frac{1}{12n^2\lg n} \quad \text{for all } n \geq 2.$$

Hence $\lim_{n\to\infty} \frac{\alpha_n}{n\lg n} = 0$.

Now

$$\lg n! = \frac{1}{2}\lg(2\pi) + \frac{1}{2}\lg n + n\lg n - n\lg e + \alpha_n \lg e.$$

Therefore

$$\frac{\lg n!}{n\lg n} = \frac{\lg(2\pi)}{2n\lg n} + \frac{1}{2n} + 1 - \frac{\lg e}{\lg n} + \frac{\alpha_n}{n\lg n}\lg e.$$

and, hence, $\lim_{n\to\infty} \frac{\lg n!}{n\lg n} = 1$. Therefore, by Theorem 2.12, $\lg n! = \theta(n\lg n)$.

(iii) Let $k > 1$ be a real number. Then $\frac{n!}{k^n} < \frac{(n+1)!}{k^{n+1}}$ for all $n \geq k$. Therefore $\left\{\frac{n!}{k^n}\right\}$ is a strictly increasing sequence of positive terms from some stage onwards. Therefore

$$\lim_{n\to\infty}\left|\frac{n!}{k^n}\right| = \lim_{n\to\infty}\frac{n!}{k^n} = \infty$$

and, hence, $n! = \omega(k^n)$. In particular, $n! = \omega(2^n)$.

Observe that in the above proof, we nowhere used k being greater than 1. Thus (iii) is valid for any positive real number k.

Exponentials beat polynomials

Exponential functions grow much faster than polynomial functions. For this we prove the following example.

Example 2.19. If k is an integer ≥ 1 and b is a real number > 1, then $n^k = o(b^n)$ but $n^k \neq \theta(b^n)$.

Solution. Since $\lim_{n \to \infty}(1 + \frac{1}{n})^k = 1$, $b > (1 + \frac{1}{n})$ for $n >$ some n_0. Therefore, for $n > n_0$, $\frac{n^k}{b^n} > \frac{(n+1)^k}{b^{n+1}}$ and, eventually, $\{\frac{n^k}{b^n}\}$ is a strictly decreasing monotonic sequence of positive terms. Hence $\lim_{n \to \infty} \frac{n^k}{b^n} = 0$ and, therefore, $n^k = o(b^n)$. In particular, $n^k = O(b^n)$.

Suppose that there exist positive constants c and n_0, n_0 an integer, such that for all $n \geq n_0$,

$$0 \leq cb^n \leq n^k. \tag{2.13}$$

The argument leading to convergence of $\{\frac{n^k}{b^n}\}$ to zero also shows that the sequence $\left\{\frac{b^n}{n^k}\right\}$ diverges to ∞. Thus, given any $M > 0$, there exists a positive integer n_1 such that for all $n \geq n_1$, $b^n > Mn^k$. This is, in particular, true for $M = \frac{1}{c}$, i.e., for all $n \geq n_1$,

$$cb^n > n^k. \tag{2.14}$$

Let $n_2 = \max\{n_0, n_1\}$. Then (2.13) and (2.14) are simultaneously true for all $n \geq n_2$, which cannot happen. Hence $n^k \neq \Omega(b^n)$ and, therefore, $n^k \neq \theta(b^n)$.

Remark. If $0 < b < 1$ in the above example, then $n^k = \omega(b^n)$ and $n^k \neq \theta(b^n)$.

We end this section by considering the question:
What happens to the functions (i) $\lg f(n)$ and $\lg g(n)$ and (ii) $2^{f(n)}$ and $2^{g(n)}$ if $f(n) = O(g(n))$?

Theorem 2.15. *Let $f(n) \geq 1$ and $\lg g(n) \geq 1$ for sufficiently large n and $f(n)$ be in $O(g(n))$. Then $\lg f(n) \in O(\lg g(n))$.*

Proof. Since $f(n) = O(g(n))$, there exist positive constants c and n_0, n_0 an integer, such that for all $n \geq n_0$,

$$f(n) \leq cg(n). \tag{2.15}$$

Also we can always take $c > 1$. Let n_1 be a positive integer such that for all

$$n \geq n_1, \ f(n) \geq 1 \quad \text{and} \quad \lg g(n) \geq 1. \tag{2.16}$$

Let $n_2 = \max\{n_0, n_1\}$. Then (2.15) and (2.16) both hold for all $n \geq n_2$.

Taking logarithms (2.15) and (2.16) imply that for all $n \geq n_2$,

$$0 \leq \lg f(n) \leq \lg c + \lg g(n) \leq \lg c \lg g(n) + \lg g(n) = (\lg c + 1) \lg g(n).$$

Therefore $\lg f(n)$ is in $O(\lg g(n))$. $\qquad\qquad\qquad\qquad\square$

Exercise 2.2.

1. Is the result of Theorem 2.15 true if we drop one or both of the conditions that $f(n) \geq 1$, $\lg g(n) \geq 1$ for sufficiently large n?
2. What happens to the result of Theorem 2.15 if the asymptotic notation O is replace by the asymptotic notation o?
3. Find conditions on $f(n)$ and $g(n)$ so that $f(n) = \theta(g(n))$ implies that $\lg f(n) = \theta(\lg g(n))$.

Example 2.20. If $f(n)$ is in $O(g(n))$, it is not necessary that $2^{f(n)}$ is in $O(2^{g(n)})$. Take, for example, $f(n) = 3n$, $g(n) = n$. Then, for all $n \geq 1$, $f(n) \leq 4g(n)$ so that $f(n) = O(g(n))$.

If $2^{f(n)}$ is in $O(2^{g(n)})$, there exist positive constants c and n_0, n_0 an integer, such that for all $n \geq n_0$,

$$2^{3n} \leq c2^n \quad \text{or} \quad 2^{2n} \leq c. \qquad\qquad (2.17)$$

Since c is a constant, we can choose an n_1 such that $2^{2n_1} > c$. Taking $n_2 = \max\{n_0, n_1\}$, we get $2^{2n_2} > c$ which contradicts (2.17). Hence $2^{f(n)}$ is not in $O(2^{g(n)})$.

Exercise 2.3. Find necessary and sufficient conditions on $f(n)$ and $g(n)$, if some exist, which ensure that $f(n) = O(g(n))$ implies $2^{f(n)} = O(2^{g(n)})$.

2.6. Some Typical Examples

Example 2.21. If a, b are real numbers both greater than 1, then $a^n = o(a^{bn})$.

Solution. $b > 1$ implies that $b - 1 > 0$ and, so, $(b - 1)n \to \infty$ as $n \to \infty$. Therefore $\lim_{n\to\infty} a^{(b-1)n} = \infty$, as $a > 1$, and, so, $\lim_{n\to\infty} \frac{a^n}{a^{bn}} = \lim_{n\to\infty} \frac{1}{a^{(b-1)n}} = 0$. Hence $a^n = o(a^{bn})$. In particular, $a^n = O(a^{bn})$. However, $a^n \neq \theta(a^{bn})$.

Example 2.22. If x is a positive real variable, then an application of L'Hospital's rule shows that $\lim_{x \to \infty} \frac{\lg x}{x} = 0$ and, therefore, $\lg x = o(x)$. In particular, $\lg x = O(x)$ but $\lg x \neq \theta(x)$. Consequently, we have $\lg \lg \lg n = O(\lg \lg n)$, $\lg \lg n = O(\lg n)$, $\lg n = O(n)$, $n = O(2^n)$, $2^n = O(2^{2^n})$ etc.

Again, $\lg n \to \infty$ when $n \to \infty$. Therefore, we have $\lg n = O(\sqrt{\lg n})$, $(\lg n)^2 = O(\lg n)$ etc.

For any real variable x, the quotient $\frac{x}{2^{\frac{x}{2}}} \to 0$ when $x \to \infty$. Therefore, the sequence $\left\{ \frac{n}{2^{\frac{n}{2}}} \right\}$ converges to the limit 0 when $n \to \infty$. Therefore, $\lg n = o(\sqrt{n})$ and, so, $\lg n = O(\sqrt{n})$.

In all the above cases, the asymptotic notation O may be replaced by the notation o.

Example 2.23. Let $f(n) = O(g(n))$ and $g(n)$ have positive asymptotic lower bound. Then for any real number a, $f(n) + a$ is in $O(g(n))$.

Proof. Let c and n_0 be positive constants (n_0 an integer) such that for all $n \geq n_0$,

$$|f(n)| \leq c|g(n)|.$$

Since $g(n)$ is asymptotically bounded below, there exist positive constants b and n_1, n_1 an integer, such that for all $n \geq n_1$,

$$g(n) \geq b \quad \text{and,} \quad \text{so,} \quad |g(n)| \geq b.$$

Let $n_2 = \max\{n_0, n_1\}$. Now for any real number a and for all $n \geq n_2$

$$|f(n) + a| \leq |f(n)| + |a| \leq c|g(n)| + \frac{|a|}{b}b \leq (c + \frac{|a|}{b})|g(n)| = c'|g(n)|,$$

where $c' = c + \frac{|a|}{b} > 0$. Hence $f(n) + a$ is in $O(g(n))$. $\square$

Example 2.24. Show that $5n \lg n + \sqrt[4]{n}(\lg n)^4$ is in $\theta(n \lg n)$.

Solution. For all $n \geq 2$, $5n \lg n \leq 5n \lg n + \sqrt[4]{n}(\lg n)^4$.

For proving the existence of positive constants c and n_0 such that $5n \lg n + \sqrt[4]{n}(\lg n)^4 \leq cn \lg n$ for all $n \geq n_0$, we may take 2 as the base of logarithm. Let $\lg n = x$ so that $n = 2^x$. Then $\sqrt[4]{n}(\lg n)^4 \leq n \lg n$ if and only if $(\lg n)^3 \leq (\sqrt[4]{n})^3$ or $x \leq 2^{\frac{x}{4}}$ which is so for all $x \geq 17$ or $n \geq 2^{17}$. Thus, for $n \geq 2^{17}$, $5n \lg n + \sqrt[4]{n}(\lg n)^4 \leq 6n \lg n$. Therefore, we have

$c_1 n \lg n \le 5n \lg n + \sqrt[4]{n}(\lg n)^4 \le c_2 n \lg n$ for all $n \ge 2^{17}$, where $c_1 = 5$, $c_2 = 6$.

Example 2.25. Show that $n^6 3^n = \omega(n^{10})$.

Solution. A repeated application of L'Hospital's rule shows that

$$\lim_{n \to \infty} \frac{n^6 3^n}{n^{10}} = \lim_{n \to \infty} \frac{3^n}{n^4} = \lim_{n \to \infty} \frac{3^n (\lg 3)^4}{4!} = \infty.$$

Therefore $n^6 3^n = \omega(n^{10})$.

Example 2.26. Show that $80n^7 \in O(e^n)$.

Solution. For any positive numbers $a, b, b > 1$, we know that $a = b^{\log_b a}$. In particular, $n = e^{\log_e n} = e^{\ln(n)}$.

Taking upper Riemann sum using intervals of unit size for $\ln(n) = \int_1^n \frac{dx}{x}$ for $n > 1$, we get

$$\ln(n) < 1 + \frac{1}{2} + \cdots + \frac{1}{n}$$

$$= \left(1 + \frac{1}{2} + \cdots + \frac{1}{6}\right) + \frac{1}{7} + \cdots + \frac{1}{n} < 6 + \frac{n-6}{7} = \frac{n+36}{7}.$$

Thus $7 \ln(n) \le n + 36$. Therefore, for all $n > 1$,

$$80n^7 = 80e^{7 \ln(n)} \le 80e^{n+36} < e^{n+41} = e^{41}e^n,$$

which implies that $80n^7 \in O(e^n)$.

(Also refer to Example 2.18.)

Example 2.27. Find the least integer n such that $\frac{x^5 + x \lg x}{x^4 + 2}$ is in $O(x^n)$.

Solution. We consider two functions $\frac{x^5}{x^4 + 2}$ and $\frac{x \lg x}{x^4 + 2}$ separately. Observe that

$$\frac{\frac{x^5}{x^4+2}}{x} = \frac{1}{1 + \frac{2}{x^4}} \to 1 \quad \text{when } x \to \infty.$$

If $\frac{x^5}{x^4+2}$ is in $O(1)$, there exist positive constants c and x_0 such that for all $x \ge x_0$,

$$\frac{x^5}{x^4 + 2} = x - \frac{2x}{x^4 + 2} \le c \quad \text{or} \quad x \le c + \frac{2x}{x^4 + 2}.$$

When $x \to \infty$, $\frac{2x}{x^4+2} \to 0$ and, therefore, there exists $x_1 > 0$ such that $\frac{2x}{x^4+2} < c$ for all $x \geq x_1$.

Then, for all $x \geq x_0 + x_1$, $x < 2c$.

Choosing $x > \max\{2c, x_0 + x_1\}$ leads to a contradiction to $x < 2c$.

Hence $\frac{x^5}{x^4+2}$ is not in $O(1)$.

Therefore, the least value of n for which $\frac{x^5}{x^4+2}$ is in $O(x^n)$ is 1.

Again

$$\lim_{x \to \infty} \frac{x \lg x}{x^4 + 2} = \lim_{x \to \infty} \frac{1 + \lg x}{4x^3} = \lim_{x \to \infty} \frac{1}{12x^3} = 0.$$

Therefore $\frac{x \lg x}{x^4+2}$ is in $o(1)$ and, so, also in $O(1)$. But then $\frac{x \lg x}{x^4+2}$ is in $O(x)$.

It now follows from Theorem 2.3 that $\frac{x^5+x \lg x}{x^4+2}$ is in $O(x)$ but it cannot be in $O(1)$. For example, if $\frac{x^5+x \lg x}{x^4+2}$ is in $O(1)$ and as $\frac{-x \lg x}{x^4+2}$ is already in $O(1)$, their sum which is $\frac{x^5}{x^4+2}$ is in $O(1)$ which is not the case. Hence the least value of n for which $\frac{x^5+x \lg x}{x^4+2}$ is in $O(x^n)$ is 1.

We finally collect together some functions in increasing $O(big\ oh)$ order.

Prove the correctness of the statement for every pair of consecutive functions which have not been proved earlier in the chapter.

1	$\log(\log n)$	$\log n$	$\sqrt{n}$	$\sqrt{n}\log n$	n	
$\log n!$	$n \log n$	$n(\log n)^2$	n^2	n^3	3^n	$n!$

Exercise 2.4.

1. Is it always true that $f(n) + o(f(n)) = \theta(f(n))$?
2. Is it always true that $f(n) + \omega(f(n)) = \theta(f(n))$?
3. Is it always true that $f(n) = O(f(n)^3)$? How about $f(n) = \omega(f(n)^2)$?
4. Prove that higher degree polynomial beats lower degree polynomial.
5. Prove that the base of an exponential matters, i.e., prove that for real numbers a, b, $1 \leq a < b$, $a^n = o(b^n)$.
6. For which values of k, is it true that for any positive real number a, $a^n = \theta(a^{n+k})$?

7. Give an example to show that if $f(n) = O(g(n))$ and a is constant, then $f(n) + a$ need not be in $O(g(n))$.

8. Prove that $\omega(g(n)) \subseteq \Omega(g(n)) - \theta(g(n))$.

9. If $f(n) = (1 + \cos n)n$ and $g(n) = n$, prove that $f(n) \in O(g(n)) - \theta(g(n))$. Is $f(n)$ in $o(g(n))$?

10. Examine the asymptotic relationship between $f(n) = (1 + \sin n + \cos n)n$ and $g(n) = n$, if one exists.

11. Prove that $f(n) + o(f(n)) = \theta(f(n))$. What can be said about $f(n) + \omega(f(n))$?

12. Prove that $x3^x + x^2 \in O(x3^x)$.

13. Find the least n for which $\frac{x^3 + x(\log x)^2}{x^2 + 1}$ is in $O(x^n)$.

14. Prove that $o(g(n)) \cap \Omega(g(n))$ is an empty set and so is $\omega(g(n)) \cap O(g(n))$.

15. Prove that $n^k \neq \theta(n^{k+1})$ for any $k \geq 1$.

16. Prove that $2n^3 + 3n^2 + \theta(n) = 2n^3 + \theta(n^2) = \theta(n^3)$.

17. Prove that for all $n \geq 8$, $n^2 \leq 4n^2 - 24n$.

18. Is it true that $n \leq 2^{\frac{n}{4}}$ for all $n \geq 17$? Justify.

19. If $\lg(\lg g(n)) > 0$ and $f(n) \geq 1$ for sufficiently large n and $f(n) = O(g(n))$, is it necessary that $\lg f(n) = O(\lg(\lg n))$?

20. If $f(n) = \sin n$, is $f(n) = \theta(f(\frac{n}{2}))$?

21. Is $\cos n = \theta(\cos \frac{n}{2})$?

22. Show that for any real number $a \geq 2$, $(\log_a x)^2 = o(x)$.

23. If $f(x) = 1 - \frac{1}{x}$, is $f(n) = O(f(n)^2)$?

24. Arrange the following functions as a list g_1, g_2, g_3, etc. such that $g_i = O(g_{i+1})$ for every $i \geq 1$. Are there any functions in this list such that $g_i = \theta(g_{i+1})$?
$(\sqrt{2})^{\lg n}, n^2, n!, (\log n)^2, n^3, \lg n!, \ln n, n^{\lg(\lg n)}, n2^n, 2^{\lg n}, 2^{2^n}, 2^n, n,$
$4^{\lg n}, n \lg n, n(\lg n)^2, \sqrt{n} \lg n, e^n, 2^{\sqrt{\lg n}}$.

25. Prove that $\theta(m) + \theta(n) = \theta(m + n)$.

Chapter 3

Backtracking

The concept of backtracking is best explained by the literal meaning of the word. In brute-force technique, we need to check a lot of cases to arrive at a final solution of a problem but backtracking method helps in reducing the number of cases to be tried. We apply backtracking to the problem of coloring a graph, the n-queen problem and solving a Sudoku problem.

3.1. Backtracking Procedure

Sometimes we come across problems asking for optimal solution or problems solutions for which satisfy certain constraints but there are no general methods available for solving such problems. We may in such cases be forced to resort to exhaustive search or brute force. **Backtracking** then helps us in reducing, and many times considerably, the number of solutions to be tried. In a possible solution $x_1, x_2, \ldots, x_n$ of a given problem, each x_i has a number of choices. Suppose that we have come to a position $x_1, x_2, \ldots, x_k, k < n$, where the x_i satisfy the given constraints (of the problem) but find that there are no choices for $x_{k+1}, \ldots, x_n$ so that $x_1, \ldots, x_k, x_{k+1}, \ldots, x_n$ is a solution to the problem. We backtrack and make a different choice for x_k out of all possible choices and then with this changed choice of x_k check whether it is possible to extend $x_1, x_2, \ldots, x_k$ to a solution of the problem or not. If we extend $x_1, x_2, \ldots,$ by one step to $x_1, \ldots, x_k, x_{k+1}$, we then check if it is possible to extend $x_1, \ldots, x_k,$ x_{k+1} to a possible solution of the problem. If this process of step-by-step extension leads to a final solution $x_1, x_2, \ldots, x_n$, we stop. On the other hand, if we cannot extend this to a solution (by considering all the possible choices for x_k's), the problem has no solution. We illustrate this procedure by a couple of examples, but before that we formalize this procedure.

Suppose that a desired solution is expressible as an n-tuple $(x_1, x_2, \ldots, x_n)$ where the x_i are chosen from some finite set S_i, $1 \le i \le n$ and the chosen x_i have to satisfy a criterion function P. The backtracking algorithm builds the solution vector (tuple) step by step by choosing an x_i once $x_1, x_2, \ldots, x_{i-1}$ have already been chosen.

Algorithm 3.1.

---

1. Choose an $x \in S_1$ and call it x_1
2. Choose an $x \in S_2$ and see if the criterion $P(x_1, x)$ is satisfied. If yes, set $x = x_2$. Otherwise ignore this x and choose another $x \in S_2$ and see if the criterion $P(x_1, x)$ is satisfied. Continue this process of ignoring x and choosing another element from S_2 till the criterion $P(x_1, x)$ is satisfied. Call the final element $x \in S_2$ satisfying the criterion $P(x_1, x)$ as x_2. If no such element $x \in S_2$ can be chosen (i.e., $P(x_1, x)$ is not satisfied for any $x \in S_2$), a solution vector cannot be built with first component x_1. We backtrack, ignore x_1 and choose another element $x \in S_1$. Call the new element chosen x_1
3. Repeat step 2
4. Suppose that $k \ge 2$ and that we have chosen elements $x_i \in S_i$, $1 \le i \le k$, such that criterion $P(x_1, \ldots, x_k)$ is satisfied
5. Choose an $x \in S_{k+1}$ and check if the criterion $P(x_1, \ldots, x_k, x)$ is satisfied. If yes, set $x = x_{k+1}$, otherwise ignore this x and choose another element from S_{k+1}. Continue this process till we arrive at an $x \in S_{k+1}$ such that the criterion $P(x_1, \ldots, x_k, x)$ is satisfied. Set this $x = x_{k+1}$. If no such $x \in S_{k+1}$ can be found, backtrack and replace x_k by another element in S_k for which $P(x_1, \ldots, x_k)$ is satisfied
6. Repeat 5 with the new choice of x_k
7. Repeat 5 and 6 with k replaced by $k + 1$
8. Continue till we arrive at $x_1, x_2, \ldots, x_n$ such that $P(x_1, x_2, \ldots, x_n)$ is satisfied. Then $(x_1, x_2, \ldots, x_n)$ is a solution to the problem. If no such sequence $x_1, x_2, \ldots, x_n$ is obtained, i.e., the process stops with $x_1, x_2, \ldots, x_k$, $k < n$ and we cannot proceed further, then the problem has no solution with the starting choice x_1. In that case ignore x_1 and choose another $x_1 \in S_1$ and repeat the process with this x_1

3.2. Graph Coloring

As a first example of backtracking we consider the problem of coloring a graph.

Given a graph G, coloring of G consists of assigning colors to the vertices of G such that no two adjacent vertices carry the same color. Graph coloring problem consists of finding minimum number of colors that suffice to color the graph. We now explain how backtracking can be used for graph coloring problem.

Let the vertices of the graph G be denoted by $a, b, c, \ldots$ and the colors be called $1, 2, \ldots, k$.

Consider vertex a and assign color 1 to a. Next consider vertex b and assign color 1 to b if b is not adjacent to a and assign color 2 to b otherwise. Next consider the third vertex c. Assign color 1 to c if c is not adjacent to a, otherwise assign color 2 to c if c is not adjacent to b. If even this is not possible, assign color 3 to c. Continue this process of coloring. If we come to a vertex to which it is not possible to assign a given color, we backtrack to the previous vertex, change the earlier color assigned by another permissible color and then go to the next vertex from which we had to backtrack.

Example 3.1. Using backtracking color the graph of Fig. 3.1 using three colors.

Solution. Assign color 1 to a, color 2 to b, color 1 to c and color 2 to d. Now e being adjacent to b, c, d, we can only assign color 3 to e. Then f being adjacent to a, b, e none of the colors assigned to these three vertices can be assigned to f. Therefore, we need to see if a different color could be assigned to e. Colors having been already assigned to b, c, d, choice of color for e is only one. Hence we have to backtrack to d. Choice of color

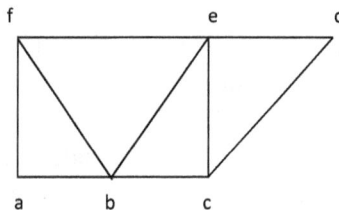

Fig. 3.1

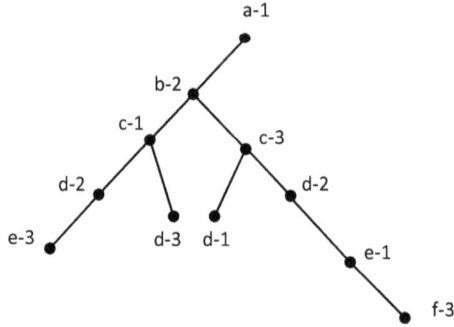

Fig. 3.2

for c having already been made, the only next possible choice for assigning a color to d is 3. But then b, c, d have been assigned all the three colors available, there is no choice of any color for e. Therefore, this choice of color for d is also not permissible. We, therefore, backtrack to c. We now assign color 3 to c the only other possibility. We may then assign color 1 to d. We find that the three colors available have been assigned to b, c, d and, so, there is no choice of a color for e. But there is another choice of a color for d, namely color 2. Only two choices namely 2 and 3 have been used to color the vertices b, c, d. Therefore, we can assign color 1 to e. Colors 1 and 2 are the only ones used for the vertices a, b, e which are adjacent to f. We can then assign color 3 to f. The final coloring of the graph G is then color 1 to a, color 2 to b, color 3 to c, color 2 to d, color 1 to e and finally color 3 to f. Pictorially, the above process of coloring is given in Fig. 3.2.

3.3. n-Queen Problem

Our next example on the application of backtracking is the n-queen problem.

In the game of chess, a queen can move any number of steps vertically (up or down), horizontally and diagonally provided there is no obstruction on the way. Two queens can attack each other if they are sitting in the same column or same row or same diagonal. Consider a checker's board (chess board) having n squares horizontally and n squares vertically (all squares being equal in size). We need to place n queens on the $n \times n$ board in such a way that no two of these attack each other, i.e., no two of these are placed in the same column or same row or same diagonal. Let (i, j) denote the (little) square that is on the intersection of the ith row and jth column. The main diagonal of the board consists of the squares

$(1, 1), (2, 2), \ldots, (n, n)$ and the off-diagonal consists of
$(n, 1), (n - 1, 2), \ldots, (2, n - 1), (1, n)$.
The diagonals parallel to the main diagonal are given by
(i, j) such that $i - j = m$ for some integer m
while the diagonals parallel to the off-diagonal are given by
(i, j) such that $i + j = m$ for some integer m.

Let Q_i, $1 \le i \le n$, be the queen that is placed in the ith row. If Q_i is placed in the jth column, the queen Q_i occupies the square (i, j). Suppose that two queens Q_i, Q_k are placed in the squares (i, j) and (k, l), respectively. Then Q_i, Q_k do not attack or hit each other if and only if

$$i \ne k, j \ne l \text{ and } i - j \ne k - l \text{ and } i + j \ne k + l.$$

Suppose that $i \ne k$ and $j \ne l$ (so that the queens Q_i and Q_k are not placed in the same row or in the same column). Then Q_i and Q_k are placed in the same diagonal if and only if $i - k = j - l$ or $i - k = l - j$ which is equivalent to saying that $|i - k| = |j - l|$. Hence Q_i and Q_k are not placed in the same diagonal if and only if

$$|i - k| \ne |j - l|. \tag{3.1}$$

Suppose that the queens Q_i are so placed that no two of them attack each other. Let x_i denote the column in which the queen Q_i is placed. Since no two queens are placed in the same column, $x_1, x_2, \ldots, x_n$ are all distinct. Hence $x_1, x_2, \ldots, x_n$ is a permutation of the integers $1, 2, \ldots, n$. Since a square in the chess board may be occupied by any one of the n queens, there are n^n possible placements to begin with. Since no two queens are placed in the same column, any placement $x_1, x_2, \ldots, x_n$ of the queens is such that no two x_i's are equal. Hence the total number of placements of the queens reduces to $n!$ i.e., the number of permutations of the integers $1, 2, \ldots, n$. The total number of these choices may be represented by the possible paths in a rooted tree from the root to the leaves of the tree which is generated giving all the permutations of $1, 2, \ldots, n$.

In this tree called **permutation tree**, the edges are labeled by the possible values of x_i.

Before pursuing the n-queen problem further, we construct the permutation tree for $n = 3$. The root vertex of the permutation tree is taken at level 1 and is named $\langle 1 \rangle$. The root now has n children which are all at level 2 and these nodes are named or numbered $\langle 2 \rangle, \langle 3 \rangle, \ldots, \langle n+1 \rangle$. The edges from the root to the nodes at level 2 are all the possible values of x_1 namely these are numbered $1, 2, \ldots, n$. For each node at level 2, there are $n - 1$ children so that the nodes at level 3 are $n(n-1)$. Each node at level 3 has $n - 2$ children and so on. In general, every node at level $i + 1$ has $n - i$ children. If $x_1, x_2, \ldots, x_i$ are the edges on the path from the root to a node at level $i + 1$, the edges leading from a node at level $i + 1$ to the $n - i$ children are named $y_1, y_2, \ldots, y_{n-i}$, where $\{y_1, y_2, \ldots, y_{n-i}\} = \{1, 2, \ldots, n\} \backslash \{x_1, x_2, \ldots, x_i\}$. Thus, for $n = 3$, the permutation tree is as given in Fig. 3.3.

As an example on the application of backtracking, we consider the n-queen problem for $n = 3, 4, 5$.

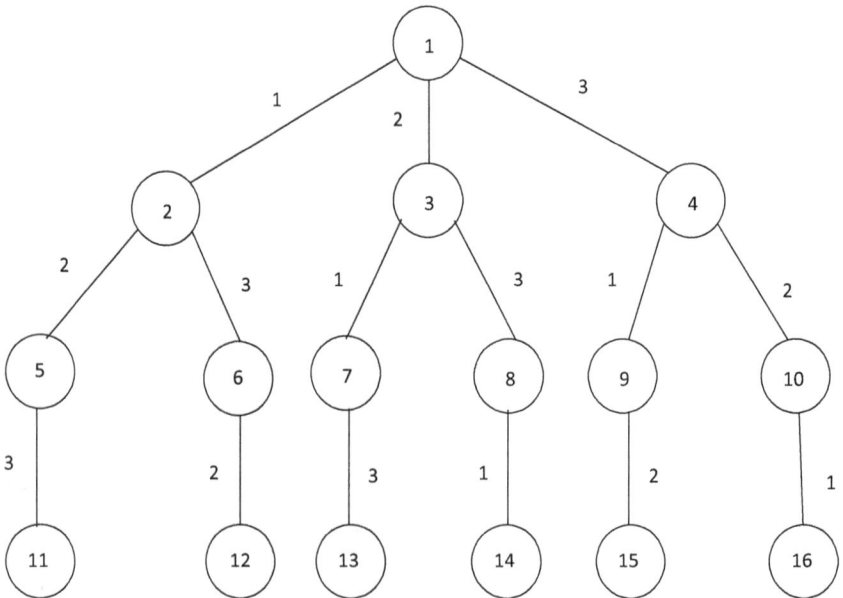

Fig. 3.3

Example 3.2.

(a) **The case** $n = 3$.

Let the queens be Q_1, Q_2, Q_3. If Q_1 is placed in the first column, the only choice possible for Q_2 is $(2, 3)$. Once Q_2 is placed in this position, the only possible available place for Q_3 is $(3, 2)$. But $(2, 3)$ and $(3, 2)$ being on the same diagonal, there is no place available for Q_3. Hence the queens cannot be placed in non-attacking positions with Q_1 sitting in the $(1, 1)$ position.

Suppose that Q_1 is placed in the $(1, 2)$ position. Then Q_2 can neither be placed in $(2, 1)$ position nor in $(2, 3)$ position. Therefore Q_1 cannot be placed in $(1, 2)$ position.

If Q_1 is placed in the $(1, 3)$ position, the only place where Q_2 can be placed is the $(2, 1)$ position. The only place available for Q_3 is the $(3, 2)$ position in which case Q_2, Q_3 happen to be in the same diagonal.

Hence there is no way to place three queens on a 3×3 board so that no two queens attack each other. (Refer Table 3.1(a)–(c).)

(b) **The case** $n = 4$.

Place Q_1 in the $(1, 1)$ cell and place Q_2 in the $(2, 3)$ position. Then there is no way to place Q_3. We backtrack and place Q_2 in the $(2, 4)$ position. Then Q_3 can only be placed in the $(3, 2)$ position and we find that there is no way to place Q_4. Hence there is no solution to the 4-queen problem with Q_1 in the $(1, 1)$ position.

We backtrack and place Q_1 in the $(1, 2)$ position. Then there is only one choice for placing Q_2 namely the position $(2, 4)$. Again there is only one way to place Q_3 which has to be placed in $(3, 1)$ position. The only place available for placing Q_4 is the $(4, 3)$ position. Hence we get a placement of the four queens on a 4×4 chess board. (Refer Table 3.2(c).)

Next we consider the case when Q_1 is placed in the $(1, 3)$ position. Then the only choice available for placing Q_2 is the $(2, 1)$ position which

Table 3.1

(a)   (b)   (c)

Table 3.2

(a)			
1			
		2	

(b)			
1			
			2
		3	

(c)			
	1		
			2
3			
		4	

(d)			
		1	
2			
			3
	4		

(e)			
			1
	2		

(f)			
			1
2			
		3	

then forces Q_3 to be placed in $(3, 4)$ position. Finally we have to place Q_4 in the $(4, 2)$ position. We thus get a valid positioning of the four queens on a 4×4 chess board.

Finally, we consider if it is possible to get a valid placement of queens with Q_1 in the $(1, 4)$ position. If we then place Q_2 in the $(2, 2)$ position, there is no way to place Q_3. So, we place Q_2 in the only other place available, namely $(2, 1)$ position. The only place available for placing Q_3 is the $(3, 3)$ position and finally we cannot place Q_4 anywhere. Hence, there is no valid placement of the four queens with Q_1 placed in the $(1, 4)$ position. (Refer Table 3.2(f)).

(c) **The case $n = 5$.**

Place Q_1 in the $(1, 1)$ position. There are three choices to place Q_2 and we place Q_2 in the $(2, 3)$ position. There being only one choice for placing Q_3, we place Q_3 in the $(3, 5)$ position. There is then only one choice for placing Q_4 and finally only one choice for placing Q_5. We thus place Q_4 and Q_5 in the $(4, 2)$ and $(5, 4)$ positions, respectively. We thus get a valid placement of the five queens on the 5×5-chess board as $(1, 3, 5, 2, 4)$, i.e., $x_1 = 1, x_2 = 3, x_3 = 5, x_4 = 2, x_5 = 4$ or Q_1 in the first column, Q_2 in the third, Q_3 in the fifth, Q_4 in the second and Q_5 in the fourth column, respectively. (Refer Table 3.3(a).)

There may be other valid placements of the 5-queen problem. For example when Q_1 is placed in the $(1, 1)$ position, there are two other choices for placing Q_2. When Q_2 is placed in $(2, 4)$ position, there is only one choice for placing Q_3 which is the $(3, 2)$ position. After placing

Table 3.3

1				
		2		
				3
	4			
			5	

1				
				2
		3		
			4	
		5		

1				
				2
			3	

    (a)   (b)   (c)

Q_3, there is only one choice, namely the $(4, 5)$ position for placing Q_4. Finally Q_5 has to be placed in the only position $(5, 3)$ available. Thus, this placement of queens is $(1, 1), (2, 4), (3, 2), (4, 5), (5, 3)$.

Again when Q_1 is placed in the $(1, 1)$ cell, the only other choice for placing Q_2 is the $(2, 5)$ cell. Then the only place for Q_3 is the $(3, 2)$. There is then no choice for placing Q_4.

Hence with Q_1 placed in the $(1, 1)$ position, there are two valid solutions for the 5-queen problem (as given in Table 3.3(a) and (b)).

3.4. *n*-Queen Problem and Permutation Tree

We next consider how the permutation tree helps in solving the n-queen problem. As already mentioned, we denote by Q_i the queen placed in the ith row for $i = 1, 2, \ldots, n$. Also x_i is the column in which the queen Q_i is placed so that $x_1, x_2, \ldots, x_n$ is a permutation of the numbers $1, 2, \ldots, n$. Thus the number of choices for the possible solutions of the n-queen problem reduces to $n!$ which is still very large. Consider now the permutation tree in which the edges from level k to level $k + 1$ nodes specify the values for x_k for $1 \leq k \leq n$. As already observed, when we come to node at level $i + 1$, then x_j for $1 \leq j \leq i$ have already been assigned values and this node will have $n - i$ edges leaving this node. If we arrive at this node along the edge with $x_i = k$, then the outgoing edges at this node (for the n-queen problem) cannot have the values $k - 1$ or $k + 1$ (for otherwise the queens Q_i and Q_{i+1} hit each other), so the branch of the tree with the present node as root with $x_{i+1} = k - 1$ or $k + 1$ need not be explored further. We call the nodes at level $i + 2$ leading along the edges (i.e., the other ends of the edges) with $x_{i+1} = k - 1$ or $k + 1$ as **bounded**. So our search for a solution is simplified.

We better explain the above procedure for $n = 4, 5, 6$.

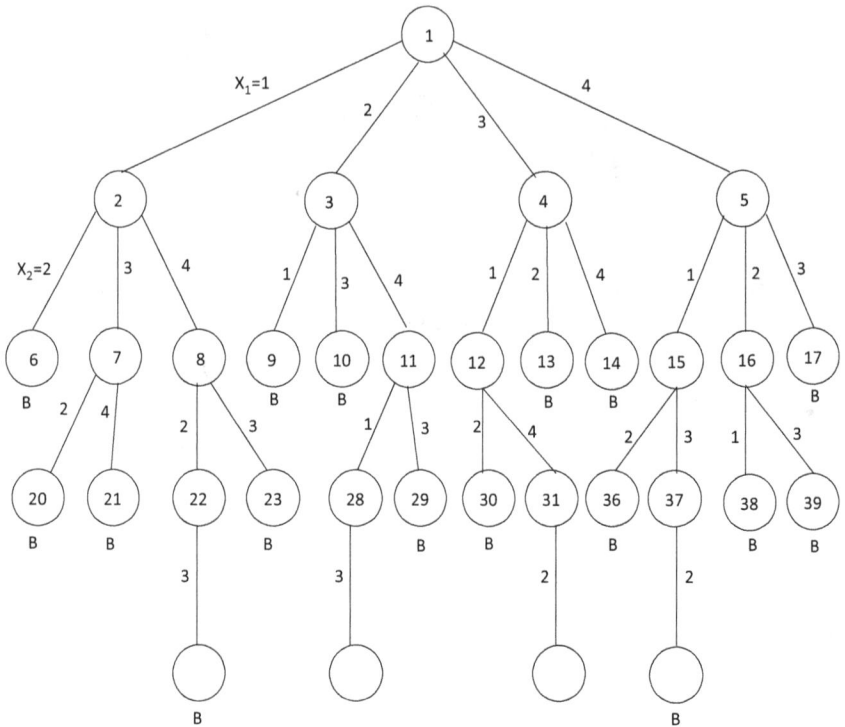

Fig. 3.4

The case $n = 4$.

In Fig. 3.4, we have not shown the whole tree and have omitted part of tree below a node named B which indicates a bound node. In the leftmost branch of the tree observe that $x_1 = 1$, $x_2 = 2$ which is not possible because if we allow this to happen, then Q_1, Q_2 sit in the $(1, 1)$, $(2, 2)$ place respectively both of which are positions on the main diagonal. Therefore, the branch of the tree with node 6 as root need not be explored. Node 6 is bounded and in our search we need to backtrack to node 2. For the branch of the tree leading from node 3, the two possible placements of the first two queens such as $(1, 2)$, $(2, 1)$ and $(1, 2)$, $(2, 3)$ are not possible. Therefore, the nodes 9 and 10 are bounded. Similarly the nodes 13, 14 and 17 are bounded. Therefore, we need to explore further the sub-tree with nodes 7, 8, 11, 12, 15 and 16 only as active nodes and the nodes 6, 9, 10, 13, 14, 17 as dead nodes. At level 4, we find that nodes 20, 21, 23, 29, 30, 36, 38 and 39 are dead nodes and nodes 22, 28, 31 and 37 are the only active nodes. At level 5, only two nodes are active nodes and the other two are dead nodes. The two active nodes provide the paths $(2, 4, 1, 3)$ and $(3, 1, 4, 2)$. Thus the positions of the queens are $(1, 2)$, $(2, 4)$, $(3, 1)$, $(4, 3)$ as per the first path and the positions of the queens provided by the second path are $(1, 3)$, $(2, 1)$, $(3, 4)$, $(4, 2)$. Hence there are exactly the two solutions as already given in Table 3.2(c) and (d).

The case $n = 5$.

We have obtained two possible solutions of the 4-queen problem which are $(1, 2)$, $(2, 4)$, $(3, 1)$, $(4, 3)$ and $(1, 3)$, $(2, 1)$, $(3, 4)$, $(4, 2)$. We may use these solutions to get possible solutions of the 5-queen problem. One way to obtain a possible solution is by replacing every (i, j) in a solution of the 4-queen problem by $(i + 1, j + 1)$ and then taking $(1, 1)$ as another ordered pair. Since for (i, j), (k, l) in a solution of the 4-queen problem $|i - k| \neq |j - l|$, we have $|i + 1 - (k + 1)| \neq |j + 1 - (l + 1)|$, we only need to check that for every (i, j) in a solution of the 4-queen problem $|i + 1 - 1| \neq |j + 1 - 1|$, i.e., $|i| \neq |j|$. For both the above solutions this is trivially true.

Another way to get a possible solution of the 5-queen problem from a solution of the 4-queen problem is to take $(5, 5)$ as an additional ordered pair. For this, we need to verify that for any (i, j) in a solution of the 4-queen problem $|i - 5| \neq |j - 5|$. Since for every (i, j) in the above two solutions $i \neq j$, we have $|i - 5| \neq |j - 5|$. Hence we get the following four solutions of the 5-queen problem:

$$(1,1),(2,3),(3,5),(4,2),(5,4); \quad (1,1),(2,4),(3,2),(4,5),(5,3);$$
$$(1,2),(2,4),(3,1),(4,3),(5,5); \quad (1,3),(2,1),(3,4),(4,2),(5,5).$$

However the above solutions need not be all the solutions of the 5-queen problem. For getting all the solutions, we need to go to the permutation tree. We may mention here that the above solutions are not obtained as an application of the backtracking approach. Rather these are obtained by divide and conquer approach which we shall discuss in a later chapter. For the sake of simplicity of drawing and due to space restrictions, we only draw the five sub-trees with roots arrived at level 2 (Figs. 3.5(a)–(e)).

Nodes that become dead as and when have been marked B which means a bound node. Only the paths to the active nodes at level 5 lead to possible solutions. There are 10 active nodes and the corresponding possible solutions are

$$(1, 3, 5, 2, 4), (1, 4, 2, 5, 3), (2, 4, 1, 3, 5), (2, 5, 3, 1, 4),$$

$$(3, 1, 4, 2, 5), (3, 5, 2, 4, 1), (4, 1, 3, 5, 2), (4, 2, 5, 3, 1),$$

$$(5, 2, 4, 1, 3), (5, 3, 1, 4, 2). \tag{3.2}$$

Since every one of the above permutations satisfy the requirement ((3.1)), namely, $|i - j| \neq |x_i - x_j|$, the above are all solutions of the 5-queen problem.

The case $n = 6$.

Recall that we have to look for permutations $(x_1, x_2, \ldots, x_6)$ of the set $X = \{1, 2, \ldots, 6\}$ for which $|i - j| \neq |x_i - x_j|$ for $i \neq j$ in X. We use this implicit requirement/restriction for the bounding function. Using this bounding function, we construct the state space tree without giving a detailed discussion. The sub-tree of this state space tree for $x_1 = 1$ is given in Fig. 3.6. Since all the leaf nodes at level 7 we have arrived at are dead nodes, there is no solution provided by the leftmost branch of the permutation tree. Hence there is no solution of the 6-queen problem with $x_1 = 1$.

We next explore the branch of the tree with $x_1 = 2$. This branch of the permutation tree is given in Fig. 3.7.

All the leaf nodes at level 7 are killed except one and, therefore, there is one solution to the 6-queen problem with $x_1 = 2$ and it is $(2, 4, 6, 1, 3, 5)$. This solution may be represented as in Table 3.4.

Thus the solution is: Q_1 in column 2, Q_2 in column 4, Q_3 in column 6, Q_4 in column 1, Q_5 in column 3 and Q_6 in column 5.

We next consider the branch of the permutation tree with $x_1 = 3$ given in Fig. 3.8. It is clear from the branch that there is only one live node at level 7 and, so, the 6-queen problem has only one solution in

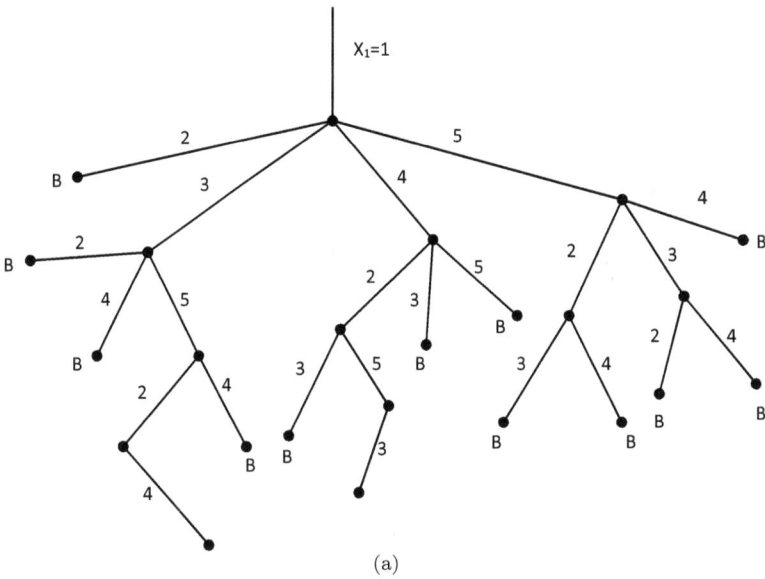

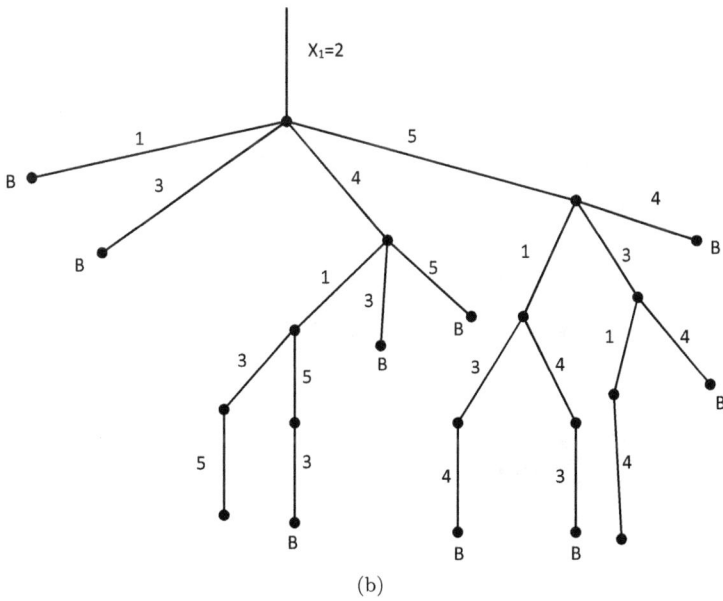

(a)

(b)

Fig. 3.5 (a) and (b)

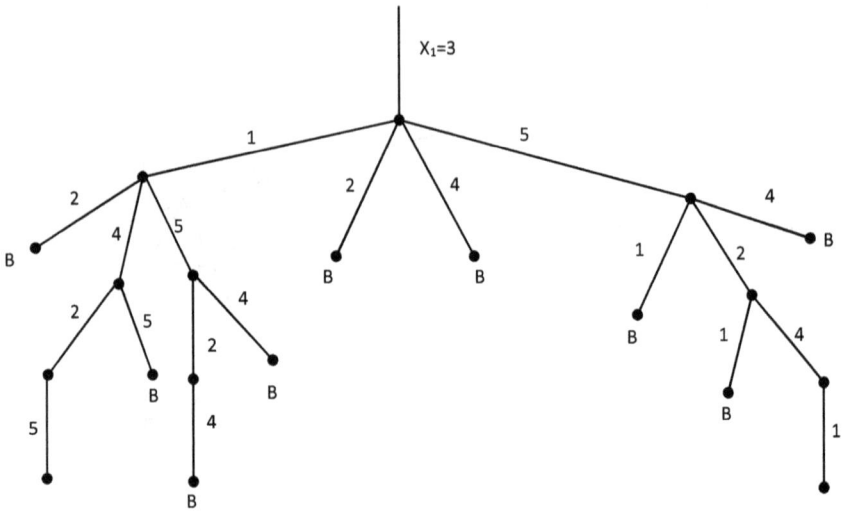

(c)

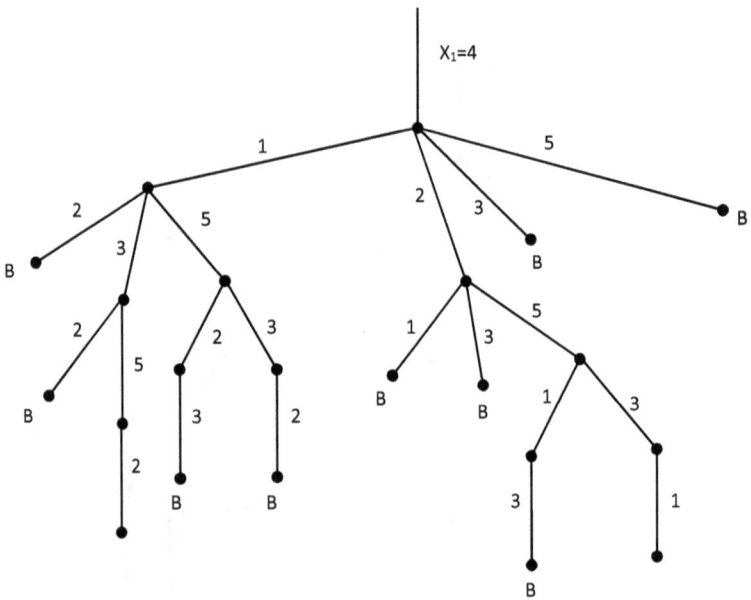

(d)

Fig. 3.5 (c) and (d)

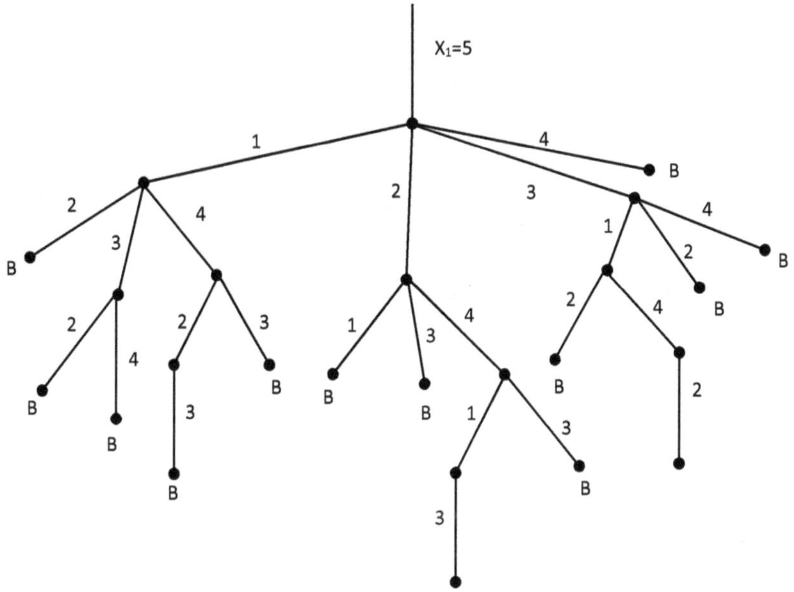

(e)

Fig. 3.5 (e)

this case, namely $(3, 6, 2, 5, 1, 4)$. This solution is $Q_1 \rightarrow 3, Q_2 \rightarrow 6,$
$Q_3 \rightarrow 2, Q_4 \rightarrow 5, Q_5 \rightarrow 1, Q_6 \rightarrow 4.$

Remark. If $(x_1, x_2, \ldots, x_n)$ is a permutation of the set $X = \{1, 2, \ldots, n\}$
which provides a solution to the n-queen problem, then the permutation
$(y_1, y_2, \ldots, y_n)$ of X where $y_i = x_{n-i+1}, 1 \leq i \leq n$, also provides a
solution to the n-queen problem. This is so because for $1 \leq i \neq j \leq n,$

$$|y_i - y_j| = |x_{n-i+1} - x_{n-j+1}|$$
$$\neq |(n - i + 1) - (n - j + 1)| = |i - j|.$$

In view of this observation, it follows that $(5, 3, 1, 6, 4, 2)$ and
$(4, 1, 5, 2, 6, 3)$ are also solutions of the 6-queen problem.

If there is a solution to the 6-queen problem with $x_1 = 6$, then there
will be a solution $(y_1, y_2, \ldots, y_6)$ to the 6-queen problem with
$y_6 = x_1 = 6$. Hence $(y_1, y_2, \ldots, y_5)$ will be a solution to the 5-queen
problem. Taking 6 as the 6th component in every solution of the 5-queen
problem as given in (3.2), we get 10 likely solutions of the 6-problem.

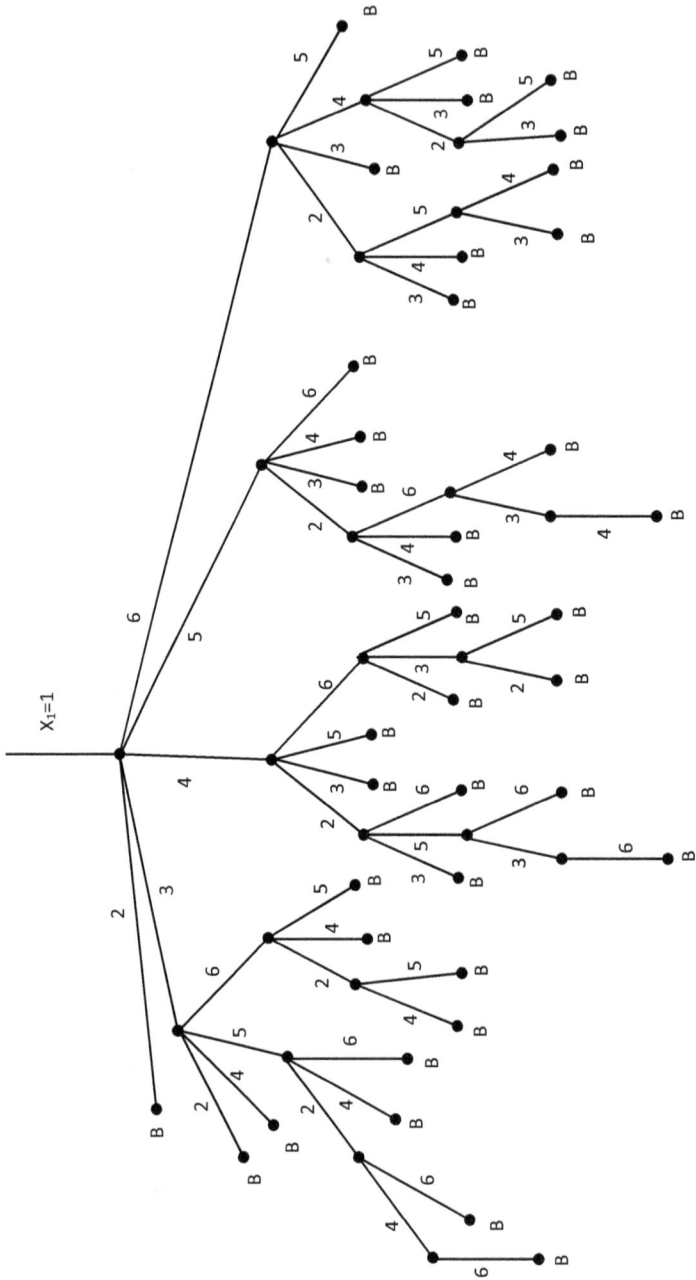

Fig. 3.6

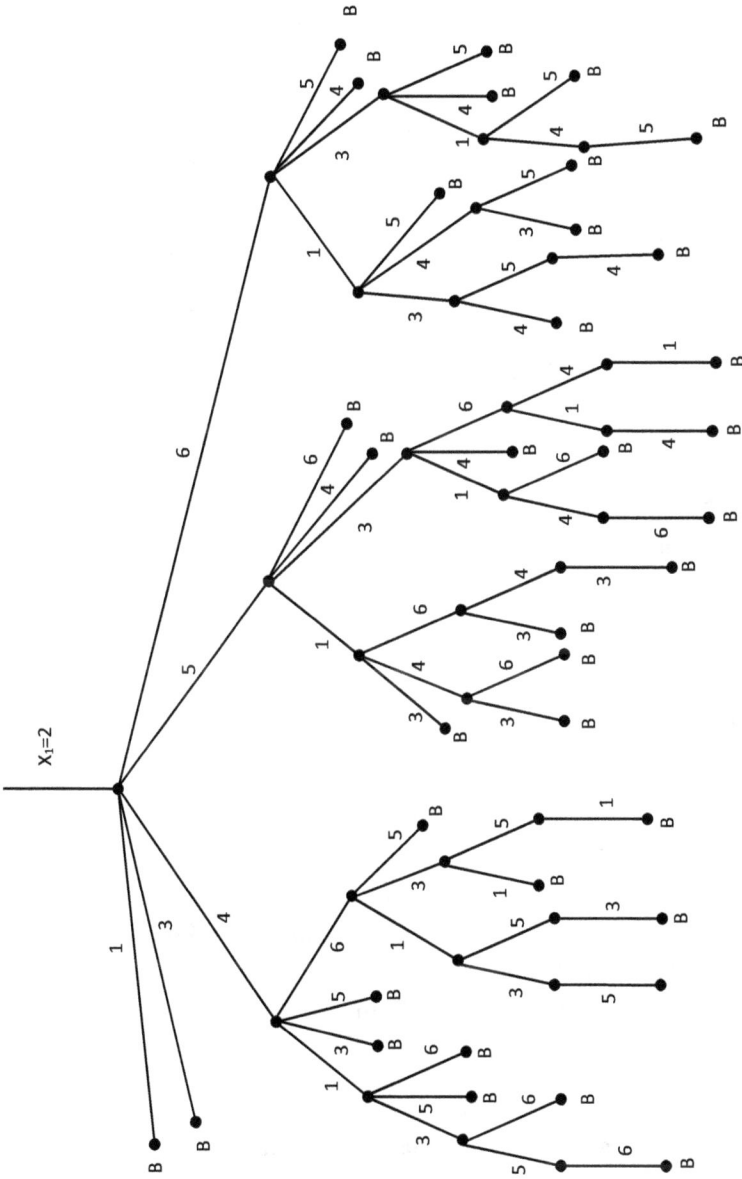

Fig. 3.7

Table 3.4

	1			
		2		
			3	
4				
		5		
			6	

However, we find that

$$|y_1 - y_6| = |1 - 6|, \ |y_1 - y_6| = |1 - 6|, \ |y_5 - y_6| = |5 - 6|,$$

$$|y_3 - y_6| = |3 - 6|, \ |y_5 - y_6| = |5 - 6|, \ |y_4 - y_6| = |4 - 6|,$$

$$|y_3 - y_6| = |3 - 6|, \ |y_2 - y_6| = |2 - 6|, \ |y_2 - y_6| = |2 - 6|,$$

$$|y_4 - y_6| = |4 - 6|$$

respectively in the 10 solutions as in (3.2) with the 6th entry 6. Thus, none of the ten solutions of the 5-queen problem leads to a solution of the 6-queen problem with $y_6 = 6$. Hence there is no solution $(x_1, x_2, \ldots, x_6)$ to the 6-queen problem with $x_1 = 6$.

The case $n = 8$.

Construction of permutation tree for the set $\{x_1, x_2, \ldots, x_8\}$ will lead to all the solutions of the 8-queen problem. In view of the constraint $|x_i - x_j| \neq |i - j|$ (cf. (3.1)), there is no solution to the n-queen problem with $x_1 = 1, x_2 = 2$ for any n. Looking at the branch of the permutation tree for the case $n = 8$ as in Fig. 3.9 and using the **branch and bound procedure** we find that there is only one solution namely $(1, 5, 8, 6, 3, 7, 2, 4)$ of the 8-queen with $x_1 = 1, x_2 = 5, x_3 = 8$. In tabular form the above solution is as given in Table 3.5.

Remark. A chess board is of size 2^k, k an integer > 1, and the 5-queen, 6-queen problems appear hypothetical. Therefore, for the n-queen problems, we may think of the queens being placed on a square board instead of a chessboard.

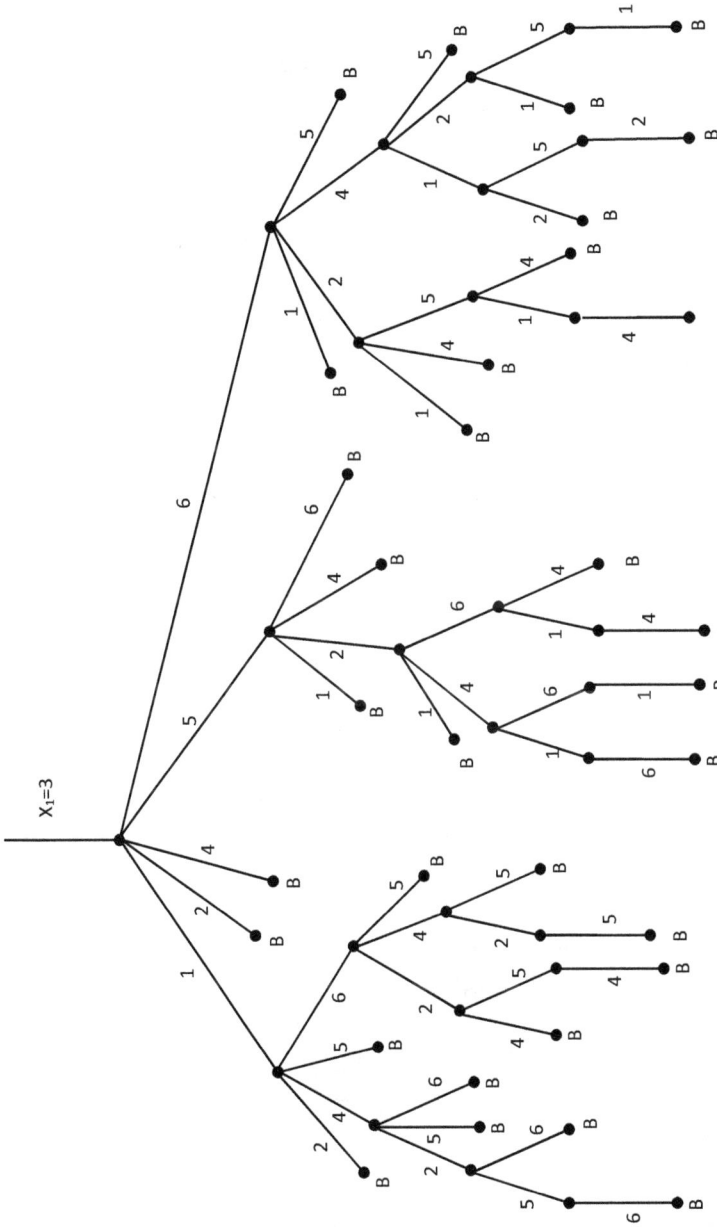

Fig. 3.8

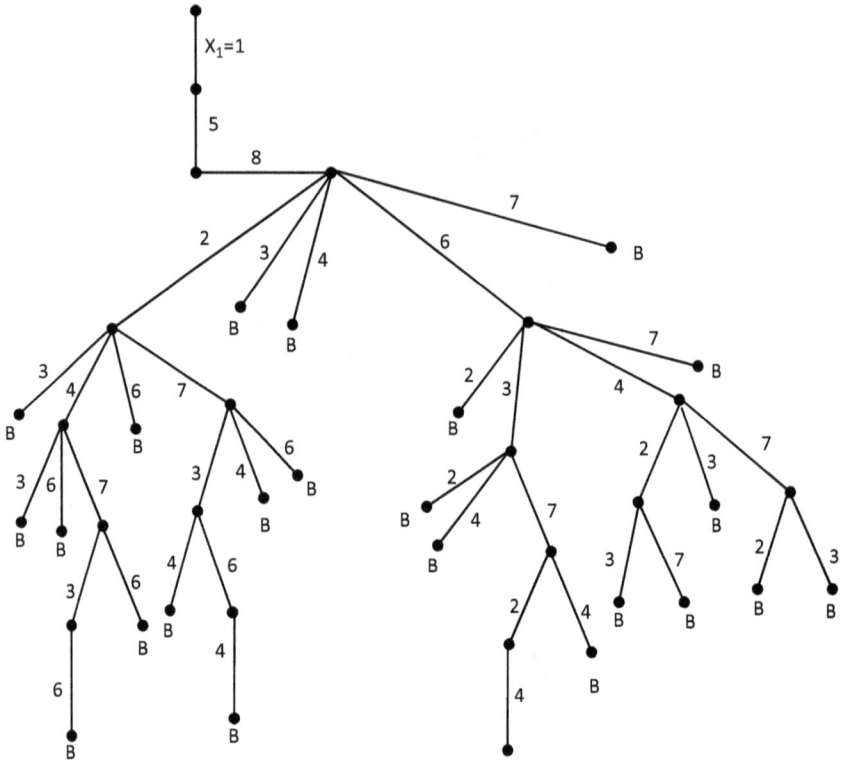

Fig. 3.9

Table 3.5

	1						
					2		
							3
					4		
			5				
						6	
		7					
				8			

Definition. Two solutions $(x_1, x_2, \ldots, x_n)$ and $(y_1, y_2, \ldots, y_n)$ of the n-queen problem with $y_i = x_{n-i+1}$ for $i = 1, 2, \ldots, n$ are called **equivalent solutions.**

Observe that $(4, 2, 7, 3, 6, 8, 5, 1)$ is also a solution of the 8-queen problem.

Algorithm 3.2. n-Queen Algorithm

---

1. Take $X = \{x_1, x_2, \ldots, x_n\}$
2. Choose x in X and set $x_1 = x$
3. For $2 \leq k \leq n$, choose $x_2, \ldots, x_k$ in $X \backslash \{x\}$ such that $|x_i - x_j| \neq |i - j|$ for $1 \leq i \neq j \leq k$
4. If $k = n$, then $(x_1, x_2, \ldots, x_n)$ is a solution to the n-queen problem with $x_1 = x$ If not, there is no solution to the n-queen problem with $x_1 = x$
5. Choose another x in X and repeat steps 2 to 4
6. Repeat step 5 till all the elements of X are exhausted

Analysis of the n-queen algorithm

As already seen, to look for a solution to the n-queen problem, we have to consider the $n!$ permutations of the set $X = \{1, 2, \ldots, n\}$. However, the constraint that there is no diagonal attack reduces the number of permutations to be considered considerably. For example, if the queen Q_1 is placed in the cell $(1, j)$, then the cells $(2, j - 1)$,$(2, j)$ and $(2, j + 1)$ are not available for placing Q_2. Thus there are n choices for placing Q_1 but there are at most $n - 2$ choices for placing Q_2 (it could be $n - 3$ if $j \neq 1, n$). Hence the number of permutations to be considered reduces to $n(n - 2)!$. This is a considerable improvement on the number of computations.

3.5. Solving Sudoku Problem Using Backtracking

As another application of backtracking we consider solving Sudoku puzzles and find that backtracking procedure is quite useful in solving these puzzles. In Sudoku puzzle we are given an $n \times n$ checkers table consisting of n^2 smaller squares all of the same size. Some of the smaller squares called cells are already filled with numbers from 1 to n. The problem is to fill the remaining empty cells again by numbers from 1 to n. The rules of the game are that no number appears more than once in any row or any column and either of the two diagonals. Sometimes, it is required that, if $n = m^2$ and the board is divided into n smaller $m \times m$ boards, then every one of the n smaller boards also has the same property.

The basic idea in finding a solution to the puzzle is as follows:

1. Scan the board to look for an empty cell that can be filled by the fewest possible number of values based on the simple constraints of the game.
2. If there is a cell that can take only one possible value, fill it with that one value and continue the algorithm.
3. If no such cell exists, place one of the numbers for that cell in the cell and repeat the process.
4. If we ever get stuck, we erase the last number placed and see if there are other possible choices for that cell and try those next.
5. Continue the above procedure till all the cells have been occupied and we have a solution to the problem or till we find that no such solution exists.

We illustrate the above procedure through a couple of examples.

Example 3.3. Fill the table given in Table 3.6 by numbers 1 to 4 in such a way that no number is repeated in any row or column or any of the four non-intersecting sub-square tables of order 2 each.

Solution. The cell $(1, 3)$ can be filled in only one way and, so, we place the number 3 there. Once this is done, the cell $(1, 2)$ has only one choice, namely 4. So place 4 in this cell. Now we find that the cell $(2, 3)$ has only one choice, namely 1. Place the number 1 in this cell. Next we observe that cell $(2, 1)$ has only one choice, namely 2. Place the number 2 in this cell. Once this is done, cell $(4, 1)$ has only one choice. Place 4 in this cell. Next the cell $(4, 2)$ has only one choice, namely 1. Place 1 in this cell. Observe that the cell $(3, 2)$ and $(4, 4)$ have one choice each. So place 2 in the cell

Table 3.6

1			2
	3		4
3		4	
		2	

Table 3.7

1		3	2
	2		4
3		4	
		2	

(a)

1	4	3	2
	3		4
3		4	
		2	

(b)

1	4	3	2
	3	1	4
3		4	
		2	

(c)

1	4	3	2
2	3	1	4
3		4	
		2	

(d)

1	4	3	2
2	3	1	4
3		4	
4		2	

(e)

1	4	3	2
2	3	1	4
3		4	
4	1	2	3

(f)

1	4	3	2
2	3	1	4
3	2	4	
4	1	2	3

(g)

1	4	3	2
2	3	1	4
3	2	4	1
4	1	2	3

(h)

Table 3.8

1			
3			
	1		
		3	

$(3, 2)$ and place the number 3 in the cell $(4, 4)$. The only remaining cell $(3, 4)$ has only one choice. So place 1 in this cell.

Observe that there being exactly one choice at each step, there is only one solution to the problem as given in Table 3.7(h). Also observe that the main diagonal does not have all entries distinct. Hence, if we had wanted a solution in which the entries on the main diagonal are also distinct, there will be no solution to the problem.

Example 3.4. Is it possible to fill Table 3.8 by the numbers 1 to 4 in such a way that no number is repeated in any row or column or any non-intersecting square sub-table of order 2 each?

Solution. Observe that the cell $(3, 1)$ has two choices. Either it takes 2 or 4. Place the number 2 in this cell. Then there is only one choice for the cell $(4, 1)$ where we need to place 4. Once this is done, there is only one choice for the cell $(4, 2)$ where we place the only possible number 2. Having done that, we find that the number 2 is repeated in the sub-square of order 2 placed in the left bottom corner (cf. Table 3.9(a)). Therefore backtrack to the cell $(3, 1)$ (once cell $(3, 1)$ is filled, there being only one choice for the

Table 3.9

1			
3			
2	1		
4	2	3	

(a)

1			
3			
4	1		
2	4	3	

(b)

Table 3.10

1				
			3	
	2			
4				5
		2		

cell (4, 1), there is no use backtracking to cell (4, 1)) and place the other possible number 4 in place of 2. Then there being only one choice for the cell (4, 1), we place the number 2 there. Again there is only one choice for the cell (4, 2) and we place the number 4 there. We find that, the number 4 is repeated in bottom left sub-square of order 2 (cf. Table 3.9(b)). Since we have exhausted the possible choices for the cell (3, 1), we cannot fill Table 3.8 under the given constraints.

Example 3.5. Fill the Table 3.10 by the numbers 1 to 5 in such a way that no number is repeated in any row or column or either of the two main diagonal.

Solution. For the cell (2, 1) there are two choices namely the numbers 2 and 5. Place the number 2 in this cell. There are two choices for the cell (3, 1), namely the numbers 3 and 5. Place the number 3 there. Then there is only one choice for the cell (5, 1). Place the number 5 there. There are two choices for the cell (4, 2). These are the numbers 1, 3. Place the number 1 in this cell. Then there is only one choice for the cell (4, 3). So, place the number 3 there. There then being only one choice for the cell (4, 4), place the number 2 there. There are two choices namely 4, 5 for the cell (1, 3). So place the number 4 in this cell. Then there is only one choice for the cell (1, 4). Place the number 5 there. There is only one choice for the cell (1, 2). We place the number 3 there and then we need to place 2 in the cell (1, 5). Now there is only one choice namely the number 4 for the

Table 3.11

(a)

1				
2			3	
	2			
4				5
		2		

(b)

1				
2			3	
3	2			
4				5
		2		

(c)

1				
2			3	
3	2			
4				5
5	2			

(d)

1				
2			3	
3	2			
4	1			5
5	2			

(e)

1				
2			3	
3	2			
4	1	3		5
5	2			

(f)

1				
2			3	
3	2			
4	1	3	2	5
5	2			

(g)

1		4		
2			3	
3	2			
4	1	3	2	5
5	2			

(h)

1		4	5	
2			3	
3	2			
4	1	3	2	5
5	2			

(i)

1	3	4	5	2
2			3	
3	2			
4	1	3	2	5
5	2			

(j)

1	3	4	5	2
2	5		3	
3	2			
4	1	3	2	5
5	4	2		

(k)

1	3	5	4	2
2			3	
3	2			
4	1	3	2	5
5		2		

(l)

1	3	5	4	2
2			3	
3	2		5	
4	1	3	2	5
5	4	2	1	3

(m)

1	3	5	4	2
2	5		3	
3	2		5	
4	1	3	2	5
5	4	2	1	3

(n)

1	3	5	4	2
2	5	1	3	4
3	2	4	5	1
4	1	3	2	5
5	4	2	1	3

cell $(5, 2)$. Place 4 there. Then we only have the choice of placing 5 in the $(2, 2)$ cell. The cell $(3, 3)$ then has two choices 1, 5. Placing either of these two numbers results in two numbers being repeated in the diagonal which violates the constraints of the problem. Hence, we need to backtrack by at least four steps (as there is only one choice in each of the last four steps), i.e., we go back to the cell $(1, 3)$ and place the number 5 in place of 4 there.

Then there is the only choice of placing 4 in the $(1, 4)$ cell, then 3 in the $(1, 2)$ cell and then 2 in the cell $(1, 5)$.

Now there is only one choice for the $(5, 2)$ cell. Place the number 4 there. Only choice available to fill the cell $(5, 4)$ is to place the number 1 there. Only choice for the cell $(3, 4)$ is to place 5 there and for the $(5, 5)$ cell is to place 3 there. Then cell $(2, 2)$ has only one choice, i.e., the number 5. For each one of the four remaining cells there are two choices namely 1 and 4. Place 1 in the $(2, 3)$ cell. Then each one of the remaining three cells has only one choice. So, place 4 in the cell $(3, 3)$, 4 in the cell $(2, 5)$ and 1 in the cell $(3, 5)$ and we get a valid solution as given in Table 3.11(n).

Observe that if we place 4 in the cell $(2, 3)$ in place of 1, then 1 is placed in the $(3, 3)$ cell which is not acceptable (as the main diagonal will have two entries each equal to 1).

The various steps involved are given in Table 3.11 (a)–(n).

Since backtracking is just an improvement on the brute-force procedure, there are several problems that may be solved by backtracking. Two simple such problems are (a) finding largest common subsequence of two sequences and (b) the sum of subsets problem. However, we consider these in a later chapter.

Exercise 3.1.

1. Find all other solutions, if any, of the 6-queen problem with $x_1 = 4$ and $x_1 = 5$.
2. Find all possible solutions of the 8-queen problem with $x_1 = 1, x_2 = 3$.
3. Prove that the number of solutions of the n-queen problem for any $n \geq 4$ is even.
4. If possible, extend $(1, 4, 7)$ to a solution of the 8-queen problem. Do the same for the triplets $(1, 4, 6)$, $(2, 4, 8)$.
5. We have five black marbles and four white marbles on a playing board consisting of 10 places with the black marbles in the five places on the left, the white marbles in the four rightmost places and the 6th place a free space. The black marbles can move to the right and white ones can only move to the left. At each move, a marble can either (a) move one space ahead, if that space is free or (b) jump ahead over exactly one marble of the opposite color, if the space just beyond that

marble is clear. Use backtrack procedure to shift the white marbles to the leftmost places/spaces on the board and the black marbles to the rightmost five spaces on the board (Matuszek (cf. [57]) for discussion on this problem).

6. Formulate, if possible, the idea as given for solving a Sudoku puzzle into an algorithm.

Chapter 4

Divide and Conquer

One of the most important and widely used techniques for solving computing/complex problems is to divide the problem into two or more problems of smaller size but of the same nature as the original problem, solving the sub-problems and then combining the solutions of the sub-problems to get a solution of the original problem. This technique of solving problems is called divide and conquer approach. We have already used this technique while discussing quick sort. Here, we consider some more applications of the technique but before that we give the divide and conquer algorithm in a formal setting.

4.1. Divide and Conquer General Algorithm

Let $T(n)$ denote the running time of the algorithm D and $C(P)$ where P is a problem of size n. In case n is small, the problem can be solved directly in a constant (finite) amount of time $c(n)$ (say). Otherwise, some time is needed for subdividing the problem into sub-problems

---

Algorithm 4.1. Procedure D and $C(P)$

---

1. P is a given problem of size n
2. if n is small, solve P directly
3. when n is large, divide P into sub-problems $P_1, P_2, \ldots, P_k$ (say)
4. apply D and C to $P_1, P_2, \ldots, P_k$
5. combine the solutions D and $C(P_1), \ldots, D$ and $C(P_k)$ to get a solution of P

$P_1, P_2, \ldots, P_k$. Suppose that sub-problems are of size $n_1, n_2, \ldots, n_k$, respectively. Solving the sub-problems using D and C algorithm takes times $T(n_1), T(n_2), \ldots, T(n_k)$, respectively. Once the sub-problems have been solved, it takes time to combine their solutions to get a solution of P. If $a(n)$ denotes the time taken (i) to subdivide the problem and (ii) combining the solutions of the sub-problems to get a solution of P, then $T(n)$ is given by

$$T(n) = \begin{cases} c(n) & \text{if } n \text{ is small,} \\ T(n_1) + \cdots + T(n_k) + a(n) & \text{otherwise.} \end{cases} \quad (4.1)$$

In case the sub-problems are all of the same size, i.e., of size, $\frac{n}{k}$, the running time $T(n)$ takes the following form:

$$T(n) = \begin{cases} c(n) & \text{if } n \text{ is small,} \\ kT\left(\frac{n}{k}\right) + a(n) & \text{otherwise.} \end{cases} \quad (4.2)$$

When n is small, the time taken $c(n)$ may be taken as $T(1)$. Presence of n in $c(n)$ and $a(n)$ indicates the dependence of the time taken on the size of the problem. The $T(n)$ as in (4.2) is a recurrence relation of a very special type and can be solved by the substitution method, i.e., by replacing $T(\frac{n}{k})$ on the right side by $kT(\frac{n}{k^2}) + ka(\frac{n}{k})$ and so on. We shall elucidate this method of solution in some of the examples we consider.

4.2. Max–Min Problem

We begin with an extremely simple application of divide and conquer. Let S be a given set of integers. We are interested in finding a maximum element as well as a minimum element of S. Partition S into two subsets S_1, S_2 of almost equal size. Apply divide and conquer to S_1 and S_2 to get maximum and minimum of S_1 and S_2. If M_i is the maximum and m_i the minimum of Si for $i = 1, 2$, then $\max\{M_1, M_2\}$ is maximum of S and $\min\{m_1, m_2\}$ is the minimum of S. For the sake of simplicity, we assume that the number n of elements of S is a power of 2.

Let $T(n)$ be the running time or the number of comparisons for finding max-min(S) for $|S| = n = 2^k$. For $n = 2$, only one comparison is needed to find the maximum and minimum of $S = \{a, b\}$. Therefore, $T(2) = 1$. If $n > 2$, then $T(n)$ is the total number of calls to max-min as at 6 and 7 of

---

Algorithm 4.2. Procedure Max–Min (S)

---

1. S is a set with $n = 2^k$ elements
2. if $n = 2$, let $S = \{a, b\}$
   if $a > b$, $\max(S) = a$, $\min(S) = b$
   else $\max(S) = b$, $\min(S) = a$
3. if $n > 2$
4. divide S into subsets S_1, S_2 with $|S_1| = |S_2| = \frac{n}{2}$
5. apply max-min to S_1, S_2
6. $(\max(S_1), \min(S_1)) = $ max-min(S_1)
7. $(\max(S_2), \min(S_2)) = $ max-min(S_2)
8. return $\max\{\max(S_1), \max(S_2)\}$, $\min\{\min(S_1), \min(S_2)\}$

---

the above procedure plus the two comparisons as at 8. Therefore

$$T(n) = \begin{cases} 1 & \text{if } n = 2, \\ 2T\left(\dfrac{n}{2}\right) + 2 & \text{if } n > 2. \end{cases}$$

(We are assuming that each comparison requires one unit of time. Also we are ignoring the constant amount of time needed for partitioning S into S_1, S_2 as at 4.)

Now, for $n = 2^k > 2$,

$$T(n) = 2T\left(\frac{n}{2}\right) + 2 = 2\left(2T\left(\frac{n}{4}\right) + 2\right) + 2 = 2^2 T\left(\frac{n}{4}\right) + (2^2 + 2)$$

and, so, in general,

$$T(n) = 2^r T\left(\frac{n}{2^r}\right) + (2^r + 2^{r-1} + \cdots + 2).$$

Since $n = 2^k$, we finally have

$$T(n) = 2^{k-1} T(2) + (2^{k-1} + 2^{k-2} + \cdots + 2)$$
$$= 2^{k-1} + \frac{2^k - 1}{2 - 1} - 1 = \frac{n}{2} + n - 2 = \frac{3}{2}n - 2.$$

4.3. Round-Robin Tennis Tournament

In Round-Robin tennis tournament, there are n players with each player playing against every other player and every player plays exactly one match every day. As each player has to play with the remaining $n - 1$ players and he is playing exactly one match a day, the tournament lasts for $n - 1$ days. Let the players be referred to as $1, 2, \ldots, n$. The schedule of the tournament can then be represented by an $n \times (n - 1)$ matrix in which the (i, j)th entry denotes the player against whom the ith player plays on the jth day. Thus, the rows of the matrix represent players and the columns represent the days on which the games are played. For the sake of simplicity, the schedule is considered when the number n of the players taking part in the tournament is a power of 2. The schedule is constructed by using a recursive application of the algorithm by first finding a schedule for half the players. Thus, we finally get down to the case when there are only two players and there is only one match on the first day itself in which the two players play each other.

Let us consider the case when $n = 2^4$, i.e., there are 16 players taking part in the tournament. We can divide the problem first into two sub-problems with eight players in each. Each of the two sub-problems is divided into two sub-problems each having four players. Thus, we get four sub-problems with four players in each. Finally, we divide every one of the four sub-problems into two sub-problems each having only two players. Thus, we have come down to eight base problems with only two players each. The tournament with two players lasts for only one day and the schedule is given in Table 4.1 in which the first player plays the second and the second player plays the first. For four players, we partition the table as in Table 4.2. On the first day of tournament, we have player 1 plays with player 2 and player 2 plays with player 1. For the other two players named 3 and 4, we can fill the last two entries of the first column by adding 2 to the first two entries, respectively. On the second and third

Table 4.1

	1
1	2
2	1

Table 4.2

	1	2	3
1	2		
2	1		
3	4		
4	3		

Table 4.3

	1	2	3
1	2	3	4
2	1	4	3
3	4	1	2
4	3	2	1

days of the tournament, players 1, 2 have to play with only players 3, 4 and vice versa. We can fill the second and third columns of row one by taking the respective entries as 3, 4. Then the second row of this portion of the table is completed by giving a cyclic shift to the relevant portion of the first row. Then we have to fill the bottom right corner of the table. In the column headed 2 we take the respective entries as $(1, 2)^t$ (= transpose of the row vector $(1, 2)$) and then the next column is filled by giving a cyclic shift to this column. Thus, a schedule of the round-robin tournament with four players is as given in Table 4.3.

Next consider the tournament with eight players. In this case we have an 8×7 table with $1, 2, \ldots, 8$ heading the rows and $1, 2, \ldots, 7$ heading the columns. We divide the problem into two sub-problems: one in which the players are $1, 2, 3, 4$ and the other in which the players are $5, 6, 7, 8$. The schedule for the first three days tournament for the second sub-problem is given in Table 4.3 by adding 4 to every number in the header column and to every entry of this table. The first three days' schedule for the eight players is given in Table 4.4, where the upper and lower 4×4 portions of the table are yet to be filled. Observe that on the fifth to eighth day the players $\{1, 2, 3, 4\}$ have to play with the players $\{5, 6, 7, 8\}$ and vice versa. The top right 4×4 portion of Table 4.4 is then obtained by taking entries of the first row as $5, 6, 7, 8$ and the next three rows are obtained by giving a cyclic shift to this and the subsequent rows. The bottom right 4×4 portion of the table is filled by taking $1, 2, 3, 4$, respectively, as the entries of the first column and then the next three columns are filled by giving a cyclic shift to the first and the subsequent

Table 4.4

	1	2	3	4	5	6	7
1	2	3	4				
2	1	4	3				
3	4	1	2				
4	3	2	1				
5	6	7	8				
6	5	8	7				
7	8	5	6				
8	7	6	5				

Table 4.5

	1	2	3	4	5	6	7
1	2	3	4	5	6	7	8
2	1	4	3	6	7	8	5
3	4	1	2	7	8	5	6
4	3	2	1	8	5	6	7
5	6	7	8	1	4	3	2
6	5	8	7	2	1	4	3
7	8	5	6	3	2	1	4
8	7	6	5	4	3	2	1

Table 4.6

	1	2	3	4	5	6	7	8	9	10	11	12	13	14	15
1	2	3	4	5	6	7	8	9	10	11	12	13	14	15	16
2	1	4	3	6	7	8	5	10	11	12	13	14	15	16	9
3	4	1	2	7	8	5	6	11	12	13	14	15	16	9	10
4	3	2	1	8	5	6	7	12	13	14	15	16	9	10	11
5	6	7	8	1	4	3	2	13	14	15	16	9	10	11	12
6	5	8	7	2	1	4	3	14	15	16	9	10	11	12	13
7	8	5	6	3	2	1	4	15	16	9	10	11	12	13	14
8	7	6	5	4	3	2	1	16	9	10	11	12	13	14	15
9	10	11	12	13	14	15	16	1	8	7	6	5	4	3	2
10	9	12	11	14	15	16	13	2	1	8	7	6	5	4	3
11	12	9	10	15	16	13	14	3	2	1	8	7	6	5	4
12	11	10	9	16	13	14	15	4	3	2	1	8	7	6	5
13	14	15	16	9	12	11	10	5	4	3	2	1	8	7	6
14	13	16	15	10	9	12	11	6	5	4	3	2	1	8	7
15	16	13	14	11	10	9	12	7	6	5	4	3	2	1	8
16	15	14	13	12	11	10	9	8	7	6	5	4	3	2	1

columns. The schedule of the tournament with eight players is thus given in Table 4.5.

Arguing on the above lines, the schedule of tournament for 16 players is as given in Table 4.6.

Remark. While obtaining Table 4.5 for tournament schedule for eight players using Table 4.3 for tournament schedule for four players, the upper right 4×4 portion of the table is filled by taking the first row of this portion as 5, 6, 7, 8 and then obtaining the next three rows by giving a cyclic shift to this row and the successive rows. On the other hand, the bottom right 4×4 portion of the table is filled by taking the entries of the first column as $(1, 2, 3, 4)^t$ and the next three columns are filled by giving a cyclic shift to the entries of this column and the successive columns. If we had filled the bottom right portion by taking the entries of the first row (instead of first column) as 1, 2, 3, 4 and then the next three rows are obtained by giving a cyclic shift to this and the successive rows, the schedule may be given by Table 4.7.

Observe that the games scheduled for the fifth day are between the players: $(1, 6)$, $(2, 7)$, $(3, 8)$, $(4, 5)$, $(5, 2)$, $(6, 3)$, $(7, 4)$ and $(8, 1)$. Thus, on this day every player has to play two games rather than only one. This contradicts the rules of the tournament. Hence for filling the top 4×4

Table 4.7

	1	2	3	4	5	6	7
1	2	3	4	5	6	7	8
2	1	4	3	6	7	8	5
3	4	1	2	7	8	5	6
4	3	2	1	8	5	6	7
5	6	7	8	1	2	3	4
6	5	8	7	2	3	4	1
7	8	5	6	3	4	1	2
8	7	6	5	4	1	2	3

portion and the bottom 4×4 portion, we cannot fill the first row of each portion (and similarly the first column of each portion) first and then completing the two portions by giving cyclic shift to the first and successive rows (or columns). Hence, if the top right corner is filled by filling the first row, the bottom corner needs to be filled by filling the first column and vice versa.

We next give the procedure for obtaining the tournament schedule for n players when n is a power of 2.

Observe that a table

$$M = \frac{\begin{array}{c|c} & b \end{array}}{\begin{array}{c|c} a & A \end{array}}$$

correctly represents a schedule of games for the tournament if all the entries in every row and every column are distinct, the ith row of M does not contain i as an entry and if, for any i, j, $a(i, j) = k$, then $a(k, j) = i$. The last condition ensures that if i plays with k on the jth day, then on this day k plays with i and no one else.

Let $T(n)$ denote the running time of the round-robin tournament with n players. Then

$$T(n) = 2T\left(\frac{n}{2}\right) + \text{the time taken for obtaining } M_3, \ M_4$$

from $\overline{M_2}$, $\overline{M_1}$, respectively.

Since M_3 is obtained from $\overline{M_2}$ by giving cyclic shifts to the rows of $\overline{M_2}$ and each cyclic shift tales a constant amount of time, say a units, the time taken for obtaining M_3 from $\overline{M_2}$ is $\frac{an}{2}$. Similarly, the time taken for

---

Algorithm 4.3. Procedure Tournament Schedule $(S, |S| = n = 2^k)$

---

1. $n = 2^k$ is the number of players
2. each player plays exactly one game a day
3. every player plays with every other player — there being exactly one game between any two players
4. the players be named $1, 2, \ldots, n$
5. the tournament lasts for exactly $n - 1$ days
6. let $a(i, j)$ denote the player with whom the player i plays on the jth day
7. then $M = (a(i, j))$ is the $n \times (n - 1)$ matrix (table) representing the schedule of the tournament
8. let M_1 be the matrix that represents the schedule of games between the players $1, 2, \ldots, \frac{n}{2}$
9. let M_2 be the matrix that represents the schedule of games between the players $\frac{n}{2} + 1, \frac{n}{2} + 2, \ldots, n$
10. let $\overline{M_1}$ be the $\frac{n}{2} \times \frac{n}{2}$ matrix which is the matrix M_1 preceded by $(1, 2, \ldots, \frac{n}{2})^t$ as the first column
11. let $\overline{M_2}$ be the $\frac{n}{2} \times \frac{n}{2}$ matrix which is the matrix M_2 preceded by $(\frac{n}{2} + 1, \frac{n}{2} + 2, \ldots, n)^t$ as the first column
12. take M_3 as the $\frac{n}{2} \times \frac{n}{2}$ matrix in which the first row is the first row of $\overline{M_2}$ and the subsequent rows are obtained by giving a cyclic shift to the previous row
13. take M_4 as the $\frac{n}{2} \times \frac{n}{2}$ matrix in which the first column is the first column of $\overline{M_1}$ and the subsequent columns are obtained by giving a cyclic shift to the previous column
14. then the schedule of the tournament for the n players is given by the table

$$M = \begin{array}{c|c|c} & \multicolumn{2}{c}{b} \\ \hline a & M_1 & M_3 \\ \hline & M_2 & M_4 \end{array}$$

where a is the column with respective entries $1, 2, \ldots, n$ and b is the row with entries being respectively $1, 2, \ldots, n - 1$

---

obtaining M_4 from $\overline{M_1}$ is $\frac{an}{2}$. Hence

$$T(n) = 2T\left(\frac{n}{2}\right) + an$$

$$= 2\left\{2T\left(\frac{n}{2^2}\right) + a\frac{n}{2}\right\} + an = 2^2 T\left(\frac{n}{2^2}\right) + 2an$$

$$= 2^2\left\{2T\left(\frac{n}{2^3}\right) + a\frac{n}{2^2}\right\} + 2an = 2^3 T\left(\frac{a}{2^3}\right) + 3an$$

$$\vdots$$

$$= 2^k T(1) + kan, \quad \text{as } n = 2^k.$$

But $T(1)$ is a constant, say b. Therefore $T(n) = bn + an \log n$. Hence $T(n) = O(n + n \log n) = O(n \log n)$.

4.4. The Problem of Counterfeit Coin

We are given a heap or collection of coins of the same kind containing at most one counterfeit coin; it being given that a counterfeit coin is not of the same weight as a genuine coin. We are also given a balancing machine which can compare the weights of two sets of coins. The problem is to identify counterfeit coin, if any, in the given collection of coins.

(a) For the sake of simplicity, assume that the given set has 2^k coins where k is a positive integer.

(i) Suppose that it is given that a counterfeit coin is lighter in weight than a genuine coin. We divide the coins into two subsets A and B each having $\frac{n}{2}$ coins. Using the balancing machine, we compare the weights of A and B. If wt(A) = wt(B), then all the coins are of equal weight and so, there is no counterfeit coin. If wt(A) < wt(B), then there is a counterfeit coin in the set A. We repeat the process of division for the collection A and continue till we arrive at a collection of just two coins one among which is the counterfeit coin. Comparing the weights of these two coins, we can identify the counterfeit coin.

In case wt(A) > wt(B), we can adopt the above procedure with B in place of A.

(ii) We are not given that a counterfeit coin is lighter than a genuine coin but only that counterfeit coin is not of the same weight as a genuine coin.

If $wt(A) < wt(B)$, we divide the set A of coins into subsets A_1 and A_2, the number of coins in A_1 being equal to the number of coins in A_2. If $wt(A_1) = wt(A_2)$, then the counterfeit coin does not belong to A but is in B and the counterfeit coin is heavier than a genuine coin. We then subdivide the subset B into subsets B_1, B_2 with $o(B_1) = o(B_2)$. Then the weights of B_1, B_2 are not equal (because there is indeed a counterfeit coin in the collection) and we continue the process of subdivision of the heavier subset till we arrive at a subset containing only two coins. The heavier of the two final coins is the counterfeit coin.

If $wt(A_1) \neq wt(A_2)$, say $wt(A_1) < wt(A_2)$, then the counterfeit coin is lighter than a genuine coin. We then arrive at a counterfeit coin as in (i) above.

(b) Suppose that the given set S of coins has 3^k coins, where k is a positive integer. In this case, we divide the set S into three subsets A, B, C with $o(A) = o(B) = o(C)$. If $wt(A) = wt(B) = wt(C)$, then there is no counterfeit coin. Otherwise $wt(A)$, $wt(B)$, $wt(C)$ cannot be all distinct. Thus weights of two of the sets A, B, C are equal. Suppose that $wt(A) < wt(B) = wt(C)$.

Then the counterfeit coin is lighter than a genuine coin and is in A. We then continue the process of subdivision dividing A into subsets A_1, A_2, A_3 with $o(A_1) = o(A_2) = o(A_3)$. Without loss of generality, we may assume that $wt(A_1) < wt(A_2) = wt(A_3)$. We then continue the process of subdivision, subdividing A_1 and at every step subdividing the lighter of the three subsets arrived at till we come to a subset having exactly three coins. Lighter of these three coins is then a counterfeit coin.

In case $wt(A) > wt(B) = wt(C)$, the counterfeit coin is heavier than a genuine coin. The process of subdivision is now carried out with the set with higher weight than the other two.

Observe that $wt(A) < wt(B) = wt(C)$ or $wt(A) > wt(B) = wt(C)$ becomes significant at every point of subdivision as comparison of weights of only two subsets needs to be considered (once it is decided whether a counterfeit coin is lighter or heavier than a genuine coin).

Algorithm 4.4. Procedure Counterfeit Coin $(S, |S| = n = 2^k)$

---

1. S is a set of n coins, $n = 2^k$, a counterfeit coin is lighter than a genuine coin, it being given that there is at most one counterfeit coin
2. if $n = 2$
3. if weights of the two coins are equal
4. there is no counterfeit coin
5. if weights of the two coins are unequal
6. lighter of the two is counterfeit coin
7. if $n > 2$
8. divide S into subsets A, B with $o(A) = o(B)$
9. if wt(A) = wt(B)
10. there is no counterfeit coin
11. otherwise if wt(A) < wt(B)
12. apply Counterfeit Coin (A)

end

In case a counterfeit coin is heavier than a genuine coin, only nominal changes are made in the above algorithm — the word lighter is replaced by heavier in 6 and 12 is replaced by

12′ apply Counterfeit Coin (B)

In case a counterfeit coin is heavier than a genuine coin, then the numbers 6 and 14 in the above algorithm get replaced by

6′. heavier of the three coins is a counterfeit coin
14′. apply Counterfeit Coin (B)

Observe that there is one comparison of weights of the three coins a, b, c (say) as at 6 is needed. If wt(a) = wt(b), then c is a counterfeit coin, while if wt(a) < wt(b), then a is a counterfeit coin.

We now consider the running time $T(n)$ of Algorithm 4.4.

When $n = 3$, comparison of weights of only two of the three coins is enough to determine a counterfeit coin. This takes only a unit time and so $T(3) = 1$. When $n > 3$, the action as at 8 takes a time proportional to n the size of the set S, i.e., cn (say), where c is a constant. Action at

---

Algorithm 4.5. Procedure Counterfeit Coin $(S, |S| = n = 3^k)$

---

1. S is a set of n coins, $n = 3^k$, a counterfeit coin is lighter than a genuine coin given that there is at most one counterfeit coin
2. if $n = 3$
3. if the weights of the three coins are equal
4. there is no counterfeit coin
5. otherwise
6. lighter of the three coins is a counterfeit coin
7. if $n > 3$
8. divide S into subsets A, B, C with $o(A) = o(B) = o(C)$
9. if $\mathrm{wt}(A) = \mathrm{wt}(B) = \mathrm{wt}(C)$
10. there is no counterfeit coin
11. otherwise, if $\mathrm{wt}(A) = \mathrm{wt}(B)$
12. apply Counterfeit Coin (C)
13. if $\mathrm{wt}(A) < \mathrm{wt}(B)$
14. apply Counterfeit Coin(A)

---

end

---

numbers 9, 11, 13 require at most two comparisons of weights of the three subsets. This may take at most two units of time. The action at 12 or 14 takes time $T\left(\frac{n}{3}\right)$. Therefore

$$T(n) = \begin{cases} 1 & \text{if } n = 3, \\ T\left(\frac{n}{3}\right) + cn & \text{if } n > 3. \end{cases}$$

Substitution method then gives $T(n)$ as follows (for $n = 3^k, k > 1$):

$$T(n) = T\left(\frac{n}{3}\right) + cn$$

$$= T\left(\frac{n}{3^2}\right) + c\frac{n}{3} + cn$$

$$= T\left(\frac{n}{3^3}\right) + c\frac{n}{3^2} + c\frac{n}{3} + cn$$

$$\vdots$$

$$= T(3) + c\frac{n}{3^{k-2}} + c\frac{n}{3^{k-3}} + \cdots + cn$$

$$= 1 + cn \left(1 + \frac{1}{3} + \cdots + \frac{1}{3^{k-2}} \right)$$

$$= 1 + cn \left(1 - \frac{1}{3^{k-1}} \right) \Big/ \left(1 - \frac{1}{3} \right)$$

$$= 1 + 3c\frac{n}{2} \left(1 - \frac{1}{3^{k-1}} \right) < 2cn = dn,$$

where $d = 2c$ is a constant.

A similar argument gives the running time $T(n) \leq cn$, where c is a constant for the Algorithm 4.4.

Algorithm 4.6. Procedure Counterfeit Coin $(S, |S| = n = 2^k, k > 1)$

---

1. S is a set of coins with $o(S) = n = 2^k, k \geq 1$, containing at most one counterfeit coin it being given that weight of a counterfeit coin is not equal to the weight of a genuine coin
2. divide S into subsets A, B with $o(A) = o(B)$
3. if $\mathrm{wt}(A) = \mathrm{wt}(B)$
4. there is no counterfeit coin
5. if $\mathrm{wt}(A) < \mathrm{wt}(B)$
6. divide A into subsets A_1, A_2 with $o(A_1) = o(A_2)$
7. if $\mathrm{wt}(A_1) = \mathrm{wt}(A_2)$
8. counterfeit coin is heavier than a genuine coin and is in B
9. apply Counterfeit Coin (B)
10. if $\mathrm{wt}(A_1) < \mathrm{wt}(A_2)$
11. the counterfeit coin is lighter than a genuine coin
12. apply Counterfeit Coin (A_1)

---

end

Remark. If it is not given when $n = 2^k$ that a counterfeit coin is lighter or heavier than a genuine coin but it is only given that a counterfeit coin is not of the same weight as a genuine coin, it is equally simple to obtain a solution without considering subcases. Let us suppose that $n = |S| > 2$. Subdivide S into subsets A, B with $o(A) = o(B)$. If $\mathrm{wt}(A) = \mathrm{wt}(B)$, there is no counterfeit coin. If $\mathrm{wt}(A) < \mathrm{wt}(B)$, subdivide A into subsets A_1, A_2 with $o(A_1) = o(A_2)$. If $\mathrm{wt}(A_1) = \mathrm{wt}(A_2)$, then the counterfeit coin is heavier than a genuine coin and is in B, while if $\text{wt}(A_1) < \text{wt}(A_2)$ the counterfeit coin is lighter than a genuine coin and is in A. Then we can proceed to find a counterfeit coin as before. Also Algorithm 4.4 gets replaced by Algorithm 4.6.

4.5. Tiling a Defective Chess Board with Exactly One Defective Square Using Triominoes

Consider a square partitioned into four squares by joining the midpoints of opposite sides of the square. Suppose one of the sub-squares removed. The resulting figure is called a **triomino**. A triomino is given in Fig. 4.1 and it consists of three equal squares.

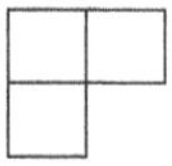

Fig. 4.1

We are given a square chess board with 2^k squares in each row and 2^k squares in each column. All the squares on the chess board are of the same size. Exactly one square on the chess board is defective. It is required to put tiles on the chess board leaving the only defective square using tiles in the shape of triominoes, it being given that each of the three squares comprising a triomino is of the same size as a square on the chess board. Is it possible to tile the chess board as above? And if yes, what is exactly the number of triominoes needed?

The total number of squares on the chess board that are to be covered is $2^k \times 2^k - 1 = (2^k - 1)(2^k + 1)$. Since 3! (factorial 3) divides a product of three consecutive positive integers, 3! divides $(2^k - 1)$ $2^k(2^k + 1)$ and, therefore, 3 divides $(2^k - 1)(2^k + 1)$. Hence it is possible to put triomino tiles on the chess board as desired. We can work by induction on k to see that this is indeed possible. The inductive argument also provides us the number $N(2^k)$ of triominoes required. For $k = 1$, the chess board consists of four squares with one defective

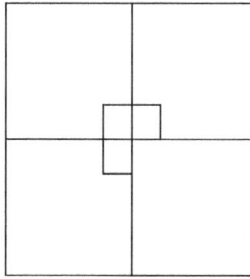

Fig. 4.2

square. Thus leaving the defective square, the chess board is covered by a single triomino. Now suppose that $k \geq 2$ and the chess board is partitioned into four square sub-boards of the same size. Each of these four square boards has $2^{k-1} \times 2^{k-1}$ squares. Since the original chess board has exactly one defective square, exactly one of the smaller chess boards has that defective square and none of the other three boards has a defective square. Place one triomino tile in the center of the original board in such a way that the three squares of the tile cover precisely one square in each of the three boards which have no defective square. Thus, the fourth board which has one defective square has no square covered by the one tile placed. The result is depicted in Fig. 4.2.

In the three boards, exactly one square is already occupied. Therefore, for the sake of putting tiles, we may regard each one of these boards as $2^{k-1} \times 2^{k-1}$ board with one defective square in each. By our induction hypothesis, every one of the four boards can be tiled by using $N(2^{k-1})$ triominoes each. Hence, the original board can be tiled and the number of triominoes needed is given by

$$N(2^k) = 4N(2^{k-1}) + 1.$$

This recurrence relation can be solved by the substitution method as follows:

$$\begin{aligned} N(2^k) &= 1 + 4N(2^{k-1}) \\ &= 1 + 4\{1 + 4N(2^{k-2})\} \\ &= 1 + 4 + 4^2\{1 + 4N(2^{k-3})\} \end{aligned}$$

$$= 1 + 4 + 4^2 + 4^3 N(2^{k-3})$$

$$\vdots$$

$$= 1 + 4 + 4^2 + \cdots + 4^{k-2} N(2^2)$$

$$= 1 + 4 + 4^2 + \cdots + 4^{k-2}\{1 + 4N(2^1)\}$$

$$= 1 + 4 + 4^2 + \cdots + 4^{k-2} + 4^{k-1} = \frac{4^k - 1}{3}.$$

4.6. Strassen's Matrix Multiplication Algorithm

Let $A = (a(i, j))$ (respectively, $B = (b(i, j))$) be a square matrix of order n, where $a(i, j)$ (respectively, $b(i, j)$) denotes the element in the ith row and jth column of the matrix A (respectively, B). Let $AB = C = (c(i, j))$. In the traditional method of multiplication of square matrices

$$c(i, j) = \sum_{1 \leq k \leq n} a(i, k)b(k, j). \tag{4.3}$$

Thus $c(i, j)$ is obtained using n multiplications and $n - 1$ additions. Since there are n^2 elements $c(i, j)$ in C, obtaining the product C of the matrices A, B requires n^3 multiplications and $n^3 - n^2$ additions. Hence, the traditional method of multiplication of matrices gives running time $T(n)$ proportional to n^3, i.e., $O(n^3)$ (because every multiplication of two numbers or of two elements of a ring over which the matrices are defined takes a constant amount of time and addition also takes a constant amount of time). Henceforth, we assume that the matrices are defined over numbers.

We now like to see if the divide and conquer approach leads to any improvement in the running time for the multiplication $AB = C$ of matrices. For this purpose, we assume that n is even. Partition the matrices A and B as follows:

$$A = \begin{pmatrix} A_{11} & A_{12} \\ A_{21} & A_{22} \end{pmatrix}, \quad B = \begin{pmatrix} B_{11} & B_{12} \\ B_{21} & B_{22} \end{pmatrix}$$

where $A_{11}, A_{12}, A_{21}, A_{22}$ and $B_{11}, B_{12}, B_{21}, B_{22}$ are all square matrices of order $\frac{n}{2}$. This partitioning of A, B is conformable for multiplication and

$$AB = \begin{pmatrix} A_{11} & A_{12} \\ A_{21} & A_{22} \end{pmatrix}\begin{pmatrix} B_{11} & B_{12} \\ B_{21} & B_{22} \end{pmatrix} = \begin{pmatrix} C_{11} & C_{12} \\ C_{21} & C_{22} \end{pmatrix}, \tag{4.4}$$

where

$$C_{11} = A_{11}B_{11} + A_{12}B_{21},$$
$$C_{12} = A_{11}B_{12} + A_{12}B_{21},$$
$$C_{21} = A_{21}B_{11} + A_{22}B_{21},$$
$$C_{22} = A_{21}B_{12} + A_{22}B_{22}.$$

(4.5)

Since $A_{ik}B_{kj}$ is a product of two square matrices of order $\frac{n}{2}$, the number of multiplications and additions needed for this product are $\frac{n^3}{8}$ and $\frac{n^3}{8} - \frac{n^2}{4}$, respectively. Also every C_{ij} is the sum of two products of the type $A_{ik}B_{kj}$. Therefore, the number of computations required for finding C_{ij} is $\frac{n^3}{4} - \frac{n^2}{2} + \frac{n^2}{4} = \frac{n^3}{4} - \frac{n^2}{4}$ additions and $\frac{n^3}{4}$ multiplications. Hence, the number of computations involved for finding AB is n^3 multiplications and $n^3 - n^2$ additions which is the same as the number of computations in obtaining C without partitioning A, B. Since the running time is proportional to the number of computations involved, divide and conquer leads to no improvement in running time over the running time by traditional method of multiplication. However, as already seen, each C_{ij} is a sum of two products of two square matrices of order $\frac{n^2}{2}$, the running time $T(n)$ is related to $T(\frac{n^2}{2})$ as follows:

$$T(n) = 8T\left(\frac{n}{2}\right) + bn^2,$$

where b is a constant (which is the time taken for addition of two numbers). Also for $n = 2$, the product AB involves only eight multiplications and four additions. Therefore $T(2) = a$ i.e., a constant. For $n = 1$, product of two square matrices of order 1 involves only one multiplication of two numbers and, so, the product takes a constant amount of time. We thus obtain the following recurrence relation for the running time.

$$T(n) = \begin{cases} a & \text{if } n \leq 2, \\ 8T\left(\frac{n}{2}\right) + bn^2 & \text{if } n > 2. \end{cases}$$

(4.6)

The general case when n is even but not a power of 2 does not provide any specific information about the running time.

Now suppose that n is a power of 2, say $n = 2^k$, where k is a positive integer. Using substitution method, we can then find the total running time $T(n)$ for the multiplication of two square matrices of order n by the traditional method. For $k > 1$,

$$T(n) = bn^2 + 8T\left(\frac{n}{2}\right)$$

$$= bn^2 + 8\left\{b\left(\frac{n}{2}\right)^2 + 8T\left(\frac{n}{2^2}\right)\right\}$$

$$= (2^2 - 1)bn^2 + 8^2 T\left(\frac{n}{2^2}\right)$$

$$= (2^2 - 1)bn^2 + 8^2\left\{b\frac{n^2}{16} + 8T\left(\frac{n}{2^3}\right)\right\}$$

$$= (2^3 - 1)bn^2 + 8^3 T\left(\frac{n}{2^3}\right)$$

$$\vdots$$

$$= (2^k - 1)bn^2 + 8^k T(1)$$

$$= b(n - 1)n^2 + n^3 a = O(n^3).$$

Since addition of two square matrices of order n takes $O(n^2)$ time and multiplication (by the traditional method of multiplication) takes $O(n^3)$ time which is much larger than $O(n^2)$ for n large, it will be interesting to explore an approach which leads to reduction in the number of matrix multiplications at the cost of matrix additions. As observed above, traditional matrix multiplication involves eight multiplications and four additions of square matrices of order $\frac{n}{2}$ for the computation of C_{ij}'s in (4.5). A method given below which was discovered by Strassen [26] shows that for the computation of C_{ij}'s in (4.5) only 7 multiplications and 18 additions of square matrices of order $\frac{n}{2}$ are enough.

Lemma 4.1. *With the notations as in (4.4) and (4.5), let $P = (A_{11} + A_{22})(B_{11} + B_{22})$, $Q = (A_{21} + A_{22})B_{11}$, $R = A_{11}(B_{12} - B_{22})$, $S = A_{22}(B_{21} - B_{11})$, $T = (A_{11} + A_{12})B_{22}$, $U = (A_{21} - A_{11})(B_{11} + B_{12})$, $V = (A_{12} - A_{22})(B_{21} + B_{22})$. Then $C_{11} = P + S - T + V$, $C_{12} = R + T$, $C_{21} = Q + S$, $C_{22} = P + R - Q + U$.*

Proof.

$$P + S - T + V = (A_{11} + A_{22})(B_{11} + B_{22}) + A_{22}(B_{21} - B_{11})$$
$$- (A_{11} + A_{12})B_{22} + (A_{12} - A_{22})(B_{21} + B_{22})$$
$$= A_{11}B_{11} + A_{11}B_{22} + A_{22}B_{11} + A_{22}B_{22}$$
$$+ A_{22}B_{21} - A_{22}B_{11} - A_{11}B_{22} - A_{12}B_{22} + A_{12}B_{21}$$
$$+ A_{12}B_{22} - A_{22}B_{21} - A_{22}B_{22} = A_{11}B_{11} + A_{12}B_{21}$$
$$= C_{11} \qquad P + R - Q + U = (A_{11} + A_{22})(B_{11} + B_{22})$$
$$+ A_{11}(B_{12} - B_{22})$$
$$- (A_{21} + A_{22})B_{11} + (A_{21} - A_{11})(B_{11} + B_{12})$$
$$= A_{11}(B_{11} + B_{22} + B_{12} - B_{22} - B_{11} - B_{12})$$
$$+ A_{22}(B_{11} + B_{22} - B_{11}) - A_{21}(B_{11} - B_{11} - B_{12})$$
$$= A_{22}B_{22} + A_{21}B_{12}$$
$$= C_{22} \qquad R + T = A_{11}(B_{12} - B_{22}) + (A_{11} + A_{12})B_{22}$$
$$= A_{11}B_{12} + A_{12}B_{22} = C_{12}$$

and

$$Q + S = (A_{21} + A_{22})B_{11} + A_{22}(B_{21} - B_{11})$$
$$= A_{21}B_{11} + A_{22}B_{21} = C_{21}. \qquad \square$$

Theorem 4.2. *If $n = 2^k$, $k > 1$ an integer, then the running time for computing the product $C = AB$ of two square matrices of order n is $T(n) = O(n^{\log_2 7})$.*

Proof. Each of $P, Q, \ldots, V$ is obtained using addition or difference of square matrices of order $\frac{n}{2}$ and then using exactly one multiplication of two matrices in each case. Then the C_{ij} are obtained using only sum or difference of $P, Q, \ldots, V$. Therefore C, the product AB, is obtained by using 7 multiplications of two square matrices of order $\frac{n}{2}$ at a time and 18 additions or differences of $\frac{n}{2} \times \frac{n}{2}$ matrices. Hence, the running time $T(n)$

for obtaining the product $C = AB$ is given by

$$T(n) = \begin{cases} a & \text{for } n \leq 2, \\ 7T\left(\dfrac{n}{2}\right) + bn^2 & \text{for } n > 2, \end{cases} \tag{4.7}$$

where a and b are constants. We solve this recurrence relation by substitution method for $n > 2$:

$$T(n) = bn^2 + 7T\left(\frac{n}{2}\right)$$

$$= bn^2\left(1 + \frac{7}{4}\right) + 7^2 T\left(\frac{n}{2^2}\right)$$

$$= bn^2\left\{1 + \frac{7}{4} + \left(\frac{7}{4}\right)^2\right\} + 7^3 T\left(\frac{n}{2^3}\right)$$

$$\vdots$$

$$= bn^2\left\{1 + \frac{7}{4} + \left(\frac{7}{4}\right)^2 + \cdots + \left(\frac{7}{4}\right)^{k-1}\right\} + 7^k T(1)$$

$$= \left(\frac{4b}{3}\right)n^2\left\{\left(\frac{7}{4}\right)^k - 1\right\} + a7^k$$

$$= \left(\frac{4b}{3}\right)n^2\left\{\left(\frac{7}{4}\right)^{\log n} - 1\right\} + a7^{\log n}. \tag{4.8}$$

If $z = x^{\log y}$, then $\log z = \log y \log x = \log y^{\log x}$ and, therefore, $z = y^{\log x}$. Thus $x^{\log y} = y^{\log x}$. Therefore, using this relation in (4.8) above, we get

$$T(n) = \left(\frac{4b}{3}\right)n^2\left\{\left(\frac{7^{\log n}}{4^{\log n}}\right) - 1\right\} + a7^{\log n}$$

$$= \left(\frac{4b}{3}\right)n^2\left\{\left(\frac{n^{\log 7}}{n^{\log 4}}\right) - 1\right\} + an^{\log 7}$$

$$\leq \left\{\left(\frac{4b}{3}\right) + a\right\}n^{\log 7}.$$

Hence $T(n) = O(n^{\log_2 7})$.   □

Remark. Since $\log_2 7 < \log_2 8 = 3$, Strassen's method of matrix multiplication improves upon the running time for matrix multiplication by traditional method.

Let us now partition the matrices A, B as

$$A = \begin{pmatrix} A_{11} & A_{12} & A_{13} \\ A_{21} & A_{22} & A_{23} \\ A_{31} & A_{32} & A_{33} \end{pmatrix}, \quad B = \begin{pmatrix} B_{11} & B_{12} & B_{13} \\ B_{21} & B_{22} & B_{23} \\ B_{31} & B_{32} & B_{33} \end{pmatrix},$$

where it is understood that A and B are square matrices of order $n = 3^k$, $k > 1$ an integer. Then $C = AB$ is partitioned as follows:

$$C = \begin{pmatrix} C_{11} & C_{12} & C_{13} \\ C_{21} & C_{22} & C_{23} \\ C_{31} & C_{32} & C_{33} \end{pmatrix}$$

and

$$C_{ij} = A_{i1}B_{1j} + A_{i2}B_{2j} + A_{i3}B_{3j} \quad \text{for } 1 \le i \le 3,\ 1 \le j \le 3.$$

The traditional method of multiplication of matrices using the above partitioning of A, B involves n^3 multiplications and $n^3 - n^2$ additions which leads to no improvement in the running time over the multiplication of matrices without using any partitioning. However, mimicking Strassen's method, if we can express C_{ij} as linear combinations of some t square matrices P_r of order $\frac{n}{3}$ and each one of the t matrices is a product of only two matrices either one of which is sum and/or difference of the A_{rs}, B_{uv}, then the running time for finding the product AB is $O(\log_3 t)$. The largest positive integer t for which $\log_3 t < \log_2 7$ is 21. Therefore, to get a true improvement in the running time over the one obtained by Strassen's method, it is necessary to find $t \le 21$ square matrices P_r of order $\frac{n}{3}$ as above.

4.7. Medians and Order Statistic

A problem closely related to sorting is finding order statistic and median of a given sequence/set. Given a set of n elements or records and k a positive integer, the kth element in the (increasing) order in the set of all given elements is called the **kth-order statistic**. The first-order statistic of the given set is the minimum element of the set while the nth-order statistic is the maximal element of the set. If n is odd, $\frac{n+1}{2}$th-order statistic is called the **median** of the set. In case n is even, **there are two medians**, namely the $\frac{n}{2}$th-order statistic and the $(\frac{n}{2}+1)$th-order statistic. As a matter of convention, we call the lower median, i.e., $(\frac{n}{2})$th-order statistic as its median. In this section, we consider the problem of finding the kth-order statistic of a given set of n elements. Using heap sort or merge sort as also quick sort, this selection of the kth-order statistic can be solved in $O(n \log n)$ time. As an application of divide and conquer, we give a worst-case linear running time procedure for selecting kth-order statistic of a given set of n distinct elements. The algorithm to achieve this is known as **Select** and determines the kth smallest element of an input array of $n > 1$ distinct elements by executing the following steps.

Algorithm 4.7. Procedure Select (order statistic)

1. divide the n elements of the input array into $\lfloor \frac{n}{7} \rfloor$ groups of seven elements each and at most one group made up of n (mod 7) elements. Then the total number of groups is $\lceil \frac{n}{7} \rceil$

2. sorting the elements (which are at most 7) of each of the $\lceil \frac{n}{7} \rceil$ groups by using insertion sort find the medians of the $\lceil \frac{n}{7} \rceil$ groups

3. recursively using Select, find the median x of the $\lceil \frac{n}{7} \rceil$ medians found in step 2 above

4. partition the given array of numbers using the median-of-medians x as pivot by the modified version of Partition. Let l be one more than the number of elements on the low side of the partition. Then x is the lth smallest element of the given array and there are $n - l$ elements on the high side of partition

5. if $k = l$, then return x (so that x is the kth smallest element of the given array). Otherwise, use Select recursively to find the kth smallest element on the low side if $k < l$ or the $(k - l)$th smallest element on the high side if $k > l$

4.7.1. Worst-case running time of select (order statistic)

For analyzing the running time of Select, we first find a lower bound on the number of elements of the given array that are greater than the partitioning element x. There being $\lceil \frac{n}{7} \rceil$ medians and x being the median of the array of medians, x is the $\lceil \frac{1}{2} \lceil \frac{n}{7} \rceil \rceil$th element of the array of medians in the sorted order. For each of the medians $> x$ found in step 2, there are at least four elements of the corresponding group which are greater than x except for one group the number of elements of which is $n \pmod 7$. Also, the group for which x is the median, the number of elements greater than x is 3 and not 4. Hence, the number of elements of the given array greater than x is at least

$$4 \left(\left\lceil \frac{1}{2} \left\lceil \frac{n}{7} \right\rceil \right\rceil - 2 \right) \geq \frac{2n}{7} - 8.$$

Similarly, the number of elements of the given array that are less than x is at least

$$4 \left(\left\lceil \frac{1}{2} \left\lceil \frac{n}{7} \right\rceil \right\rceil - 1 \right) \geq 4 \left(\left\lceil \frac{1}{2} \left\lceil \frac{n}{7} \right\rceil \right\rceil - 2 \right) \geq \frac{2n}{7} - 8.$$

Therefore, the number of elements that are greater than x is at most

$$n - \left(\frac{2n}{7} - 8 \right) = \frac{5n}{7} + 8$$

and the number of elements that are less than x is also at most $\frac{5n}{7} + 8$.

Hence, in the worst case, Select in step 5 is called recursively on at most $\frac{5n}{7} + 8$ times.

Let $T(n)$ be the worst-case running time of the algorithm Select. Step 1 involves no sorting and no swapping. It only involves counting of seven elements of the given array from left to right and putting a semicolon (;) to separate each set of seven elements to put it in one group. Step 2 consists of $\lceil \frac{n}{7} \rceil$ calls of insertion sort for finding the median of each group of just at most seven elements. Step 4 partitions the set of $\lceil \frac{n}{7} \rceil$ elements which are medians of the groups of seven elements each (except for possibly one) around the median x of the $\lceil \frac{n}{7} \rceil$ medians. Thus steps 1, 2, 4 taken together take $O(n)$ time. Step 3 recursively uses Select to find the median x of the $\lceil \frac{n}{7} \rceil$ elements which are the medians of the groups of at most seven elements each. Thus, this step involves $T(\lceil \frac{n}{7} \rceil)$ running time. In step 5, the number of recursive calls for Select depends on the number of elements of the given array which are less than or equal to the partitioning element x or which are greater than x. But the number of elements which are less than or equal to x is at most $\frac{5n}{7} + 8$ and the number of elements which are greater than x is also at most $\frac{5n}{7} + 8$. Since we are looking for the kth element in the sorted order of the given elements and either kth element $\leq x$ or kth element $> x$, the total running time is $\leq T(\frac{5n}{7} + 8)$. Hence, we have

$$T(n) \leq O(n) + T\left(\left\lceil\frac{n}{7}\right\rceil\right) + T\left(\frac{5n}{7} + 8\right), \qquad (4.9)$$

which happens from some stage, say m, onwards. For $n \leq m$, the given set of elements is small in number, and using sorting algorithm, we may find the kth element in $\leq an$ time where a is some positive constant. Also the term $O(n)$ in (4.9) is $\leq bn$ for some constant b. Thus, the recursive relation for $T(n)$ becomes

$$T(n) \leq \begin{cases} an & \text{for } n \leq m, \\ bn + T\left(\left\lceil\frac{n}{7}\right\rceil\right) + T\left(\frac{5n}{7} + 8\right) & \text{for } n > m. \end{cases}$$

To prove that $T(n) = O(n)$, let us assume that $T(n) \leq cn$ for all $n > m$. Then

$$T(n) \leq bn + c\left(\left\lceil\frac{n}{7}\right\rceil\right) + c\left(\frac{5n}{7} + 8\right)$$

$$< bn + c\left(\frac{n}{7} + 1\right) + c\left(\frac{5n}{7} + 8\right)$$

$$= cn + \left(\left(b - \frac{c}{7}\right)n + 9c\right) \leq cn$$

provided

$$bn - \left(c\frac{(n - 63)}{7}\right) \leq 0 \quad \text{or} \quad c \geq \frac{7bn}{n - 63}.$$

If $n > 126$, then $\frac{n}{n-63} = 1 + \frac{63}{n-63} < 2$ so that $\frac{7bn}{n-63} < 14b$. Hence, if we choose $c \geq 14b$, then $T(n) \leq cn$ for all $n > 126$, so that the number m assumed above is 126. Letting $d = \max\{a, c\}$, we then get $T(n) \leq dn$ for all n, i.e., $T(n) = O(n)$.

Hence the worst-case running time of the procedure Select is $O(n)$, i.e., it is a linear running time.

When we partition the given array around the median-of-medians x, it does not imply that the number of the given elements less than x is $\leq \frac{n}{2}$ and that the number of elements greater than x is $\geq \frac{n}{2}$.

Example 4.1. Let us consider the array (with groups of seven elements with the corresponding medians)

28, 5, 7, 33, 41, 3, 48;	28
40, 2, 25, 17, 34, 6, 44;	25
31, 15, 37, 49, 1, 18, 46;	31
32, 4, 45, 14, 51, 42, 26;	32
19, 21, 53, 16, 35, 11, 24;	21
52, 47, 8, 13, 38, 54, 29;	38
10, 23, 39, 50, 9, 36, 43;	36
12, 22, 27, 30, 20;	22

The medians of the seven groups of seven elements each and the last group of five elements are, respectively,

$$28, 25, 31, 32, 21, 38, 36, 22.$$

The median-of-medians x is then 28 (by our convention for an array of even length).

The number of elements less than $x (= 28)$ is $\geq 4 \left(\left\lceil \frac{1}{2} \left\lceil \frac{n}{7} \right\rceil \right\rceil - 1 \right) > 4 \left(\left\lceil \frac{1}{2} \left\lceil \frac{n}{7} \right\rceil \right\rceil - 2 \right) = 8$ (remember that $n = 54$ and so $\left\lceil \frac{n}{7} \right\rceil = 8$) and the number of elements greater than x is $\geq 4 \left(\left\lceil \frac{1}{2} \left\lceil \frac{n}{7} \right\rceil \right\rceil - 2 \right) = 8$.

Then $n - 8 = 54 - 8 = 46 = n - 4 \left(\left\lceil \frac{1}{2} \left\lceil \frac{n}{7} \right\rceil \right\rceil - 2 \right)$ and the number of elements $\leq x$ is 28 which is < 46. Similarly, the number of elements greater than $x = 28$ is 26 which is again < 46.

Example 4.2. Let us next consider $n = 49$ with the numbers being the numbers in the first seven groups of seven elements each in the above example. Then the median-of-medians x is 31. Now $4 \left(\left\lceil \frac{1}{2} \left\lceil \frac{n}{7} \right\rceil \right\rceil - 2 \right) = 8$ again and $n - 4 \left(\left\lceil \frac{1}{2} \left\lceil \frac{n}{7} \right\rceil \right\rceil - 2 \right) = 41$. The elements less than 31 are 25 and those greater than 31 are 23 and both these numbers are less than 41.

Algorithm 4.8. Procedure Select (order statistic with groups of size 3)

1. divide the n elements of the given input array into $\lfloor \frac{n}{3} \rfloor$ groups of three elements each and at most one group of one or two elements. Thus, the total number of groups into which the given array of n elements is partitioned is $\lceil \frac{n}{3} \rceil$

2. sorting the elements which are at most three of each of the $\lceil \frac{n}{3} \rceil$ groups by using insertion sort find the medians of the $\lceil \frac{n}{3} \rceil$ groups

3. recursively using Select, find the median x of the $\lceil \frac{n}{3} \rceil$ medians found in step 2 above

4. partition the given array of n numbers using median-of-medians x as pivot by the modified version of Partition. Let l be one more than the number of elements on the lower side of the partition so that x is the lth smallest element of the given array and there are $n - l$ elements on the right side of partition

5. if $k = l$, then return x (so that x is the kth smallest element of the given array). Otherwise, use Select recursively to find the kth smallest element on the low side of partition if $k < l$ or the $(k - l)$th smallest element on the high side if $k > l$

We next prove that the algorithm Select does not run in linear time if groups of three elements each are used instead of the groups of seven elements each.

There being $\lceil \frac{n}{3} \rceil$ medians and x being median of the array of medians, x is the $\left\lceil \frac{1}{2} \lceil \frac{n}{3} \rceil \right\rceil$th element of the array of medians in the sorted order. For each of the median greater than x found in step 2, there are at least two elements of the corresponding group which are greater than x except for possibly one group the number of elements of which is 1 or 2. Also, the group for which x is the median, the number of elements greater than x is 1 and not 2. Hence, the number of elements of the given array $> x$ is at least

$$2 \left(\left\lceil \frac{1}{2} \left\lceil \frac{n}{3} \right\rceil \right\rceil - 2 \right) \geq \frac{n}{3} - 4.$$

Similarly, the number of elements of the given array that are less than x is at least

$$2 \left(\left\lceil \frac{1}{2} \left\lceil \frac{n}{3} \right\rceil \right\rceil - 1 \right) > 2 \left(\left\lceil \frac{1}{2} \left\lceil \frac{n}{3} \right\rceil \right\rceil - 2 \right) = \frac{n}{3} - 4.$$

Therefore, the number of elements of the given array that are greater than x is at most

$$n - \left(\frac{n}{3} - 4\right) \leq \frac{2n}{3} + 4$$

and the number of elements of the given array that are less than x is also at most

$$n - \left(\frac{n}{3} - 4\right) \leq \frac{2n}{3} + 4.$$

Suppose that the running time $T(n)$ in this case is $O(n)$. Then there exist positive constants c and m such that

$$T(n) \leq cn \quad \text{for all } n \geq m.$$

Step 1 of the algorithm Select involves no swapping. It only involves counting three elements of the given array from left to right and putting a semicolon (;) after every count to separate each set of three elements to put it in one group. Step 2 consists of $\lceil \frac{n}{3} \rceil$ calls of insertion sort to find the median of each group of at most three elements. Step 4 partitions the set of $\lceil \frac{n}{3} \rceil$ elements which are the medians of the $\lceil \frac{n}{3} \rceil$ groups of at most three elements. Thus steps 1, 2, 4 taken together take $O(n)$ time. Step 3 recursively uses Select to find the median x of the set of $\lceil \frac{n}{3} \rceil$ medians. Thus, this step involves $T(\lceil \frac{n}{3} \rceil)$ running time. In step 5, the number of recursive calls for Select depends on the number of elements of the given array which are less than or equal to the partitioning element x or which are greater than x. But the number of elements which are less than or equal to x is at most $\frac{2n}{3} + 4$ and the number of elements which are greater than x is also at most $\frac{2n}{3} + 4$. Since we are looking for the kth element in the sorted order of the given elements and either the kth element is $\leq x$ or the kth element is $> x$, the total running time is $\leq T(\frac{2n}{3} + 4)$. Hence, we have

$$T(n) \leq O(n) + T\left(\left\lceil \frac{n}{3} \right\rceil\right) + T\left(\frac{2n}{3} + 4\right), \tag{4.10}$$

which happens from some stage, say $n \geq m$, onwards. For $n \leq m$, the given set of elements is small in number and using some sorting algorithm, we may find the kth element in $\leq an$ time, where a is some positive constant. Also, the term $O(n)$ in (4.10) is $\leq bn$ for some positive constant

b for all $n > m_1$, for some positive m_1. Thus, we have (for m larger than m_1 and the above m)

$$T(n) \leq \begin{cases} an & \text{for } n \leq m, \\ bn + T\left(\left\lceil \dfrac{n}{3} \right\rceil\right) + T\left(\dfrac{2n}{3} + 4\right) & \text{for } n > m. \end{cases}$$

Since we are assuming $T(n) \leq cn$ for $n > m$, we have

$$T(n) \leq bn + c\left(\left\lceil \frac{n}{3} \right\rceil\right) + c\left(\frac{2n}{3} + 4\right)$$

$$\leq bn + c\left(\frac{n}{3} + 1\right) + c\left(\frac{2n}{3} + 4\right)$$

$$= cn + (bn + 5c)$$

$$\leq cn$$

provided $bn + 5c \leq 0$. Since b, c are positive and $n > m$ (a positive integer), $bn + 5c$ can never be ≤ 0. Hence, we cannot have $T(n) \leq cn$, for all $n > m$ where c is a positive constant. This proves that the algorithm Select when the given elements are divided into groups of three elements each together with at most one group of just one or two elements does not have a linear running time.

Exercise 4.1.

1. Prove that for n odd, it is not possible to find a schedule of round-robin tournament for n players if the restrictions of the tournament are as already stated. However, if the restriction that every player plays exactly one game every day is replaced by "every player plays at most one game every day", then it is possible to find a schedule of tournament for n players, with n odd. Also prove that minimum number of days of tournament is n.
2. Using divide and conquer or otherwise, find a schedule of the revised round-robin tournament for five players that lasts for five days.
3. Prove that it is always possible to find a schedule of round-robin tournament for n players when n is even.

4. Using divide and conquer or otherwise, find a schedule of round-robin tournament for (a) 6 players (b) 10 players.

5. In the algorithm Select, the input elements are divided into groups of seven elements each. Prove that the algorithm works in linear time if the input elements are divided into groups of (i) 5, (ii) 9, (iii) 11 and (iv) 13 elements each.

{Hint:

$$(ii) \ T(n) \leq T\left(\left\lceil \frac{n}{9} \right\rceil\right) + T\left(\frac{13n}{18} + 10\right) + O(n)$$

$$\leq c\left(\frac{n}{9} + 1\right) + c\left(\frac{13n}{18} + 10\right) + an$$

$$= \frac{5nc}{6} + (11c + an) \leq cn$$

provided $\frac{-cn}{6} + 11c + an \leq 0$ or $c \geq \frac{6an}{n-66}$. Thus $T(n) \leq cn$ for all $n \geq 132$ if we choose $c \geq 12a$.

$$(iii) \ T(n) \leq \begin{cases} O(1) & \text{for } n \leq m, \\ T\left(\left\lceil \frac{n}{11} \right\rceil\right) + T\left(\frac{8n}{11} + 12\right) + O(n) & \text{for } n > m. \end{cases}$$

Suppose that $O(n) \leq an$ and that $T(n) \leq cn$ for some positive constants a, c for all $n > m$. Then, for $n > m$,

$$T(n) \leq c\left(\left\lceil \frac{n}{11} \right\rceil\right) + c\left(\frac{8n}{11} + 12\right) + an$$

$$\leq c\left(\frac{n}{11} + 1\right) + c\left(\frac{8n}{11} + 12\right) + an$$

$$= cn + \left(\frac{-2cn}{11} + 13c + an\right) \leq cn$$

provided $\frac{-2cn}{11} + 13c + an \leq 0$ or $c \geq \frac{11an}{2n-143}$, which is possible for all $n \geq 143$ when we choose $c \geq 11a$.

(iv) Here $T(n) \leq T\left(\left\lceil \frac{n}{13} \right\rceil\right) + T\left(\frac{19n}{26} + 14\right) + an$ for all $n > m$.

If we assume that $T(n) \leq cn$ for $n > m$, then

$$T(n) \leq c\left(\left\lceil \frac{n}{13} \right\rceil\right) + c\left(\frac{19n}{26} + 14\right) + an$$

$$\leq c\left(\frac{n}{13} + 1\right) + c\left(\frac{19n}{26} + 14\right) + an$$

$$= cn + \left(\frac{-5cn}{26} + 15c + an\right) \leq cn$$

provided $\frac{-5cn}{26} + 15c + an \leq 0$ or $c \geq \frac{26an}{5n-390} = \frac{26an}{5(n-78)}$.

Observe that for $n \geq 156$, $c \geq \frac{52a}{5}$, so if we choose $m = 156$ and $c \geq 11a > \frac{52a}{5}$, then $T(n) \leq cn$ holds.}

6. In the algorithm Select, the input elements are divided into groups of seven elements each with possible one exception. Prove or disprove that the algorithm Select works in linear time if the input elements are divided into groups of (i) 4, (ii) 6, (iii) 8 and (iv) 10 elements each.

Chapter 5

Greedy Algorithms

Greedy approach is an attempt to solve an optimization problem by using a sequence of decisions so that the choice made at each decision point is optimal at the moment of choice. The process of choices made may or may not lead to a final optimal solution. The greedy choice is easy to implement and in many cases leads to an optimal solution. In this chapter, we consider some problems which are solved using greedy approach. We also give a couple of examples where the greedy method does not lead to an optimal solution.

5.1. Some Examples

Our first example of greedy algorithm is that of "**Change Making**".

Example 5.1. A customer buys some biscuits and pays the cashier a Rs. 100 note and the cashier has to return Rs. 67. The cashier has an unlimited supply of only Re. 1, Rs. 5, Rs. 10 and Rs. 20 notes and likes to give the change using the least number of notes. He does it by paying 3 notes of Rs. 20, 1 note of Rs. 5 and 2 notes of Re. 1 each-six notes in all. There is no way that he could pay by less than six notes. In case the money to be returned to the customer is Rs. 59, the cashier pays it by 2 notes of Rs. 20, 1 note of Rs. 10, 1 note of Rs. 5 and 4 notes of Re. 1 each.

The greedy algorithm for change making is as given in Algorithm 5.1.

The greedy algorithm works well because of the special nature of values of notes. However, if the notes available were of values Re. 1, Rs. 8 and Rs. 13 and the change to be given were Rs. 63 (respectively, Rs. 64),

---

Algorithm 5.1. Change Making Algorithm

---

1. Let n be the number of coins to be paid
2. Let a_1 be the value of the highest coin which is less than n
3. Let n_1 be the least remainder on dividing n by a_1 and q_1 be the quotient
4. Let a_2 be the value of the highest coin which is less than n_1
5. Repeat step 3 with $n = n_1$ and $a_1 = a_2$. Continue the process till the remainder becomes 0
6. Then the number of coins is the sum of the quotients $q_1, q_2, \ldots$

---

then the greedy algorithm shows that the amount may be paid by using 4 notes of Rs. 13 each, 1 note of Rs. 8 and 3 notes of Re. 1 each, i.e., a total of 8 notes (respectively, 4 notes of Rs. 13 each, 1 note of Rs. 8 and 4 notes of Re. 1 each, i.e., a total of 9 notes) whereas the amount could be paid by 3 notes of Rs. 13 each and 3 notes of Rs. 8 each, i.e., a total of 6 notes (respectively, 3 notes of Rs. 13 each, 3 notes of Rs. 8 each and 1 note of Re. 1 i.e., a total of 7 notes). If the notes available were of value Re. 1, Rs. 7 and Rs. 11 and the change to be given were Rs. 43, the greedy algorithm shows that the change may be given by using 3 notes of Rs. 11 each, 1 note of Rs. 7 and 3 notes of Re. 1 each, i.e., a total of 7 notes whereas we find that the change could be given by using 2 notes of Rs. 11 each and 3 notes of Rs. 7 each, i.e., a total of 5 notes.

Our second example of greedy method is that of "**Machine-Scheduling**" or "**Activity-Assignment**" **Problem**.

Example 5.2. There are n activities each activity a_i having a start time s_i and finish time f_i, $s_i < f_i$, for $i = 1, 2, \ldots, n$ and $[s_i, f_i]$ is the period of activity a_i (or the processing time of a_i). We also indicate this by writing activity $a_i = [s_i, f_i]$. Two activities a_i and a_j are said to be **compatible** or **non-overlapping** if $[s_i, f_i] \cap [s_j, f_j] = \varphi$ (or $f_i \leq s_j$ or $f_j \leq s_i$). For example, the activities $a_i = [1, 5]$ and $a_j = [3, 6]$ are overlapping or non-compatible whereas the activities $a_i = [1, 4]$, $a_j = [4, 6]$ are compatible. The **optimal activity-selection problem** is to find a largest subset of mutually compatible activities. The activity-selection problem has a simple extension. A **feasible activity-to-facility assignment problem** is an assignment of activities to facilities available in which no facility is assigned to two overlapping activities. Therefore,

Table 5.1

i	1	2	3	4	5	6	7	8	9	10	11
s_i	1	3	0	3	5	5	6	9	2	8	12
f_i	4	5	6	7	8	9	10	11	12	12	14

in a feasible assignment each facility is being used by at most one activity at any given time. An **optimal assignment** is a feasible assignment that utilizes the least number of facilities.

We consider an example.

Example 5.3. In a conference in computer science and applications, talks and lectures are divided into 11 areas/subareas. According to the availability of experts and their choices of time slots, the start and finish time of the areas are as given in Table 5.1.

Here the starting time of the conference being 8 a.m., the times in Table 5.1 are adjusted accordingly.

(a) Also the finishing times of the subareas are written in the increasing order. To choose a largest subset A of the areas which could be assigned the same lecture hall, we start by selecting the subarea with the earliest finish time, i.e., we always start by selecting a_1. Then we select the first a_i for which $s_i \geq f_1$. This condition is met by a_5. Thus, we next select a_5. Next we consider the first a_i with $s_i \geq f_5$ — a condition which is met by a_8. After selecting a_8, we consider the first a_i with $s_i \geq f_8 = 11$. This condition is met by a_{11}. Hence, an optimal solution of activity-selection problem is $A = \{a_1, a_5, a_8, a_{11}\}$.

Once having chosen a_i in A, instead of choosing the first a_j with $s_j \geq f_i$, we may choose a_j with the smallest s_j with $s_j \geq f_i$. If we do this, then after a_5, we need to choose a_{10} and then a_{11}. Thus, with this choice we get $A = \{a_1, a_5, a_{10}, a_{11}\}$.

Thus the activities in A can take place simultaneously.

As a matter of convention, we always adopt the first choice.

(b) We next consider the same data but this time the starting times of the activities in non-decreasing order. So, timings of the activities are given in Table 5.2.

Whenever two activities have the same start time, we arrange these in the increasing order of their finish time. That is, activity $[3, 5]$ comes

Table 5.2

i	1	2	3	4	5	6	7	8	9	10	11
s_i	0	1	2	3	3	5	5	6	8	9	12
f_i	6	4	12	5	7	8	9	10	12	11	14

Table 5.3

i	2	3	4	6	7	9	10
s_i	2	0	3	5	6	2	8
f_i	5	6	7	9	10	12	12

before activity [3, 7] and activity [5, 8] comes before activity [5, 9]. This time we start building a largest subset A of activities by taking the a_i with largest start time, i.e., we put a_{11} in A. Next, we consider a_i with largest i for which $f_i < s_{11}$. Suppose a_i has been put in A, then we place a_j in A which has $f_j < s_i$ and continue the process. Thus, in the present case, after a_{11} we need to put a_{10} in A, then a_7, then a_4 and we cannot continue further. Thus $A = \{a_4, a_7, a_{10}, a_{11}\}$, again having four elements as it should.

(c) In case the data of activities is given as in Table 5.2 and we start building A by first taking the activity with lowest start time, i.e., we take a_1 in A and once a_i is chosen, we put a_j in A with the lowest j such that $f_i < s_j$, then we find that $A = \{a_1, a_8, a_{11}\}$ having only three elements. Thus, this procedure does not yield a largest subset of compatible activities.

(d) We next go back to case (a) above and find the least number of lecture halls that will be needed for the conference. We write A_1 for $A = \{a_1, a_5, a_8.a_{11}\}$, and ignore the four entries of A_1 from Table 5.1. The remaining activities are given in Table 5.3.

We build an optimal set of compatible activities as given in Table 5.3 by first selecting a_2 (the activity with least finish time). By the procedure obtained in (a), we next select a_i with $s_i \geq f_2$ which gives a_6 as the next element in A_2. Since there is no j with $s_j \geq f_6$, the process stops and we get $A_2 = \{a_2, a_6\}$. Ignoring a_2, a_6 from the data in Table 5.3, we are left with the activities given by the data as in Table 5.4.

The procedure adopted gives $A_3 = \{a_3, a_7\}$. Ignoring a_3, a_7 from Table 5.4, we are left with the activities given in Table 5.5.

Table 5.4

i	3	4	7	9	10
s_i	0	3	6	2	8
f_i	6	7	10	12	12

Table 5.5

i	4	9	10
s_i	3	2	8
f_i	7	12	12

Table 5.6

i	1	2	3	5	6	8	9
s_i	0	1	2	3	5	6	8
f_i	6	4	12	7	8	10	12

Starting with a_4 we build $A_4 = \{a_4, a_{10}\}$ and we are left with only one activity a_9 so that $A_5 = \{a_9\}$. We have, therefore, five subsets of compatible assignments and, hence, we need five lecture halls for the conference to be held as scheduled.

(e) We like to see the number of lecture halls required if we had opted for the procedure adopted in (b). In this case $A_1 = A = \{a_4, a_7, a_{10}, a_{11}\}$. Ignoring these entries from Table 5.2, the data of remaining entries is given in Table 5.6.

Ignoring the details we find that $A_2 = \{a_9, a_6, a_2\}$, $A_3 = \{a_8, a_1\}$, $A_4 = \{a_5\}$, $A_5 = \{a_3\}$ which again shows that we require five lecture halls for the conference to be held as scheduled.

The algorithms used in (a) and (b) for finding an optimal set of compatible activities are given below.

Algorithm 5.2. Algorithm for Optimal Activities Set with Finish Times in Non-decreasing Order

---

1. let n activities a_i with start and finish time s_i, f_i be given
2. arrange the activities in non-decreasing order of finish times
3. put a_1 in an optimal set A to be constructed
4. for i, $1 \le i < n$, if a_i is already put in A, choose the smallest j with $f_i \le s_j$. Put a_j in A
5. continue the process as given in 4 above till all the activities are exhausted

---

Algorithm 5.3. Algorithm for Optimal Activities Set with Start Times in Non-decreasing Order

---

1. given n activities a_i with start and finish times s_i, f_i and start times arranged in non-decreasing order
2. put a_n in an optimal set A to be constructed
3. for i, $1 < i \leq n$, if a_i is already put in A, choose the largest j with $f_j \leq s_i$. Put a_j in A
4. continue the process as in 3 above till all the activities are exhausted

---

Arranging activities as at step 3 of Algorithm 5.2 requires $\frac{n(n-1)}{2}$ comparisons and possible interchanges. Again $\frac{n(n-1)}{2}$ possible comparisons and placements in A are required. Therefore, running time of Algorithm 5.2 is $O(n^2)$. Similar reasoning shows that running time of Algorithm 5.3 is also $O(n^2)$.

Example 5.4. Consider eight tasks the start and finish times of which are given in Table 5.7. Find the minimum number of machines to which these tasks are assigned.

Solution. Arranging the finish times in non-decreasing order, the data are as given in Table 5.8.

Now, using the process as given in (a) or Algorithm 5.2 and omitting the details, we get a first optimal set A_1 of tasks as $A_1 = \{a_1, a_3, a_6, a_8\}$. Next sets A_2, A_3 of optimal tasks are $A_2 = \{a_2, a_5\}$, $A_3 = \{a_4, a_7\}$. Therefore, the set of tasks is given by $S = A_1 \cup A_2 \cup A_3$, and we need to assign these tasks to three machines.

Table 5.7

i	1	2	3	4	5	6	7	8
s_i	0	3	4	9	7	1	6	7
f_i	2	7	7	11	10	5	8	9

Table 5.8

i	1	2	3	4	5	6	7	8
s_i	0	1	3	4	6	7	7	9
f_i	2	5	7	7	8	9	10	11

Observe that the data of tasks as in Table 5.8 is also arranged in the order of non-decreasing start times. Using Algorithm 5.3, we find the optimal sets of compatible tasks to be $B_1 = \{a_8, a_6, a_4, a_1\}$, $B_2 = \{a_7, a_3\}$ and $B_3 = \{a_5, a_2\}$ which also shows that a minimum of three machines are needed to schedule the tasks.

5.2. Knapsack Problem

We next consider applying greedy method to the knapsack problem. Given n items having weights $w_i, 1 \le i \le n$, and each item when sold gives a profit $p_i, 1 \le i \le n$. A hawker wants to carry in his knapsack fractions $x_i, 1 \le i \le n$, of these items so as to maximize his profit $\sum_{1 \le i \le n} p_i x_i$. The knapsack that he has can accommodate a total weight W so that $\sum_{1 \le i \le n} w_i x_i \le W$. To sum up, the knapsack problem may be stated as follows.

Given n items with weights w_i, $(w_i > 0)$ and profits $p_i, 1 \le i \le n$, and W, we have to find fractions $x_i, 0 \le i \le 1$ for $1 \le i \le n$ such that

$$P = \sum_{1 \le i \le n} p_i x_i \qquad (5.1)$$

is maximized subject to the constraints

$$\sum_{1 \le i \le n} w_i x_i \le W, \qquad (5.2)$$

$$0 \le x_i \le 1, \quad \text{for} \quad 1 \le i \le n. \qquad (5.3)$$

A solution $(x_1, x_2, \ldots, x_n)$ that satisfies the constraints (5.2) and (5.3) is called a **feasible solution** and a feasible solution that maximizes P is called an **optimal solution**. There are three ways in which we may get a feasible solution using greedy methods.

(a) Suppose that the profits p_i are arranged in non-increasing order, i.e., $p_1 \ge p_2 \ge \cdots \ge p_n$. Consider $(x_1, x_2, \ldots, x_n)$ such that

$$x_j = 1, 1 \le j \le i - 1,$$

$$x_i = \frac{W - \sum_{1 \le j < i} w_j x_j}{w_i} \le 1, x_j = 0 \quad \text{for } j = i + 1, \ldots, n.$$

Then conditions (5.2) and (5.3) are satisfied and we get a feasible solution. In this case our concern is based on increasing the profits at each step without taking into account the weights of the items. This greedy strategy may or may not lead to an optimal solution.

(b) Suppose that the weights w_i of the items are arranged in non-decreasing order, i.e., $w_1 \le w_2 \le \cdots \le w_n$. Again, we build a feasible solution by considering

$$x_j = 1 \text{ for } 1 \le j \le i - 1,$$

$$x_i = \frac{W - \sum_{1 \le j < i} w_j x_j}{w_i} \le 1, x_j = 0 \quad \text{for } j = i+1, \ldots, n.$$

This time we are filling knapsack as slowly as possible. However, this strategy also may or may not lead to an optimal solution.

(c) The next method is based on the profits per unit of weight/space utilized. In this case, we consider the ratios $\frac{p_i}{w_i}$ and arrange these in non-increasing order, i.e., $\frac{p_1}{w_1} \ge \frac{p_2}{w_2} \ge \cdots \ge \frac{p_n}{w_n}$ and build a solution $(x_1, x_2, \ldots, x_n)$ accordingly, i.e., $x_j = 1$, for $1 \le j < i$, $x_i = \frac{W - \sum_{1 \le j < i} w_j x_j}{w_i} \le 1$ and $x_j = 0$ for $j = i+1, \ldots, n$.

Thus there are at least three different greedy strategies for solving the knapsack problem depending upon three different measures. These measures are profit, capacity utilized and the profit per unit of space/weight utilized. To show that the first two strategies may not always yield an optimal solution, we consider an example. At least for this example, we find that the third strategy leads to an optimal solution.

Remark. If $(x_1, x_2, \ldots, x_n)$ is a solution to the knapsack problem and $\sum_{1 \le i \le n} w_i x_i < W$, then we may increase some of the values of x_i, $1 \le i \le n$ so as to have $\sum_{1 \le i \le n} w_i x_i = W$. Thus an optimal solution to the knapsack problem will always have $\sum_{1 \le i \le n} w_i x_i = W$.

Example 5.5. Suppose that $n = 4$, $W = 25$, $(p_1, p_2, p_3, p_4) = (25, 24, 15, 12)$ and $(w_1, w_2, w_3, w_4) = (18, 15, 10, 8)$. Consider the following solutions:

(x_1, x_2, x_3, x_4)	$\sum w_i x_i$	$\sum p_i x_i$
1. $(\frac{1}{2}, \frac{1}{3}, \frac{1}{4}, \frac{1}{4})$	18.5	$\frac{25}{2} + 8 + \frac{15}{4} + 3 = 27.25$
2. $(1, \frac{2}{15}, \frac{3}{10}, \frac{1}{4})$	25	$25 + \frac{48}{15} + \frac{45}{10} + 3 = 35.7$
3. $(0, 1, \frac{1}{2}, \frac{5}{8})$	25	$24 + \frac{15}{2} + \frac{60}{8} = 39$
4. $(0, \frac{2}{3}, 1, \frac{5}{8})$	25	$16 + 15 + \frac{15}{2} = 38.5$
5. $(1, \frac{2}{15}, 0, \frac{5}{8})$	25	$25 + \frac{48}{15} + \frac{60}{8} = 35.7$
6. $(1, \frac{2}{15}, \frac{1}{2}, 0)$	25	$25 + \frac{48}{15} + \frac{15}{2} = 35.7$
7. $(0, 1, 1, 0)$	25	$24 + 15 = 39$
8. $(1, \frac{7}{15}, 0, 0)$	25	$25 + \frac{56}{5} = 36.2$
9. $(0, \frac{7}{15}, 1, 1)$	25	$\frac{56}{5} + 15 + 12 = 38.2$

Observe that of these nine solutions 3 and 7 yield the maximum profit. An optimal solution must have $\sum w_i x_i = W$. So, in the present case, we must have $18x_1 + 15x_2 + 10x_3 + 8x_4 = 25$.

This gives $x_4 = \frac{25 - 18x_1 - 15x_2 - 10x_3}{8}$ so that

$$P = 25x_1 + 24x_2 + 15x_3 + 12x_4$$

$$= 25x_1 + 24x_2 + 15x_3 + \frac{3(25 - 18x_1 - 15x_2 - 10x_3)}{2}$$

$$= \frac{75}{2} - 2x_1 + \frac{3x_2}{2} \tag{5.4}$$

is maximized subject to $x_1, x_2, x_3 \geq 0$ and also

$$25 - 18x_1 - 15x_2 - 10x_3 \geq 0. \tag{5.5}$$

Since the coefficient of x_1 in P is negative, P will be maximal when $x_1 = 0$. Thus we need to maximize

$$P = \frac{75}{2} + \frac{3x_2}{2} \tag{5.6}$$

subject to the conditions $x_2 \geq 0$, $x_3 \geq 0$ and

$$25 - 15x_2 - 10x_3 \geq 0. \tag{5.7}$$

Now P is maximal for $x_2 = 1$. Also $x_2 = x_3 = 1$ satisfy (5.7). Hence $P_{max} = 39$ and an optimal solution is $(0, 1, 1, 0)$ the optimal profit being 39. It also follows that solution $(0, 1, \frac{1}{2}, \frac{5}{8})$ as at number 3 is also an optimal solution.

Observe that the greedy strategy (a) gives a feasible solution as at number 8 but is not an optimal solution. Greedy solution of strategy (b) is also as given in number 9 and is not an optimal solution.

We next consider the profit per unit of space/weight of the knapsack to be occupied by the four items. These are

$$\frac{p_1}{w_1} = \frac{25}{18}, \quad \frac{p_2}{w_2} = \frac{8}{5}, \quad \frac{p_3}{w_3} = \frac{3}{2}, \quad \frac{p_4}{w_4} = \frac{3}{2}.$$

Arranging these in non-increasing order, we get

$$\frac{p_2}{w_2} = 1.6, \quad \frac{p_3}{w_3} = 1.5, \quad \frac{p_4}{w_4} = 1.5, \quad \frac{p_1}{w_1} = 1.389.$$

Now, applying greedy method using these ratios, we put as much amount of the item with the largest profit per unit of space utilized, then as much amount of the item with the next largest profit per unit space and then the next and the next till the maximum capacity of the knapsack is reached. Thus, in the present case, we take $x_2 = 1$, and then $x_3 = 1$ and the total capacity of the knapsack is reached. Hence, a solution with largest profit is $(0, 1, 1, 0)$ with largest profit equal to $24 + 15 = 39$.

Since $\frac{p_3}{w_3} = \frac{p_4}{w_4} = 1.5$, we may not use the whole of the material of item 3 but may use the whole of the material of 4. Thus, we may consider $x_2 = 1$, $x_4 = 1$. Then some space, namely 2 units, is still left in the knapsack and this space may be used by some material from item 3, i.e., we may take $x_3 = \frac{2}{10} = \frac{1}{5}$. The profit in this case is $24 + 12 + \frac{15}{5} = 39$ again. Thus, solution $(0, 1, \frac{1}{5}, 1)$ also leads to a maximal profit of 39, i.e., this solution is an optimal solution. Another solution leading to 39 as profit is $(0, 1, \frac{1}{2}, \frac{5}{8})$ as given in number 3.

We next prove that the greedy knapsack algorithm based on the ratios $\frac{p_i}{w_i}$ of profit per unit of space utilized always leads to an optimal solution. This algorithm may be written as follows:

Algorithm 5.4. Greedy Knapsack Algorithm (W, n)

1. $p[1 : n]$ and $w[1 : n]$ are the profits and weights respectively of the n objects ordered such that $\frac{p[i]}{w[i]} \geq \frac{p[i+1]}{w[i+1]}$. W is the total capacity of the knapsack
2. $x[1 : n]$ is the solution vector
3. initially $x[i] = 0$ for $i = 1$ to n
4. choose i such that $w[j] > W$ for $1 \leq j \leq i - 1$ but $w[i] \leq W$
5. set $x[i] = 1$, $x[j] = 0$, $1 \leq j < i$, $W = W - w[i]$
6. continue this process till we get $W - w[i] \leq w[i + 1]$
7. set $x[i + 1] = \frac{W - w[i]}{w[i+1]}$ and $x[j] = 0$ for $j = i + 2$ to n

Theorem 5.1. *If $\frac{p_1}{w_1} \geq \frac{p_2}{w_2} \geq \cdots \geq \frac{p_n}{w_n}$, and an instance of the knapsack problem is given, then the greedy knapsack algorithm (Algorithm 5.4) generates an optimal solution to the given instance of the knapsack problem.*

Proof. Observe that if $\sum_{1 \leq i \leq n} w_i \leq W$, then $x_i = 1$ for $1 \leq i \leq n$ is a solution to the knapsack problem. Therefore, for the knapsack problem to make a sense, we assume that $\sum_{1 \leq i \leq n} w_i > W$. Also, if $(x_1, x_2, \ldots, x_n)$ is a solution to the knapsack problem and $\sum_{1 \leq i \leq n} x_i w_i < W$, then we may increase some of the values of x_i, $1 \leq i \leq n$ so as to have $\sum_{1 \leq i \leq n} x_i w_i = W$. Thus, an optimal solution $(x_1, x_2, \ldots, x_n)$ of the knapsack problem will always have $\sum_{1 \leq i \leq n} x_i w_i = W$.

Let $x = (x_1, x_2, \ldots, x_n)$ be the solution of the knapsack problem generated by the Greedy Knapsack Algorithm (Algorithm 5.4). If $x_i = 1$ for all i, $1 \leq i \leq n$, then surely x is an optimal solution. Suppose that this is not the case and let j be the least index such that $x_j \neq 1$. Then, it follows from the Greedy Knapsack Algorithm that $x_i = 1$ for $i = 1, \ldots, j - 1, 0 \leq x_j < 1$ and $x_i = 0$ for $i = j + 1, \ldots, n$. Also, we have $\sum_{1 \leq i \leq j} x_i w_i = W$. Let $y = (y_1, y_2, \ldots, y_n)$ be an optimal solution. If $y_i = x_i$ for every i, $1 \leq i \leq n$, then $x = y$ is an optimal solution. So, suppose that $x \neq y$. Let r be the least index such that $y_r \neq x_r$.

Claim $y_r < x_r$.

To prove this claim, we need to consider three cases, namely $r < j$, $r = j$ and $r > j$.

If $r < j$, then $x_r = 1$ and, so, $x_r \neq y_r$ implies that $y_r < x_r$.

If $r = j$, then $W = \sum_{1 \leq i \leq n} x_i w_i = \sum_{1 \leq i < j} x_i w_i + x_j w_j = \sum_{1 \leq i < j} y_i w_i + x_j w_j$. If we assume that $y_j > x_j$, then $W = \sum_{1 \leq i < j} y_i w_i + x_j w_j < \sum_{1 \leq i \leq j} y_i w_i \leq \sum_{1 \leq i \leq n} y_i w_i = W$, which is a contradiction. Hence $y_j < x_j$.

If $r > j$, then $W = \sum_{1 \leq i \leq n} y_i w_i = \sum_{1 \leq i \leq j} y_i w_i + y_r w_r + \sum_{j < i \leq n, i \neq r} y_i w_i = \sum_{1 \leq i \leq j} x_i w_i + y_r w_r + \sum_{j < i \leq n, i \neq r} y_i w_i = W + y_r w_r + \sum_{j < i \leq m, i \neq r} y_i w_i$.

Now $y_r \neq x_r = 0$ (as $j > r$), $y_r > 0$. Therefore, $W = W + y_r w_r + \sum_{j < i \leq n, i \neq r} y_i w_i > W$, which is a contradiction. Thus the case $r > j$ does not arise.

This completes the proof of our claim.

Suppose, we increase y_r to x_r and suitably decrease as many of $y_{r+1}, \ldots, y_n$ so that the total capacity used is still W. We thus get a new solution $z = (z_1, \ldots, z_n)$ with $z_i = x_i, 1 \leq i \leq r$, and

$$\sum_{r < i \leq n} (y_i - z_i) w_i = (z_r - y_r) w_r. \tag{5.8}$$

Then, for z we have

$$\sum_{1 \leq i \leq n} z_i p_i = \sum_{1 \leq i < r} y_i p_i + z_r p_r + \sum_{r < i \leq n} z_i p_i$$

$$= \sum_{1 \leq i \leq n} y_i p_i + (z_r - y_r) p_r + \sum_{r < i \leq n} (z_i - y_i) p_i$$

$$= \sum_{1 \leq i \leq n} y_i p_i + (z_r - y_r) w_r \left(\frac{p_r}{w_r} \right) + \sum_{r < i \leq n} (z_i - y_i) w_i \left(\frac{p_i}{w_i} \right)$$

$$\geq \sum_{1 \leq i \leq n} y_i p_i + (z_r - y_r) w_r \left(\frac{p_n}{w_n} \right) + \sum_{r < i \leq n} (z_i - y_i) w_i \left(\frac{p_n}{w_n} \right)$$

$$\left(\text{as } \frac{p_i}{w_i} \text{ are non-increasing} \right)$$

$$= \sum_{1 \leq i \leq n} y_i p_i \text{ (because of (5.8)).}$$

Since y is an optimal solution and we have $\sum_{1 \le i \le n} z_i p_i = \sum_{1 \le i \le n} y_i p_i$, z is an optimal solution.

Thus, starting with an optimal solution $y = (y_1, y_2, \ldots, y_n)$ with $y_i = x_i$ for $1 \le i \le r - 1$, we have obtained an optimal solution $z = (z_1, z_2, \ldots, z_n)$ with $z_i = x_i$, $1 \le i \le r$. Repeating the above argument a finite number of times, we arrive at an optimal solution $t = (t_1, t_2, \ldots, t_n)$ such that $t_i = x_i$, for $1 \le i \le n$, i.e., $t = x$. This completes the proof that x is an optimal solution. $\qquad\square$

Example 5.6. Using Knapsack Algorithm, find an optimal solution to the knapsack problem with $n = 7$,

$$W = 13, \ (p_1, p_2, \ldots, ..p_7) = (8, 5, 12, 7, 6, 15, 3)$$

and

$$(w_1, w_2, \ldots, w_7) = (2, 3, 5, 6, 2, 3, 3).$$

Solution. Here

$$\frac{p_1}{w_1} = 4, \ \frac{p_2}{w_2} = \frac{5}{3}, \ \frac{p_3}{w_3} = \frac{12}{5}, \ \frac{p_4}{w_4} = \frac{7}{6}, \ \frac{p_5}{w_5} = 3, \ \frac{p_6}{w_6} = 5, \ \frac{p_7}{w_7} = 1.$$

Arranging these ratios of profits per unit of space utilized in non-decreasing order, we have $\frac{p_6}{w_6} > \frac{p_1}{w_1} > \frac{p_5}{w_5} > \frac{p_3}{w_3} > \frac{p_2}{w_2} > \frac{p_4}{w_4} > \frac{p_7}{w_7}$.

Therefore, an optimal solution is given by $x_6 = 1$, $x_1 = 1$, $x_5 = 1$, $x_3 = 1$. Then $w_1 + w_3 + w_5 + w_6 = 2 + 5 + 2 + 3 = 12$, so that capacity of 1 unit remains unutilized. Therefore, we need to take $x_2 = \frac{1}{3}$. Hence, an optimal solution is $(1, \frac{1}{3}, 1, 0, 1, 1, 0)$ and the optimal profit is $8 + \frac{5}{3} + 12 + 6 + 15 = \frac{128}{3}$.

The knapsack problem that we have considered so far has a solution $x = (x_1, x_2, \ldots, x_n)$ satisfying $0 \le x_i \le 1$ for all i, $1 \le i \le n$ and $\sum_{1 \le i \le n} x_i w_i \le W$. This knapsack problem may be called **Knapsack Problem (rational case)**. There is a related problem called the $\frac{0}{1}$**-Knapsack Problem** which may be stated as follows.

Given n items having weights $w_1, w_2, \ldots, w_n$, the items yielding profits $p_1, p_2, \ldots, p_n$ and the knapsack having capacity W, to find a solution $x = (x_1, x_2, \ldots, x_n)$ with $x_i = 0$ or 1 for every i, $1 \le i \le n$, satisfying $\sum_{1 \le i \le n} x_i w_i \le W$ that maximizes the profit $P = \sum_{1 \le i \le n} x_i p_i$.

In the case of $\frac{0}{1}$-knapsack problem, either an item is taken in full or is ignored altogether. Therefore it is not essential that the knapsack may be filled to its capacity. When the sum $\sum_{1 \le i \le n} w_i$ of all the weights is more than the capacity W of the knapsack, the problem has at most $2^n - 2$ solutions (the number of feasible solutions may be much less). Theorem 5.1 shows that the greedy strategy based on the ratios $\frac{p_i}{w_i}$ of profit per unit of weight/capacity utilized always yields an optimal solution. We show that the same is not true in the case of $\frac{0}{1}$-knapsack problem.

Example 5.7. Consider the $\frac{0}{1}$-knapsack problem with $n = 5$, $W = 29$, $p = (25, 24, 15, 10, 8)$ and $w = (18, 15, 9, 8, 6)$. Working with the greedy strategy by considering profits in non-increasing order, we get the solution

$$x_1 = 1 = x_3, \quad x_2 = x_4 = x_5 = 0,$$

$$\sum x_i w_i = 27 \quad \text{and} \quad P = 25 + 15 = 40.$$

Next, working with the greedy strategy by considering weights in non-decreasing order, the solution is

$$x_5 = x_4 = x_3 = 1, \quad x_1 = x_2 = 0,$$

$$\sum x_i w_i = 23 \quad \text{and} \quad P = 8 + 10 + 15 = 33.$$

Now using the greedy strategy using profit per unit of space utilized, we have

$$\frac{p_1}{w_1} = \frac{25}{18} = 1.389, \quad \frac{p_2}{w_2} = \frac{8}{5} = 1.6, \quad \frac{p_3}{w_3} = \frac{5}{3} = 1.667,$$

$$\frac{p_4}{w_4} = \frac{5}{4} = 1.25, \quad \frac{p_5}{w_5} = \frac{4}{3} = 1.333$$

so that the solution given is $x_3 = 1 = x_2$, $x_1 = x_4 = x_5 = 0$ with $w_3 + w_2 = 15 + 9 = 24$ and profit $P = 15 + 24 = 39$.

This proves that the solution obtained in the last case is not optimal. The profit obtained in this case is even smaller than the one obtained in the first case. Observe that the solution obtained in the first case is not optimal. For example, $w_2 + w_4 + w_5 = 29 = W$ so that we take $x_2 = x_4 = x_5 = 1$, $x_1 = x_3 = 0$ giving the profit $P = 24 + 10 + 8 = 42$ which is optimal.

A very special case of the $\frac{0}{1}$-knapsack problem admits an optimal solution using greedy algorithm. This special case is when $p_1 = p_2 = \cdots = p_n$. In this case $\frac{p_1}{w_1} \geq \frac{p_2}{w_2} \geq \cdots \geq \frac{p_n}{w_n}$ is equivalent to $w_1 \leq w_2 \leq \cdots \leq w_n$. The p_i being all equal, greedy criterian using profits alone does not make sense. Hence the greedy strategies considered earlier reduce to only one greedy strategy which is based on the weights of the items in non-decreasing order. A practical example that results may be called **Container Loading Problem** and may be stated as follows.

An exporter wanting to export his goods has put his cargo into containers which are all of the same size. However, the weights of the containers may be different, say $w_1, w_2, \ldots, w_n$. The cargo (i.e., the containers) are loaded on a large ship with capacity (load) C (say). The shipping company charges for the whole ship. In order to save on the shipping charges, the exporter likes to load a maximum number of containers with larger weights on the ship. We write $x_i = 0$ if the container with weight w_i is not loaded on the ship and write $x_i = 1$ if this container with weight w_i is loaded on the ship. Therefore, the total number of containers loaded is $\sum_{1 \leq i \leq n} x_i$ and this is the number we need to maximize. Since, either the container with weight w_i is loaded or not loaded at all, the total weight of the cargo loaded is $\sum_{1 \leq i \leq n} x_i w_i$. To sum up, we have the following problem.

Given n, capacity C and weights w_i, $1 \leq i \leq n$, we need to find a vector $x = (x_1, x_2, \ldots, x_n)$ that maximizes

$$\sum_{1 \leq i \leq n} x_i \tag{5.9}$$

subject to the constraints

$$\sum_{1 \leq i \leq n} x_i w_i \leq C, x_i = 0 \text{ or } 1 \quad \text{for } 1 \leq i \leq n. \tag{5.10}$$

A vector $x = (x_1, x_2, \ldots, x_n)$ that satisfies the constraints (5.10) is called a **feasible solution** and a feasible solution that maximizes the objective function, i.e., (5.9) is called an **optimal solution**.

In the greedy strategy for loading the containers we arrange the containers in non-decreasing order of their weights. If $w_1 \leq C$, then first container is loaded on the ship and we take $x_1 = 1$. Once we have loaded the container with weight w_i, the container with weight w_{i+1} is loaded if $\sum_{1 \le j \le i+1} w_j \le C$ and we take $x_{i+1} = 1$, otherwise the container with weight w_{i+1} is not loaded onto the ship, we take $x_{i+1} = 0$ and the process stops.

Algorithm 5.5. Container Loading Greedy Algorithm

1. given n containers with weights $w[1 : n]$, capacity C of the ship
2. arrange the weights of the containers in non-decreasing order: $w_1 \le w_2 \le \cdots \le w_n$
3. call the container with weight w_i as the ith container
4. build the solution $x = (x_1, x_2, \ldots, x_n)$ step by step
5. if $w_1 \le C$, set $x_1 = 1$
6. for $i = 1$ to $n - 1$, if $\sum_{1 \le j \le i+1} w_j \le C$, set $x_{i+1} = 1$ otherwise $x_{i+1} = 0$ and stop

---

Step 2 in the algorithm requires $\frac{n(n-1)}{2}$ comparisons and resulting switching (if necessary). Step 6 needs at most $n - 1$ additions and at most n comparisons. Therefore, the running time of the algorithm is $O(n^2)$.

Theorem 5.2. *The greedy algorithm generates an optimal solution to the container loading problem.*

Proof. Let $x = (x_1, x_2, \ldots, x_n)$ be a solution to the container loading problem generated by the greedy algorithm. Then there exists a t such that $x_i = 1$ for $1 \le i \le t$ and $x_i = 0$ for $i > t$. Then, we need to prove that $\sum_{1 \le i \le n} x_i \ge \sum_{1 \le i \le n} y_i$ for every feasible solution $y = (y_1, y_2, \ldots, y_n)$ of the problem. We prove this by induction on the number k of i for which $x_i \ne y_i$. Suppose that $y = (y_1, y_2, \ldots, y_n)$ is a feasible solution with $x_i \ne y_i$ for exactly one i, i.e., $k = 1$. If this $i > t$, then $x_j = y_j$ for $1 \le j \le t$ and $\sum_{1 \le j \le n} x_j w_j = \sum_{1 \le j \le t} w_j = \sum_{1 \le j \le t} y_j w_j < \sum_{1 \le j \le t} y_j w_j + w_{t+1} \le \sum_{1 \le j \le t} y_j w_j + w_i = \sum_{1 \le j \le n} y_j w_j$ and it follows from the greedy strategy that x can be extended by atleast one more step — a contradiction. Hence $i \le t$ and since $x_i = 1$, $y_i = 0$. Thus, $\sum_{1 \le j \le n} x_j > \sum_{1 \le j \le n} y_j$.

Suppose that for every feasible solution y with $x_i \ne y_i$ for the number of positions i being m, we have $\sum_{1 \le i \le n} x_i \ge \sum_{1 \le i \le n} y_i$. Consider a feasible solution y with the number of positions i with $x_i \neq y_i$ being $m+1$. Let j be the least index for which $x_j \neq y_j$. If $j > t$, then the argument used above for $k = 1$ shows that y is not a feasible solution. Hence $j \leq t$. Then $x_j = 1$ and $y_j = 0$. Replacing y_j by 1, we get a y in which the number of positions i for which $x_i \neq y_i$ is m. The new y is either feasible or not feasible. If feasible, we put the new $y = z$ and $1 + \sum y_i = \sum z_i$ which, by induction hypothesis, is $\leqslant \sum x_i$, i.e., $\sum y_i < \sum x_i$.

If the new $y(= z)$ is not a feasible solution, there must exist an l, $t + 1 \leq l \leq n$, such that $z_l = 1$. Replace this z_l by 0 and the z so obtained be called u. Since u is obtained from y by replacing $y_j = 0$ by 1 and $y_l = z_l = 1$ by 0 and $w_j \leq w_l$, we get $\sum_{1 \leq i \leq n} u_i w_i \leq \sum_{1 \leq i \leq n} y_i w_i$. Since y is a feasible solution, so is u. Since u is obtained from y by making two components y_i which were not equal to the corresponding x_i equal to x_i, the number of positions i for which $x_i \neq u_i$ is equal to $m-1$. Therefore, by induction hypothesis (for both z and u), $\sum_{1 \leq i \leq n} x_i \geq \sum_{1 \leq i \leq n} u_i = \sum_{1 \leq i \leq n} y_i$ and $\sum_{1 \leq i \leq n} x_i \geq \sum_{1 \leq i \leq n} z_i = 1 + \sum_{1 \leq i \leq n} y_i > \sum_{1 \leq i \leq n} y_i$. This completes induction and, hence, x is an optimal solution. $\qquad\square$

5.3. Job Sequencing with Deadlines

We are given a set S of n jobs, each job needing to be processed on a machine for one unit of time. Associated with the jobs are integer deadlines $d_1, d_2, \ldots, d_n$ with $1 \leq d_i \leq n$ for every i and n non-negative numbers $p_1, p_2, \ldots, p_n$ called profits. Each job i earns a profit p_i if and only if the job is completed by the deadline d_i. Only one machine is available for processing the jobs and once processing of a job begins, it cannot be interrupted until the job is completely processed. A feasible solution of this problem is a subset J of S such that every job $i \in J$ is completed by the deadline d_i. Associated with each feasible solution is the profit function $\sum_{i \in J} p_i$. An optimal solution J is a feasible solution for which the associated profit is maximal (optimized). Our problem is to find an optimal solution J and a corresponding scheduling of the problems in J. To find an optimal solution, we adopt the following greedy strategy.

List the jobs in S as $a_1, a_2, \ldots, a_n$ in such a fashion that the associated profits are ordered in a non-increasing fashion, i.e., $p_1 \geq p_2 \geq \cdots \geq p_n$. Initially the solution J that we are wanting to find is empty. We put 1 in J (here we are omitting a from the jobs a_i). Once i has been chosen and put in J, we consider the least $j > i$ for which deadline d_j of j is $\geq$ ($1 +$ the number of jobs already put in J) and put this j in J. We continue the process till all the jobs in S have been examined. Observe that if $m = \max\{d_1, d_2, \ldots, d_n\}$, then the number of jobs in any feasible solution is $\leq m$.

Let J be a feasible solution having k jobs. The jobs in J can be scheduled in $k!$ (k factorial) ways (corresponding to the permutations of the set J) and not all the ways are such that the deadlines of the jobs are not violated. In the hit and trial method, we need to check these $k!$ scheduling to arrive at one in which the deadlines of jobs are not violated. This process is very cumbersome and impractical. However, there is an easier way of getting a scheduling in which the deadlines of the jobs are not violated.

Theorem 5.3. *Let J be a set of k jobs and $\sigma = (i_1, i_2, \ldots, i_k)$ be a permutation of the jobs in J such that $d_{i_1} \leq d_{i_2} \leq \cdots \leq d_{i_k}$. Then J is a feasible solution if and only if the jobs in J can be processed in the order σ without violating any deadlines.*

Proof. If the jobs in J can be processed in the order σ without violating deadlines, it follows from the definition of feasibility that J is a feasible solution. So, suppose that J is a feasible solution. Then there exists a permutation $\tau = (r_1, r_2, \ldots, r_k)$ of the jobs in J such that the jobs in J are processed in the order τ and $d_{r_j} \geq j$, $1 \leq j \leq k$, i.e., job r_1 is processed in the time interval $[0, 1]$, job r_2 is processed in the time interval $[1, 2]$, $\ldots$, and finally the job r_k is processed in the time interval $[k-1, k]$. If $\sigma = \tau$, we are through. So, suppose that $\sigma \neq \tau$. Let a be the least index such that $r_a \neq i_a$. Since σ and τ both are permutations of the set J, there exist indices b and c such that $r_b = i_a$ and $r_a = i_c$. By the choice of a, it follows that both $b > a$ and $c > a$. Then $d_{r_b} = d_{i_a} \leq d_{i_c} = d_{r_a}$. Therefore, if r_a and r_b are interchanged in τ, there results a permutation $\vartheta = (s_1, s_2, \ldots, s_k)$ of the jobs in J and the jobs in J can be processed in the order ϑ without violating any deadlines. However, ϑ is a permutation which agrees with σ in one more position than τ does. Continuing the above argument for a finite number of times, we arrive at the permutation σ and the jobs in J can be performed in the order σ without violating any deadlines. $\square$

We next prove that the greedy strategy leads to an optimal solution.

Theorem 5.4. *The greedy method as described above always provides an optimal solution to the job scheduling problem.*

Proof. Let $p_1 \geq p_2 \geq \cdots \geq p_n \geq 0$ be the profits and d_i, $1 \leq i \leq n$, the integer deadlines of an instant of the job scheduling (or sequencing) problem. Let I be the set of jobs selected by the greedy method. If I' is a feasible solution of the job sequencing problem which properly contains I, let x be a job in I' which is not in I. Then $I \cup \{x\}$ being a subset of I' is a feasible solution which contradicts the choice of I (by the greedy strategy).

Hence I is not a proper subset of any feasible solution I'.

Let J be an optimal solution to the problem. If $I = J$ then I is an optimal solution and we have nothing to prove. So, suppose that $I \neq J$. By the observation made above, $I \not\subseteq J$. Again J being an optimal solution $J \not\subseteq I$. Therefore, there exist jobs a, b with $a \in I$ but $a \notin J$ and $b \in J$ but $b \notin I$. Among all the jobs in I which are not in J choose a to be highest profit job. This choice is possible because the set I is finite and the set of jobs in I which are not in J is non-empty. If $p_b > p_a$, then in the greedy method, job b would have been considered before the job a for inclusion in I. But the job b is not in I. Therefore, we have $p_a \geq p_b$.

Let S_I and S_J be the schedules for the feasible solutions I and J, respectively. Let c be a job which is in both I as well as in J. As a job in I, let c be scheduled in the time slot t to $t + 1$ and as a job in J, let it be scheduled in the time slot t' to $t' + 1$. If $t < t'$, and c' is a job in I scheduled in the time slot t' to $t' + 1$ in S_I, shifting c' to the earlier time slot t to $t + 1$ creates no problem. As c in S_J is scheduled in the time slot t' to $t' + 1$, the deadline $d_c \geq t' + 1$. Therefore the job c can be shifted to the time slot t' to $t' + 1$ as a job in I. Thus, if $t < t'$, interchange the job (if any) scheduled in the time slot t' to $t' + 1$ in S_I with c. If there is no such c'' in I, simply shift c to the time slot t' to $t' + 1$. There then results a feasible scheduling S_I' for I. If on the other hand $t' < t$, a similar transformation can be made to the scheduling S_J. Thus, we obtain feasible scheduling S_I' and S_J' for the feasible solutions I and J, respectively, with the property that all jobs common to both I and J are scheduled to the same time slot in S_I' as well as in S_J'.

Now consider the job $a \in I$ chosen above. Consider the time interval $[t_a, t_a + 1]$ in S'_I in which the job a is assigned. Let $b \in J$ be the job (if any) scheduled in the time interval $[t_a, t_a + 1]$ in S'_J. Since a is not in J, it follows immediately from the above choice of S'_I and S'_J that b is not in I. Then, by the choice of a, we have $p_a \geq p_b$. Schedule the job a in the time slot $[t_a, t_a + 1]$ in S'_J and discard the job b from J. We get a feasible solution $J' = (J - \{b\}) \cup \{a\}$ with feasible schedule $S'_{J'}$. Since J' is obtained from J by inserting a job with higher profit and discarding a job with lower profit value, the profit value of J' is no less than the profit value of J. Hence J' is also an optimal solution. Moreover, J' differs from I in one less job than J does.

A repeated use of the above process of transforming J to J' can transform J to I in a finite number of steps. Hence I is an optimal solution to the job scheduling problem.   □

Having proved the above theorem, the greedy strategy of obtaining I may be formalized as an algorithm for an optimal solution of the job scheduling problem.

Algorithm 5.6. Greedy Algorithm (d, p, J, n)

---

1. given n jobs with integer deadlines d_i, profits p_i, a subset J of jobs that can be completed by their deadlines and which achieve a maximum profit is to be constructed. The profits p_i are given ordered in non-increasing order. The jobs are numbered say, $1, 2, \ldots, n$
2. set $J = \{1\}$
3. for $i = 2$ to n do
4. if all jobs in $J \cup \{i\}$ can be completed by their deadlines, set $J = J \cup \{i\}$
5. end

Using the above algorithm, construction of the optimal set J of jobs that can be completed by their deadlines is achieved in a step-by-step manner. As it is, it does not give a scheduling of the jobs in J. Once an intermediary set J is constructed and a job i is to be added to J, in order to check the feasibility of $J \cup \{i\}$, we need to sort the elements of $J \cup \{i\}$ in non-decreasing order of their deadlines. Thus, sorting of the elements in J is to be done at every step. However, this additional work of sorting may be avoided by adopting the following procedure. We can keep the jobs in

J ordered by their deadlines, i.e., instead of building J as a subset, it is written as a sequence $J[1:k]$ such that $J[r]$, $1 \leq r \leq k$ are the jobs in J ordered by their deadlines. Thus, we have jobs $J[1]$, $J[2]$, ..., $J[k]$ in J with $d(J[1]) \leq d(J[2]) \leq \cdots \leq d(J[k])$. The set J being a feasible solution, we have $d(J[r]) \geq r$, for $1 \leq r \leq k$. Observe that there is a change of notation here. We are writing $d(j[r])$ for the deadline $d_{J[r]}$ associated with the job $J[r]$. When a job i is to be considered for inclusion in J, we look for an r, $1 \leq r \leq k$ such that $d(J[r]) \leq d_i \leq d(J[r+1])$ where it is understood that for $r = k$ this simply means $d(J[r]) \leq d_i$. In case, $d_i \leq d(J[1])$, we may think of it as $d(J[0]) \leq d_i \leq d(J[1])$ where an additional job $J[0]$ with $d(J[0]) = 0$ is considered as a part of J. Then the job i is considered inserted between $J[r]$ and $J[r+1]$. Thus, we get a sequence $J[1:k+1]$ of jobs $J[r]$, $1 \leq r \leq k+1$ with deadlines satisfying $d(J[1]) \leq d(J[2]) \leq \cdots \leq d(J[k+1])$. Thus, in order to check whether the enlarged set is a feasible solution we only need to check that $d(J[r]) \geq r$ for all r, $1 \leq r \leq k+1$. If the job i is inserted at the position $r+1$, then we only need to check that the jobs $J[r+1]$, ..., $J[k+1]$ do not violate the deadlines, i.e., we need only check that $d(J[q]) \geq q$, for $r+1 \leq q \leq k+1$ (as the jobs up to $J[r]$ already do not violate their deadlines). When this procedure terminates, it not only gives an optimal solution to the job scheduling problem but also gives their scheduling.

We now consider a couple of examples in all the three of which we need to find an optimal solution along with the associated scheduling of jobs.

Example 5.8. Let $n = 6$, $(p_1, p_2, \ldots, p_6) = (20, 15, 12, 8, 5, 1)$ and $(d_1, d_2, \ldots, d_6) = (2, 2, 1, 3, 3, 4)$.

Solution. Let J denote the optimal subset of jobs to be constructed. Using the feasibility law of Theorem 5.3, we have

J	assigned time slot	job	action	profit
φ	none	1	assign to slot [1,2]	0
{1}	[1,2]	2	assign to slot [0,1]	20
{2, 1}	[0,1],[1,2]	3	cannot fit, reject	20 + 15
{2, 1}	[0,1],[1,2]	4	assign to slot [2,3]	35
{2, 1, 4}	[0,1],[1,2],[2,3]	5	cannot fit, reject	35 + 8
{2, 1, 4}	[0,1],[1,2],[2,3]	6	assign to slot [3,4]	43
{2, 1, 4, 6}	[0,1],[1,2],[2,3],[3,4]	—	—	43 + 1 = 44

Thus the jobs are scheduled in the order 2, 1, 4, 6 and the profit is 44.

Example 5.9. Let $n = 6$, $(p_1, p_2, \ldots, p_6) = (20, 15, 12, 8, 5, 1)$ and $(d_1, d_2, \ldots, d_6) = (4, 3, 1, 1, 2, 4)$.

Solution. Let J denote the optimal subset of jobs to be constructed. Using the feasibility criterion of Theorem 5.3, we have

J	assigned time slot	job	action	profit
φ	none	1	assign to slot [3,4]	0
$\{1\}$	[3,4]	2	assign to slot [2,3]	20
$\{2, 1\}$	[2,3],[3,4]	3	assign to slot [0,1]	20 + 15
$\{3, 2, 1\}$	[0,1],[2,3],[3,4]	4	cannot fit, reject	35 + 12
$\{3, 2, 1\}$	[0,1],[2,3],[3,4]	5	assign to slot [1,2]	47
$\{3, 5, 2, 1\}$	[0,1],[1,2],[2,3],[3,4]	6	cannot fit, reject	47 + 5

Thus an optimal solution has four jobs and the scheduling of these jobs is 3, 5, 2, 1 with a profit of 52.

In case we write the column under J as φ, $\{1\}$, $\{1, 2\}$, $\{1, 2, 3\}$, $\{1, 2, 3, 5\}$, the column under "assigned time slot" needs to be written as

None,

[3,4]

[3,4],[2,3]

[3,4],[2,3],[0,1]

[3,4],[2,3],[0,1]

[3,4],[2,3],[0,1],[1,2]

Then, for getting the scheduling of the jobs, we need to look at the two columns simultaneously. Alternatively, we may get the scheduling by looking at the columns "job considered" and "action" simultaneously. Thus, job 1 goes to the last slot, job 2 to the slot previous to that, job 3 to the first slot and finally job 5 to the second slot. Therefore, job 3 is followed by job 5 which is followed by job 2 and that is followed by job 1.

Example 5.10. Let $n = 7$, $(p_1, p_2, \ldots, p_7) = (65, 60, 55, 45, 35, 25, 15)$ and $(d_1, d_2, \ldots, d_7) = (2, 4, 3, 4, 1, 4, 5)$.

Solution. Again, let J denote the subset of jobs to be constructed. Using, once again, the feasibility criterion of Theorem 5.3, we have

J	assigned time slot	job	action	profit
φ	none	1	assign to slot [1,2]	0
{1}	[1,2]	2	assign to slot [3,4]	65
{1, 2}	[1,2],[3,4]	3	assign to slot [2,3]	65 + 60
{1, 3, 2}	[1,2],[2,3],[3,4]	4	assign to slot [0,1]	125 + 55
{4, 1, 3, 2}	[0,1],[1,2],[2,3],[3,4]	5	cannot fit, reject	180
{4, 1, 3, 2}	[0,1],[1,2],[2,3],[3,4]	6	cannot fit, reject	180
{4, 1, 3, 2}	[0,1],[1,2],[2,3],[3,4]	7	assign to slot [4,5]	180
{4, 1, 3, 2, 7}	[0,1],[1,2],[2,3],[3,4],[4,5]	—	—	180 + 15

Therefore, an optimal solution consists of five jobs {4, 1, 3, 2, 7} and the optimal job scheduling is: job 4, followed by job 1, followed by job 3, followed by job 2, followed by job 7 (this scheduling we abbreviate as $(4 \rightarrow 1 \rightarrow 3 \rightarrow 2 \rightarrow 7)$ and the optimal profit is $180 + 15 = 195$).

A similar, as a matter of fact equivalent, problem is that of scheduling unit time jobs/tasks with integer deadlines and penalties for jobs completed after the corresponding deadline.

A company has given n tasks to one of its programmers. Each job requires one unit of time for processing on the only machine available to him/her. The jobs carry integer deadlines $d_1, d_2, \ldots, d_n, 1 \leq d_i \leq n$, meaning thereby that the ith job is to be completed on or before d_i. The jobs carry non-negative penalties $p_1, p_2, \ldots, p_n$ where a penalty p_i is to be paid by the programmer if the job i is completed after the deadline d_i. This penalty is charged from his/her bonuses and incentives. A penalty saved by the programmer may be considered his/her gain or profit. The problem is how the jobs may be scheduled so that he/she has to pay a minimum penalty. Equivalently, how should the jobs be scheduled so as to save maximum penalty, i.e., the "profit/gain" is maximized.

A job completed by the deadline is called an **early job** and a job is called **late** if the same is completed after the deadline. Consider a scheduling of the n jobs (observe that all the jobs have to be completed as the programmer is a paid employee of the company). If an early job i comes immediately after a late job j, then the positions of the two jobs may be interchanged. After the interchange i having moved to an earlier time slot still remains early and the job j which has been moved to a later time slot remains a late job. In this swapping relative positions of early jobs remain unchanged. Therefore, early jobs remain early. Repeating this process of swapping a finite number of times, we get a scheduling of jobs in which all the early jobs come before the late jobs. In any scheduling relative positions of the late jobs do not matter because they all lead to the same penalty to be paid. Since we have to maximize the penalty avoided, we can still use Algorithm 5.6 (the greedy algorithm for maximum profit) to solve this problem. To illustrate this, we consider Examples 5.8–5.10 regarding the profits in these examples as penalties. While in these examples, jobs that could not be completed by the deadline were ignored, this time we will include these in J as $\langle k \rangle$. for the late job k.

Example 5.11. Let $n = 6$, penalties $(p_1, p_2, \ldots, p_6) = (20, 15, 12, 8, 5, 1)$ and deadlines $(d_1, d_2, \ldots, d_6) = (2, 2, 1, 3, 3, 4)$.

Solution. Using the feasibility law of Theorem 5.3, we have

J	assigned time slot	job	action	penalty
φ	none	1	assign to slot [1,2]	0
$\{1\}$	[1,2]	2	assign to slot [0,1]	0
$\{2, 1\}$	[0,1],[1,2]	3	late	0
$\{2, 1, \langle 3 \rangle\}$	[0,1],[1,2]	4	assign to slot [2,3]	12
$\{2, 1, \langle 3 \rangle, 4\}$	[0,1],[1,2],[2,3]	5	late	12
$\{2, 1, \langle 3 \rangle, 4, \langle 5 \rangle\}$	[0,1],[1,2],[2,3]	6	assign to slot [3,4]	$12 + 5$
$\{2, 1, \langle 3 \rangle, 4, \langle 5 \rangle, 6\}$	[0,1],[1,2],[2,3],[3,4]	—	—	17

Thus an optimal assignment of the jobs is $2 \to 1 \to 4 \to 6 \to \langle 3 \rangle \to \langle 5 \rangle$ with the jobs 3,5 being late incurring a penalty of 17.

Example 5.12. Let $n = 6$, penalties $(p_1, p_2, \ldots, p_6) = (20, 15, 12, 8, 5, 1)$ and deadlines $(d_1, d_2, \ldots, d_6) = (4, 3, 1, 1, 2, 4)$.

Solution. Using the feasibility criterion of Theorem 5.3, we have

J	assigned time slot	job	action	penalty
φ	none	1	assign to slot [3,4]	0
$\{1\}$	[3,4]	2	assign to slot [2,3]	0
$\{2, 1\}$	[2,3],[3,4]	3	assign to slot [0,1]	0
$\{3, 2, 1\}$	[0,1],[2,3],[3,4]	4	late	0
$\{3, 2, 1, \langle 4 \rangle\}$	[0,1],[2,3],[3,4]	5	assign to slot [1,2]	8
$\{3, 5, 2, 1, \langle 4 \rangle\}$	[0,1],[1,2],[2,3],[3,4]	6	late	8
$\{3, 5, 2, 1, \langle 4 \rangle, \langle 6 \rangle\}$	[0,1],[1,2],[2,3],[3,4]	—	—	$8 + 1$

Therefore an optimal assignment is $3 \to 5 \to 2 \to 1 \to \langle 4 \rangle \to \langle 6 \rangle$ and attracts a penalty of $8 + 1 = 9$.

Example 5.13. Let $n = 7$, penalties $(p_1, p_2, \ldots, p_7) = (65, 60, 55, 45, 35, 25, 15)$ and deadlines $(d_1, d_2, \ldots, d_7) = (2, 4, 3, 4, 1, 4, 5)$.

Solution. Using the feasibility criterion of Theorem 5.3, we have

J	assigned time slot	job	action	penalty
φ	none	1	assign to slot [1,2]	0
$\{1\}$	[1,2]	2	assign to slot [3,4]	0
$\{1, 2\}$	[1,2],[3,4]	3	assign to slot [2,3]	0
$\{1, 3, 2\}$	[1,2],[2,3],[3,4]	4	assign to slot [0,1]	0
$\{4, 1, 3, 2\}$	[0,1],[1,2],[2,3],[3,4]	5	late	0
$\{4, 1, 3, 2, \langle 5 \rangle\}$	[0,1],[1,2],[2,3],[3,4]	6	late	35
$\{4, 1, 3, 2, \langle 5 \rangle, \langle 6 \rangle\}$	[0,1],[1,2],[2,3],[3,4]	7	assign to slot [4,5]	$35 + 25$
$\{4, 1, 3, 2, 7, \langle 5 \rangle, \langle 6 \rangle\}$	[0,1],[1,2],[2,3],[3,4],[4,5]	—	—	60

Thus, the optimal assignment is $4 \to 1 \to 3 \to 2 \to 7 \to \langle 5 \rangle \to \langle 6 \rangle$ and attracts a penalty of 60.

Example 5.14. Let $n = 7$, penalties $(p_1, p_2, \ldots, p_7) = (10, 20, 30, 40, 50, 60, 70)$ and deadlines $(d_1, d_2, \ldots, d_7) = (4, 2, 4, 3, 1, 4, 6)$.

Solution. Arranging the penalties in decreasing order and renaming the p's and deadlines d's accordingly, we have

$$(p_1, p_2, \ldots, p_7) = (70, 60, 50, 40, 30, 20, 10) \quad \text{and}$$

$$(d_1, d_2, \ldots, d_7) = (6, 4, 1, 3, 4, 2, 4).$$

Now using the feasibility criterion of Theorem 5.3, we get

J	assigned time slot	job	action	penalty
φ	none	1	assign to slot [5,6]	0
$\{1\}$	[5,6]	2	assign to slot [3,4]	0
$\{2, 1\}$	[3,4],[5,6]	3	assign to slot [0, 1]	0
$\{3, 2, 1\}$	[0,1],[3,4],[5,6]	4	assign to slot [2,3]	0
$\{3, 4, 2, 1\}$	[0,1],[2,3],[3,4],[5,6]	5	assign to slot [1,2]	0
$\{3, 5, 4, 2, 1\}$	[0,1],[1,2],[2,3],[3,4],[5,6]	6	late	0
$\{3, 5, 4, 2, 1, \langle 6 \rangle\}$	[0,1],[1,2],[2,3],[3,4],[5,6]	7	late	20
$\{3, 5, 4, 2, 1, \langle 6 \rangle, \langle 7 \rangle\}$	[0,1],[1,2],[2,3],[3,4],[5,6]	—	—	$20 + 10$

Therefore, an optimal assignment is $3 \to 5 \to 4 \to 2 \to 1 \to \langle 6 \rangle \to \langle 7 \rangle$ and attracts a penalty $20 + 10 = 30$. The time slot [4, 5] is not assigned to any job. In order that machine is not kept idle, we may assign this slot to the late job 6 and, so, the assignment becomes $3 \to 5 \to 4 \to 2 \to \langle 6 \rangle \to 1 \to \langle 7 \rangle$, the penalty remains unchanged.

Alternatively, and better so, we may assign the vacant time slot [4, 5] to job 1 and then the late jobs $\langle 6 \rangle$ and $\langle 7 \rangle$ in that order to the next two slots, we get the scheduling $3 \rightarrow 5 \rightarrow 4 \rightarrow 2 \rightarrow 1 \rightarrow \langle 6 \rangle \rightarrow \langle 7 \rangle$. Now going to the numbering of the jobs as given, the above scheduling of jobs becomes $5 \rightarrow 3 \rightarrow 4 \rightarrow 6 \rightarrow 7 \rightarrow \langle 2 \rangle \rightarrow \langle 1 \rangle$.

5.4. Huffman Code

Consider the problem of sending messages from base station to a satellite or from one station to another. The messages are written in the English language which has 26 alphabets or characters. Messages, for transmitting through transmission lines, are first to be converted into binary sequences or strings of 0s and 1s. For this purpose, to each English character is associated a small string of 0s and 1s, i.e., to each character is associated a code word. Then using this code, the original file or message is converted into a binary sequence. There are two problems associated with the conversion of the file into a binary sequence. One is the storage capacity needed for the binary sequence and the other is the transmission time. Both these depend on the length of the binary sequence obtained. In order to save both the storage capacity and transmission time, we need as short a sequence as possible for the given file. This can be achieved through choosing a suitable code for converting the characters. There being 26 characters and $2^4 < 26 < 2^5$, each character may be represented as a binary sequence of length 5 (getting fixed length code) and a message/file having n characters gets converted into a binary sequence of length $5n$. There are variable length codes possible and if k is the average length of a code word, the file having n characters gets converted into a binary sequence of length kn. In case we can choose code of average length $k < 5$, such a code is better than a fixed length code. There are some characters such as a, c, e which occur much more frequently than some others such as q, x, z. To achieve minimization of average code length, the alphabets which occur more frequently are represented by binary strings of smaller length and the ones which occur less frequently by larger binary strings.

In the case of fixed length code where every character is associated with a binary string of length 5 (say), on the receiving end, the sequence received is divided into subsequences of length 5 each and then each such substring is replaced by the associated character to get the message in ordinary language. In the case of variable length code, conversion of the received binary sequence into English language is not so easy. Also, when such a conversion is achieved, this conversion may not be unique. For example, if the letters a, e, d, m are represented by the sequences $00, 1, 01, 000$ respectively and the sequence received is 00011, then it could be translated as *ade* and also as *mee*. This problem of decoding can be overcome by using a procedure called prefix codes.

A set of binary sequences is called a **prefix code** if no sequence in the set is a prefix (the initial part) of another sequence in the set. For example $\{1, 00, 01, 000, 001\}$ is not a prefix code while $\{000, 001, 01, 11\}$ and $\{000, 001, 10, 01, 111\}$ are prefix codes.

Let a binary tree T be given. (Recall that a tree is called a binary tree if every node in it has at most two children.) Assign values to the edges in the tree as follows: If a vertex a in T has two children, assign the value 0 to the edge leading to the left child of a and assign the value 1 to the edge leading to the right child of a. Label each leaf of the tree with the binary sequence that is sequence of labels of the edges in the path from the root to the leaf. Since no leaf is an ancestor of another leaf, no sequence which is the label of a leaf is a prefix of another sequence which is the label of another leaf. Thus, the set of all sequences which are labels of the leaves of the binary tree is a prefix code. For example, the prefix codes given by the binary trees of Figs. 5.1(a) and 5.1(b) are, respectively, $\{000, 001, 010, 10, 11\}$ and $\{000, 001, 010, 011, 101, 110, 111\}$.

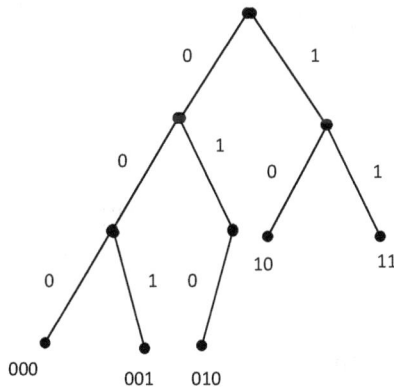

Fig. 5.1(a)

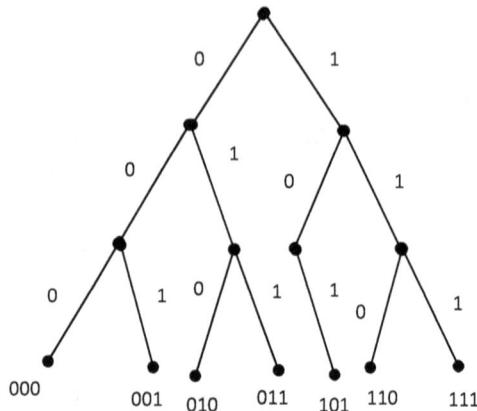

Fig. 5.1(b)

We next prove that given any prefix code, we can find a binary tree the leaves of which carry the binary sequences in the given prefix code as labels. Let h be the length of a largest possible code word. Construct a full regular binary tree of height h. The number of sequences in the given prefix code is at most $2 + 2^2 + \cdots + 2^h$ and the number of vertices in the full regular binary tree of height h is $1 + 2 + 2^2 + \cdots + 2^h$. Assign values to all the edges in this tree as done earlier (i.e., assign value 0 to the edge leading to the left child of a node a (say) and 1 to the edge leading to the right child of the node a). Also assign to each vertex a of the tree a binary sequence that is the sequence of labels of edges in the path from the root to the node a. Thus all binary sequences of length 1 to h appear as labels of the nodes. Identify the nodes which carry as labels the sequences of the prefix code. Then delete all the descendants along with the edges involved of these identified nodes and also delete all nodes that do not have an identified node as descendant together with subtrees with those nodes as roots. There results a binary tree the leaves of which are the binary sequences of the given prefix code.

We have thus proved that to every binary tree corresponds a unique prefix code consisting of the labels of the leaves of the tree and to every prefix code corresponds a unique binary tree the leaves of which have labels which are the sequences of the prefix code. Also observe that the length of any label of a node (or vertex) of a binary tree equals the depth of that node.

Let C be a given set of characters and T a binary tree which corresponds to a prefix code for the set C of characters. For $c \in C$, let $f(c)$ denote the probability of occurrence of c in the given file. Then average length of the code for C is

$$B(T) = \sum_{c \in C} f(c) D_T(c), \qquad (5.11)$$

where $D_T(c)$ denotes the depth of the leaf corresponding to c in T.

In case $f(c)$ denotes the frequency of occurrence of c in the file, $B(T)$ is called the cost of the given tree. When the code corresponding to C is a fixed length code, say $D_T(c) = k$ for every $c \in C$, then whatever be the probabilities $f(c)$, $B(T) = k$. This is so because the sum of probabilities of occurrence of the c's in C is always 1. If all the elements of C are equally likely to occur in the file, then $B(T) = \frac{\sum_{c \in C} D_T(c)}{|C|}$, where $|C|$ is the number of elements in C. If $f(c)$ is constant for the code of Fig. 5.1(a) and also for the code of Fig. 5.1(b), then $B(T) = \frac{13}{5} = 2.6$ and $B(T) = 3$, respectively. Since code of Fig. 5.1(b) is a fixed length code, whatever be the probabilities of occurrence of the characters of C in the file, $B(T) = 3$. On the other hand, if the probabilities of occurrence of the words in the code given by Fig. 5.1(a) are 0.3, 0.25, 0.20, 0.10, 0.15 respectively, then

$$B(T) = 3 \times 0.3 + 3 \times 0.25 + 3 \times 0.20 + 2 \times 0.10 + 2 \times 0.15$$
$$= 0.9 + 0.75 + 0.60 + 0.20 + 0.30 = 2.75.$$

If n is the number of characters in the file that we want to convert into a binary sequence using the code for C, length of the binary sequences becomes $nB(T)$. On the other hand, if $f(c)$ denotes the frequency of occurrence of the character c in the given file, then $B(T)$ itself (which we called the cost of the tree T) gives the length of the sequence which is obtained by using the code for C.

The problem is the construction of a binary tree T for C such that the average length $B(T)$ of the code is minimum. Any tree T that minimizes $B(T)$ is called an **optimal tree**. Huffman suggested the following technique for the construction of such a tree. Using this technique there results a prefix code which is called the **Huffman code**.

Algorithm 5.7. Huffman Algorithm

---

1. given a set C of characters with given probabilities $f(c)$, $c \in C$
2. choose $x, y \in C$ with $f(x), f(y)$ the least possible i.e., there is no z in C with $f(z) < f(x)$ and no t in C with $f(t) < f(y)$ except that we may have $f(x) < f(y)$ or $f(y) < f(x)$
3. merge the two elements x, y into a new element z and set $f(z) = f(x) + f(y)$
4. replace C by $C' = (C - \{x, y\}) \cup z$
5. repeat step 2 for C' to get C'' as in steps 3 and 4
6. continue till we arrive at a single point λ with value $\sum_{c \in C} f(c)$
7. we are then lead to a binary tree in which the leaves are the elements of the set C
8. assign values 0 and 1 to the edges in the tree in the usual way i.e., assign 0 to an edge leading to left child of a given vertex and 1 to the edge leading to right child of a vertex
9. label the leaves in the tree by binary sequences which are the sequences formed by the edges of a path from the root of the tree to the leaf under consideration

---

We illustrate this process through a couple of examples in which we may replace the probabilities $f(c)$ by arbitrarily assigned numbers $f(c)$ (for simplicity).

Example 5.15. Using the Huffman algorithm, construct the binary tree and the corresponding Huffman code from the data $a : 1, b : 1, c : 2, d : 3, e : 5, f : 8$.

Solution. The steps needed for the encoding are given in Figs. 5.2(a)–5.2(f).

Thus, prefix code for the given data is

$$a = 11100, b = 11101, c = 1111, d = 110, e = 10, f = 0.$$

a : 1   b : 1   c : 2   d : 3   e : 5   f : 8

Fig. 5.2(a)

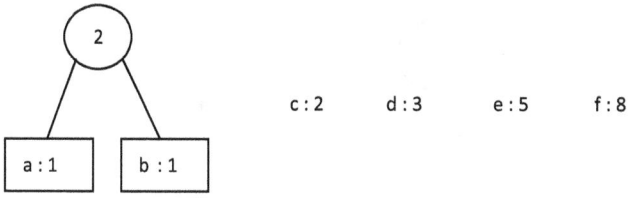

c : 2   d : 3   e : 5   f : 8

Fig. 5.2(b)

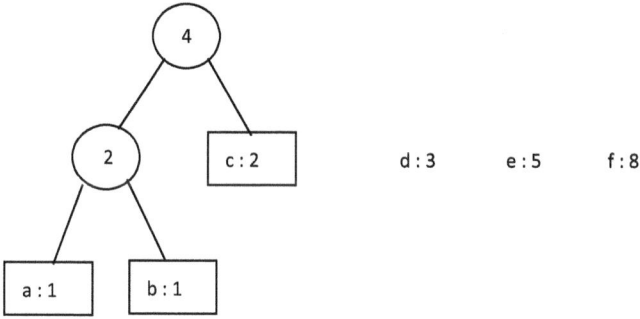

d : 3   e : 5   f : 8

Fig. 5.2(c)

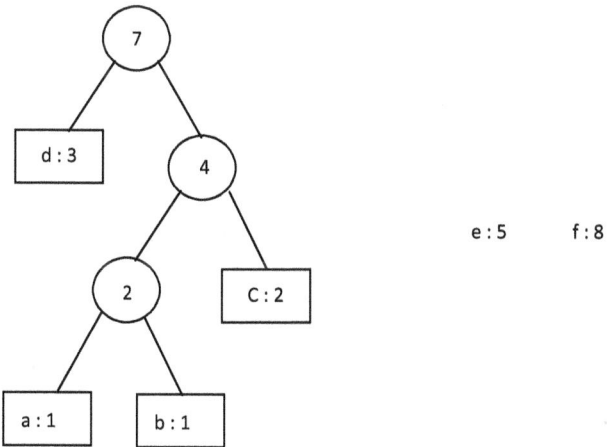

e : 5   f : 8

Fig. 5.2(d)

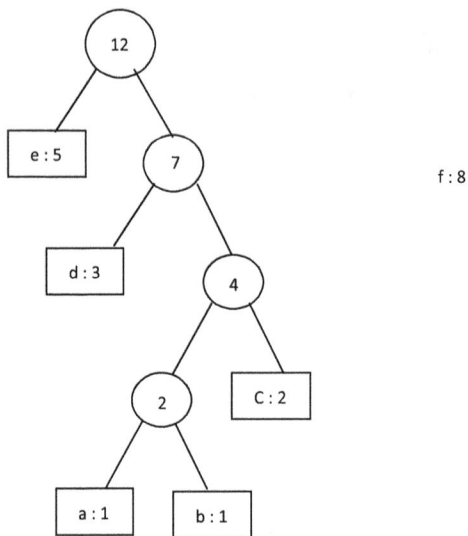

Fig. 5.2(e)

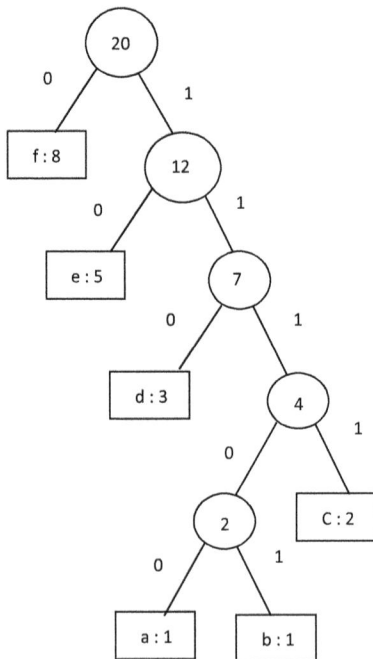

Fig. 5.2(f)

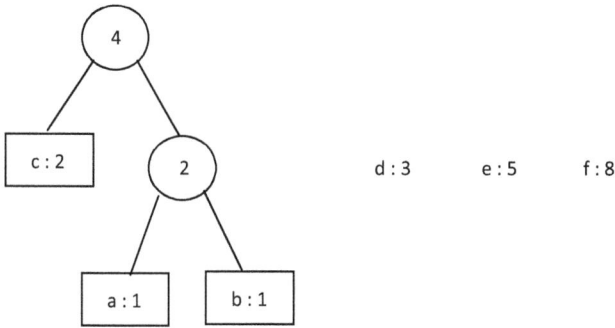

Fig. 5.2(c′)

A few observations are in order. When we merge two elements with the same weight or assigned value, it is immaterial which one of the two is taken as the left child and which one as right child of the new node created. However the two codes obtained for the two assignments will differ slightly. Also the leaves at every step are enclosed in a rectangle while the internal nodes, say x, are placed within brackets as $\langle x \rangle$.

In Fig. 5.2(c) where we merged $c : 2$ and the subtree with node $\langle 2 \rangle$, if we interchange the positions of these two merged subtrees, Fig. 5.2(c) gets replaced by Fig. 5.2(c′).

With this change, the final tree becomes the one given in Fig. 5.3.

The code given by this tree is

$$a = 11110, b = 11111, c = 1110, d = 110, e = 10, f = 0.$$

Had we also interchanged a and b which are of equal weight in the graph of Fig. 5.2(b), the resulting final tree will give the code

$$a = 11111, b = 11110, c = 1110, d = 110, e = 10, f = 0.$$

Example 5.16. Using Huffman coding, find the prefix code with the data

$$a : 1, b : 1, c : 2, d : 3, e : 5, f : 8, g : 13, h : 21, i : 34$$

which are the first nine Fibonacci numbers.

Solution. The steps needed for the encoding are given in Fig. 5.3. Observe that the forest for the first six Fibonacci numbers has already been constructed. Therefore, omitting the first six steps, we have (Fig. 5.4(a)).

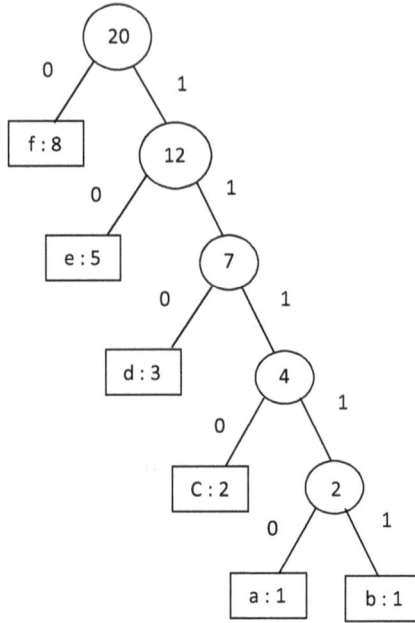

Fig. 5.3

Now merging the sub-tree with $g : 13$, then the resulting subtree with $h : 21$ and finally that result with $i : 34$, we get the tree (see Fig. 5.4(b)).
Thus the Huffman code is

$$a = 11111111, b = 11111110, c = 1111110, d = 111110,$$

$$e = 11110, f = 1110, g = 110, h = 10, i = 0$$

We may point out here that in the code the last number has code 0, the number before that has code 10, the one earlier than that has code 110 and so on with the last 2 (each equal to 1) have codes 11111110 and 11111111, respectively. Thus starting with the last number the code for which is 0, once a code for a point is obtained the code for the following number is obtained by increasing the number of 1's by one and the process continues right up to the second number. Then the code for the first number is obtained by replacing the only 0 in the code for the second number by 1. Therefore, the Huffman code for the n Fibonacci numbers $a_1, a_2,$

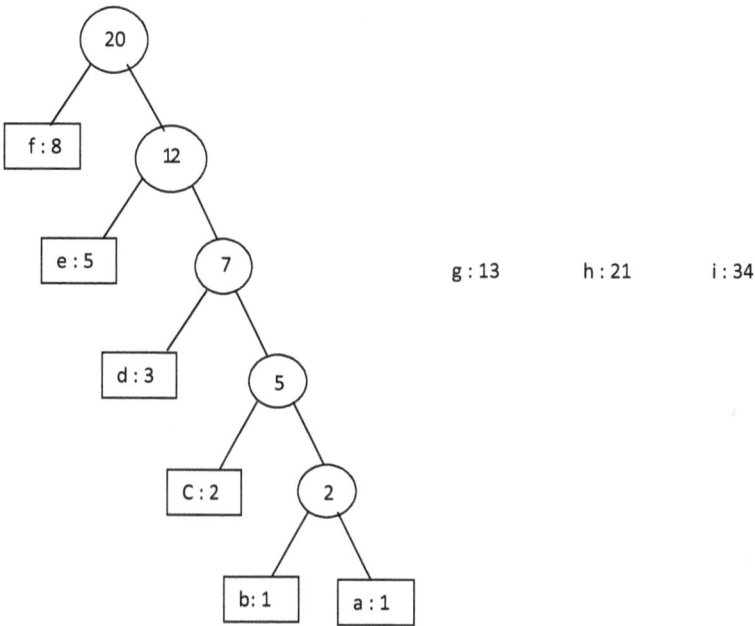

Fig. 5.4(a)

$a_3, \ldots, a_n$, where $a_{i-1} + a_i = a_{i+1}$ for $i \geq 2$ and $a_1 = a_2 = 1$, is

$$a_1 = 11 \ldots 1, a_2 = 11 \ldots 10, a_3 = 11 \ldots 10, \ldots, a_{n-1} = 10, a_n = 0$$

with $n - 1$ 1's in a_1, $n - 2$ 1's in a_2, $n - 3$ 1's in a_3 and finally only one 1 in a_{n-1} and no 1's in a_n.

The Huffman algorithm builds the tree from bottom upwards starting with characters of lowest weight thus ensuring that characters with the lowest weights have the longest code words. Also at an intermediate step, two nodes with smallest weights at that stage are merged. Since at every step two nodes with smallest weights are merged, the algorithm is a greedy algorithm. However, not all Huffman codes are as easy to obtain as the one for Fibonacci numbers.

Example 5.17. Construct the Huffman code for the data given below.

$$a : 3, b : 4, c : 5, d : 6, e : 9, f : 13.$$

Solution. The steps needed in the code are given in Figs. 5.5(a)–5.5(e).

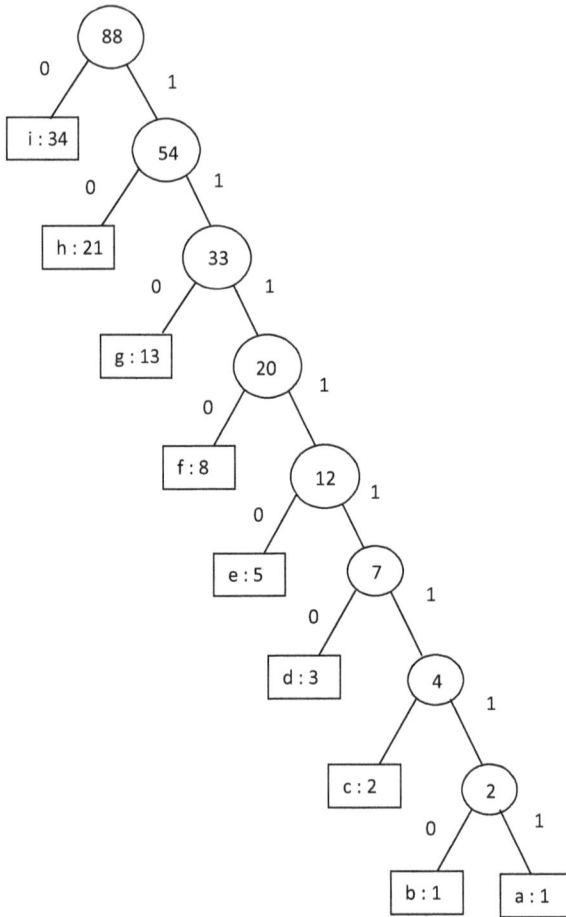

Fig. 5.4(b)

The Huffman code for the given data is

$$a = 000, b = 001, c = 100, d = 101, e = 01, f = 11.$$

Our aim is now to prove that Huffman code for an alphabet C is an optimal code.

Let C be an alphabet having only finite number of characters. Let k be the least positive integer such that $2^{k-1} \le |C| \le 2^k$. Then every character c in C may be represented as a binary sequence of length $\le k$. Therefore,

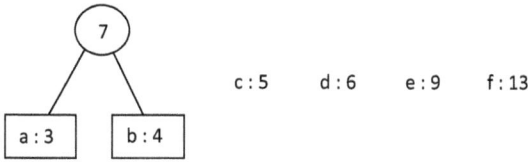

c : 5   d : 6   e : 9   f : 13

Fig. 5.5(a)

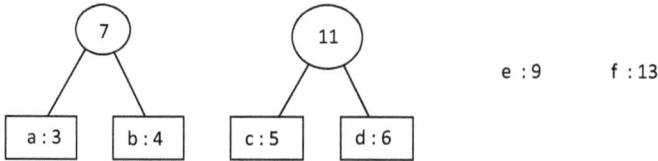

e : 9   f : 13

Fig. 5.5(b)

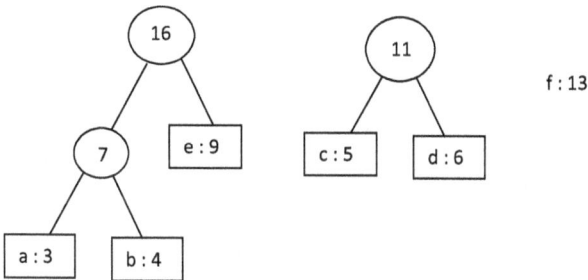

f : 13

Fig. 5.5(c)

the number of choices of codes for c in C is $1 + 2! + \cdots + k! = n$ (say). Then the number of choices of codes for all the characters in C is $n(n-1)(n-2)\ldots(n-|C|+1) = m$ (say), i.e., the number of codes for C is m (finite). The number of prefix codes for C and, so, the number of binary trees representing prefix codes for C is finite. Hence, there always exists a binary tree T for which $B(T)$ is minimum, i.e., **there always exists a binary tree T representing an optimal prefix code for C.**

Let C an alphabet with frequencies $f(c)$ for $c \in C$ be given. Then, we have the following lemma.

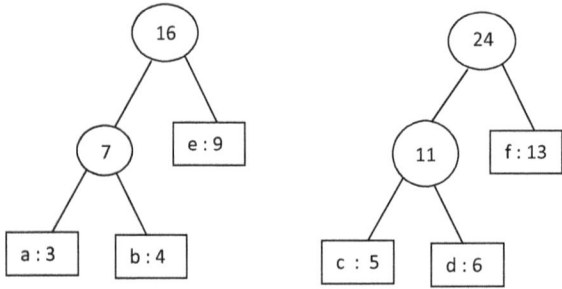

Fig. 5.5(d)

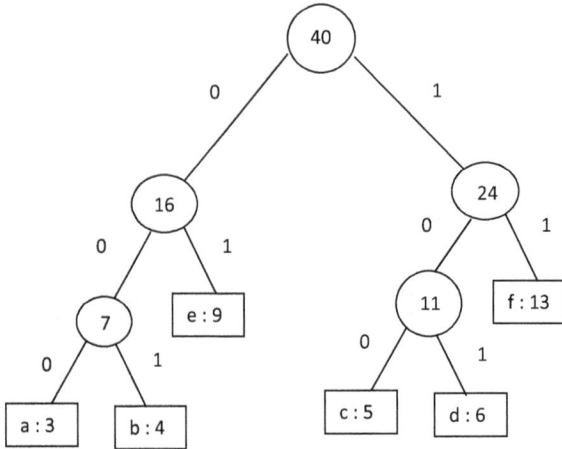

Fig. 5.5(e)

Lemma 5.5. *Every binary tree T representing an optimal prefix code for C is regular.*

Proof. Let T be a binary tree representing an optimal prefix code for C. If T is not regular, let v be a node having only one child x (say). Coalesce v and x and use x for the resulting node. We get a binary tree T' representing C which is of weight $B(T') < B(T)$. This is so because x and all its descendants y have depth $D_{T'}(y) = D_T(y) - 1$. This contradiction proves that T is regular. $\qquad\square$

Lemma 5.6. *If $a, c \in C$ and $f(a) \le f(c)$, then there exists a binary tree T representing a prefix code for C with $D_T(c) \le D_T(a)$.*

Proof. Let T be a binary tree representing an optimal prefix code for C. Suppose that $D_T(c) > D_T(a)$. Let T' be the tree obtained from T by interchanging the positions of a and c. Then $D_{T'}(a) = D_T(c)$, $D_{T'}(c) = D_T(a)$, $D_{T'}(x) = D_T(x)$ for every other x in C. Then

$$
\begin{aligned}
B(T) - B(T') &= \sum_{x \in C} f(x) D_T(x) - \sum_{x \in C} f(x) D_{T'}(x) \\
&= f(a) D_T(a) + f(c) D_T(c) - f(a) D_{T'}(a) \\
&\quad - f(c) D_{T'}(c) \\
&= (f(c) - f(a)) D_T(c) + (f(a) - f(c)) D_T(a) \\
&= (f(c) - f(a))(D_T(c) - D_T(a)). \tag{5.12}
\end{aligned}
$$

If $f(c) > f(a)$, then $B(T) - B(T') > 0$ which contradicts the hypothesis that T is optimal. On the other hand, if $f(c) = f(a)$, it follows from (5.12) that $B(T) = B(T')$, i.e., T' is also a binary tree representing an optimal prefix code for C. Also, in this case

$$
D_{T'}(c) = D_T(a) < D_T(c) = D_{T'}(a)
$$

Hence, in either case, we get a binary tree representing an optimal code for C with $D_T(c) \le D_T(a)$. $\qquad\square$

Lemma 5.7. *If $a, b, c, d \in C$ and $f(a) \le f(c)$, $f(b) \le f(d)$, then there exists a binary tree T representing an optimal prefix code for C such that $D_T(c) \le D_T(a)$ and $D_T(d) \le D_T(b)$.*

Proof. By Lemma 5.6 above, there exists a binary tree T representing an optimal code for C for which $D_T(c) \le D_T(a)$. The argument used in the proof of Lemma 5.6 then shows that either $D_T(d) \le D_T(b)$ or there exists a binary tree T' representing an optimal prefix code for C for which $D_{T'}(d) \le D_{T'}(b)$. Since T' is obtained from T by just interchanging b and d, the depths of a and c in T remain unchanged in T'. Hence we get a binary tree T representing an optimal prefix code for C such that $D_T(c) \le D_T(a)$ and $D_T(d) \le D_T(b)$. $\qquad\square$

Lemma 5.8. *Let a, b be two characters in C having the lowest frequencies. Then there exists an optimal prefix code for C in which a, b have the same length and differ only in their last bits.*

Proof. Let T be a binary tree representing an optimal prefix code for C. Let v be a branch node of maximum depth in T. Since T is regular, let x, y be sons of v. Then x, y are leaves of maximum depth with $D_T(x) = D_T(y)$ and

$$D_T(x) \geq D_T(a), \ D_T(x) \geq D_T(b) \tag{5.13}$$

Case (i). Suppose that $f(a) < f(x)$, $f(b) < f(y)$. Then, by Lemma 5.6,

$$D_T(x) \leq D_T(a), \ D_T(x) = D_T(y) \leq D_T(b) \tag{5.14}$$

It follows from (5.13) and (5.14) that $D_T(x) = D_T(y) = D_T(a) = D_T(b)$.

Let T' be the tree obtained from T by interchanging x, y with a, b, respectively. Then T' also represents C and $D_{T'}(c) = D_T(c)$ for all $c \in C - \{a, b, x, y\}$. Therefore

$$\begin{aligned}
B(T) - B(T') &= f(x)D_T(x) + f(y)D_T(y) + f(a)D_T(a) \\
&\quad + f(b)D_T(b) - \{f(x)D_{T'}(x) + f(y)D_{T'}(y) \\
&\quad + f(a)D_{T'}(a) + f(b)D_{T'}(b)] \\
&= (f(x) + f(y) + f(a) + f(b))D_T(a) \\
&\quad - \{f(x)D_T(a) + f(y)D_T(b) + f(a)D_T(x) \\
&\quad + f(b)D_T(y)\} \\
&= (f(x) + f(y) + f(a) + f(b))D_T(a) \\
&\quad - (f(x) + f(y) + f(a) + f(b))D_T(a) \\
&= 0.
\end{aligned}$$

Hence T' also represents an optima prefix code for C.

Case (ii). $f(a) = f(x)$, $f(b) < f(y)$. Then, as in case (i) above, we have $D_T(y) = D_T(b)$.

Again, let T' be the tree obtained from T by interchanging x, y with a, b, respectively. Then

$$
\begin{aligned}
B(T) - B(T') &= f(x)D_T(x) + f(y)D_T(y) + f(a)D_T(a) \\
&\quad + f(b)D_T(b) - \{f(x)D_T(a) + f(y)D_T(b) \\
&\quad + f(a)D_T(x) + f(b)D_T(y)\} \\
&= \{f(x)D_T(x) + f(a)D_T(a) - f(x)D_T(a) \\
&\quad - f(a)D_T(x)\} + \{f(y) + f(b) - f(y) - f(b)\}D_T(b) \\
&= 0.
\end{aligned}
$$

Hence T' again represents an optimal prefix code for C.

The case $f(a) < f(x)$ but $f(b) = f(y)$ follows on similar lines.

Case (iii). $f(a) = f(x)$, $f(b) = f(y)$. In this case also, interchanging a, x and b, y, we get a tree T' for which

$$
\begin{aligned}
B(T) - B(T') &= f(x)D_T(x) + f(y)D_T(y) + f(a)D_T(a) \\
&\quad + f(b)D_T(b) - \{f(x)D_T(a) + f(y)D_T(b) \\
&\quad + f(a)D_T(x) + f(b)D_T(y)\} \\
&= (f(x) - f(a))D_T(x) + (f(a) - f(x))D_T(a) \\
&\quad + (f(y) - f(b))D_T(y) + (f(b) - f(y))D_T(b) \\
&= 0
\end{aligned}
$$

and T' represents an optimal prefix code for C.

Hence, in all cases, there exists a binary tree T (say) which represents an optimal prefix code for C and in this tree a, b are sibling leaves of maximum depth. Then the paths from the root of T to the leaves a, b are of equal length. Since the path from the root of T to the branch node v which is the parent of a and b is common to the paths from the root to a, b, the codes for a, b given by T are not only of equal length but also differ only in the last bit. □

Lemma 5.9. *Let a and b be two characters in C with minimum frequency and $C' = (C - \{a, b\}) \cup \{z\}$ where z is a new character not already in C. For the characters in C' define frequencies f as for C except that $f(z) =$*

$f(a) + f(b)$. *Let T' be any binary tree representing an optimal prefix code for C' and T be the tree obtained from T' by making the leaf node z an internal node having a, b as sons. Then T represents an optimal prefix code for C.*

Proof. That T represents a prefix code for C is clear from its construction. Let $B(T)$, $B(T')$ denote the cost of the tree T, T', respectively. For $c \in C - \{a, b\}$, $D_T(c) = D_{T'}(c)$ and $D_T(a) = D_T(b) = D_{T'}(z) + 1$. Therefore,

$$B(T) - B(T') = f(a)D_T(a) + f(b)D_T(b) - f(z)D_{T'}(z)$$
$$= f(a)(D_{T'}(z) + 1) + f(b)(D_{T'}(z) + 1)$$
$$- (f(a) + f(b))D_{T'}(z)$$
$$= f(a) + f(b)$$

or

$$B(T) = B(T') + f(a) + f(b). \tag{5.15}$$

Equivalently

$$B(T') = B(T) - f(a) - f(b) \tag{5.16}$$

Observe that relations (5.15) and (5.16) are true even if T' does not represent an optimal prefix code for C'.

Suppose that T does not represent an optimal prefix code for C. Then there exists a tree T'' that represents an optimal prefix code for C, having a, b as siblings and $B(T'') < B(T)$. Let v be the branch node in T'' having a, b as children. Let T''' be the tree obtained from T'' by deleting a, b and the edges from v leading to a, b. Rename v as z and let the frequency of z be $f(z) = f(a) + f(b)$. Then T''' is a binary tree representing a prefix code for C' and (by (5.16))

$$B(T''') = BT'') - f(a) - f(b)$$
$$< B(T) - f(a) - f(b) = B(T').$$

This contradicts the hypothesis that T' is a tree representing an optimal prefix code for C'. Hence T must represent an optimal prefix code for C. $\qquad\qquad\square$

Fig. 5.6

Theorem 5.10. *Huffman code is an optimal prefix code for C.*

Proof. Let T be the tree that represents the Huffman code for C. It is clear from the construction of Huffman code that it is a prefix code. We need to prove its optimality which we do by induction on the number of characters in C. If C has only two characters, say a,b, then T is the tree (Fig. 5.6) with $f(v) = f(a) + f(b)$. Since depth of a, b in T is 1 for each of the two characters, $B(T) = f(a) + f(b)$ which is minimal. Hence T represents an optimal prefix code in this case. Suppose that C has $k + 1$ characters, $k \geq 2$, and that the Huffman code for any alphabet C' having at most k characters is an optimal code. Let a, b be characters of C with minimum frequency. Let $C' = (C - \{a, b\}) \cup \{z\}$ where z is a character not already in C. Define frequencies for elements of C' as defined already for elements of C except that $f(z) = f(a) + f(b)$. It is clear from the construction of T that a, b are siblings. Let v be the branch node in T having a, b as children. Let T' be the tree obtained from T by deleting a, b and the edges connecting a, b to v. The frequency of v is $f(a) + f(b)$. We rename v as z. Then T' becomes a tree which represents the Huffman code for C'. By induction hypothesis, the Huffman code for C' is optimal. Hence T' represents an optimal prefix code for C'. It then follows from Lemma 5.9 that T represents an optimal prefix code for C. This completes induction. Therefore Huffman code is an optimal prefix code for any alphabet C. $\qquad\square$

There are several other areas such as Kruskal'a algorithm and Prim's algorithm for finding minimum spanning trees, Dijkstra's algorithm for shortest path problem, Traveling sales person problem and many more where greedy approach is used. We discuss some of these in later chapters.

Exercise 5.1.

1. In a Huffman code,

    (i) if some character occurs with frequency more than $\frac{2}{5}$, then there is a code word of length 1;

    (ii) if all characters occur with frequency $< \frac{1}{3}$, then there is no code word of length 1.

2. Is Theorem 5.2 valid if the weights of containers in the container loading problem are arranged in non-increasing order? Justify.

Remark. The proof of Theorem 5.10 along with its lemmas is modeled on the proof of the theorem as in [5].[1]

---

[1]The Lemma, Theorem and Proof of this chapter are reproduced with permissions from "Cormen, Thomas H., Charles E. Leiserson, and Ronald L. Rivest, *Introduction to Algorithms*, © 1990 by the Massachusetts Institute of Technology, published by the MIT Press."

Chapter 6

Dynamic Programming

Dynamic programming solves problems by combining the solutions to sub-problems. In this procedure, a given problem is partitioned into some sub-problems which are not independent but share the characteristics of the original problem, i.e., the sub-problems are similar in nature to the problem itself. Moreover, some sub-problems of the original problem occur in some other sub-problems. Every sub-problem is solved just once, its answer is saved in a table and avoids the work of re-computing the answer every time.

Dynamic programming is applied to optimization problems for which there can be more possible solutions than just one. Each solution to a problem or a sub-problem has a value (cost or profit or distance) and we wish to find a solution with the optimal value. Such a solution is called an **optimal solution**.

As opposed to simplex method and dual simplex method for solving linear programming problem, there is no simple unique method(s) for solving some typical dynamic problems and a solution is to be devised for every class of some typical problems. However, one thing is common in most problems solved using dynamic programming: Given the constraints and other available data of the problem a difference equation is formed which is then solved in a step-by-step method. We explain the process of solutions using dynamic programming for some classes of typical problems.

Following are the essential steps to develop a dynamic programming algorithm for optimization problems.

155

1. Characterize the structure of an optimal solution to the problem.
2. Define the value of an optimal solution in a recursive manner (i.e., obtain a difference equation for an optimal value).
3. Compute the value of an optimal solution (i.e., an optimal value) in a bottom up fashion.
4. From the computed information construct an optimal solution, if required.

6.1. Matrix-Chain Multiplication

Let a sequence or chain $(A_1, A_2, \ldots, A_n)$ of n matrices compatible for multiplication be given. Recall that matrices A, B are compatible with respect to multiplication if the number of columns of A equals the number of rows of B and the result AB is a matrix the number of rows of which is equal to the number of rows of A and the number of columns of AB equals the number of columns of B. Also the product is obtained by using pqr scalar multiplications when A is a $p \times q$ matrix and B is a $q \times r$ matrix. The time taken for the multiplication is proportional to the number pqr of scalar multiplications and the time taken on addition of scalars is negligible and, so, is ignored. We thus say that the time taken for the multiplication AB is pqr time units. If we have a chain A_1, A_2, A_3, we could multiply these matrices in two ways, namely $(A_1 A_2)A_3$ and $A_1(A_2 A_3)$ while if we have a chain of four matrices A_1, A_2, A_3, A_4, we could multiply this chain in the following five ways:

$$A_1(A_2(A_3 A_4)), (A_1 A_2)(A_3 A_4),$$

$$((A_1 A_2)A_3)A_4, (A_1(A_2 A_3))A_4, A_1((A_2 A_3)A_4).$$

Let A_i be a matrix of order $p_{i-1} \times p_i$ for $1 \le i \le n$, Then the time taken for the product $(A_1 A_2)A_3$ is

$$p_0 p_1 p_2 + p_0 p_2 p_3 = 2 \times 3 \times 4 + 2 \times 4 \times 5 = 24 + 40 = 64,$$

and for the product $A_1(A_2 A_3)$ is

$$p_0 p_1 p_3 + p_1 p_2 p_3 = 2 \times 3 \times 5 + 3 \times 4 \times 5 = 30 + 60 = 90$$

when $p_0 = 2$, $p_1 = 3$, $p_2 = 4$, $p_3 = 5$ (say).

The above shows that the time taken for the product of a sequence of matrices depends upon the parenthesizing for obtaining the product. The minimum of the times taken for product of a sequence over all parenthesizing is called the minimum time for product of a sequence of matrices. Choose i and j with $1 \leq i < j \leq n$ and choose a fixed k with $i \leq k < j$ and consider a subsequence $(A_i, \ldots, A_j)$ of the given sequence $(A_1, \ldots, A_i, \ldots, A_j, \ldots, A_n)$ of matrices. We denote by $m(i, j)$ the minimum time for multiplying the chain $(A_i, \ldots, A_j)$ of matrices. Write, for convenience, $A_{i\ldots j}$ for the matrix product $A_i A_{i+1} \ldots A_j$. With this notation, observe that we may first compute $A_{i\ldots k}$ and $A_{k+1\ldots j}$ and then multiply the resulting matrices to obtain $A_{i\ldots j}$. Then the cost of this parenthesizing is the cost of computing the matrix $A_{i\ldots k}$, plus the cost of computing the matrix $A_{k+1\ldots j}$, plus the cost of multiplying the two matrices $A_{i\ldots k}$ and $A_{k+1\ldots j}$.

The optimal sub-problem of the problem is as follows.

Suppose that an optimal parenthesizing of $A_i A_{i+1} \ldots A_j$ splits the product between A_k and A_{k+1}, i.e., it splits as the product of $A_i A_{i+1} \ldots A_k$ and $A_{k+1} \ldots A_j$. Then the parenthesization of the sub-chain $A_i A_{i+1} \ldots A_k$ within this optimal parenthesization of $A_i A_{i+1} \ldots A_j$ must be an optimal parenthesization of $A_i A_{i+1} \ldots A_k$. If this were not the case and there is a less costly parenthesization of $A_i A_{i+1} \ldots A_k$, substituting this parenthesization of $A_i A_{i+1} \ldots A_k$ in the given parenthesization of $A_i A_{i+1} \ldots A_j$ leads to a parenthesization of $A_i A_{i+1} \ldots A_j$ which is less costly than the optimal parenthesization we started with and we get a contradiction. Similarly, the parenthesization of $A_{k+1} \ldots A_j$ within the optimal parenthesization of $A_i A_{i+1} \ldots A_j$ must be an optimal parenthesization of $A_{k+1} \ldots A_j$. The cost of optimal parenthesization of $A_i A_{i+1} \ldots A_j$ then must equal

$$m(i, j) = m(i, k) + m(k + 1, j) + p_{i-1} p_k p_j,$$

where obviously $p_{i-1} p_k p_j$ is the cost of multiplying optimal parenthesization of $A_i A_{i+1} \ldots A_k$ and optimal parenthesization of $A_{k+1} \ldots A_j$. Since an optimal parenthesization of $A_i A_{i+1} \ldots A_j$ may split between A_k and A_{k+1} for any k with $i \leq k < j$, we must have

$$m(i, j) = \begin{cases} 0 & \text{if } i = j, \\ \min_{i \leq k < j} \{m(i, k) + m(k + 1, j) + p_{i-1} p_k p_j\} & \text{if } i < j. \end{cases}$$

We also record (and denote) $s(i, j) = k$, the value of k for which $m(i, j)$ happens to be the minimum of $m(i, k) + m(k + 1, j) + p_{i-1}p_k p_j$. By recording the values of $m(i, j)$ in a table and $s(i, j)$ in an adjoining table, we can construct the optimal parenthesization of $A_i \ldots A_j$.

The total number $P(n)$ of all possible parenthesizations of a chain of n matrices is given by the difference equation

$$P(n) = \begin{cases} 1 & \text{if } n = 1, \\ \displaystyle\sum_{1 \leq k < n} P(k)P(n-k) & \text{if } n \geq 2, \end{cases}$$

and $P(n)$ becomes large for large n. However, to find $m(i, j)$ we do not need to find the cost of all the parenthesizations of the given chain. This is so because many sub-problems of multiplying a given sequence of matrices may have to be computed several times. Since we have recorded the computed values in a table, we may use the resulting answer from the table whenever this has to be recomputed.

The recursive equation for computing the minimum cost $m(i, j)$ for multiplication of $j - i + 1$ matrices, we need to compute minimum cost for multiplication of less than $j - i + 1$ matrices — once for $k - i + 1$ matrices and the other time for $j - k$ matrices.

The procedure for writing the two tables for $m(i, j)$ and $s(i, j)$ is best explained through a couple of examples.

Example 6.1. Find the optimal cost and an optimal parenthesization of a matrix-chain product whose sequence of dimensions is $(30, 35, 15, 5, 10, 20, 25)$.

Solution. Since the length of the sequence of dimensions is 7, there are six matrices say $A_1, A_2, A_3, A_4, A_5, A_6$ and if A_i is a $p_{i-1}p_i$ matrix, then we have $p_0 = 30$, $p_1 = 35$, $p_2 = 15$, $p_3 = 5$, $p_4 = 10$, $p_5 = 20$, $p_6 = 25$.

We have to find $m(1, 6)$. Observe that

$$m(1, 6) = \min \begin{cases} (m(1, 1) + m(2, 6) + p_0 p_1 p_6, \\ m(1, 2) + m(3, 6) + p_0 p_2 p_6, \\ m(1, 3) + m(4, 6) + p_0 p_3 p_6, \\ m(1, 4) + m(5, 6) + p_0 p_4 p_6, \\ m(1, 5) + m(6, 6) + p_0 p_5 p_6. \end{cases}$$

Since $m(i, i) = 0$ and $m(i, i + 1)$ is the cost of multiplying $A_i A_{i+1}$, i.e., $m(i, i+1) = p_{i-1} p_i p_{i+1}$, we only need to compute $m(1, 3)$, $m(1, 4)$, $m(1, 5)$ and $m(2, 6)$, $m(3, 6)$, $m(4, 6)$. Now

$$
m(1, 3) = \min \begin{cases} m(1, 1) + m(2, 3) + p_0 p_1 p_3 = p_1 p_2 p_3 + p_0 p_1 p_3 \\ \quad = 45 \times 35 \times 5 = 7875, \\ m(1, 2) + m(3, 3) + p_0 p_2 p_3 = p_0 p_1 p_2 + p_0 p_2 p_3 \\ \quad = 30 \times 40 \times 15 = 18000. \end{cases}
$$

So $m(1, 3) = 7875$, $s(1, 3) = 1$.

$$
m(4, 6) = \min \begin{cases} m(4, 4) + m(5, 6) + p_3 p_4 p_6 = p_4 p_5 p_6 + p_3 p_4 p_6 \\ \quad = 250 \times 25 = 6250, \\ m(4, 5) + m(6, 6) + p_3 p_5 p_6 = p_3 p_4 p_5 + p_3 p_5 p_6 \\ \quad = 100 \times 35 = 3500. \end{cases}
$$

Thus $m(4, 6) = 3500$, $s(4, 6) = 5$.

$$
m(2, 4) = \min \begin{cases} m(2, 2) + m(3, 4) + p_1 p_2 p_4 = p_2 p_3 p_4 + p_1 p_2 p_4 \\ \quad = 40 \times 150 = 6000, \\ m(2, 3) + m(4, 4) + p_1 p_3 p_4 = p_1 p_2 p_3 + p_1 p_3 p_4 \\ \quad = 35 \times 5 \times 25 = 4375. \end{cases}
$$

Therefore $m(2, 4) = 4375$, $s(2, 4) = 3$.

$$
m(3, 5) = \min \begin{cases} m(3, 3) + m(4, 5) + p_2 p_3 p_5 = p_3 p_4 p_5 + p_2 p_3 p_5 \\ \quad = 25 \times 100 = 2500, \\ m(3, 4) + m(5, 5) + p_2 p_4 p_5 = p_2 p_3 p_4 + p_2 p_4 p_5 \\ \quad = 25 \times 150 = 3750. \end{cases}
$$

So $m(3, 5) = 2500$, $s(3, 5) = 3$.
Now

$$
m(1, 4) = \min \begin{cases} m(1, 1) + m(2, 4) + p_0 p_1 p_4 \\ \quad = 4375 + 10500 = 14875, \\ m(1, 2) + m(3, 4) + p_0 p_2 p_4 \\ \quad = p_0 p_1 p_2 + p_2 p_3 p_4 + p_0 p_2 p_4 \\ \quad = 15750 + 150 \times 35 = 21000, \\ m(1, 3) + m(4, 4) + p_0 p_3 p_4 \\ \quad = 7875 + p_0 p_3 p_4 = 7875 + 1500 = 9375. \end{cases}
$$

Thus $m(1, 4) = 9375$, $s(1, 4) = 3$.

Next, we have

$$m(2, 5) = \min \begin{cases} m(2, 2) + m(3, 5) + p_1 p_2 p_5 \\ \quad = 2500 + 35 \times 15 \times 20 = 13000, \\ m(2, 3) + m(4, 5) + p_1 p_3 p_5 \\ \quad = p_1 p_2 p_3 + p_3 p_4 p_5 + p_1 p_3 p_5 \\ \quad = 35 \times 15 \times 5 + 45 \times 100 = 7125, \\ m(2, 4) + m(5, 5) + p_1 p_4 p_5 \\ \quad = 4375 + 35 \times 10 \times 20 = 11375. \end{cases}$$

Hence $m(2, 5) = 7125$, $s(2, 5) = 3$.

Now

$$m(3, 6) = \min \begin{cases} m(3, 3) + m(4, 6) + p_2 p_3 p_6 \\ \quad = 3500 + 15 \times 5 \times 25 = 5375, \\ m(3, 4) + m(5, 6) + p_2 p_4 p_6 \\ \quad = p_2 p_3 p_4 + p_4 p_5 p_6 + p_2 p_4 p_6 \\ \quad = 750 + 5000 + 3750 = 9500, \\ m(3, 5) + m(6, 6) + p_2 p_5 p_6 \\ \quad = 2500 + p_2 p_5 p_6 = 2500 + 7500 = 10000. \end{cases}$$

Therefore, $m(3, 6) = 5375$, $s(3, 6) = 3$.

Now,

$$m(1, 5) = \min \begin{cases} m(1, 1) + m(2, 5) + p_0 p_1 p_5 \\ \quad = 7125 + 30 \times 35 \times 20 = 7125 + 21000 = 28125, \\ m(1, 2) + m(3, 5) + p_0 p_2 p_5 \\ \quad = p_0 p_1 p_2 + 2500 + p_0 p_2 p_5 \\ \quad = 2500 + 24750 = 27250, \\ m(1, 3) + m(4, 5) + p_0 p_3 p_5 \\ \quad = 7875 + p_3 p_4 p_5 + p_0 p_3 p_5 \\ \quad = 7875 + 40 \times 100 = 11875, \\ m(1, 4) + m(5, 5) + p_0 p_4 p_5 = 9375 + 6000 = 15375. \end{cases}$$

Hence $m(1, 5) = 11875$, $s(1, 5) = 3$.

Next

$$m(2, 6) = \min \begin{cases} m(2, 2) + m(3, 6) + p_1 p_2 p_6 \\ \quad = 5375 + 35 \times 15 \times 25 = 5375 + 13125 = 18500, \\ m(2, 3) + m(4, 6) + p_1 p_3 p_6 \\ \quad = p_1 p_2 p_3 + 3500 + p_1 p_3 p_6 \\ \quad = 3500 + 40 \times 35 \times 5 = 10500, \\ m(2, 4) + m(5, 6) + p_1 p_4 p_6 \\ \quad = 4375 + p_4 p_5 p_6 + p_1 p_4 p_6 \\ \quad = 4375 + 13750 = 18125, \\ m(2, 5) + m(6, 6) + p_1 p_5 p_6 = 7125 + p_1 p_5 p_6 \\ \quad = 7125 + 17500 = 24625. \end{cases}$$

Thus $m(2, 6) = 10500$, $s(2, 6) = 3$.

Finally, all the requirements for $m(1, 6)$ have been computed and

$$m(1, 6) = \min \begin{cases} 10500 + p_0 p_1 p_6 = 10500 + 26250 = 36750, \\ p_0 p_1 p_2 + 5375 + p_0 p_2 p_6 \\ \quad = 5375 + 60 \times 30 \times 15 = 32375, \\ 7875 + 3500 + p_0 p_3 p_6 = 11375 + 125 \times 30 \\ \quad = 11375 + 3750 = 15125, \\ 9375 + p_4 p_5 p_6 + p_0 p_4 p_6 = 9375 + 50 \times 250 \\ \quad = 9375 + 12500 = 21875, \\ 11875 + p_0 p_5 p_6 = 11875 + 15000 = 26875. \end{cases}$$

Hence $m(1, 6) = 15125$, $s(1, 6) = 3$.

Therefore, optimal cost for matrix multiplication of the chain $A_1 \ldots A_6$ is 15125 and the optimal parenthesization is $(A_1(A_2 A_3))$ $((A_4 A_5) A_6)$.

Example 6.2. Find the optimal cost and an optimal parenthesization of a matrix-chain product whose sequence of dimensions is $(5, 10, 3, 12, 5, 50, 6, 8)$.

Solution. The length of the dimensions sequence being 8, there is a matrix-chain $A_1 \ldots A_7$ of length 7 and $p_0 = 5$, $p_1 = 10$, $p_2 = 3$, $p_3 = 12$,

$p_4 = 5$, $p_5 = 50$, $p_6 = 6$, $p_7 = 8$. We need to compute

$$m(1, 7) = \min \begin{cases} (m(1, 1) + m(2, 7) + p_0 p_1 p_7, \\ m(1, 2) + m(3, 7) + p_0 p_2 p_7, \\ m(1, 3) + m(4, 7) + p_0 p_3 p_7, \\ m(1, 4) + m(5, 7) + p_0 p_4 p_7, \\ m(1, 5) + m(6, 7) + p_0 p_5 p_7, \\ m(1, 6) + m(7, 7) + p_0 p_6 p_7. \end{cases}$$

Therefore, we have to compute $m(1, 3)$, $m(1, 4)$, $m(1, 5)$, $m(1, 6)$ and $m(5, 7)$, $m(4, 7)$, $m(3, 7)$ and $m(2, 7)$. We compute the required $m(i, j)$ as follows:

(i)

$$m(1, 3) = \min \begin{cases} m(1, 1) + m(2, 3) + p_0 p_1 p_3 \\ \quad = p_1 p_2 p_3 + p_0 p_1 p_3 = 360 + 600 = 960, \\ m(1, 2) + m(3, 3) + p_0 p_2 p_3 \\ \quad = p_0 p_1 p_2 + p_0 p_2 p_3 = 150 + 180 = 330, \end{cases}$$

so $m(1, 3) = 330$, $s(1, 3) = 2$.

(ii)

$$m(2, 4) = \min \begin{cases} m(2, 2) + m(3, 4) + p_1 p_2 p_4 \\ \quad = p_2 p_3 p_4 + p_1 p_2 p_4 = 180 + 150 = 330, \\ m(2, 3) + m(4, 4) + p_1 p_3 p_4 \\ \quad = p_1 p_2 p_3 + p_1 p_3 p_4 = 360 + 600 = 960, \end{cases}$$

so $m(2, 4) = 330$, $s(2, 4) = 2$.

(iii)

$$m(3, 5) = \min \begin{cases} m(3, 3) + m(4, 5) + p_2 p_3 p_5 \\ \quad = p_3 p_4 p_5 + p_2 p_3 p_5 = 3000 + 1800 = 4800, \\ m(3, 4) + m(5, 5) + p_2 p_4 p_5 \\ \quad = p_2 p_3 p_4 + p_2 p_4 p_5 = 180 + 750 = 930, \end{cases}$$

so $m(3, 5) = 930$, $s(3, 5) = 4$.

(iv)

$$m(4, 6) = \min \begin{cases} m(4, 4) + m(5, 6) + p_3 p_4 p_6 \\ \quad = p_4 p_5 p_6 + p_3 p_4 p_6 = 1500 + 360 = 1860, \\ m(4, 5) + m(6, 6) + p_3 p_5 p_6 \\ \quad = p_3 p_4 p_5 + p_3 p_5 p_6 = 3000 + 3600 = 6600, \end{cases}$$

so $m(4, 6) = 1860$, $s(4, 6) = 4$.

(v)

$$m(5, 7) = \min \begin{cases} m(5, 5) + m(6, 7) + p_4 p_5 p_7 \\ \quad = p_5 p_6 p_7 + p_4 p_5 p_7 = 2400 + 2000 = 4400, \\ m(5, 6) + m(7, 7) + p_4 p_6 p_7 \\ \quad = p_4 p_5 p_6 + p_4 p_6 p_7 = 1500 + 240 = 1740, \end{cases}$$

so $m(5, 7) = 1740$, $s(5, 7) = 6$.

(vi)

$$m(1, 4) = \min \begin{cases} m(1, 1) + m(2, 4) + p_0 p_1 p_4 \\ \quad = 330 + 250 = 580, \\ m(1, 2) + m(3, 4) + p_0 p_2 p_4 \\ \quad = p_0 p_1 p_2 + p_2 p_3 p_4 + p_0 p_2 p_4 \\ \quad = 150 + 180 + 75 = 405, \\ m(1, 3) + m(4, 4) + p_0 p_3 p_4 \\ \quad = 330 + 300 = 630, \end{cases}$$

so $m(1, 4) = 405$, $s(1, 4) = 2$.

(vii)

$$m(2, 5) = \min \begin{cases} m(2, 2) + m(3, 5) + p_1 p_2 p_5 \\ \quad = 930 + 1500 = 2430, \\ m(2, 3) + m(4, 5) + p_1 p_3 p_5 \\ \quad = p_1 p_2 p_3 + p_3 p_4 p_5 + p_1 p_3 p_5 \\ \quad = 360 + 3000 + 6000 = 9360, \\ m(2, 4) + m(5, 5) + p_1 p_4 p_5 \\ \quad = 330 + 2500 = 2830, \end{cases}$$

so $m(2, 5) = 2430$, $s(2, 5) = 2$.

(viii)

$$m(3, 6) = \min \begin{cases} m(3, 3) + m(4, 6) + p_2 p_3 p_6 \\ \quad = 1860 + 216 = 2076, \\ m(3, 4) + m(5, 6) + p_2 p_4 p_6 \\ \quad = p_2 p_3 p_4 + p_4 p_5 p_6 + p_2 p_4 p_6 \\ \quad = 180 + 1500 + 90 = 1770, \\ m(3, 5) + m(6, 6) + p_2 p_5 p_6 \\ \quad = 930 + 900 = 1830, \end{cases}$$

so $m(3, 6) = 1770, s(3, 6) = 4$.

(ix)

$$m(4, 7) = \min \begin{cases} m(4, 4) + m(5, 7) + p_3 p_4 p_7 \\ \quad = 1740 + 480 = 2220, \\ m(4, 5) + m(6, 7) + p_3 p_5 p_7 \\ \quad = p_3 p_4 p_5 + p_5 p_6 p_7 + p_3 p_5 p_7 \\ \quad = 3000 + 2400 + 4800 = 10200, \\ m(4, 6) + m(7, 7) + p_3 p_6 p_7 \\ \quad = 1860 + 576 = 2436, \end{cases}$$

so $m(4, 7) = 2220, s(4, 7) = 4$.

(x)

$$m(1, 5) = \min \begin{cases} m(1, 1) + m(2, 5) + p_0 p_1 p_5 \\ \quad = 2430 + 2500 = 4930, \\ m(1, 2) + m(3, 5) + p_0 p_2 p_5 \\ \quad = p_0 p_1 p_2 + 930 + 750 \\ \quad = 150 + 1680 = 1830, \\ m(1, 3) + m(4, 5) + p_0 p_3 p_5 \\ \quad = 330 + p_3 p_4 p_5 + p_0 p_3 p_5 \\ \quad = 330 + 3000 + 3000 = 6330, \\ m(1, 4) + m(5, 5) + p_0 p_4 p_5 \\ \quad = 405 + 1250 = 1655, \end{cases}$$

so $m(1, 7) = 1655, s(1, 5) = 4$.

(xi)

$$m(2, 6) = \min \begin{cases} m(2, 2) + m(3, 6) + p_1 p_2 p_6 \\ \quad = 1770 + 180 = 1950, \\ m(2, 3) + m(4, 6) + p_1 p_3 p_6 \\ \quad = p_1 p_2 p_3 + 1860 + 720 = 2940, \\ m(2, 4) + m(5, 6) + p_1 p_4 p_6 \\ \quad = 330 + p_4 p_5 p_6 + p_1 p_4 p_6 \\ \quad = 330 + 1500 + 300 = 2130, \\ m(2, 5) + m(6, 6) + p_1 p_5 p_6 \\ \quad = 2430 + 3000 = 5430, \end{cases}$$

so $m(2, 6) = 1950$, $s(2, 6) = 2$.

(xii)

$$m(3, 7) = \min \begin{cases} m(3, 3) + m(4, 7) + p_2 p_3 p_7 \\ \quad = 2220 + 288 = 2508, \\ m(3, 4) + m(5, 7) + p_2 p_4 p_7 \\ \quad = p_2 p_3 p_4 + 1740 + 120 \\ \quad = 180 + 1740 + 120 = 2040, \\ m(3, 5) + m(6, 7) + p_2 p_5 p_7 \\ \quad = 930 + p_5 p_6 p_7 + p_2 p_5 p_7 \\ \quad = 930 + 2400 + 1200 = 4530, \\ m(3, 6) + m(7, 7) + p_2 p_6 p_7 \\ \quad = 1770 + 144 = 1914, \end{cases}$$

so $m(3, 7) = 1914$, $s(3, 7) = 6$.

(xiii)

$$m(1, 6) = \min \begin{cases} m(1, 1) + m(2, 6) + p_0 p_1 p_6 \\ \quad = 1950 + 300 = 2250, \\ m(1, 2) + m(3, 6) + p_0 p_2 p_6 \\ \quad = p_0 p_1 p_2 + 1770 + 90 \\ \quad = 150 + 1860 = 2010, \\ m(1, 3) + m(4, 6) + p_0 p_3 p_6 \\ \quad = 330 + 1860 + 360 = 2550, \\ m(1, 4) + m(5, 6) + p_0 p_4 p_6 \\ \quad = 405 + p_4 p_5 p_6 + 150 \\ \quad = 555 + 1500 = 2055, \\ m(1, 5) + m(6, 6) + p_0 p_5 p_6 \\ \quad = 1655 + 1500 = 3155, \end{cases}$$

so $m(1, 6) = 2010$, $s(1, 6) = 2$.

(xiv)

$$m(2, 7) = \min \begin{cases} m(2, 2) + m(3, 7) + p_1 p_2 p_7 \\ \quad = 1914 + 240 = 2154, \\[4pt] m(2, 3) + m(4, 7) + p_1 p_3 p_7 \\ \quad = p_1 p_2 p_3 + 2220 + 960 \\ \quad = 360 + 3180 = 3540, \\[4pt] m(2, 4) + m(5, 7) + p_1 p_4 p_7 \\ \quad = 330 + 1740 + 400 = 2470, \\[4pt] m(2, 5) + m(6, 7) + p_1 p_5 p_7 \\ \quad = 2430 + p_5 p_6 p_7 + 4000 \\ \quad = 6430 + 2400 = 8830, \\[4pt] m(2, 6) + m(7, 7) + p_1 p_6 p_7 \\ \quad = 1950 + 480 = 2430, \end{cases}$$

so $m(2, 7) = 2154$, $s(2, 7) = 2$.

We now have all the information available for computing optimal matrix-chain multiplication and the corresponding parenthesization of matrix-chain multiplication. We have

$$m(1, 7) = \min \begin{cases} m(1, 1) + m(2, 7) + p_0 p_1 p_7 \\ \quad = 2154 + 400 = 2554, \\[4pt] m(1, 2) + m(3, 7) + p_0 p_2 p_7 \\ \quad = p_0 p_1 p_2 + 1914 + 120 = 2184, \\[4pt] m(1, 3) + m(4, 7) + p_0 p_3 p_7 \\ \quad = 330 + 2220 + 480 = 3030, \\[4pt] m(1, 4) + m(5, 7) + p_0 p_4 p_7 \\ \quad = 405 + 1740 + 200 = 2345, \\[4pt] m(1, 5) + m(6, 7) + p_0 p_5 p_7 \\ \quad = 1655 + p_5 p_6 p_7 + 2000 \\ \quad = 3655 + 2400 = 6055, \\[4pt] m(1, 6) + m(7, 7) + p_0 p_6 p_7 \\ \quad = 2010 + 240 = 2250. \end{cases}$$

Hence the optimal matrix-chain multiplication cost in this case is 2184. Also the corresponding optimal parenthesization of the matrix-chain multiplication being provided by

$$m(1, 7) = 2, m(3, 7) = 6, m(3, 6) = 4, \text{ is } (A_1 A_2)(((A_3 A_4)(A_5 A_6))A_7).$$

Observation. For matrix-chain multiplication for a chain of n matrices we need to compute

$$\left. \begin{aligned} &m(1, 2), m(2, 3), \ldots, m(n-1, n) \\ &m(1, 3), m(2, 4), \ldots, m(n-2, n) \\ &m(1, 4), m(2, 5), \ldots, m(n-3, n) \\ &\qquad\qquad\vdots \\ &m(1, n-1), m(2, n) \\ &m(1, n) \end{aligned} \right\} \tag{6.1}$$

The total number of computations is $1 + 2 + \cdots + (n-1) = \frac{n(n-1)}{2}$ whereas the total number of parenthesization is $P(n) = \sum_{1 \le k \le n-1} P(k)P(n-k)$ for $n \ge 2$ which is much larger than $\frac{n(n-1)}{2}$. To see this, we compare the values of $\frac{n(n-1)}{2}$ and $P(n)$ for some small values of n.

n	$\frac{n(n-1)}{2}$	$P(n)$
2	1	$P(1)P(1) = 1 \times 1 = 1$
3	3	$P(1)P(2) + P(2)P(1)) = 2$
4	6	$P(1)P(3) + P(2)P(2) + P(3)P(1) = 2 + 1 + 2 = 5$
5	10	$P(1)P(4) + P(2)P(3) + P(3)P(2) + P(4)P(1)$
		$\quad = 5 + 2 + 2 + 5 = 14$
6	15	$P(1)P(5) + P(2)P(4) + P(3)P(3) + P(4)P(2) + P(5)P(1)$
		$\quad = 2(14 + 5) + 2 \times 2 = 42$
7	21	$P(1)P(6) + P(2)P(5) + P(3)P(4) + P(4)P(3)$
		$\quad + P(5)P(2) + P(6)P(1) = 2(42 + 14 + 2 \times 5) = 132$

so that $P(n)$ grows extremely faster than $\frac{n(n-1)}{2}$ as n grows larger. Also observe that once the $n-1$ quantities in the first row of (6.1) are computed, all the information for the computation of the second row are available. Once the quantities in the first and second row are computed, all the information for the computation of the third row are available. The process continues.

6.2. Multistage Graphs

In multistage dynamic problem, each problem consists of some stages, say k and every stage consists of certain options called states from which decision for choosing one option, keeping in mind the optimality of the objective function and the constraints of the problem, is taken. As an example of such a dynamic problem, we consider directed weighted graph $G = (V, E)$ in which the set V of vertices is partitioned into subsets V_i, $1 \leq i \leq k$, i.e., $V = \bigcup_{1 \leq i \leq k} V_i$, $V_i \cap V_j = \emptyset$ for $i \neq j$ and $V_i \neq \emptyset$ for every i, $1 \leq i \leq k$. The subsets V_i are such that $o(V_1) = o(V_k) = 1$, and if $e = (u, v)$ is an edge in G, then $u \in V_i$ and $v \in V_{i+1}$ for some i. Thus, no two vertices in any V_i are adjacent and no vertex in V_i is adjacent to any vertex in V_j if $j \geq i + 2$. Each V_i determines a stage in the graph and the elements of V_i are states from which one state is to be chosen, If $V_1 = \{a\}$ and $V_k = \{b\}$, **the problem is to obtain a shortest length path from a to b and shortest path length**. A shortest length path is also called **shortest cost path** and the length of this path is called the shortest cost. For the sake of simplicity we list the elements of V as $1, 2, \ldots, n$ in order, such that for any i, $1 \leq i \leq k - 1$, the elements of V_{i+1} are listed only after the elements of V_i have been listed. Then $V_1 = \{1\}$, $V_k = \{n\}$. If j is a node in V_i, let $p(i, j)$ denote the shortest cost path from j to n and $\text{cost}(i, j)$ denote the shortest cost associated with this shortest cost path. For nodes $j, l \in V$, we write $c(j, l)$ for the cost or weight associated with the edge (j, l) if j, l are adjacent and ∞ otherwise. Once $\text{cost}(i + 1, j)$ have been computed for all j in V_{i+1}, we observe that

$$\text{cost}(i, j) = \min_{l \in V_{i+1}, (j,l) \in E} \{c(j, l) + \text{cost}(i + 1, l)\}. \tag{6.2}$$

Note that j is a vertex in V_i and every edge incident with j has to terminate at a vertex l (say) in V_{i+1}. Also, as per our convention, $c(j, l) = \infty$ if (j, l) is not an edge in E and since we are interested in the minimum of $c(j, l) + \text{cost}(i + 1, l)$, we need not consider the case where $(j, l) \notin E$. Let $d(i, j)$ denote the value of the node l in V_{i+1} for which the minimum is obtained in (6.2) i.e.,

$$c(j, d(i, j)) + \text{cost}(i + 1, d(i, j)) \leq c(j, l) + \text{cost}(i + 1, l)$$

for all l in V_{i+1} for which $(j, l) \in E$. With this notation, a minimum cost path will be of the form $a = v_1 = 1, v_2, \ldots, v_{k-1}, v_k = b$ where

$$v_2 = d(1, 1), v_3 = d(2, v_2), v_4 = d(3, v_3), \ldots, v_{k-1} = d(k - 2, v_{k-2}).$$
$$(6.3)$$

Example 6.3. As an example, we consider the case of graph as given in Fig. 6.1.

This is 6-stage graph with $V_1 = \{1\}$, $V_2 = \{2, 3, 4\}$, $V_3 = \{5, 6, 7, 8, 9\}$, $V_4 = \{10, 11, 12, 13\}$, $V_5 = \{14, 15, 16\}$ and $V_6 = \{17\}$. Using the result (6.2) for cost(i, j) we have (we do not consider cost$(5, j)$

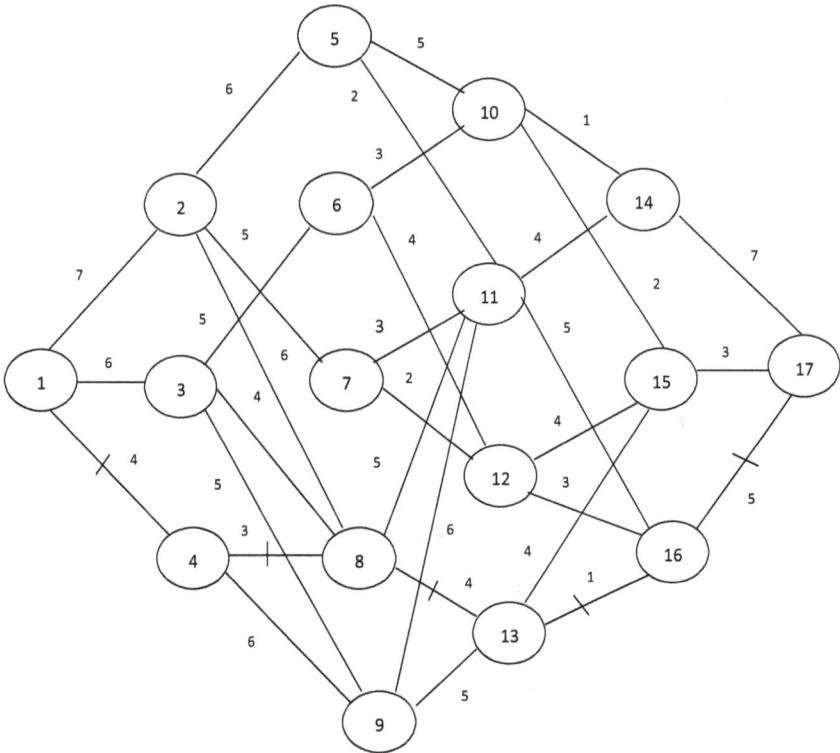

Fig. 6.1

which are simply $c(j, 17))$

$$\begin{aligned}
\text{cost}(4, 10) &= \min\{c(10, 14) + \text{cost}(5, 14), c(10, 15) + \text{cost}(5, 15)\} \\
&= \min\{1 + 7, 2 + 3\} = 5 \text{ with } d(4, 10) = 15,
\end{aligned}$$

$$\begin{aligned}
\text{cost}(4, 11) &= \min\{c(11, 14) + \text{cost}(5, 14), c(11, 16) + \text{cost}(5, 16)\} \\
&= \min\{4 + 7, 5 + 5\} = 10 \text{ with } d(4, 11) = 16,
\end{aligned}$$

$$\begin{aligned}
\text{cost}(4, 12) &= \min\{c(12, 15) + \text{cost}(5, 15), c(12, 16) + \text{cost}(5, 16)\} \\
&= \min\{4 + 3, 3 + 5\} = 7 \text{ with } d(4, 12) = 15,
\end{aligned}$$

$$\begin{aligned}
\text{cost}(4, 13) &= \min\{c(13, 15) + \text{cost}(5, 15), c(13, 16) + \text{cost}(5, 16)\} \\
&= \min\{4 + 3, 1 + 5\} = 6 \text{ with } d(4, 13) = 16,
\end{aligned}$$

$$\begin{aligned}
\text{cost}(3, 5) &= \min\{c(5, 10) + \text{cost}(4, 10), c(5, 11) + \text{cost}(4, 11)\} \\
&= \min\{5 + 5, 2 + 10\} = 10 \text{ with } d(3, 5) = 10,
\end{aligned}$$

$$\begin{aligned}
\text{cost}(3, 6) &= \min\{c(6, 10) + \text{cost}(4, 10), c(6, 12) + \text{cost}(4, 12)\} \\
&= \min\{3 + 5, 4 + 7\} = 8 \text{ with } d(3, 6) = 10,
\end{aligned}$$

$$\begin{aligned}
\text{cost}(3, 7) &= \min\{c(7, 11) + \text{cost}(4, 11), c(7, 12) + \text{cost}(4, 12)\} \\
&= \min\{3 + 10, 2 + 7\} = 9 \text{ with } d(3, 7) = 12,
\end{aligned}$$

$$\begin{aligned}
\text{cost}(3, 8) &= \min\{c(8, 11) + \text{cost}(4, 11), c(8, 13) + \text{cost}(4, 13)\} \\
&= \min\{5 + 10, 4 + 6\} = 10 \text{ with } d(3, 8) = 13,
\end{aligned}$$

$$\begin{aligned}
\text{cost}(3, 9) &= \min\{c(9, 11) + \text{cost}(4, 11), c(9, 13) + \text{cost}(4, 13)\} \\
&= \min\{6 + 10, 5 + 6\} = 11 \text{ with } d(3, 9) = 13,
\end{aligned}$$

$$\begin{aligned}
\text{cost}(2, 2) &= \min\{c(2, 5) + \text{cost}(3, 5), c(2, 7) + \text{cost}(3, 7), c(2, 8) \\
&\quad + \text{cost}(3, 8)\} \\
&= \min\{6 + 10, 5 + 9, 6 + 10\} = 14 \text{ with } d(2, 2) = 7,
\end{aligned}$$

$$\begin{aligned}
\text{cost}(2, 3) &= \min\{c(3, 6) + \text{cost}(3, 6), c(3, 8) + \text{cost}(3, 8), c(3, 9) \\
&\quad + \text{cost}(3, 9)\} \\
&= \min\{5 + 8, 4 + 10, 5 + 11\} = 13 \text{ with } d(2, 3) = 6,
\end{aligned}$$

$$\begin{aligned}
\text{cost}(2, 4) &= \min\{c(4, 8) + \text{cost}(3, 8), c(4, 9) + \text{cost}(3, 9)\} \\
&= \min\{3 + 10, 6 + 11\} = 13 \text{ with } d(2, 4) = 8,
\end{aligned}$$

$$\text{cost}(1, 1) = \min\{c(1, 2) + \text{cost}(2, 2), c(1, 3) + \text{cost}(2, 3), c(1, 4)$$
$$+ \text{cost}(2, 4)\}$$
$$= \min\{7 + 14, 6 + 13, 4 + 13\} = 17 \text{ with } d\{1, 1\} = 4.$$

Hence the minimum cost of the minimum cost path from node I to node 17 is 17 and the minimum cost path is

$$1 \to d(1, 1) = 4 \to d(2, 4) = 8 \to d(3, 8)$$
$$= 13 \to d(4, 13) = 16 \to d(5, 16) = 17,$$

i.e.,

$$1 \to 4 \to 8 \to 13 \to 16 \to 17.$$

Observe that for any node $j \neq n$ in V, there is an i, $1 \leq i \leq k-1$, such that $j \in V_i$ and all l for which $(j, l) \in E$ have to be in V_{i+1}. Therefore, we may write $\text{cost}(j)$ for $\text{cost}(i, j)$ and $\text{cost}(l)$ for $\text{cost}(i + 1, l)$. We may then rewrite (6.2) simply as

$$\text{cost}(j) = \min_{l \in V_{i+1}, (j,l) \in E} \{c(j, l) + \text{cost}(l)\}. \tag{6.2'}$$

We may then also write $d(j)$ in place of $d(i, j)$. With this simplified notation the shortest cost path (6.3) becomes

$$v_1 = 1, v_2 = d(1), v_3 = d(v_2), v_4 = d(v_3), \ldots, v_{k-1}$$
$$= d(v_{k-2}), v_k = n. \tag{6.3'}$$

This simplified notation will be of great help in writing the algorithm for shortest cost and the associated shortest cost path. Also observe that the shortest cost path is obtained in a step-by-step method moving forward.

In (6.2), we observe that cost (i, j) is determined once cost $(i + 1, l)$ are determined for all l in V_{i+1}, i.e., first we determine cost (k, n) which is 0.

6.3. A Business/Industry Oriented Problem

We next consider a business and industry related multistage problem. The number of states in each stage being not so small, drawing a graph becomes quite cumbersome and the computations made at each stage are recorded in a table.

Example 6.4. A company has a surplus money of seven crore rupees which it likes to invest (no company lets any money remain idle). A total of four investment options are available to it. When x_j crores of rupees are invested in option j, then a net present value, i.e., the maturity amount as at present is $v_j(x_j)$ crores of rupees and the $v_j(x_j)$ are as given below:

$$v_1(x_1) = 6x_1 + 5, \; v_2(x_2) = 3x_2 + 7,$$

$$v_3(x_3) = 4x_3 + 3, \; v_4(x_4) = 5x_4 + 2,$$

where $x_1, x_2, x_3, x_4 \geq 0$ are integers and $v_1(0) = v_2(0) = v_3(0) = v_4(0) = 0$. The amount placed in each investment has to be an integral multiple of one crore so that each x_j is an integer (as already stated). How should the company invest the whole amount in order to maximize the present value obtained from the investment?

Solution. Observe that there are four decisions to be taken which are to assign the values to each x_i. First choose x_1 and once this is done the balance amount $S - x_1 = S_2$, where $S = S_1$ is the total amount available for investment, is to be invested in the other three options. Then we need to choose x_2. Once this is done, the amount available for the remaining two investment options is $S_3 = S_2 - x_2$. Next we choose x_3 and the balance $S_4 = S_3 - x_3$ is the amount available for the fourth investment option. Thus $x_4 = S_4$. As such we would like to solve the problem as a 4-stage dynamic problem. In each stage i, the states are from 0 to S_i, i.e., possible values that x_i can take.

If we denote by $f_n^*(S_n)$ the optimal pay off at stage n for state S_n (i.e., the amount available for investment at the state S_n), then $f_n^*(S_n)$ is the maximum of {the maximum pay off at stage $n + 1$ i.e., $f_{n+1}^*(S_{n+1} - x_n)$ plus the payoff from the current investment} or that

$$f_n^*(S_n) = \max_{x_n} \{v_n(x_n) + f_{n+1}^*(S_{n+1} - x_n)\}.$$

The x_n for which this maximum is attained is denoted by x_n^*. We also write

$$f_n(S_n) = v_n(x_n) + f_{n+1}^*(S_{n+1} - x_n).$$

With the notations introduced, we now consider the tables for the payoffs for the four stages.

Table 6.1

State S_4	$f_4(S_4) = v_4(x_4) = 5x_4 + 2$								$f_4^*(S_4)$	x_4^*
	$x_4 = 0$	$x_4 = 1$	$x_4 = 2$	$x_4 = 3$	$x_4 = 4$	$x_4 = 5$	$x_4 = 6$	$x_4 = 7$		
0	0								0	0
1	0	7							7	1
2	0	7	12						12	2
3	0	7	12	17					17	3
4	0	7	12	17	22				22	4
5	0	7	12	17	22	27			27	5
6	0	7	12	17	22	27	32		32	6
7	0	7	12	17	22	27	32	37	37	7

Table 6.2

State S_3	$f_3(S_3) = v_3(x_3) + f_4^*(S_4 - x_3) = 4x_3 + 3 + f_4^*(S_4 - x_3)$								$f_3^*(S_3)$	x_3^*
	$x_3 = 0$	$x_3 = 1$	$x_3 = 2$	$x_3 = 3$	$x_3 = 4$	$x_3 = 5$	$x_3 = 6$	$x_3 = 7$		
0	0								0	0
1	7	7							7	1
2	12	14	11						14	1
3	17	19	18	15					19	1
4	22	24	23	22	19				24	1
5	27	29	28	27	26	23			29	1
6	32	34	33	32	31	30	27		34	1
7	37	39	38	37	36	35	34	31	39	1

Stage 4. The payoffs at this stage are as given in Table 6.1.

As the amount S_4 available at this stage has to be invested in full, the entries below the main diagonal in this table are unnecessary and the payoffs are given by the matrix

$$\text{diag}(0, 7, 12, 17, 22, 27, 32, 37).$$

Stage 3. The payoffs at this stage are shown in Table 6.2. The payoffs for each option (i.e., state) are also indicated.

Stage 2. The payoffs at this stage along with payoffs for each option (i.e., state) are given in Table 6.3.

Stage 1. Since the full amount of 7 crore rupees is available for investment at this stage, the table of payoffs in this case consists of a single row and is given in Table 6.4.

Table 6.3

State S_2	$f_2(S_2) = v_2(x_2) + f_3^*(S_3 - x_2) = 3x_2 + 7 + f_3^*(S_3 - x_2)$									
	$x_2 = 0$	$x_2 = 1$	$x_2 = 2$	$x_2 = 3$	$x_2 = 4$	$x_2 = 5$	$x_2 = 6$	$x_2 = 7$	$f_2^*(S_2)$	x_2^*
0	0								0	0
1	7	10							10	1
2	14	17	13						17	1
3	19	24	20	16					24	1
4	24	29	27	23	19				29	1
5	29	34	32	30	26	22			34	1
6	34	39	37	35	33	29	25		39	1
7	39	44	42	40	38	36	32	28	44	1

Table 6.4

State S_1	$f_1(S_1) = v_1(x_1) + f_2^*(S_2 - x_1) = 6x_1 + 5 + f_2^*(S_2 - x_1)$									
	$x_1 = 0$	$x_1 = 1$	$x_1 = 2$	$x_1 = 3$	$x_1 = 4$	$x_1 = 5$	$x_1 = 6$	$x_1 = 7$	$f_1^*(S_1)$	x_1^*
7	44	50	51	52	53	52	51	47	53	4

Optimal solution. As is clear from Table 6.4 for stage 1, optimal payoff is 53 crore rupees and $x_1 = 4$, i.e., 4 crores of rupees are to be invested in the first option. Then 3 crores of rupees are available for investment in stage 2. The maximum payoff at this stage is 24 with $x_2^* = 1$ so that 1 crore rupees are to be invested in second option. We are then left with 2 crore rupees for investment in the remaining two options. Thus, looking at the payoff table for stage 3, we find that the maximal payoff is Rs.14 crores with $x_3 = 1$, i.e., 1 crore rupees is invested in the third option. We are thus left with 1 crore rupees which is invested in the fourth option. Thus, the optimal solution is $x_1 = 4$, $x_2 = 1 = x_3 = x_4$ and the payoff $= (4 \times 6 + 5) + (3 \times 1 + 7) + (4 \times 1 + 3) + (5 \times 1 + 2) = 29 + 10 + 7 + 7 = 53$, which is as it should be.

6.4. Largest Common Subsequence

Given sequence $x = (x_1, x_2, \ldots, x_n)$ with n entries $x_1, x_2, \ldots, x_n$ (which may be integers or other well recognized elements i.e., elements of certain set), a sequence $a = (a_1, a_2, \ldots, a_k)$ is called a **subsequence** of x if a can be obtained from x by deleting some (even none) entries from x. There are two things inherent in it namely every sequence is a subsequence of itself and every a_i equals some x_j without changing their relative positions, i.e., there exist integers $1 \le i_1 < i_2 < \cdots < i_k \le n$ such that $a_j = x_{i_j}$, $1 \le j \le k$. Also, we call n **the length of** x. Given two sequences x, y (not necessarily of the same length), a sequence a is called a **common subsequence** of x and y if a is a subsequence of x as well as of y. Also a is called a **longest (or largest) common subsequence** of x and y if a is a common subsequence of x and y and there is no common subsequence b of x and y the length of which is strictly greater than the length of a. When we come to consider examples, we shall see that the sequences x and y may have more than one longest common subsequences.

The longest common subsequence problem states that given sequences $x = (x_1, x_2, \ldots, x_n)$, $y = (y_1, y_2, \ldots, y_m)$ to find a longest possible common subsequence of x and y. In this section, we use dynamic programming (in tabular form) to solve this problem. It is however easier to find the largest length of a common subsequence than finding an actual common subsequence (LCS). To define and solve sub-problems of this problem, we need to define a prefix of a given sequence. Given $x = (x_1, x_2, \ldots, x_n)$, ith prefix of x is the subsequence $(x_1, x_2, \ldots, x_i)$. To avoid confusion, we write X for the sequence x and then X_i for the ith prefix of x, i.e., $X_i = (x_1, x_2, \ldots, x_i)$. Observe that nth prefix of X is X itself and X_0 is the empty sequence. Also observe that if X is an empty sequence or Y is an empty sequence then LCS of X and Y is the empty sequence and the length of the LCS is zero. The following observation is immediate from the definition of LCS.

Theorem 6.1. *Any two LCS of sequences X and Y have the same length.*

The following simple result is extremely useful in finding length of an LCS of two sequences X and Y.

Theorem 6.2. *Let $Z = (z_1, z_2, \ldots, z_k)$ be an LCS of sequences $X = (x_1, x_2, \ldots, x_m)$ and $Y = (y_1, y_2, \ldots, y_n)$. Then the following conditions hold.*

(1) If $x_m = y_n$, then $z_k = x_m = y_n$ and Z_{k-1} is an LCS of X_{m-1} and Y_{n-1}.

(2) If $x_m \neq y_n$, then Z is an LCS of X_{m-1} and Y and also an LCS of X and Y_{n-1}.

Proof. (1) If $z_k \neq x_m$, then $Z' = (z_1, \ldots, z_k, x_m)$ is a common subsequence of X and Y and is larger than Z which contradicts the hypothesis that Z is an LCS of X and Y. Hence $z_k = x_m = y_n$.

(2) Suppose that $x_m \neq y_n$. Then $z_k \neq y_n$ or $z_k \neq x_m$. If $z_k \neq x_m$, then Z is a subsequence of X_{m-1}. As it is already a subsequence of Y, Z is a subsequence of X_{m-1} and Y. If it is not an LCS of X_{m-1} and Y, an LCS of X_{m-1} and Y has length longer than that of Z (as any subsequence of two sequences can be extended to an LCS of the two sequences) which contradicts the hypothesis that Z is an LCS of X and Y. Hence Z is an LCS of X_{m-1} and Y.

Also if $z_k \neq y_n$, by an argument similar to the one used above Z is an LCS of X and Y_{n-1}. $\square$

Remark. The result as at (2) of the above theorem is quite useful for obtaining a recursive formula for obtaining length of an LCS of X and Y.

In view of Theorem 6.2(1), for finding an LCS of $X = (x_1, x_2, \ldots, x_m)$ and $Y = (y_1, y_2, \ldots, y_n)$ when $x_m = y_n$, we need to find an LCS of X_{m-1} and Y_{n-1}, and then append $x_m (= y_n)$ to this LCS. Thus length of LCS of X and Y equals $(1 +$ length LCS of X_{m-1} and $Y_{n-1})$. In the case of $x_m \neq y_n$ (i.e., case (2) of Theorem 6.2), we need to find an LCS of X_{m-1} and Y and an LCS of X and Y_{n-1}. Thus we need to solve three sub-problems. Then we iterate backwards solving three sub-problems at each step. We thus have the case of a dynamic programming problem. When $x_m \neq y_n$, we find that LCS of X and Y is the larger of the LCS of X_{m-1} and Y and of the LCS of X and Y_{n-1}. If we write $l(X)$ for the length of the sequence X and $c(i, j)$ for the length of an LCS of X_i and Y_j the above observations lead to (with $l(X) = m, l(Y) = n$)

$$c(l(X), l(Y))$$
$$= \begin{cases} 0 & \text{if } l(X) = 0 \text{ or } l(Y) = 0, \\ 1 + c(m-1, n-1) & \text{if } l(X), l(Y) > 0 \text{ and } x_m = y_n, \\ \max\{c(m-1, n), c(m, n-1)\} & \text{if } l(X), l(Y) > 0 \text{ and } x_m \neq y_n. \end{cases}$$

Then moving backwards, we get the following recurrence relation for the length of an LCS of X and Y (since for any i, j, $l(X_i) = i$ and $L(Y_j) = j$)

$$c(i, j) = \begin{cases} 0 & \text{if } i = 0 \text{ or } j = 0, \\ 1 + c(i-1, j-1) & \text{if } i, j > 0 \text{ and } x_i = y_j, \\ \max\{c(i-1, j), c(i, j-1)\} & \text{if } i, j > 0 \text{ and } x_i \neq y_j. \end{cases}$$
(6.4)

Thus, for finding $c(i, j)$ we need to compute $c(i, j-1)$, $c(i-1, j)$ and $c(i-1, j-1)$. We record the computation of $c(i, j)$ in an $(m+1) \times (n+1)$ table. The table is completed by computing all the rows in order. The first row which corresponds to $i = 0$ consists of all zeros. Also the first column which corresponds to $j = 0$ consists of all zeros. Once the first row is completed, the second row corresponding to $i = 1$ is completed. Then the third row and so on. The computation of the $(i+1)$th row only needs the ith row and the entry $c(i+1, j)$ for the computation of the $c(i+1, j+1)$ entry. In this table, we enclose the entry $c(i, j)$ if it equals $1 + c(i-1, j-1)$ as $\langle c(i, j) \rangle$. This enclosing the entries helps in writing down an LCS of X and Y. We explain the $c = (c(i, j))$ table through a couple of examples but before that we record the algorithm for length of LCS.

Algorithm 6.1. Algorithm LCS Length (X, Y)

---

1. given sequences $X = (x_1, x_2, \ldots, x_m)$, $Y = (y_1, y_2, \ldots, y_n)$ so that $l(X) = m$, $l(Y) = n$ and $x_1, x_2, \ldots, x_m, y_1, y_2, \ldots, y_n$ are integers (or elements of a set)
2. if $m = 0$ or $n = 0$ then $c(m, n) = 0$
3. with both $m > 0$, $n > 0$; for $i = 1$ to m, $j = 1$ to n
4. $c(i, j) = 1 + c(i-1, j-1)$ if $x_i = y_j$; write $c(i, j)$ as $\langle c(i, j) \rangle$ (writing $\langle c(i, j) \rangle$ is not essential for finding $c(i, j)$)
5. but $c(i, j) = \max\{c(i-1, j), c(i, j-1)\}$ if $x_i \neq y_j$

Observe that there are mn comparisons involved plus another at most mn comparisons for computing $\max\{c(i-1, j), c(i, j-1)\}$ Therefore, the running time of the algorithm is $O(mn)$.

Table 6.5

j		0	1	2	3	4	5	6	7	8	9	10
i	y_j		0	1	0	1	1	0	1	1	0	1
0	x_i	0	0	0	0	0	0	0	0	0	0	0
1	1	0	0	⟨1⟩	1	⟨1⟩	⟨1⟩	1	⟨1⟩	⟨1⟩	1	⟨1⟩
2	0	0	⟨1⟩	1	⟨2⟩	2	2	⟨2⟩	2	2	⟨2⟩	2
3	0	0	⟨1⟩	1	⟨2⟩	2	2	⟨3⟩	3	3	⟨3⟩	3
4	1	0	1	⟨2⟩	2	⟨3⟩	⟨3⟩	3	⟨4⟩	⟨4⟩	4	⟨4⟩
5	0	0	⟨1⟩	2	⟨3⟩	3	3	⟨4⟩	4	4	⟨5⟩	5
6	1	0	1	⟨2⟩	3	⟨4⟩	⟨4⟩	4	⟨5⟩	⟨5⟩	5	⟨6⟩
7	0	0	⟨1⟩	2	⟨3⟩	4	4	⟨5⟩	5	5	⟨6⟩	6
8	1	0	1	⟨2⟩	3	⟨4⟩	⟨5⟩	5	⟨6⟩	⟨6⟩	6	⟨7⟩
9	0	0	⟨1⟩	2	⟨3⟩	4	5	⟨6⟩	6	6	⟨7⟩	7

Example 6.5. Find an LCS and the length of an LCS and an LCS of X and Y, where $X = 1, 0, 0, 1, 0, 1, 0, 1, 0$ and $Y = 0, 1, 0, 1, 1, 0, 1, 1, 0, 1$.

Solution. Here $m = 9$ and $n = 10$ so that $c(= c(i, j))$ is a 10×11 table (see Table 6.5). We now compute the table c by using the recursive relation (6.4).

Observe that none of the entries is larger than i ($i = 0, 1, 2, \ldots, m$). Once an entry in a row becomes i, none of the entries to its right in the row is less than i. Similar observations apply about the columns of the table. Highest entry enclosed in ⟨ ⟩ in the table gives the length of an LCS of X and Y. For obtaining an LCS, we start with highest entry within ⟨ ⟩. The largest such entry in the present case is 7 so that the length of an LCS is 7. This entry in ⟨ ⟩ occurs in two places (x_9, y_9) and (x_8, y_{10}). Suppose we start with (x_9, y_9). Next we look for an entry 6 in ⟨ ⟩ in the row immediately above the 9th row. There are two such entries in the 8th row. Both of these occur in columns numbered less than 9. We may choose either of the two. So, the branching starts here. So, we have

$$(x_8, y_8), (x_9, y_9),$$

$$(x_8, y_7), (x_9, y_9).$$

Then, we look for the entry $\langle 5 \rangle$ in the 7th row. This occurs in the position (x_7, y_6). So, we have

$$(x_7, y_6), (x_8, y_8), (x_9, y_9),$$
$$(x_7, y_6), (x_8, y_7), (x_9, y_9),$$

Then look for the entry $\langle 4 \rangle$ in the 6th row. It occurs in two positions $(x_6, y_4), (x_6, y_5)$. Thus, we get

$$(x_6, y_5), (x_7, y_6), (x_8, y_8), (x_9, y_9),$$
$$(x_6, y_4), (x_7, y_6), (x_8, y_8), (x_9, y_9),$$
$$(x_6, y_5), (x_7, y_6), (x_8, y_7), (x_9, y_9),$$
$$(x_6, y_4), (x_7, y_6), (x_8, y_7), (x_9, y_9).$$

There is only one entry $\langle 3 \rangle$ in the 5th row and the above sequences yield the same common subsequence and, so, we get

$$(x_5, y_3), (x_6, y_5), (x_7, y_6), (x_8, y_8), (x_9, y_9).$$

The next entry within $\langle \ \rangle$ to the left and above (x_5, y_3) occurs in the position (x_4, y_2) and, so, there results the next common subsequence as

$$(x_4, y_2), (x_5, y_3), (x_6, y_5), (x_7, y_6), (x_8, y_8), (x_9, y_9).$$

Finally the entry $\langle 1 \rangle$ to the left and top of the entry in $\langle \ \rangle$ at (x_4, y_2) occurs in the position (x_3, y_1). Thus an *LCS* that results is

$$(x_3, y_1), (x_4, y_2), (x_5, y_3), (x_6, y_5), (x_7, y_6), (x_8, y_8), (x_9, y_9)$$

or

$$x_3, x_4, x_5, x_6, x_7, x_8, x_9 \text{ i.e., } 0, 1, 0, 1, 0, 1, 0.$$

When we start with the other entry $\langle 7 \rangle$ in the position (x_8, y_{10}), and move up and left every time when we need to move, we get common

Table 6.6

j	0	1	2	3	4	5	6
i	y_j	2	4	3	1	2	1
0 x_i	0	0	0	0	0	0	0
1 1	0	0	0	0	$\langle 1\rangle$	1	$\langle 1\rangle$
2 2	0	$\langle 1\rangle$	1	1	1	$\langle 2\rangle$	2
3 3	0	1	1	$\langle 2\rangle$	2	2	2
4 2	0	$\langle 1\rangle$	1	2	2	$\langle 3\rangle$	3
5 4	0	1	$\langle 2\rangle$	2	2	3	3
6 1	0	1	2	2	$\langle 3\rangle$	3	$\langle 4\rangle$
7 2	0	$\langle 1\rangle$	2	2	3	$\langle 4\rangle$	4

subsequences

$$(x_2, y_1), (x_3, y_3), (x_4, y_5), (x_5, y_6), (x_6, y_8), (x_7, y_9), (x_8, y_{10}),$$

$$(x_2, y_1), (x_3, y_3), (x_4, y_5), (x_5, y_6), (x_6, y_7), (x_7, y_9), (x_8, y_{10}),$$

$$(x_2, y_1), (x_3, y_3), (x_4, y_4), (x_5, y_6), (x_6, y_8), (x_7, y_9), (x_8, y_{10}),$$

$$(x_2, y_1), (x_3, y_3), (x_4, y_4), (x_5, y_6), (x_6, y_7), (x_7, y_9), (x_8, y_{10}),$$

all of which yield the same LCS $x_2, x_3, x_4, x_5, x_6, \quad x_7, x_8$ or $0, 0, 1.0, 1, 0, 1$. Observe that there result two distinct LCS which are $0, 1, 0.1, 0, 1, 0$ and $0, 0, 1, 0, 1, 0, 1$ both of length 7 (as it should be).

Example 6.6. Find an LCS and length of an LCS of X, Y, where $X = (1, 2, 3, 2, 4, 1, 2)$, $Y = (2, 4, 3, 1, 2, 1)$.

Solution. Here $m = 7$ and $n = 6$ so that c is an 8×7 table (see Table 6.6). We now compute the table c by using the recursive relation (6.4) putting the entries $c(i, j)$ as $\langle c(i, j)\rangle$ which are of the form $c(i, j) = 1 + c(i - 1, j - 1)$.

As mentioned, in general, the first row consists of all zeros and so does the first column. Then the second row is completed step-by-step, then the third row and so on. Observe that the highest entry $\langle 4\rangle$ occurs in two positions (x_7, y_5) and (x_6, y_6) so there will be at least two LCS of length 4. We first begin with the entry in (x_7, y_5) and look for the entries $\langle 3\rangle, \langle 2\rangle, \langle 1\rangle$ in succession moving at least one step up and at least one step to the left from the entry already arrived at. In this fashion we get an LCS as (x_4, y_1), (x_5, y_2), (x_6, y_4), (x_7, y_5), i.e., the sequence x_4, x_5, x_6, x_7 or $2, 4, 1, 2$.

Next we start with the entry (x_6, y_6) and get an LCS as (x_2, y_1), (x_3, y_3), (x_4, y_5), (x_6, y_6), i.e., the sequence x_2, x_3, x_4, x_6 or $2, 3, 2, 1$.

Thus, there are exactly two LCS namely $2, 4, 1, 2$ and $2, 3, 2, 1$.

Increasing or decreasing subsequence of a sequence

The procedure for finding an LCS may be modified slightly to obtain an increasing subsequence or a decreasing subsequence of a given sequence.

Let X be a given sequence, the elements of which are from a linearly ordered set. In order to find a strictly increasing subsequence of X, we first obtain a maximal element of X, say n; then obtain a maximal element of the remaining sequence of $n - 1$ elements and call this maximal element $n - 1$. Continue the process till we have found all such maximal elements. Then consider the sequence $Y = (1, 2, 3, \ldots, n)$ $(Y = (0, 1, 2, \ldots, n)$ if 0 is present among the elements of X). An LCS of X and Y will then be a strictly increasing subsequence of X. If we are interested in a strictly decreasing subsequence of X, we need to take $Y = (n, n - 1, \ldots, 1)$ (or $Y = (n, n - 1, \ldots, 1, 0)$ in case 0 occurs in the sequence X). Then an LCS of X and Y is a strictly decreasing subsequence of X.

Algorithm 6.2. Algorithm: $\max X$, X A Sequence

1. $X = (x_1, x_2, \ldots, x_n)$ is a sequence with the entries $x_1, x_2, \ldots, x_n$ in a linearly ordered set
2. set $x = \max\{x_1, x_2\}$
3. for i from 3 to n set $x = \max\{x, x_i\}$
4. end

Observe that $n - 1$ comparisons of two elements at a time are needed to get the largest element of X and, therefore, the running time of the algorithm is $O(n)$. Changing "max" to "min" in the

Table 6.7

j		0	1	2	3	4
i	y_j	1	2	3	4	
0	x_i	0	0	0	0	0
1	2	0	0	⟨1⟩	1	1
2	4	0	0	1	1	⟨2⟩
3	3	0	0	1	⟨2⟩	2
4	1	0	⟨1⟩	1	2	2
5	2	0	1	⟨2⟩	2	2
6	1	0	⟨1⟩	2	2	2

above algorithm gives an algorithm for finding the least element of the sequence.

We now consider a couple of examples.

Example 6.7. Find a strictly increasing and a strictly decreasing subsequence of $X = (2, 4, 3, 1, 2, 1)$.

Solution. Observe that maximal element of X is 4. Therefore, we consider the following.

(a) $Y = (1, 2, 3, 4)$. We then find an *LCS* of X and Y. We obtain this by writing the c table for which we omit the details. Now, the c table is as given in Table 6.7.

The largest entry ⟨a⟩ in Table 6.7 being ⟨2⟩, the length of a largest subsequence is 2 and largest increasing subsequences of X have coordinate (x_4, y_1), (x_5, y_2); (x_1, y_2), (x_3, y_3); (x_1, y_2), (x_2, y_4) so that the subsequences are x_4, x_5; x_1, x_3; x_1, x_2 which are 1, 2; 2, 3 and 2, 4, respectively.

(b) Here $Y = (4, 3, 2, 1)$. The c table for a largest decreasing subsequence of X is as given in Table 6.8.

The largest entry in Table 6.8 being ⟨4⟩, the length of a largest decreasing subsequence of X is 4 and the only such subsequence has coordinates (x_2, y_1), (x_3, y_2), (x_5, y_3), (x_6, y_4). Thus the subsequence is x_2, x_3, x_5, x_6, i.e., 4, 3, 2, 1.

Observe that there is no relationship between a largest increasing and a largest decreasing subsequence of a sequence. On the one hand, one may feel that once a largest strictly increasing subsequence is obtained, a largest strictly decreasing subsequence may be obtained by just taking the terms

Table 6.8

j		0	1	2	3	4
i		y_j	4	3	2	1
0	x_i	0	0	0	0	0
1	2	0	0	0	⟨1⟩	1
2	4	0	⟨1⟩	1	1	1
3	3	0	1	⟨2⟩	2	2
4	1	0	1	2	2	⟨3⟩
5	2	0	1	2	⟨3⟩	3
6	1	0	1	2	3	⟨4⟩

Table 6.9

j		0	1	2	3	4	5
i		y_j	1	2	3	4	5
0	x_i	0	0	0	0	0	0
1	2	0	0	⟨1⟩	1	1	1
2	3	0	0	1	⟨2⟩	2	2
3	1	0	⟨1⟩	1	2	2	2
4	4	0	1	1	2	⟨3⟩	3
5	2	0	1	⟨2⟩	2	3	3
6	4	0	1	2	2	⟨3⟩	3
7	3	0	1	2	⟨3⟩	3	3
8	5	0	1	2	3	3	⟨4⟩

of an increasing subsequence in the reverse order. But this is not the case as shown in the above example.

Example 6.8. Find a strictly increasing and a strictly decreasing subsequence of $X = (2, 3, 1, 4, 2, 4, 3, 5)$.

Solution. Observe that $\max(X) = 5$. Therefore, we take the following condition.

(a) $Y = (1, 2, 3, 4, 5)$. For finding an LCS of X and Y, the c table is as given in Table 6.9.

Table 6.10

j	0	1	2	3	4	5
i	y_j	5	4	3	2	1
0 x_i	0	0	0	0	0	0
1 2	0	0	0	0	$\langle 1 \rangle$	1
2 3	0	0	0	$\langle 1 \rangle$	1	1
3 1	0	0	0	1	1	$\langle 2 \rangle$
4 4	0	0	$\langle 1 \rangle$	1	1	2
5 2	0	0	1	1	$\langle 2 \rangle$	2
6 4	0	0	$\langle 1 \rangle$	1	2	2
7 3	0	0	1	$\langle 2 \rangle$	2	2
8 5	0	$\langle 1 \rangle$	1	2	2	2

Table 6.9 shows that length of a largest strictly increasing subsequence is 4 and largest strictly increasing subsequences are with coordinates:

$$(x_3, y_1), (x_5, y_2), (x_7, y_3), (x_8, y_5);$$
$$(x_3, y_1), (x_5, y_2), (x_6, y_4), (x_8, y_5);$$
$$(x_1, y_2), (x_2, y_3), (x_4, y_4), (x_8, y_5);$$
$$(x_1, y_2), (x_2, y_3), (x_6, y_4), (x_8, y_5).$$

Then, the sequences are

$$x_3, x_5, x_7, x_8 \text{ or } y_1, y_2, y_3, y_5; \quad x_3, x_5, x_6, x_8 \text{ or } y_1, y_2, y_4, y_5;$$
$$x_1, x_2, x_4, x_8 \text{ or } y_2, y_3, y_4, y_5; \quad x_1, x_2, x_6, x_8 \text{ or } y_2, y_3, y_4, y_5.$$

Thus the strictly increasing largest subsequences are $1, 2, 3, 5$; $1, 2, 4, 5$; $2, 3, 4, 5$ and $2, 3, 4, 5$ so that there are three such subsequences.

(b) Again $Y = \{5, 4, 3, 2, 1\}$. For finding a longest decreasing subsequence of X, we need to find an LCS of X and Y and for this the c table is as given in Table 6.10.

Thus the length of a longest decreasing subsequence is 2 and the longest decreasing subsequences of X have coordinates

$$(x_6, y_2), (x_7, y_3); \ (x_4, y_2), (x_7, y_3); \ (x_2, y_3), (x_5, y_4);$$
$$(x_4, y_2), (x_5, y_4); \ (x_2, y_3), (x_3, y_5); \ (x_1, y_4), (x_3, y_5),$$

so that the subsequences are x_6, x_7; x_4, x_7; x_2, x_5; x_4, x_5; x_2, x_3; x_1, x_3 or that the subsequences are y_2, y_3; y_2, y_3; y_3, y_4; y_2, y_4; y_3, y_5; y_4, y_5 or that the subsequences are 4, 3; 3, 2; 4, 2; 3, 1; 2, 1. Thus, we have five longest strictly decreasing subsequences.

6.5. Optimal Binary Search Tree

Let a finite set S of linearly ordered elements be given. Recall that a binary search tree on the set S is a binary tree such that every element that appears in the left subtree of a given node is less than the given node and every element that appears in the right subtree of the node is greater than the node. Given S, we can always construct a binary search tree starting with any element of S as a root. If the set S has n elements, since any one of these elements may be chosen as a root, several binary search trees are possible with the elements of the set S. For example, if $S = \{a, b, c\}$ with $a < b < c$, we can have five binary search trees (Fig. 6.2).

If $S = \{a, b, c, d\}$ with the natural ordering, i.e., $a < b < c < d$, we have 14 binary search trees given in Figs. 6.3(a)–6.3(n) and the number of binary search trees goes on increasing with the increase in the order of S.

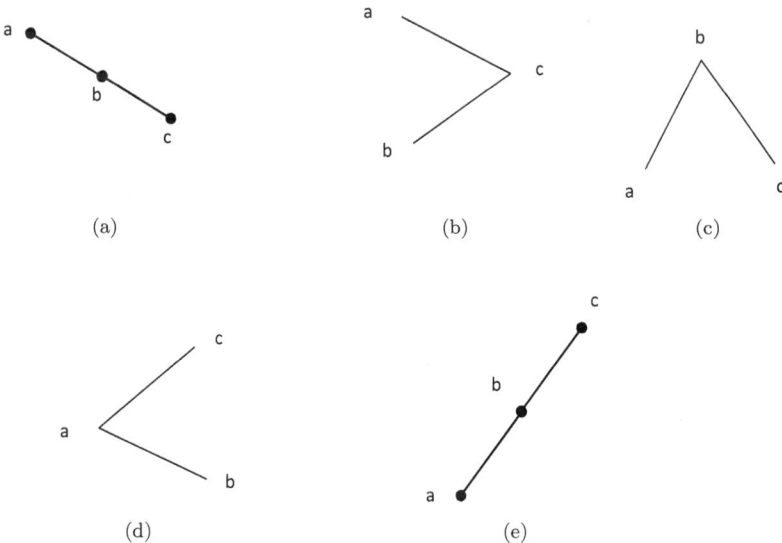

Fig. 6.2

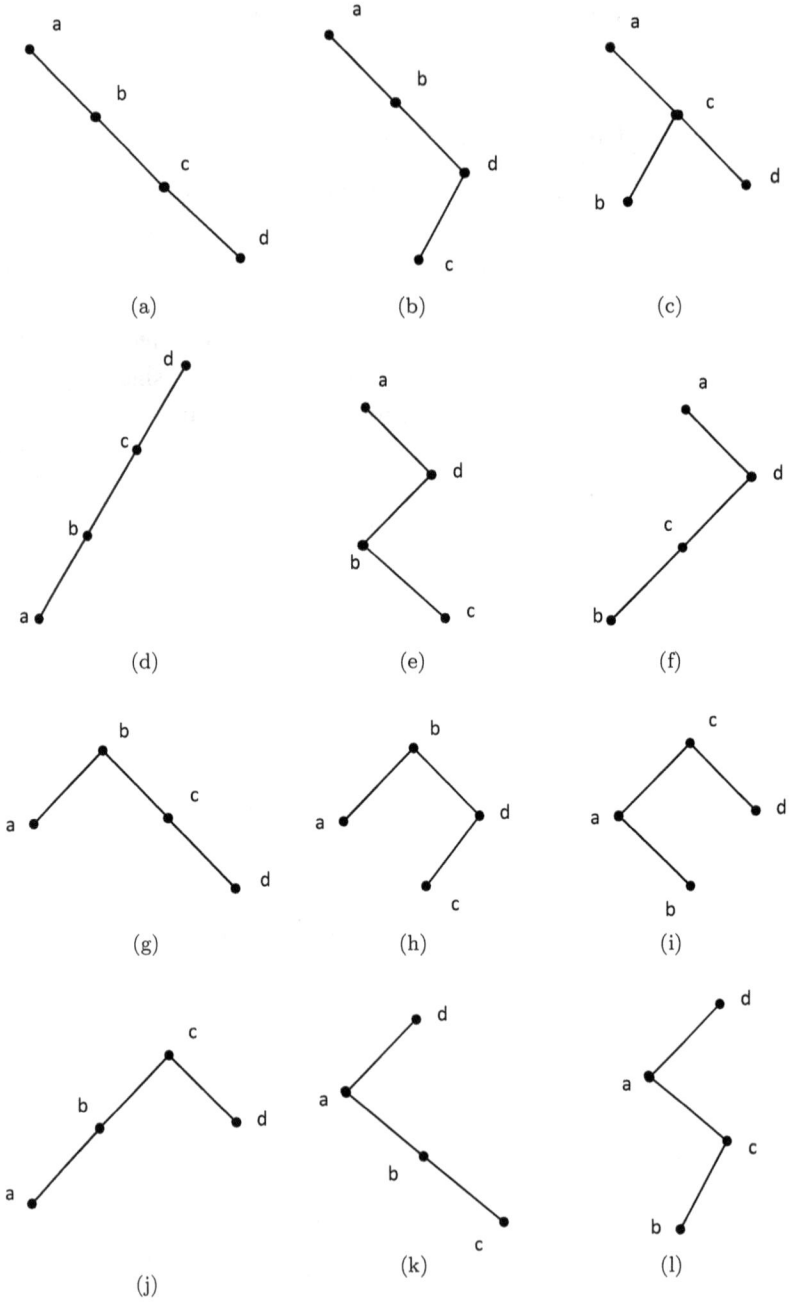

Fig. 6.3

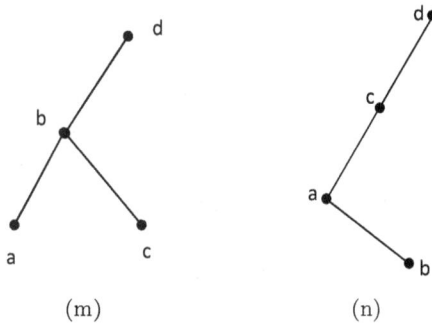

Fig. 6.3 (*Continued*)

As the name suggests, one of the main purpose of a search tree is to look for the existence and location of and otherwise of a particular element in the tree. Two other purposes of the binary search tree are (a) the insertion of an element in the tree, if it is not already there and (b) the deletion of an element from the tree. In the case of information retrieval, we need to search for a document which contains a sample collection of words. Each sample word occurs several times, i.e., it occurs with a frequency. Also, there may be several words which are in documents that are not chosen in the sample set of words. Such words also occur with some frequency and these may occur between two consecutive chosen words, i.e., are greater than one chosen word and are less than a consecutive chosen word. The chosen words may be represented by a binary search tree with particular frequencies (or probabilities) for the words. If we are looking for a word x, we start looking for it from node to node. We first compare it with the root of the tree. There are three possibilities. Either x equals the root r (say) of the tree or $x < r$ or $x > r$. If x equals r, the search for x terminates. Otherwise, if $x < r$, then we need to search for x in the left subtree for r while if $x > r$, we need to look for x in the right subtree of r. Continuing the search, there result two cases. Either x is located as an element in the left subtree of r or x is located in the right subtree of r. If one of the two things happens, we say that the search is successful and x equals one of the nodes of the tree. If neither of the two things happens, we say that search is unsuccessful and x does not appear as a node in the tree. In this case we indicate x by introducing a new leaf in the tree calling it an external node. In a binary search tree formed by the given words external nodes are so added that none of the given words is a leaf in the newly formed binary search tree, one of the given nodes which now is an internal node has only one sibling and every external node is a leaf. We call the new binary search tree arrived at a binary search tree with the given words (keys) along with the words not given (also called dummy words or keys). Thus all the given words together with the words (or collection of words) not given among the chosen words are represented by a binary search tree with the non-given words as leaves or external nodes (and indicated that way). In this tree when searching for a document/word x we come to a leaf b (i.e., a non-given word) which is a son of a given word a, the number of comparisons required is exactly equal to the number of comparisons involved for coming to a. Thus, the work in coming to b is equal to the work involved in coming to a which equals the level of b in the tree (relative to the root)-1.

While scanning a document, if a is a given chosen word which is encountered first of all, we consider a binary search tree with a as root. If $a_1, a_2, \ldots, a_n$ are the chosen words (called keys) with their frequency or probability $p(1), p(2), \ldots, p(n)$ of occurrence, we let $b_0, b_1, \ldots, b_n$ be the probable words (or collection of words) with probabilities $q(0), q(1), \ldots, q(n)$ of their occurrence, we define the **cost of the binary search tree** T by

$$\sum_{1 \leq i \leq n} p(i)\text{level}(a_i) + \sum_{0 \leq i \leq n} q(i)(\text{level}(b_i) - 1), \qquad (6.5)$$

where level(x) denotes the level of x in the binary search tree with the level of the root being 1. Thus, the level of x is the same as $1 + \text{depth of } x$. Also, since every search is either successful or unsuccessful, we must have

$$\sum_{1 \leq i \leq n} p(i) + \sum_{0 \leq i \leq n} q(i) = 1. \qquad (6.6)$$

In terms of depth of nodes in the tree T, the cost of the binary search tree T may be written as

$$\text{cost}(T) = \sum_{1 \leq i \leq n} (1 + \text{depth}_T(a_i))p(i) + \sum_{0 \leq i \leq n} \text{depth}_T(b_i)q(i).$$

The cost of the binary search tree T measures how close the document is to the document we are looking for. Therefore, smaller the cost of the binary search tree, the closer we are to the actual document we are looking for. Given words (or keys) $a_1, a_2, \ldots, a_n$ with the probabilities $p(1), p(2), \ldots, p(n)$ of occurrence and probable words (or dummy keys) $b_0, b_1, \ldots, b_n$ with probabilities $q(0), q(1), \ldots, q(n)$ a binary search tree T (say) with minimum cost, cost(T), is called an **optimal binary search tree**. Different cost binary search trees correspond to different choices of the root.

A subtree of a tree T, in general, is a subgraph of T obtained by removing some of the nodes together with the associated edges from T such that the resulting subgraph is connected (and, so, is a tree). In this section and elsewhere in the book we use a rather restrictive definition (as given below) of a subtree of a tree. If a is a node of a tree T, then a together with all its descendants in T and the edges in T connecting a and all its descendants is a subtree of T with a as root. This subtree is the one that is obtained from T by omitting all nodes that are not descendants of a and by removing all edges associated with the nodes removed. A subtree in the restrictive sense is indeed a subtree in the general sense but the converse is not always true. For example, the tree of Fig. 6.4(b) is a subtree of the tree of Fig. 6.4(a) in the general sense but is not a subtree in the restrictive sense.

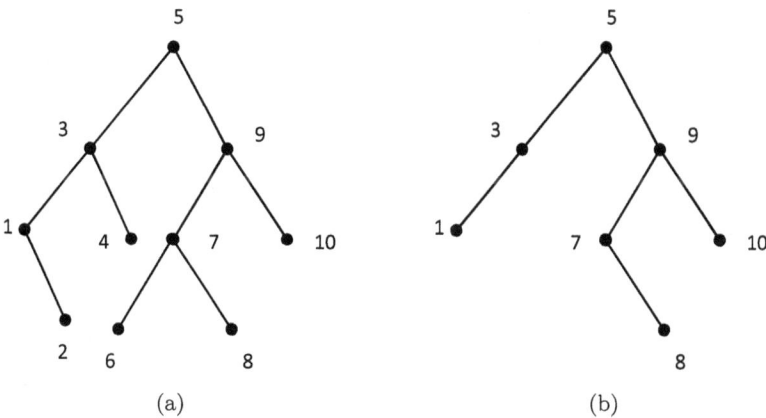

Fig. 6.4

If T is a tree with root r (say), by left subtree of T denoted by $ltree(T)$ we mean the subtree of T obtained by deleting r and all right successors of r along with the associated edges. Right subtree, $rtree(T)$, is defined similarly. We can then formalize search for an x in a binary search tree.

Algorithm 6.3. Algorithm Search (T, x)

1. T is a binary search tree with root r
2. if $x = r$, search terminates
3. if $x < r$, search $(ltree(T), x)$
4. else search $(rtree(T), x)$
5. end

Theorem 6.3. *Every subtree of a binary search tree is again a binary search tree on a certain subset of the set of keys along with corresponding dummy keys.*

Proof. Let T' be a subtree of the binary search tree T with keys $a_1, a_2, \ldots, a_n$ and dummy keys $b_0, b_1, \ldots, b_n$. If T' is a tree with root which is a left son (or sibling) of a_{j+1}, then every node in T' is $< a_{j+1}$. If a_{i+1} is a key with least index in T', then every node with key a_k in T' is such that $i + 1 \leq k < j + 1$. Consider a node in T with key a_k such that $i + 1 \leq k \leq j$. Then $a_k < a_{j+1}$ so that a_k occurs in the left subtree of the subtree of T with root a_{j+1}. T being a binary tree and T' being a left subtree of a subtree of T with root a_{j+1}, a_k must occur in T'. Hence T' is a binary search tree on the keys $a_{i+1}, \ldots, a_j$ and the corresponding dummy keys $b_i, b_{i+1}, \ldots, b_j$. We may also write $T(i, j)$ for the subtree T'.

If T' is a subtree with root which is a right son (or sibling) of a_{j-1}, then every node in T' is $> a_j$. If a_{i+1} is a key in T' with largest index, then every node with key a_k in T' is such that $j - 1 < k \leq i + 1$. Consider a node in T with key a_k such that $j \leq k \leq i+1$. Then $a_{j-1} < a_j \leq a_k$. Thus a_k belongs to the right subtree of T with root a_{j-1}. Now T being a binary search tree and T' being a right subtree of a subtree of T with root a_{j-1}, a_k is in T'. Hence T' is a binary search tree on the keys $a_j, \ldots, a_{i+1}$ and the corresponding dummy keys $b_{j-1}, b_j, \ldots, b_{i+1}$. Thus in either case T' is a binary search tree.

Suppose that T is an optimal binary search tree and T' is a subtree of T with keys $a_{i+1}, \ldots, a_j$ and dummy keys $b_i, b_{i+1}, \ldots, b_j$. If T' is not an optimal binary search tree, let T'' be an optimal binary search tree with keys on which T' is defined. Then cutting out (or removing) T' from T and pasting T'' we obtain a binary search tree T_1 with keys $a_1, a_2, \ldots, a_n$ and dummy keys $b_0, b_1, \ldots, b_n$ and $\mathrm{cost}(T_1) < \mathrm{cost}(T)$ which contradicts that T is optimal. Hence T' is optimal. We have thus proved the following result. $\square$

Theorem 6.4. *Every subtree of an optimal binary search tree is again an optimal binary search tree.*

This observation helps us in applying dynamic programming for finding the optimal cost and also an optimal binary search tree on a given set of keys and a given set of dummy keys with given frequencies (or probabilities).

A recursive characterization of optimal cost

Let i, j with $1 \le i \le j \le n$ be given. Consider the subtree $T(i, j)$ defined by the keys $a_{i+1}, \ldots, a_j$ and dummy keys $b_i, b_{i+1}, \ldots, b_j$ with their respective probabilities (frequencies). We write $c(i, j)$ for the cost of an optimal subtree of T on these keys so that $c(0, n)$ becomes the cost of an optimal binary search tree with keys $a_1, a_2, \ldots, a_n$ and dummy keys $b_0, b_1, \ldots, b_n$ with their given frequencies $p(1), p(2), \ldots, p(n)$ and $q(0), q(1), \ldots, q(n)$ respectively. As in (6.5) for T, we define the cost of $T(i, j)$ by

$$\mathrm{cost}(T(i, j)) = \sum_{i < l \le j} p(l)\mathrm{level}(a_l) + \sum_{i \le l \le j} q(i)(\mathrm{level}(b_i) - 1),$$

where $\mathrm{level}(a_l)$ and $\mathrm{level}(b_l)$ denote the level of a_l and b_l, respectively, in the binary search tree $T(i, j)$. Let a_k be the root of $T(i, j)$ and let L denote the left subtree and R the right subtree of $T(i, j)$. Then

$$
\begin{aligned}
\text{cost}(T(i, j)) = {}& p(k) + \sum_{i<l<k} p(l)\text{level}(a_l) + \sum_{k<l\leq j} p(l)\text{level}(a_l) \\
& + \sum_{i\leq l<k} q(l)(\text{level}(b_l) - 1) + \sum_{k\leq l\leq j} q(l)(\text{level}(b_l) - 1) \\
= {}& p(k) + \left\{ \sum_{i<l<k} p(l)\text{level}(a_l) + \sum_{i\leq l<k} q(l)(\text{level}(b_l) - 1) \right\} \\
& + \left\{ \sum_{k<l\leq j} p(l)\text{level}(a_l) + \sum_{k\leq l\leq j} q(l)(\text{level}(b_l) - 1) \right\}
\end{aligned}
$$

For $i \leq l < k$, $\text{level}(a_l) = 1 + \text{level}_L(a_l)$, $\text{level}(b_l) = 1 + \text{level}_L(b_l)$ and for $k < l \leq j$, $\text{level}(a_l) = 1 + \text{level}_R(a_l)$, $\text{level}(b_l) = 1 + \text{level}_R(b_l)$. Also $\text{level}(b_k) = 1 + \text{level}_R(b_k)$.

Therefore, we have

$$
\begin{aligned}
& \text{cost}(T(i, j)) \\
& = p(k) + \left\{ \sum_{i<l<k} p(l)\text{level}_L(a_l) + \sum_{i\leq l<k} q(l)(\text{level}_L(b_l) - 1) \right\} \\
& \quad + \left\{ \sum_{k<l\leq j} p(l)\text{level}_R(a_l) + \sum_{k\leq l\leq j} q(l)(\text{level}_R(b_l) - 1) \right\} \\
& \quad + \sum_{i<l<k} p(l) + \sum_{i\leq l<k} q(l) + \sum_{k<l\leq j} p(l) + \sum_{k\leq l\leq j} q(l).
\end{aligned}
$$

Let us write $w(r, s) = q(r) + \sum_{r<l\leq s}(p(l) + q(l))$ so that

$$
\sum_{i<l<k} p(l) + \sum_{i\leq l<k} q(l) = w(i, k - 1) \quad \text{and}
$$

$$
\sum_{k<l\leq j} p(l) + \sum_{k\leq l\leq j} q(l) = w(k, j).
$$

Therefore,

$$\text{cost}(T(i, j)) = p(k) + \text{cost}(L) + \text{cost}(R) + w(i, k - 1) + w(k, j).$$

(6.7)

Since different binary search trees on the keys $a_{i+1}, \ldots, a_j$ and dummy keys $b_i, b_{i+1}, \ldots, b_j$ correspond to different choices of the root a_k,

$$\begin{aligned}
c(i, j) &= \min_{i < k \le j} \{p(k) + \min \text{cost}(L) \\
&\quad + \min \text{cost}(R) + w(i, k - 1) + w(k, j)\} \\
&= \min_{i < k \le j} \{p(k) + c(i, k - 1) + c(k, j) + w(i, k - 1) + w(k, j)\} \\
&= \min_{i < k \le j} \{c(i, k - 1) + c(k, j)\} + w(i, j).
\end{aligned}$$

(6.8)

Also the root of this optimal binary search tree on $a_{i+1}, \ldots, a_j$ and $b_i, b_{i+1}, \ldots, b_j$ is a_k where k is the index in $i < k \le j$ for which the minimum in (6.8) is achieved.

For $i = 0$ and $j = n$, the optimal cost of the binary search tree is given by

$$\begin{aligned}
c(0, n) &= \min_{1 \le k \le n} \{p(k) + c(0, k - 1) + c(k, n) + w(0, k - 1) + w(k, n)\} \\
&= \min_{1 \le k \le n} \{c(0, k - 1) + c(k, n)\} + w(0, n),
\end{aligned}$$

(6.9)

and root of corresponding binary search tree is a_k, where k is the index in $1 \le k \le n$ for which the minimum in (6.9) is achieved,

Observe that for finding $c(0, n)$, we need to compute

$$c(0, 0), c(1, n); c(0, 1), c(2, n); c(0, 2), c(3, n); \ldots;$$

$$c(0, n - 1), c(n, n).$$

Similarly for obtaining an optimal binary search tree, we need to obtain vertices of various optimal binary search subtrees. To apply dynamic programming to this problem, we compute $c(i, j)$ and the corresponding k which we denote by $r(i, j)$ (for the root of the subtree) and record these in a table. The computations made in one case may be needed in the next computation. The method is best explained through a couple of examples.

Before considering these examples, we record the following observations about $c(i, j)$, $w(i, j)$ and $r(i, j)$ which are helpful in later computations.

Remarks.

1. Since $T(i, i)$ is a tree defined by the keys $a_{i+1}, \ldots, a_i$ there are no internal nodes and, therefore, $c(i, i) = 0$ and $r(i, i) = 0$ for all $i, 0 \le i \le n$.
2. $c(i, i + 1) = w(i, i + 1), r(i, i + 1) = i + 1$ for all $i, 0 \le i < n$.
3. $w(i, i + 1) = q(i) + p(i + 1) + q(i + 1)$ for all $i, 0 \le i < n$.
4. $w(i, j) = p(j) + q(j) + w(i, j - 1)$ for all $i, j, 0 \le i < j \le n$.

The above observations give the initial step in the following algorithm for finding optimal cost $c(i, j)$ of an optimal binary search tree $T(i, j)$.

Algorithm 6.4. Algorithm Optimal Cost $BST(p, q, n)$

---

1. given keys $a_1 < a_2 < \cdots < a_n$ with probabilities $p(i)$, $1 \le i \le n$, and probabilities $q(i)$, $0 \le i \le n$ of dummy keys. We need to compute the optimal cost $c(i, j)$ of the binary search subtree $T(i, j)$ and the root $r(i, j)$ of the BST T. For this auxiliary sums $w(i, j)$ are first computed
2. initially $w(i, i) = q(i), c(i, i) = 0, r(i, i) = 0$ for all $i, 0 \le i \le n$ and for all $i, 0 \le i < n$, $w(i, i+1) = q(i)+p(i+1)+q(i+1)$, $c(i, i+1) = w(i, i + 1), r(i, i + 1) = i + 1$
3. for $0 \le i < j \le n$, set $w(i, j) = p(j) + q(j) + w(i, j - 1)$
4. for $0 \le i < j \le n$, choose k with $i < k \le j$ which minimizes $p(l) + c(i, l - 1) + c(l, j) + w(i, j), i < l \le j$
5. set $c(i, j) = p(k) + c(i, k - 1) + c(k, j) + w(i, j), r(i, j) = k$
6. return $c(0, n), r(0, n)$

Observe that $w(i, j)$ are recursively determined by using only a finite number of additions (no comparisons are involved). Also $p(l) + c(i, l - 1) + c(l, j) + w(i, j)$ are computed using only additions. However, for finding k, which is finally the root $r(i, j)$ we need to compare $j - i$ quantities. Hence the running time for the computation of $c(0, n)$ and the corresponding $OBST$ is $\sum_{1 \le j \le n} \sum_{0 \le i < j} (j - i) = \sum_{1 \le j \le n} \frac{j(j+1)}{2}$ which is $O(n^3)$ (by ignoring the time involved for additions of certain numbers).

Example 6.9. Find the optimal cost and an optimal binary search tree with keys a_1, a_2, a_3, a_4 and dummy keys b_0, b_1, b_2, b_3, b_4 where the frequencies $p(i), q(j)$ are $p(1:4) = (3, 2, 2, 1); q(0:4) = (1, 3, 2, 1, 1)$.

Solution. We first compute $c(i, j)$, $w(i, j)$ and $r(i, j)$ in a step-by-step manner as follows (for $j - i = 1, 2, 3, 4$):

(i) $w(0, 1) = p(1) + q(1) + q(0) = 3 + 3 + 1 = 7;$

$c(0, 1) = w(0, 1) = 7; \quad r(0, 1) = 1.$

(ii) $w(1, 2) = q(1) + p(2) + q(2) = 3 + 2 + 2 = 7;$

$c(1, 2) = w(1, 2) = 7; \quad r(1, 2) = 2.$

(iii) $w(2, 3) = q(2) + p(3) + q(3) = 2 + 2 + 1 = 5;$

$c(2, 3) = w(2, 3) = 5; \quad r(2, 3) = 3.$

(iv) $w(3, 4) = q(3) + p(4) + q(4) = 1 + 1 + 1 = 3;$

$c(3, 4) = w(3, 4) = 3; \quad r(3, 4) = 4.$

(v) $w(0, 2) = p(2) + q(2) + w(0, 1) = 2 + 2 + 7 = 11;$

$c(0, 2) = w(0, 2) + \min_{0 < k \leq 2} \{c(0, 0) + c(1, 2), c(0, 1) + c(2, 2)\}$

$\qquad = 11 + \min\{7, 7\} = 18;$

$r(0, 2) = 1 \text{ or } 2.$

(vi) $w(1, 3) = p(3) + q(3) + w(1, 2) = 2 + 1 + 7 = 10;$

$c(1, 3) = w(1, 3) + \min_{1 < k \leq 3} \{c(1, 1) + c(2, 3), c(1, 2) + c(3, 3)\}$

$\qquad = 10 + \min\{5, 7\} = 15;$

$r(1, 3) = 2.$

(vii) $w(2, 4) = p(4) + q(4) + w(2, 3) = 1 + 1 + 5 = 7;$

$c(2, 4) = w(2, 4) + \min_{2 < k \leq 4} \{c(2, 2) + c(3, 4), c(2, 3) + c(4, 4)\}$

$\qquad = 7 + \min\{3, 5\} = 10;$

$r(2, 4) = 3.$

(viii) $w(0, 3) = p(3) + q(3) + w(0, 2) = 2 + 1 + 11 = 14;$

$$c(0, 3) = w(0, 3) + \min_{0 < k \leq 3} \{c(0, 0) + c(1, 3), c(0, 1)$$

$$+ c(2, 3), c(0, 2) + c(3, 3)\}$$

$$= 14 + \min\{15, 7 + 5, 18\} = 26;$$

$r(0, 3) = 2.$

(ix) $w(1, 4) = p(4) + q(4) + w(1, 3) = 1 + 1 + 10 = 12;$

$$c(1, 4) = w(1, 4) + \min_{1 < k \leq 4} \{c(1, 1) + c(2, 4), c(1, 2)$$

$$+ c(3, 4), c(1, 3) + c(4, 4)\}$$

$$= 12 + \min\{10, 7 + 3, 15\} = 22;$$

$r(1, 4) = 2$ or $3.$

(x) $w(0, 4) = p(4) + q(4) + w(0, 3) = 1 + 1 + 14 = 16;$

$$c(0, 4) = w(0, 4) + \min_{0 < k \leq 4} \{c(0, 0) + c(1, 4), c(0, 1)$$

$$+ c(2, 4), c(0, 2) + c(3, 4), c(0, 3) + c(4, 4)\}$$

$$= 16 + \min\{22, 7 + 10, 18 + 3, 26\} = 33;$$

$r(0, 4) = 2.$

Hence optimal cost of an optimal binary search tree ($OBST$) is 33. In case we need the cost in terms of probability, we need to divide the $p(i)$'s and $q(j)$'s by $3 + 2 + 2 + 1 + 1 + 3 + 2 + 1 + 1 = 16$ and then the optimal cost is $\frac{33}{16} = 2.0625$. Also root of any optimal binary search tree is a_2 with $T(0, 1)$ as its left subtree and $T(2, 4)$ as its right subtree. The left subtree has only one node namely a_1. The $OBST$ on a_2, a_3, a_4 has root a_3 with left subtree having only one node a_2 and the right subtree is a tree on a_3, a_4. Hence binary search tree is as given in Fig. 6.5.

Observe that cost of this binary search tree is

$$p(2) + 2 \times p(1) + 2 \times q(0) + 2 \times q(1) + 2 \times p(3) + 2 \times q(2)$$

$$+ 3 \times p(4) + 3 \times q(3) + 3 \times q(4)$$

$$= 2 + 3 \times 2 + 1 \times 2 + 3 \times 2 + 2 \times 2 + 2 \times 2 + 1 \times 3 + 1 \times 3 + 1 \times 3$$

$$= 2 + 6 + 2 + 6 + 4 + 4 + 3 + 3 + 3 = 33$$

as it should be.

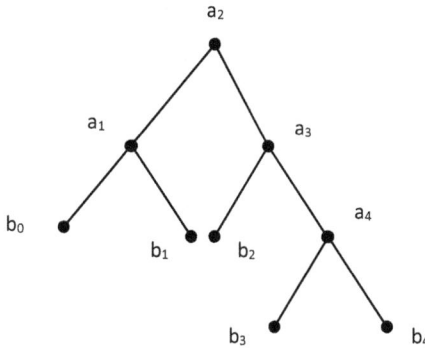

Fig. 6.5

Example 6.10. Find the optimal cost and an optimal binary search tree on the keys a_1, a_2, a_3, a_4, a_5 and the dummy keys $b_0, b_1, b_2, b_3, b_4, b_5$ where the frequencies of the a's and b's are given by $p(1 : 5) = (3, 3, 2, 1, 1)$ and $q(0 : 5) = (2, 3, 3, 1, 1, 1)$.

Solution. Making a repeated use of the remark above Example 6.9 we compute the $c(i, j)$, $w(i, j)$ and $r(i, j)$ for $j - i = 1, 2, 3, 4, 5$ as follows:

(i) $w(0, 1) = p(1) + q(1) + q(0) = 3 + 3 + 2 = 8$;

$\quad c(0, 1) = w(0, 1) = 8$; $\quad r(0, 1) = 1$.

(ii) $w(1, 2) = p(2) + q(2) + q(1) = 3 + 3 + 3 = 9$;

$\quad c(1, 2) = w(1, 2) = 9$; $\quad r(1, 2) = 2$.

(iii) $w(2, 3) = p(3) + q(3) + q(2) = 2 + 1 + 3 = 6$;

$\quad c(2, 3) = w(2, 3) = 6$; $\quad r(2, 3) = 3$.

(iv) $w(3, 4) = p(4) + q(4) + q(3) = 1 + 1 + 1 = 3$;

$\quad c(3, 4) = w(3, 4) = 3$; $\quad r(3, 4) = 4$.

(v) $w(4, 5) = p(5) + q(5) + q(4) = 1 + 1 + 1 = 3$;

$\quad c(4, 5) = w(4, 5) = 3$; $\quad r(4, 5) = 5$.

(vi) $w(0, 2) = p(2) + q(2) + w(0, 1) = 3 + 3 + 8 = 14;$

$\quad c(0, 2) = w(0, 2) + \min_{0 < k \le 2} \{c(0, 0) + c(1, 2), c(0, 1) + c(2, 2)\};$

$\quad\quad = 14 + \min\{9, 8\} = 22;$

$\quad r(0, 2) = 2.$

(vii) $w(1, 3) = p(3) + q(3) + w(1, 2) = 2 + 1 + 9 = 12;$

$\quad c(1, 3) = w(1, 3) + \min_{1 < k \le 3} \{c(1, 1) + c(2, 3), c(1, 2) + c(3, 3)\}$

$\quad\quad = 12 + \min\{6, 9\} = 18;$

$\quad r(1, 3) = 2.$

(viii) $w(2, 4) = p(4) + q(4) + w(2, 3) = 1 + 1 + 6 = 8;$

$\quad c(2, 4) = w(2, 4) + \min_{2 < k \le 4} \{c(2, 2) + c(3, 4), c(2, 3) + c(4, 4)\}$

$\quad\quad = 8 + \min\{3, 6\} = 11;$

$\quad r(2, 4) = 3.$

(ix) $w(3, 5) = p(5) + q(5) + w(3, 4) = 1 + 1 + 3 = 5;$

$\quad c(3, 5) = w(3, 5) + \min_{3 < k \le 5} \{c(3, 3) + c(4, 5), c(3, 4) + c(5, 5)\}$

$\quad\quad = 5 + \min\{3, 3\} = 8;$

$\quad r(3, 5) = 4 \text{ or } 5.$

(x) $w(0, 3) = p(3) + q(3) + w(0, 2) = 2 + 1 + 14 = 17;$

$\quad c(0, 3) = w(0, 3) + \min_{0 < k \le 3} \{c(0, 0) + c(1, 3), c(0, 1)$

$\quad\quad\quad + c(2, 3), c(0, 2) + c(3, 3)\}$

$\quad\quad = 17 + \min\{18, 8 + 6, 22\} = 31;$

$\quad r(0, 3) = 2.$

(xi) $w(1, 4) = p(4) + q(4) + w(1, 3) = 1 + 1 + 12 = 14;$

$c(1, 4) = w(1, 4) + \min_{1 < k \leq 4} \{c(1, 1) + c(2, 4), c(1, 2)$

$\qquad + c(3, 4), c(1, 3) + c(4, 4)\}$

$\qquad = 14 + \min\{11, 9 + 3, 18\} = 25;$

$r(1, 4) = 2.$

(xii) $w(2, 5) = p(5) + q(5) + w(2, 4) = 1 + 1 + 8 = 10;$

$c(2, 5) = w(2, 5) + \min_{2 < k \leq 5} \{c(2, 2) + c(3, 5), c(2, 3)$

$\qquad + c(4, 5), c(2, 4) + c(5, 5)\}$

$\qquad = 10 + \min\{8, 6 + 3, 11\} = 18;$

$r(2, 5) = 3.$

(xiii) $w(0, 4) = p(4) + q(4) + w(0, 3) = 1 + 1 + 17 = 19$

$c(0, 4) = w(0, 4) + \min_{0 < k \leq 4} \{c(0, 0) + c(1, 4), c(0, 1)$

$\qquad + c(2, 4), c(0, 2) + c(3, 4), c(0, 3) + c(4, 4)\}$

$\qquad = 19 + \min\{25, 8 + 11, 22 + 3, 31\} = 38;$

$r(0, 4) = 2.$

(xiv) $w(1, 5) = p(5) + q(5) + w(1, 4) = 1 + 1 + 14 = 16$

$c(1, 5) = w(1, 5) + \min_{1 < k \leq 5} \{c(1, 1) + c(2, 5), c(1, 2)$

$\qquad + c(3, 5), c(1, 3) + c(4, 5), c(1, 4) + c(5, 5)\}$

$\qquad = 16 + \min\{18, 9 + 8, 18 + 3, 25\} = 33;$

$r(1, 5) = 3.$

(xv) $w(0, 5) = p(5) + q(5) + w(0, 4) = 1 + 1 + 19 = 21$

$c(0, 5) = w(0, 5) + \min_{0 < k \leq 5} \{c(0, 0) + c(1, 5), c(0, 1)$

$\qquad + c(2, 5), c(0, 2) + c(3, 5), c(0, 3)$

$\qquad + c(4, 5), c(0, 4) + c(5, 5)\}$

$\qquad = 21 + \min\{33, 8 + 18, 22 + 8, 31 + 3, 38\} = 47;$

$r(0, 5) = 2.$

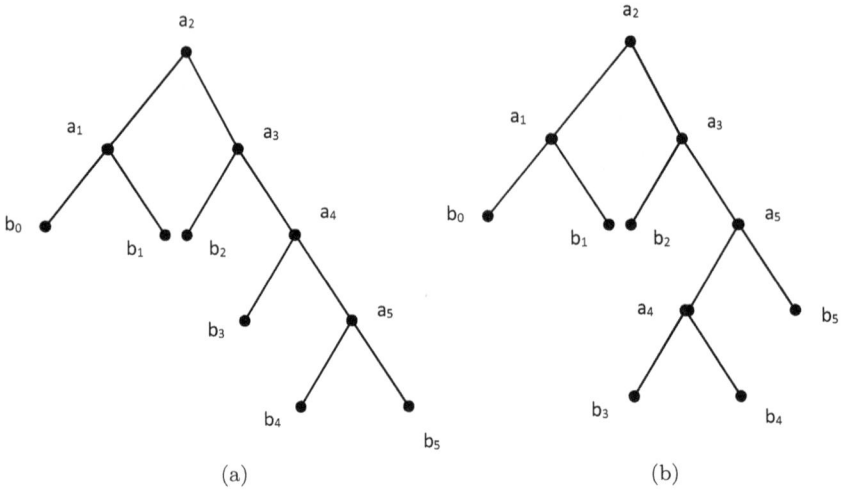

Fig. 6.6

Hence the optimal cost is 47 and a_2 is root of an optimal binary search tree. Also, the left subtree of an $OBST$ is $T(0, 1)$ consisting of a single node a_1 and the right subtree is $T(2, 5)$. The left subtree of $OBST$ $T(2, 5)$ consists of no node, has a_3 as its root and the right subtree is $T(3, 5)$. The $OBST$ $T(3, 5)$ has either root a_4 or a_5 and so there are two choices for this subtree. Hence there are two optimal binary search trees in this case and are given in Figs. 6.6(a) and 6.6(b).

Exercise 6.1. Verify that each of the trees in Figs. 6.6 (a) and 6.6 (b) is an optimal binary search tree on a's and b's with frequencies as in Example 6.10.

6.6. Optimal Triangulation of a Polygon

Let a polygon of n sides with vertices $v_0, v_1, \ldots, v_{n-1}$ in a cyclic order placed in clockwise direction in a plane be given. The problem is to draw chords (v_i, v_j), $j \neq i + 1$, $0 \leq i < j \leq n - 1$, in such a way that no two chords intersect each other except, if necessary, at vertices of the polygon and the polygon area is divided into triangles. Sum of the lengths (or weights assigned to) of all such chords is called the **cost of triangulation of the polygon**. A triangulation of the polygon for which this sum of

---

Algorithm 6.5. Algorithm Optimal Cost $c(0, n)$ for Binary Search Tree

---

1. T is a binary search tree on keys $a_1, a_2, \ldots, a_n$ with probabilities $p(1), p(2), \ldots, p(n)$ and dummy keys $b_0, b_1, \ldots, b_n$ with
   probabilities $q(0), q(1), \ldots, q(n)$. $w(i, j) = \sum_{i < l \leq j} p(l) +$
   $\sum_{i \leq l \leq j} q(l)$, $c(i, j)$ is the optimal cost of the sub tree $T(i, j)$
2. for $0 \leq i \leq j \leq n$, compute $w(i, j)$: for $0 \leq i \leq n$, $w(i, i) = q(i)$
   for $0 \leq i < n$, $w(i, i + 1) = q(i) + p(i + 1) + q(i + 1)$
   for $0 \leq i < j \leq n$, $w(i, j) = p(j) + q(j) + w(i, j - 1)$
3. for $0 \leq i \leq n$, $c(i, i) = w(i, i)$; for $0 \leq i < n$, $c(i, i + 1) = w(i, i + 1)$
4. for $0 \leq i \leq j \leq n$, compute $c(i, j) = \min_{i < k \leq j}\{c(i, k - 1) + c(k, j)\} +$
   $w(i, j)$ Then $c(0, n)$ is the optimal cost of T

---

the lengths of the chords is minimum is called **optimal triangulation**.
In case $n = 3$, the polygon is already a triangle and no chords are
necessary for its triangulation. Therefore optimal cost of triangulation is
zero. In case $n = 4$, the polygon is a quadrilateral and each diagonal
provides a triangulation. Smaller of the two diagonals provides an optimal
triangulation and its length gives the optimal cost. Thus the problem of
optimal triangulation is interesting only in the case when $n \geq 5$. By
drawing a chord of a polygon of n sides, we can divide the problem
of triangulation of the polygon into two sub-problems. We continue
the process of subdividing till we get a triangulation of the original
problem.

Given a vertex v_i, we can choose a vertex v_{i+k} (we can consider
$i + k$ reduced modulo n, if it is not already less than n) and draw
the chord (v_i, v_{i+k}) so that the given polygon is subdivided into two
polygons $(v_i, v_{i+1}, \ldots, v_{i+k})$ and $(v_{i+k}, v_{i+k+1}, \ldots, v_0, \ldots, v_i)$ which
are of length $k + 1$ and $n - k + 1$. The cost of triangulation of the
given polygon is the sum of the costs of triangulation of the two smaller
polygons plus the length of the chord (v_i, v_{i+k}). This is so because of
(v_i, v_{i+k}) being a side of each of the two smaller polygons does not
contribute to the cost of triangulation of either of the two polygons but
it does contribute as a chord to the triangulation of the given polygon. In
case costs of triangulations of the two polygons are replaced by optimal
costs of triangulations of the two polygons, we may not still get optimal

cost of triangulation of the given polygon. This is so because the cost that we obtain depends on the choice of k for a given i and then on the choice of the starting vertex v_i. For a given v_i, the number of choices of v_{i+k} is $n-3$, since the choices v_{i+1} and v_{i-1} do not lead to a subdivision of the polygon into two polygons. Thus, if for a given v_i, we take the minimum of the sum of minimal costs of the two polygons plus length of the chord (v_i, v_{i+k}) for k varying between 2 to $n-3$, we get the minimal cost of triangulation of the polygon starting with vertex v_i. Finally to get the optimal cost of triangulation of the polygon, we need to get the minimum of the optimal costs obtained when i varies from 0 to $n-1$.

In the above we are concerned with finding optimal costs of triangulation of the smaller polygons $(v_i, v_{i+1}, \ldots, v_{i+k})$ of length $k+1$ and $(v_{i+k}, v_{i+k+1}, \ldots, v_0, \ldots, v_i)$ of length $n-k+1$. Either of these two polygons is of the form $(v_i, v_{i+1}, \ldots, v_{i+s-1})$ of length s. So, we now consider triangulation of this polygon of length s. Let $c(i, s)$ denote the optimal cost of triangulation of this polygon of length s starting with the vertex v_i. For any $j \neq k$, we let $D(j, k)$ be the length of the chord (v_j, v_k). In case $k = j + 1$, (v_j, v_{j+1}) is a side of the polygon and, so, it does not contribute to cost of triangulation of the polygon. Therefore, we set $D(j, j+1) = 0$. For triangulation, choose a k with $i < i + k < i + s - 1$. Joining v_{i+k} with v_i and v_{i+s-1} leads to division of the polygon into two polygons $(v_i, v_{i+1}, \ldots, v_{i+k})$ of length $k+1$, $(v_{i+k}, v_{i+k+1}, \ldots, v_{i+s-1})$ of length $s-k$ and a triangle $(v_i, v_{i+k}, v_{i+s-1})$. Therefore, cost of triangulation of the polygon of length s starting with v_i = cost of triangulation of the polygon $(v_i, v_{i+1}, \ldots, v_{i+k})$+ cost of triangulation of the polygon $(v_{i+k}, v_{i+k+1}, \ldots, v_{i+s-1}) + D(i, i+k) + D(i+k, i+s-1)$.

Here we have not added the length $D(i, i+s-1)$ of the third side of the triangle $(v_i, v_{i+k}, v_{i+s-1})$ because it is a side of the polygon under consideration and, so, it does not contribute to the cost of triangulation. Also note that we are writing $j + s$ for the sum of j and s reduced modulo n, if it is not already less than n. Then

$$c(i, s) = \min_{1 \leq k \leq s-2} \{c(i, k+1) + c(i+k, s-k)$$
$$+ D(i, i+k) + D(i+k, s+i-1)\}. \tag{6.10}$$

Diagrammatically, the division of the polygon is shown in Fig. 6.7.

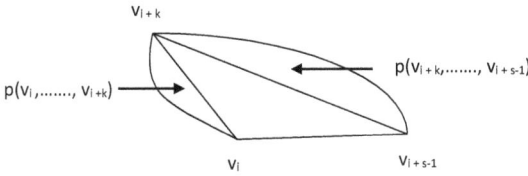

Fig. 6.7

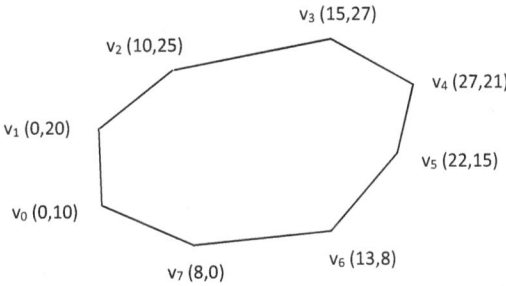

Fig. 6.8

Observe that $c(i, s) = 0$ for $s = 2$, $c(i, s) = D(i, i+2)$ for $s = 3$ and, therefore, we need to compute $c(i, s)$ for all s, $4 \leq s \leq n$ and all i, $0 \leq i \leq n-1$. It is clear from (6.10) that for the computation of $c(i, s)$ we need $c(j, t)$ for $t < s$. Therefore, the $c(i, s)$ computed are recorded in a table for use in the next $c(i, s)$ etc. Also, we need to compute the lengths of all chords (v_i, v_j). Although no $D(i, j)$ is needed for the computation of other $D(r, t)$, these need to be computed and recorded in a table for convenience of use in the computation of $c(i, s)$. Once all $c(i, s)$ for the values of i and s as mentioned have been computed, we need to look for the least among $c(i, n)$, $0 \leq i \leq n-1$ which will be the optimal cost of triangulation of the given polygon. In order to get corresponding triangulation, we let $r(i, s)$ be the value of k for which the right-hand side of (6.9) is the minimum and record these also in the table for $c(i, s)$.

We illustrate the above procedure through an example.

Example 6.11. Find the optimal cost and also the corresponding triangulation of the polygon of length 8 on vertices $v_0, v_1, \ldots, v_7$ as given in Fig. 6.8 with the Cartesian coordinates of the vertices as given. The cost is to be determined in terms of Euclidean distances between pairs of points.

Solution. First we compute $D(i, j)$ keeping in mind that $D(i, j) = D(j, i)$. Thus

$$D(0, 2) = \sqrt{100 + 256} = \sqrt{356} = 18.867,$$

$$D(0, 3) = \sqrt{225 + 289} = \sqrt{514} = 22.671,$$

$$D(0, 4) = \sqrt{729 + 121} = \sqrt{850} = 29.154,$$

$$D(0, 5) = \sqrt{484 + 25} = \sqrt{509} = 22.561,$$

$$D(0, 6) = \sqrt{169 + 4} = \sqrt{173} = 13.152,$$

$$D(1, 3) = \sqrt{225 + 49} = \sqrt{274} = 16.552,$$

$$D(1, 4) = \sqrt{729 + 1} = \sqrt{730} = 27.018,$$

$$D(1, 5) = \sqrt{484 + 25} = \sqrt{509} = 22.561,$$

$$D(1, 6) = \sqrt{169 + 144} = \sqrt{313} = 17.691,$$

$$D(1, 7) = \sqrt{64 + 400} = \sqrt{464} = 21.540,$$

$$D(2, 4) = \sqrt{289 + 25} = \sqrt{314} = 17.720,$$

$$D(2, 5) = \sqrt{144 + 121} = \sqrt{265} = 16.278,$$

$$D(2, 6) = \sqrt{9 + 324} = \sqrt{333} = 18.248,$$

$$D(2, 7) = \sqrt{4 + 676} = \sqrt{680} = 26.076,$$

$$D(3, 5) = \sqrt{49 + 144} = \sqrt{193} = 12.892,$$

$$D(3, 6) = \sqrt{4 + 361} = \sqrt{365} = 19.104,$$

$$D(3, 7) = \sqrt{49 + 729} = \sqrt{778} = 27.892,$$

$$D(4, 6) = \sqrt{196 + 169} = \sqrt{365} = 19.104,$$

$$D(4, 7) = \sqrt{361 + 441} = \sqrt{802} = 28.319,$$

$$D(5, 7) = \sqrt{196 + 225} = \sqrt{421} = 20.518,$$

and $D(i, i + 1) = 0$ for all i.

Next we compute $c(i, s)$ for $s \geq 4$ and all i, $0 \leq i \leq 7$. Now, using (6.10)

(i) $c(0, 4) = \min_{1 \leq k \leq 2} \{c(0, 2) + c(1, 3) + D(0, 1)$

$\qquad + D(1, 3), c(0, 3) + c(2, 2) + D(0, 2) + D(2, 3)\}$

$\qquad = \min\{D(1, 3), D(0, 2)\} = \min\{16.552, 18.867\} = 16.552,$

$r(0, 4) = 1.$

$c(1, 4) = \min\{c(1, 2) + c(2, 3) + D(1, 2) + D(2, 4), c(1, 3)$

$\qquad + c(3, 2) + D(1, 3) + D(3, 4)\}$

$\qquad = \min\{D(2, 4), D(1, 3)\} = \min\{17.720, 16.552\} = 16.552,$

$r(1, 4) = 2.$

$c(2, 4) = \min\{D(2, 3) + D(3, 5), D(2, 4) + D(4, 5)\}$

$\qquad = \min\{D(3, 5), D(2, 4)\} = \min\{13.892, 17.720\} = 13.892,$

$r(2, 4) = 1.$

$c(3, 4) = \min\{D(3, 4) + D(4, 6), D(3, 5) + D(5, 6)\}$

$\qquad = \min\{D(4, 6), D(3, 5)\} = \min\{19.104, 13.892\} = 13.892,$

$r(3, 4) = 2.$

$c(4, 4) = \min\{D(4, 5) + D(5, 7), D(4, 6) + D(6, 7)\}$

$\qquad = \min\{D(5, 7), D(4, 6)\} = \min\{20.518, 19.104\} = 19.104,$

$r(4, 4) = 2.$

$c(5, 4) = \min\{D(5, 6) + D(6, 0), D(5, 7) + D(7, 0)\}$

$\qquad = \min\{D(6, 0), D(5, 7)\} = \min\{13.152, 20.518\} = 13.152,$

$r(5, 4) = 1.$

$c(6, 4) = \min\{D(6, 7) + D(7, 1), D(6, 0) + D(0, 1)\}$

$\qquad = \min\{D(7, 1), D(6, 0)\} = \min\{21.540, 13.152\} = 13.152,$

$r(6, 4) = 2.$

$c(7, 4) = \min\{D(7, 0) + D(0, 2), D(7, 1) + D(1, 2)\}$

$\qquad = \min\{D(0, 2), D(7, 1)\} = \min\{18.867, 21.540\} = 18.867,$

$r(7, 4) = 1.$

(ii) $c(0, 5) = \min\{c(0, 2) + c(1, 4) + D(0, 1) + D(1, 4), c(0, 3)$
$$+ c(2, 3) + D(0, 2) + D(2, 4), c(0, 4) + c(3, 2)$$
$$+ D(0, 3) + D(3, 4)\}$$
$$= \min\{c(1, 4) + D(1, 4), D(0, 2)$$
$$+ D(2, 4), c(0, 4) + D(0, 3)\}$$
$$= \min\{16.552 + 27.018, 18.867$$
$$+ 17.720, 16.552 + 22.671\} = 36.587,$$

$r(0, 5) = 2.$

$c(1, 5) = \min\{c(1, 2) + c(2, 4) + D(1, 2) + D(2, 5), c(1, 3)$
$$+ c(3, 3) + D(1, 3) + D(3, 5), c(1, 4\} + c(4, 2)$$
$$+ D(1, 4) + D(4, 5)\}$$
$$= \min\{c(2, 4) + D(2, 5), D(1, 3)$$
$$+ D(3, 5), c(1, 4) + D(1, 4)\}$$
$$= \min\{13.892 + 16.278, 16.552 + 13.892, 16.552$$
$$+ 27.018\} = 30.170,$$

$r(1, 5) = 1.$

$c(2, 5) = \min\{c(2, 2) + c(3, 4) + D(2, 3) + D(3, 6), c(2, 3)$
$$+ c(4, 3) + D(2, 4) + D(4, 6), c(2, 4\} + c(5, 2)$$
$$+ D(2, 5) + D(5, 6)\}$$
$$= \min\{c(3, 4) + D(3, 6), D(2, 4)$$
$$+ D(4, 6), c(2, 4) + D(2, 5)\}$$
$$= \min\{13.892 + 19.104, 17.720 + 19.104, 13.892$$
$$+ 16.278\} = 30.170,$$

$r(2, 5) = 3.$

$c(3, 5) = \min\{c(3, 2) + c(4, 4) + D(3, 4) + D(4, 7), c(3, 3)$
$$+ c(5, 3) + D(3, 5) + D(5, 7), c(3, 4\} + c(6, 2)$$
$$+ D(3, 6) + D(6, 7)\}$$
$$= \min\{c(4, 4) + D(4, 7), D(3, 5)$$
$$+ D(5, 7), c(3, 4) + D(3, 6)\}$$

$$= \min\{19.104 + 28.319, 13.892 + 20.518, 13.892$$
$$+ 19.104\} = 32.996,$$

$r(3, 5) = 3.$

$$c(4, 5) = \min\{c(4, 2) + c(5, 4) + D(4, 5) + D(5, 0), c(4, 3)$$
$$+ c(6, 3) + D(4, 6) + D(6, 0), c(4, 4) + c(7, 2)$$
$$+ D(4, 7) + D(7, 0)\}$$
$$= \min\{c(5, 4) + D(5, 0), D(4, 6) + D(6, 0), c(4, 4)$$
$$+ D(4, 7)\}$$
$$= \min\{13.152 + 22.561, 19.104 + 13.152, 19.104$$
$$+ 28.319\} = 32.256,$$

$r(4, 5) = 2.$

$$c(5, 5) = \min\{c(5, 2) + c(6, 4) + D(5, 6) + D(6, 1), c(5, 3)$$
$$+ c(7, 3) + D(5, 7) + D(7, 1), c(5, 4) + c(0, 2)$$
$$+ D(5, 0) + D(0, 1)\}$$
$$= \min\{c(6, 4) + D(6, 1), D(5, 7)$$
$$+ D(7, 1), c(5, 4) + D(5, 0)\}$$
$$= \min\{13.152 + 17.691, 20.518 + 21.540, 13.152$$
$$+ 22.561\} = 30.843,$$

$r(5, 5) = 1.$

$$c(6, 5) = \min\{c(6, 2) + c(7, 4) + D(6, 7) + D(7, 2), c(6, 3)$$
$$+ c(0, 3) + D(6, 0) + D(0, 2), c(6, 4) + c(1, 2)$$
$$+ D(6, 1) + D(1, 2)\}$$
$$= \min\{c(7, 4) + D(7, 2), D(6, 0)$$
$$+ D(0, 2), c(6, 4) + D(6, 1)\}$$
$$= \min\{18.867 + 26.076, 13.152 + 18.867, 13.152$$
$$+ 17.691\} = 30.843,$$

$r(6, 5) = 3.$

$$c(7, 5) = \min\{c(7, 2) + c(0, 4) + D(7, 0) + D(0, 3), c(7, 3)$$
$$+ c(1, 3) + D(7, 1) + D(1, 3), c(7, 4) + c(2, 2)$$
$$+ D(7, 2) + D(2, 3)\}$$
$$= \min\{c(0, 4) + D(0, 3), D(7, 1)$$
$$+ D(1, 3), c(7, 4) + D(7, 2)\}$$
$$= \min\{16.552 + 22.671, 21.540 + 16.552, 18.867$$
$$+ 26.076\} = 38.092,$$

$$r(7, 5) = 2.$$

(iii) $c(0, 6) = \min\{c(0, 2) + c(1, 5) + D(0, 1) + D(1, 5), c(0, 3)$
$$+ c(2, 4) + D(0, 2) + D(2, 5), c(0, 4) + c(3, 3)$$
$$+ D(0, 3) + D(3, 5), c(0, 5) + c(4, 2)$$
$$+ D(0, 4) + D(4, 5)\}$$
$$= \min\{c(1, 5) + D(1, 5), c(2, 4) + D(0, 2)$$
$$+ D(2, 5), c(0, 4) + D(0, 3) + D(3, 5), c(0, 5)$$
$$+ D(0, 4)\}$$
$$= \min\{30.170 + 22.561, 13.892 + 18.867 + 16.278, 16.552$$
$$+ 22.671 + 13.892, 36.587 + 29.154\}$$
$$= 49.037,$$

$$r(0, 6) = 2.$$

$$c(1, 6) = \min\{c(1, 2) + c(2, 5) + D(1, 2) + D(2, 6), c(1, 3)$$
$$+ c(3, 4) + D(1, 3) + D(3, 6), c(1, 4) + c(4, 3)$$
$$+ D(1, 4) + D(4, 6), c(1, 5) + c(5, 2)$$
$$+ D(1, 5) + D(5, 6)\}$$
$$= \min\{c(2, 5) + D(2, 6), c(3, 4) + D(1, 3)$$
$$+ D(3, 6), c(1, 4) + D(1, 4) + D(4, 6), c(1, 5) + D(1, 5)\}$$

$$= \min\{30.170 + 18.248, 13.892 + 16.552 + 19.104, 16.552$$
$$+ 27.018 + 19.104, 30.170 + 22.561\}$$
$$= 48.418,$$

$r(1, 6) = 1.$

$$c(2, 6) = \min\{c(2, 2) + c(3, 5) + D(2, 3) + D(3, 7), c(2, 3)$$
$$+ c(4, 4) + D(2, 4) + D(4, 7), c(2, 4) + c(5, 3)$$
$$+ D(2, 5) + D(5, 7), c(2, 5) + c(6, 2)$$
$$+ D(2, 6) + D(6, 7)\}$$
$$= \min\{c(3, 5) + D(3, 7), c(4, 4) + D(2, 4)$$
$$+ D(4, 7), c(2, 4) + D(2, 5) + D(5, 7), c(2, 5) + D(2, 6)\}$$
$$= \min\{32.996 + 27.892, 19.104 + 17.720$$
$$+ 28.319, 13.892 + 16.278 + 20.518, 30.170 + 18.248\}$$
$$= 48.418,$$

$r(2, 6) = 4.$

$$c(3, 6) = \min\{c(3, 2) + c(4, 5) + D(3, 4) + D(4, 0), c(3, 3)$$
$$+ c(5, 4) + D(3, 5) + D(5, 0), c(3, 4) + c(6, 3)$$
$$+ D(3, 6) + D(6, 0), c(3, 5) + c(7, 2)$$
$$+ D(3, 7) + D(7, 0)\}$$
$$= \min\{c(4, 5) + D(4, 0), c(5, 4) + D(3, 5)$$
$$+ D(5, 0), c(3, 4) + D(3, 6) + D(6, 0), c(3, 5) + D(3, 7)\}$$
$$= \min\{32.256 + 29.15, 13.152 + 13.882 + 22.561, 13.892$$
$$+ 19.104 + 13.152, 32.996 + 27.892\}$$
$$= 46.148,$$

$r(3, 6) = 3.$

$$c(4, 6) = \min\{c(4, 2) + c(5, 5) + D(4, 5) + D(5, 1), c(4, 3)$$
$$+ c(6, 4) + D(4, 6) + D(6, 1), c(4, 4) + c(7, 3)$$
$$+ D(4, 7) + D(7, 1), c(4, 5) + c(0, 2)$$
$$+ D(4, 0) + D(0, 1)\}$$
$$= \min\{c(5, 5) + D(5, 1), c(6, 4) + D(4, 6)$$
$$+ D(6, 1), c(4, 4) + D(4, 7) + D(7, 1), c(4, 5) + D(4, 0)\}$$
$$= \min\{30.843 + 22.561, 13.152 + 19.104 + 17.691, 19.104$$
$$+ 28.319 + 21.540, 32.256 + 29.154\}$$
$$= 49.947,$$

$$r(4, 6) = 2.$$

$$c(5, 6) = \min\{c(5, 2) + c(6, 5) + D(5, 6) + D(6, 2), c(5, 3)$$
$$+ c(7, 4) + D(5, 7) + D(7, 2), c(5, 4) + c(0, 3)$$
$$+ D(5, 0) + D(0, 2), c(5, 5) + c(1, 2)$$
$$+ D(5, 1) + D(1, 2)\}$$
$$= \min\{c(6, 5) + D(6, 2), c(7, 4) + D(5, 7)$$
$$+ D(7, 2), c(5, 4) + D(5, 0) + D(0, 2), c(5, 5) + D(5, 1)\}$$
$$= \min\{30.843 + 18.248, 18.867 + 20.518 + 26.076, 13.152$$
$$+ 22.561 + 18.867, 30.843 + 22.561\}$$
$$= 49.091,$$

$$r(5, 6) = 1.$$

$$c(6, 6) = \min\{c(6, 2) + c(7, 5) + D(6, 7) + D(7, 3), c(6, 3)$$
$$+ c(0, 4) + D(6, 0) + D(0, 3), c(6, 4) + c(1, 3)$$
$$+ D(6, 1) + D(1, 3), c(6, 5) + c(2, 2)$$
$$+ D(6, 2) + D(2, 3)\}$$
$$= \min\{c(7, 5) + D(7, 3), c(0, 4) + D(6, 0)$$
$$+ D(0, 3), c(6, 4) + D(6, 1) + D(1, 3), c(6, 5) + D(6, 2)\}$$

$$= \min\{38.092 + 27.892, 16.552 + 13.152 + 22.671, 13.152$$
$$+ 17.691 + 16.552, 30.843 + 18.248\}$$
$$= 47.395,$$

$r(6, 6) = 3.$

$$c(7, 6) = \min\{c(7, 2) + c(0, 5) + D(7, 0) + D(0, 4), c(7, 3)$$
$$+ c(1, 4) + D(7, 1) + D(1, 4), c(7, 4) + c(2, 3)$$
$$+ D(7, 2) + D(2, 4), c(7, 5) + c(3, 2)$$
$$+ D(7, 3) + D(3, 4)\}$$
$$= \min\{c(0, 5) + D(0, 4), c(1, 4) + D(7, 1)$$
$$+ D(1, 4), c(7, 4) + D(7, 2)$$
$$+ D(2, 4), c(7, 5) + D(7, 3)\}$$
$$= \min\{36.587 + 29.154, 16.552 + 21.540$$
$$+ 27.018, 18.867 + 26.076 + 17.720, 38.092 + 27.892\}$$
$$= 62.663,$$

$r(7, 6) = 3.$

$$c(0, 7) = \min\{c(0, 2) + c(1, 6) + D(0, 1) + D(1, 6), c(0, 3)$$
$$+ c(2, 5) + D(0, 2) + D(2, 6), c(0, 4) + c(3, 4)$$
$$+ D(0, 3) + D(3, 6), c(0, 5) + c(4, 3) + D(0, 4)$$
$$+ D(4, 6), c(0, 6)$$
$$+ c(5, 2) + D(0, 5) + D(5, 6)\}$$
$$= \min\{c(1, 6) + D(1, 6), c(2, 5) + D(0, 2)$$
$$+ D(2, 6), c(0, 4) + c(3, 4) + D(0, 3) + D(3, 6), c(0, 5)$$
$$+ D(0, 4) + D(4, 6), c(0, 6) + D(0, 5)\}$$
$$= \min\{48.418 + 17.691, 30.170 + 18.867$$
$$+ 18.248, 16.552 + 13.892 + 22.671 + 19.104, 36.587$$
$$+ 29.154 + 19.104, 49.037 + 22.561\}$$
$$= 66.109,$$

$r(0, 7) = 1.$

(iv) $c(1, 7) = \min\{c(1, 2) + c(2, 6) + D(1, 2) + D(2, 7), c(1, 3)$

$\quad\quad\quad + c(3, 5) + D(1, 3) + D(3, 7), c(1, 4) + c(4, 4)$

$\quad\quad\quad + D(1, 4) + D(4, 7), c(1, 5) + c(5, 3) + D(1, 5)$

$\quad\quad\quad + D(5, 7), c(1, 6) + c(6, 2) + D(1, 6) + D(6, 7)\}$

$\quad\quad = \min\{c(2, 6) + D(2, 7), c(3, 5) + D(1, 3)$

$\quad\quad\quad + D(3, 7), c(1, 4) + c(4, 4) + D(1, 4) + D(4, 7), c(1, 5)$

$\quad\quad\quad + D(1, 5) + D(5, 7), c(1, 6) + D(1, 6)\}$

$\quad\quad = \min\{48.418 + 26.076, 32.996 + 16.552 + 27.892, 16.552$

$\quad\quad\quad + 19.104 + 27.018 + 28.319, 30.170 + 22.561$

$\quad\quad\quad + 20.518, 48.418 + 17.691\}$

$\quad\quad = 66.109,$

$r(1, 7) = 5.$

$c(2, 7) = \min\{c(2, 2) + c(3, 6) + D(2, 3) + D(3, 0), c(2, 3)$

$\quad\quad\quad + c(4, 5) + D(2, 4) + D(4, 0), c(2, 4) + c(5, 4)$

$\quad\quad\quad + D(2, 5) + D(5, 0), c(2, 5) + c(6, 3) + D(2, 6)$

$\quad\quad\quad + D(6, 0), c(2, 6) + c(7, 2) + D(2, 7) + D(7, 0)\}$

$\quad\quad = \min\{c(3, 6) + D(3, 0), c(4, 5) + D(2, 4)$

$\quad\quad\quad + D(4, 0), c(2, 4) + c(5, 4) + D(2, 5) + D(5, 0), c(2, 5)$

$\quad\quad\quad + D(2, 6) + D(6, 0), c(2, 6) + D(2, 7)\}$

$\quad\quad = \min\{46.148 + 22.671, 32.256 + 17.720 + 29.154, 13.892$

$\quad\quad\quad + 13.152 + 16.278 + 22.561, 30.170 + 18.248$

$\quad\quad\quad + 13.152, 48.418 + 26.076\}$

$\quad\quad = 61.570,$

$r(2, 7) = 4.$

$c(3, 7) = \min\{c(3, 2) + c(4, 6) + D(3, 4) + D(4, 1), c(3, 3)$

$\quad\quad\quad + c(5, 5) + D(3, 5) + D(5, 1), c(3, 4) + c(6, 4)$

$\quad\quad\quad + D(3, 6) + D(6, 1), c(3, 5) + c(7, 3) + D(3, 7)$

$\quad\quad\quad + D(7, 1), c(3, 6) + c(0, 2) + D(3, 0) + D(0, 1)\}$

$$\begin{aligned}
&= \min\{c(4,6) + D(4,1), c(5,5) + D(3,5) \\
&\quad + D(5,1), c(3,4) + c(6,4) + D(3,6) + D(6,1), c(3,5) \\
&\quad + D(3,7) + D(7,1), c(3,6) + D(3,0)\} \\
&= \min\{49.947 + 27.018, 30.843 + 13.892 + 22.561, 13.892 \\
&\quad + 13.152 + 19.104 + 17.691, 32.996 + 27.892 \\
&\quad + 21.540, 46.148 + 22.671\} \\
&= 63.839,
\end{aligned}$$

$r(3,7) = 3.$

$$\begin{aligned}
c(4,7) &= \min\{c(4,2) + c(5,6) + D(4,5) + D(5,2), c(4,3) \\
&\quad + c(6,5) + D(4,6) + D(6,2), c(4,4) + c(7,4) \\
&\quad + D(4,7) + D(7,2), c(4,5) + c(0,3) + D(4,0) \\
&\quad + D(0,2), c(4,6) + c(1,2) + D(4,1) + D(1,2)\} \\
&= \min\{c(5,6) + D(5,2), c(6,5) + D(4,6) \\
&\quad + D(6,2), c(4,4) + c(7,4) + D(4,7) + D(7,2), c(4,5) \\
&\quad + D(4,0) + D(0,2), c(4,6) + D(4,1)\} \\
&= \min\{49.091 + 16.278, 30.843 + 19.104 + 18.248, 19.104 \\
&\quad + 18.867 + 28.319 + 26.076, 32.256 + 29.154 \\
&\quad + 18.867, 49.947 + 27.018\} \\
&= 65.369,
\end{aligned}$$

$r(4,7) = 1.$

$$\begin{aligned}
c(5,7) &= \min\{c(5,2) + c(6,6) + D(5,6) + D(6,3), c(5,3) \\
&\quad + c(7,5) + D(5,7) + D(7,3), c(5,4) + c(0,4) \\
&\quad + D(5,0) + D(0,3), c(5,5) + c(1,3) + D(5,1) \\
&\quad + D(1,3), c(5,6) + c(2,2) + D(5,2) + D(2,3)\} \\
&= \min\{c(6,6) + D(6,3), c(7,5) + D(5,7) \\
&\quad + D(7,3), c(5,4) + c(0,4) + D(5,0) \\
&\quad + D(0,3), c(5,5) + D(5,1) + D(1,3), c(5,6) + D(5,2)\}
\end{aligned}$$

$$= \min\{47.395 + 19.104, 38.092 + 20.518 + 27.892, 13.152$$
$$+ 16.552 + 22.561 + 22.671, 30.843 + 22.561$$
$$+ 16.552, 49.091 + 16.278\}$$
$$= 65.369,$$

$r(5, 7) = 5.$

$$c(6, 7) = \min\{c(6, 2) + c(7, 6) + D(6, 7) + D(7, 4), c(6, 3)$$
$$+ c(0, 5) + D(6, 0) + D(0, 4), c(6, 4) + c(1, 4)$$
$$+ D(6, 1) + D(1, 4), c(6, 5) + c(2, 3) + D(6, 2)$$
$$+ D(2, 4), c(6, 6) + c(3, 2) + D(6, 3) + D(3, 4)\}$$
$$= \min\{c(7, 6) + D(7, 4), c(0, 5) + D(6, 0)$$
$$+ D(0, 4), c(6, 4) + c(1, 4) + D(6, 1) + D(1, 4), c(6, 5)$$
$$+ D(6, 2) + D(2, 4), c(6, 6) + D(6, 3)\}$$
$$= \min\{62.663 + 28.319, 36.587 + 13.152 + 29.154, 13.152$$
$$+ 16.552 + 17.691 + 27.018, 30.843 + 18.248$$
$$+ 17.720, 47.395 + 19.104\}$$
$$= 66.499,$$

$r(6, 7) = 5.$

$$c(7, 7) = \min\{c(7, 2) + c(0, 6) + D(7, 0) + D(0, 5), c(7, 3)$$
$$+ c(1, 5) + D(7, 1) + D(1, 5), c(7, 4) + c(2, 4)$$
$$+ D(7, 2) + D(2, 5), c(7, 5) + c(3, 3) + D(7, 3)$$
$$+ D(3, 5), c(7, 6) + c(4, 2) + D(7, 4) + D(4, 5)\}$$
$$= \min\{c(0, 6) + D(0, 5), c(1, 5) + D(7, 1)$$
$$+ D(1, 5), c(7, 4) + c(2, 4) + D(7, 2)$$
$$+ D(2, 5), c(7, 5) + D(7, 3) + D(3, 5), c(7, 6) + D(7, 4)\}$$
$$= \min\{49.037 + 22.561, 30.170 + 21.540 + 22.561, 18.867$$
$$+ 13.892 + 26.076 + 16.278, 38.092 + 27.892$$
$$+ 13.892, 62.663 + 28.319\}$$
$$= 71.598,$$

$r(7, 7) = 1.$

(v) $c(0, 8) = \min\{c(0, 2) + c(1, 7) + D(0, 1) + D(1, 7), c(0, 3)$

$\qquad + c(2, 6) + D(0, 2) + D(2, 7), c(0, 4) + c(3, 5)$

$\qquad + D(0, 3) + D(3, 7), c(0, 5) + c(4, 4) + D(0, 4)$

$\qquad + D(4, 7), c(0, 6) + c(5, 3) + D(0, 5) + D(5, 7), c(0, 7)$

$\qquad + c(6, 2) + D(0, 6) + D(6, 7)\}$

$\quad = \min\{c(1, 7) + D(1, 7), c(2, 6) + D(0, 2)$

$\qquad + D(2, 7), c(0, 4) + c(3, 5) + D(0, 3) + D(3, 7), c(0, 5)$

$\qquad + c(4, 4) + D(0, 4) + D(4, 7), c(0, 6) + D(0, 5)$

$\qquad + D(5, 7)\}, c(0, 7) + D(0, 6)\}$

$\quad = \min\{66.109 + 21.540, 48.418 + 18.867 + 26.076, 16.552$

$\qquad + 32.996 + 22.671 + 27.892, 36.587 + 19.104 + 29.154$

$\qquad + 28.319, 49.037 + 22.561 + 20.518, 66.109 + 13.152\}$

$\quad = 79.261,$

$r(0, 8) = 6.$

$c(1, 8) = \min\{c(1, 2) + c(2, 7) + D(1, 2) + D(2, 0), c(1, 3)$

$\qquad + c(3, 6) + D(1, 3) + D(3, 0), c(1, 4) + c(4, 5)$

$\qquad + D(1, 4) + D(4, 0), c(1, 5) + c(5, 4) + D(1, 5)$

$\qquad + D(5, 0), c(1, 6) + c(6, 3) + D(1, 6) + D(6, 0), c(1, 7)$

$\qquad + c(7, 2) + D(1, 7) + D(7, 0)\}$

$\quad = \min\{c(2, 7) + D(2, 0), c(3, 6) + D(1, 3)$

$\qquad + D(3, 0), c(1, 4) + c(4, 5) + D(1, 4) + D(4, 0), c(1, 5)$

$\qquad + c(5, 4) + D(1, 5) + D(5, 0), c(1, 6) + D(1, 6)$

$\qquad + D(6, 0), c(1, 7) + D(1, 7)\}$

$\quad = \min\{61.570 + 18.867, 46.148 + 16.552 + 22.671, 16.552$

$\qquad + 32.256 + 27.018 + 29.154, 30.170 + 13.152$

$\qquad + 22.561 + 22.561, 48.418 + 17.691$

$\qquad + 13.152, 66.109 + 21.540\}$

$\quad = 79.261,$

$r(1, 8) = 5.$

$$c(2, 8) = \min\{c(2, 2) + c(3, 7) + D(2, 3) + D(3, 1), c(2, 3)$$
$$+ c(4, 6) + D(2, 4) + D(4, 1), c(2, 4) + c(5, 5)$$
$$+ D(2, 5) + D(5, 1), c(2, 5) + c(6, 4) + D(2, 6)$$
$$+ D(6, 1), c(2, 6) + c(7, 3) + D(2, 7) + D(7, 1), c(2, 7)$$
$$+ c(0, 2) + D(2, 0) + D(0, 1)\}$$
$$= \min\{c(3, 7) + D(3, 1), c(4, 6) + D(2, 4)$$
$$+ D(4, 1), c(2, 4) + c(5, 5) + D(2, 5) + D(5, 1), c(2, 5)$$
$$+ c(6, 4) + D(2, 6) + D(6, 1), c(2, 6) + D(2, 7)$$
$$+ D(7, 1), c(2, 7) + D(2, 0)\}$$
$$= \min\{63.839 + 16.552, 49.947 + 17.720 + 27.018, 13.892$$
$$+ 30.843 + 16.278 + 22.561, 30.170 + 13.152 + 18.248$$
$$+ 17.691, 48.418 + 26.076 + 21.540, 61.570 + 18.867\}$$
$$= 79.261,$$

$$r(2, 8) = 4.$$

$$c(3, 8) = \min\{c(3, 2) + c(4, 7) + D(3, 4) + D(4, 2), c(3, 3)$$
$$+ c(5, 6) + D(3, 5) + D(5, 2), c(3, 4) + c(6, 5)$$
$$+ D(3, 6) + D(6, 2), c(3, 5) + c(7, 4) + D(3, 7)$$
$$+ D(7, 2), c(3, 6) + c(0, 3) + D(3, 0) + D(0, 2), c(3, 7)$$
$$+ c(1, 2) + D(3, 1) + D(1, 2)\}$$
$$= \min\{c(4, 7) + D(4, 2), c(5, 6) + D(3, 5)$$
$$+ D(5, 2), c(3, 4) + c(6, 5) + D(3, 6) + D(6, 2), c(3, 5)$$
$$+ c(7, 4) + D(3, 7) + D(7, 2), c(3, 6) + D(3, 0)$$
$$+ D(0, 2), c(3, 7) + D(3, 1)\}$$
$$= \min\{66.369 + 17.720, 49.091 + 13.892 + 16.278, 13.892$$
$$+ 30.843 + 19.104 + 18.248, 32.996 + 18.867 + 27.892$$
$$+ 26.076, 46.148 + 22.671 + 18.867, 63.839 + 16.552\}$$
$$= 79.261,$$

$$r(3, 8) = 2.$$

$$c(4, 8) = \min\{c(4, 2) + c(5, 7) + D(4, 5) + D(5, 3), c(4, 3)$$
$$+ c(6, 6) + D(4, 6) + D(6, 3), c(4, 4) + c(7, 5)$$
$$+ D(4, 7) + D(7, 3), c(4, 5) + c(0, 4) + D(4, 0)$$
$$+ D(0, 3), c(4, 6) + c(1, 3) + D(4, 1) + D(1, 3), c(4, 7)$$
$$+ c(2, 2) + D(4, 2) + D(2, 3)\}$$
$$= \min\{c(5, 7) + D(5, 3), c(6, 6) + D(4, 6)$$
$$+ D(6, 3), c(4, 4) + c(7, 5) + D(4, 7) + D(7, 3), c(4, 5)$$
$$+ c(0, 4) + D(4, 0) + D(0, 3), c(4, 6) + D(4, 1)$$
$$+ D(1, 3), c(4, 7) + D(4, 2)\}$$
$$= \min\{65.369 + 13.892, 49.395 + 19.104 + 19.104, 19.104$$
$$+ 38.092 + 28.319 + 27.892, 32.256 + 16.552 + 29.154$$
$$+ 22.671, 49.947 + 27.018 + 16.552, 65.369 + 17.720\}$$
$$= 79.261,$$
$$r(4, 8) = 1.$$
$$c(5, 8) = \min\{c(5, 2) + c(6, 7) + D(5, 6) + D(6, 4), c(5, 3)$$
$$+ c(7, 6) + D(5, 7) + D(7, 4), c(5, 4) + c(0, 5)$$
$$+ D(5, 0) + D(0, 4), c(5, 5) + c(1, 4) + D(5, 1)$$
$$+ D(1, 4), c(5, 6) + c(2, 3) + D(5, 2) + D(2, 4), c(5, 7)$$
$$+ c(3, 2) + D(5, 3) + D(3, 4)\}$$
$$= \min\{c(6, 7) + D(6, 4), c(7, 6) + D(5, 7)$$
$$+ D(7, 4), c(5, 4) + c(0, 5) + D(5, 0) + D(0, 4), c(5, 5)$$
$$+ c(1, 4) + D(5, 1) + D(1, 4), c(5, 6) + D(5, 2)$$
$$+ D(2, 4), c(5, 7) + D(5, 3))$$
$$= \min\{66.499 + 19.104, 62.663 + 20.518 + 28.319, 13.152$$
$$+ 36.587 + 22.561 + 29.154, 30.843 + 16.552 + 22.561$$
$$+ 27.018, 49.091 + 16.278 + 17.720, 65.369 + 13.892\}$$
$$= 79.261,$$
$$r(5, 8) = 6.$$

$$\begin{aligned}
c(6,8) = \min\{ &c(6,2) + c(7,7) + D(6,7) + D(7,5), c(6,3) \\
&+ c(0,6) + D(6,0) + D(0,5), c(6,4) + c(1,5) \\
&+ D(6,1) + D(1,5), c(6,5) + c(2,4) + D(6,2) \\
&+ D(2,5), c(6,6) + c(3,3) + D(6,3) + D(3,5), c(6,7) \\
&+ c(4,2) + D(6,4) + D(4,5)\} \\
= \min\{ &c(7,7) + D(7,5), c(0,6) + D(6,0) \\
&+ D(0,5), c(6,4) + c(1,5) + D(6,1) + D(1,5), c(6,5) \\
&+ c(2,4) + D(6,2) + D(2,5), c(6,6) + D(6,3) \\
&+ D(3,5), c(6,7) + D(6,4)\} \\
= \min\{ &71.598 + 20.518, 49.037 + 13.152 + 22.561, 13.152 \\
&+ 30.170 + 17.691 + 22.561, 30.843 + 13.892 + 18.248 \\
&+ 16.278, 47.395 + 19.104 + 13.892, 66.499 + 19.104\} \\
= {} &79.261,
\end{aligned}$$

$$r(6,8) = 4.$$

$$\begin{aligned}
c(7,8) = \min\{ &c(7,2) + c(0,7) + D(7,0) + D(0,6), c(7,3) \\
&+ c(1,6) + D(7,1) + D(1,6), c(7,4) + c(2,5) \\
&+ D(7,2) + D(2,6), c(7,5) + c(3,4) + D(7,3) \\
&+ D(3,6), c(7,6) + c(4,3) + D(7,4) + D(4,6), c(7,7) \\
&+ c(5,2) + D(7,5) + D(5,6)\} \\
= \min\{ &c(0,7) + D(0,6), c(1,6) + D(7,1) \\
&+ D(1,6), c(7,4) + c(2,5) + D(7,2) + D(2,6), c(7,5) \\
&+ c(3,4) + D(7,3) + D(3,6), c(7,6) + D(7,4) \\
&+ D(4,6), c(7,7) + D(7,5)\} \\
= \min\{ &66.109 + 13.152, 48.418 + 21.540 + 17.691, 18.867 \\
&+ 30.170 + 26.076 + 18.248, 38.092 + 13.892 + 27.892 \\
&+ 19.104, 62.663 + 28.319 + 19.104, 71.598 + 20.518\} \\
= {} &79.261,
\end{aligned}$$

$$r(7,8) = 1.$$

Observe that $c(i, 8) = 79.261$ for every i, $0 \leq i \leq 7$. Hence the minimal cost of triangulation is 79.261. We could consider any one of the $c(i, 8)$ to get an optimal triangulation of the polygon.

We now consider optimal triangulation corresponding to the optimal cost $c(0, 8)$. Since $r(0, 8) = 6$, as a first step we need to join v_6 to v_0 and to v_7 and this leads to a polygon of seven sides and a triangle (v_0, v_6, v_7). The polygon $(v_0, v_1, \ldots, v_6)$ is the next to be triangulated. Again $r(0, 7) = 1$ so we need to connect v_1 to v_0 and v_6. This leads to a triangle (v_0, v_1, v_6) and a polygon $(v_1, v_2, \ldots, v_6)$ of six sides for which we need to consider $c(1, 6)$. Again $r(1, 6) = 1$ so that for triangulation of $(v_1, v_2, \ldots, v_6)$, we need to connect v_2 to v_1 and v_6. This leads to a triangle (v_1, v_2, v_6) and polygon $(v_2, \ldots, v_6)$ of five sides. Also, in the computation of $c(1, 6)$ we are led to $c(2, 5)$. Now $r(2, 5) = 3$ so that we need to connect v_5 to v_2 and to v_6. This leads to a polygon (v_2, v_3, v_4, v_5) of four sides and a triangle (v_2, v_5, v_6). Since $c(2, 5) = c(2, 4) + D(2, 5)$, we need to consider $c(2, 4)$ which is equal to $c(2, 4) = D(3, 5)$ which then shows that optimal triangulation of the quadrilateral (v_2, v_3, v_4, v_5) is obtained by connecting v_3 and v_5. Hence the optimal triangulation of the given polygon is as given in Fig. 6.9.

Next, we consider optimal triangulation of the polygon corresponding to $c(2, 8)$, say. Now

$$c(2, 8) = c(2, 5) + c(6, 4) + D(2, 6) + D(6, 1)$$

$$= c(2, 4) + D(2, 5) + D(6, 0) + D(2, 6) + D(6, 1)$$

$$= D(3, 5) + D(2, 5) + D(6, 0) + D(2, 6) + D(6, 1),$$

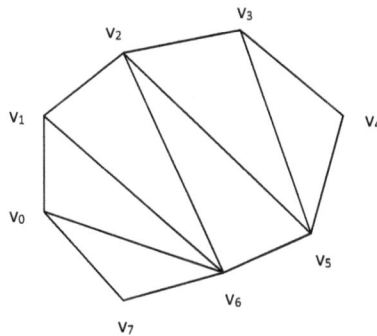

Fig. 6.9

which shows that v_6 is to be connected to v_0, v_1, v_2 and v_5 is to be connected to v_2, v_3. This is the same triangulation as given in Fig. 6.9 which was obtained from $c(0, 8)$.

Exercise 6.2. Prove that every $c(i, 8)$, $0 \leq i \leq 7$, leads to the same triangulation of the given polygon of Example 6.11.

Optimal cost of triangulation of a polygon is independent of the choice of starting vertex for triangulation.

Theorem 6.5. *Given a polygon $(v_0, v_1, \ldots, v_{n-1})$ of n sides with the vertices taken in the clockwise direction, the optimal cost of triangulation satisfies $c(i, n) = c(j, n)$ for all i, j, $0 \leq i, j \leq n - 1$.*

Proof. Exercise.   □

Example 6.12. Consider the pentagon as given in Fig. 6.10. Then optimal cost of triangulation is 7.728. Also, there is only one optimal triangulation and is as given in Fig. 6.11.

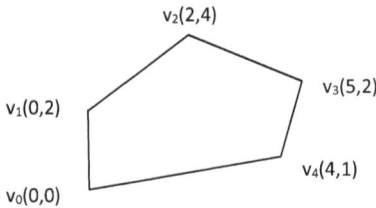

Fig. 6.10

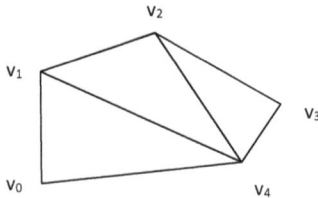

Fig. 6.11

Solution.

$$D(0, 2) = \sqrt{20} = 4.472, \quad D(0, 3) = \sqrt{29} = 5.385, \quad D(1, 3) = 5,$$

$$D(1, 4) = \sqrt{17} = 4.123, \quad D(2, 4) = \sqrt{13} = 3.605,$$

$$c(0, 4) = \min\{D(1, 3), D(0, 2)\} = 4.472, r(0, 4) = 2,$$

$$c(1, 4) = \min\{c(1, 2) + c(2, 3) + D(1, 2) + D(2, 4), c(1, 3)$$

$$+ c(3, 2) + D(1, 3) + D(3, 4)\}$$

$$= \min\{D(2, 4), D(1, 3)\} = 3.605,$$

$$r(1, 4) = 1.$$

$$c(2, 4) = \min\{c(2, 2) + c(3, 3) + D(2, 3) + D(3, 0), c(2, 3)$$

$$+ c(4, 2) + D(2, 4) + D(4, 0)\}$$

$$= \min\{D(3, 0), D(2, 4)\} = 3.605,$$

$$r(2, 4) = 2.$$

$$c(3, 4) = \min\{c(3, 2) + c(4, 3) + D(3, 4) + D(4, 1), c(3, 3)$$

$$+ c(3, 2) + D(3, 0) + D(0, 1)\}$$

$$= \min\{D(4, 1), D(3, 0)\} = 4.123,$$

$$r(3, 4) = 1.$$

$$c(4, 4) = \min\{c(4, 2) + c(0, 3) + D(4, 0) + D(0, 2), c(4, 3)$$

$$+ c(1, 2) + D(4, 1) + D(1, 2)\}$$

$$= \min\{D(0, 2), D(4, 1)\} = 4.123,$$

$$r(4, 4) = 2.$$

$$c(0, 5) = \min\{c(0, 2) + c(1, 4) + D(0, 1) + D(1, 4), c(0, 3)$$

$$+ c(2, 3) + D(0, 2) + D(2, 4), c(0, 4) + c(3, 2)$$

$$+ D(0, 3) + D(3, 4)\}$$

$$= \min\{c(1, 4) + D(1, 4), D(0, 2) + D(2, 4), c(0, 4)$$

$$+ D(0, 3)\}$$

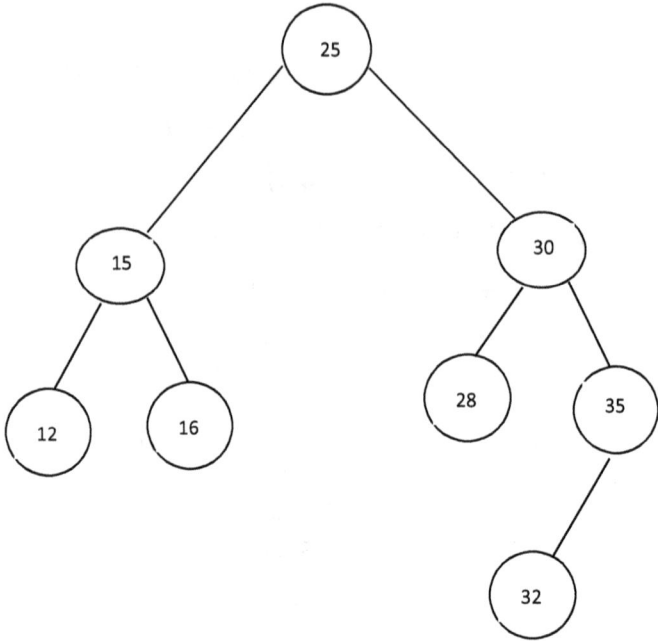

Fig. 6.12

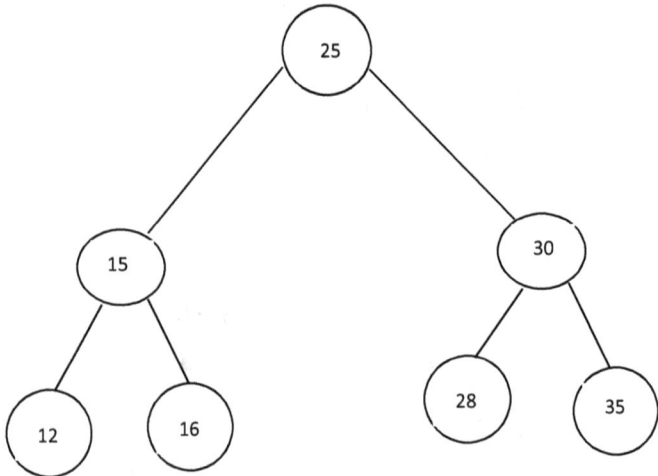

Fig. 6.13

$$= \min\{3.605 + 4.123, 4.472 + 3.605, 4.472 + 5.385\}$$
$$= 7.728,$$
$$r(0, 5) = 1.$$

It is easy to see that

$$c(1, 5) = 7.728, r(1, 5) = 3; c(2, 5) = 7.728, r(2, 5) = 2,$$
$$c(3, 5) = 7.728, r(3, 5) = 1; c(4, 5) = 7.728, r(4, 5) = 3.$$

Exercise 6.3.

1. Give the missing details in the solution of Example 6.12.
2. Prove Theorem 6.5.
3. Delete (i) 25, (ii) 30 from the binary search tree of Fig. 6.12.
4. Insert (i) 20, (ii) 32 in the binary search tree given in Fig. 6.13.

Chapter 7

Elementary Graph Algorithms

The aim of the present chapter is to study methods for representing a graph and for searching a graph. Searching a graph means visiting the vertices of the graph by systematically going over the edges of the graph. Much information about the structure of a graph can be discovered by a graph-searching algorithm.

7.1. Representations of Graphs

Given a graph $G = (V, E)$, where V is the set of vertices and E the set of edges, there are two standard ways to represent it. One is to represent it by adjacency lists and the other by its adjacency matrix. In general, representing a graph by adjacency lists is preferred over its representation by adjacency matrix-normally for graphs in which $|E|$ is much less than $|V|^2$. However, in the adjacency matrix representation of a graph it is much quicker to tell whether vertices, say u, v, are adjacent or not.

Let a graph $G = (V, E)$ be given. The **adjacency list representation** of G consists of lists $Adj[u]$ for every $u \in V$ where

$$Adj[u] = \{v \in V \mid v \text{ is adjacent to } u \text{ in } G\}.$$

The vertices in each $Adj[u]$ are stored in an arbitrary manner. Observe that the sum of the lengths or orders of all adjacency lists is $|E|$ when G is a directed graph but is $2|E|$ if G is not a directed graph. This representation is also known as **successor** or **neighbor listing representation**.

When G is a weighted graph with weight function w defined on the edge set E of G, the adjacency lists $Adj[u]$ in the adjacency list representation of G are defined by

$Adj[u] = \{v(k)|v$ is adjacent to u in G and the weight of the edge (u, v) is $k\}$.

Adjacency list representation is also possible for graphs which are not simple.

For matrix representation of G, we consider the vertices as $1, 2, \ldots, n$ rather than $v_1, v_2, \ldots, v_n$. Then G is represented by a square matrix $A = (a_{ij})$ of order n, where

$$a_{ij} = \begin{cases} 1 & \text{if } i \text{ is adjacent to } j, \\ 0 & \text{otherwise.} \end{cases}$$

In case G is a weighted graph, the entries a_{ij} of the matrix are given (or defined) by

$$a_{ij} = \text{weight of the edge } (i, j),$$

where it is understood that $a_{ij} = 0$ if j is not adjacent to i. The matrix $A = (a_{ij})$ is called the **adjacency matrix** of the graph G and the representation is known as **adjacency matrix representation**.

When the graph G is not directed, then $(i, j) \in E$ if and only if $(j, i) \in E$. Therefore, the adjacency matrix of G is symmetric. However, if parallel edges are allowed in the graph, i.e., the graph G is not simple, it is not possible to represent the graph using adjacency matrix.

There are other methods of representation of graphs, for example, incidence matrix representation, edge listing representation, etc. but we do not discuss these here.

Example 7.1. Consider the graphs as given in Figs. 7.1(a) and 7.2(a).

The adjacency list representation of Fig. 7.1(a) is given below while the adjacency matrix is given in Fig. 7.1(b).

$$Adj[1] = 2, 3, \quad Adj[2] = 1, 3, 4, 5, \quad Adj[3] = 1, 2, 4,$$

$$Adj[4] = 2, 3, 5, \quad Adj[5] = 2, 4, 6, Adj[6] = 5.$$

Sum of the lengths of the adjacency lists $= 2 + 4 + 3 + 3 + 3 + 1 = 16 = 2 \times 8 = $ twice the number of edges in the graph.

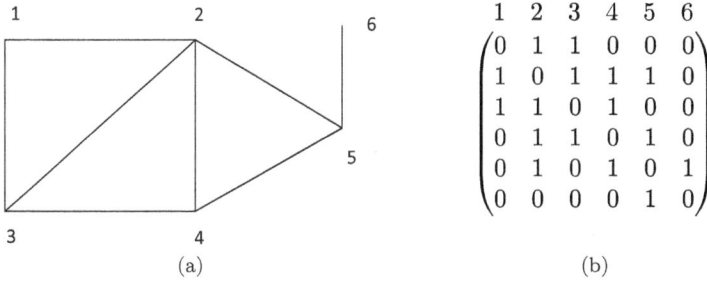

Fig. 7.1

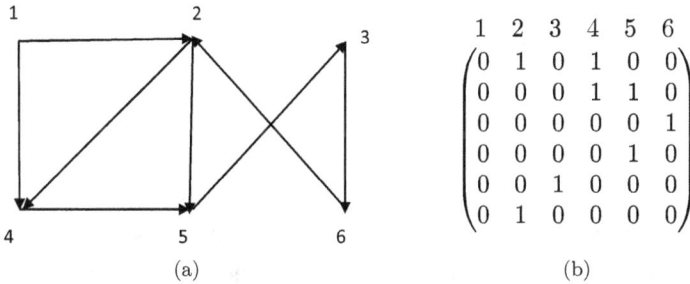

Fig. 7.2

The adjacency lists of the adjacency list representation of Fig. 7.2(a) are as given below while its adjacency matrix is given in Fig. 7.2(b)

$$Adj[1] = 2, 4, \quad Adj[2] = 4, 5, \quad Adj[3] = 6,$$

$$Adj[4] = 5, \quad Adj[5] = 3, \quad Adj[6] = 2.$$

Sum of the lengths of adjacency lists $= 2 + 2 + 1 + 1 + 1 + 1 = 8 =$ the number of edges in the graph.

7.2. Queues

A **queue** is a dynamic set in which elements can be removed one at a time by Delete operation and the element removed is pre-specified. The element deleted is always the one that has been in the set for the longest time. Thus an element which enters the queue first will be the first to leave, i.e., the queue adopts the policy of **First-in-First-out (FIFO)**. A line of cars waiting at a Toll Barrier is an example of a queue. A car that entered the line first is at the head of the line and will be the first to pay the tax and go out. Thus, this car will be the first to get served and leave. Then the car immediately behind will take its place on the head of the line and the process continues. Any car wanting to get past the barrier has to take its place at the other end of the line and wait for being served. Since the number of cars in the line continually increases or decreases, the cars in the line form a dynamic set (i.e., a set in which the number of elements does not remain fixed).

As is clear from the above example, every queue has a front or head and a tail or rear. While an element leaving a queue is the one at the head of the queue an element entering it enters at the tail or rear of the queue. An insert operation (i.e., the operation of entering the queue) is called **ENQUEUE** and the process of deleting (i.e., of leaving the queue) is called **DEQUEUE**. A queue may be represented as an array $Q[1 \ldots n - 1]$ if it consists of at most $n - 1$ elements, i.e., if at no time number of elements in the queue can exceed $n - 1$. Since an element entering the queue occupies the position at the rear, the rear is always the position one after the position of the last element already in the queue. The head position is always the one occupied by the first element in the queue. The elements in the queue Q occupy the positions $Head[Q], Head[Q]+1, \ldots, Rear[q]-1$. Since the number of elements in the queue at a given time may not be the maximum allowable, the position of the last element present is written as $Rear[q]-1$ rather than $n-1$. We may also think as if it is not the elements in queue that are moving but it is the service provider that moves. Since elements entering the queue are not stopped, the queue may be thought of as operating in a circular manner. While a new element x entering the queue Q is represented by $Enqueue(Q, x)$, the process of removing an element is represented simply by $Dequeue(Q)$. This is so because we know that it is the element at the head of the queue that is to be removed.

To illustrate an array representation of a queue, let us consider an array $Q[1 \ldots 15]$ so that the maximum number of elements allowed in the queue is 14. Suppose that at a given time elements in the queue are $4, 7, 6, 11, 3, 9, 14$ in the positions $7, 8, \ldots, 13$. Then we allow $Enqueue(Q, 10)$, $Enqueue(Q, 5)$, $Enqueue(Q, 12)$, $Enqueue(Q, 18)$, $Enqueue(Q, 16)$. After that there are two Dequeue operations. This queue implementation is given in Fig. 7.3.

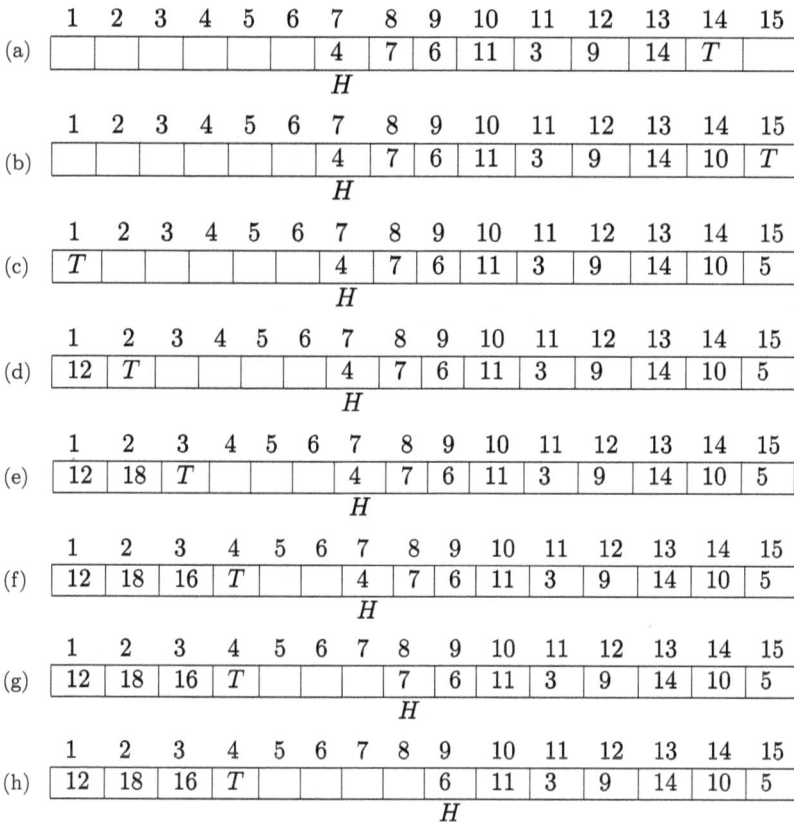

(a)

1	2	3	4	5	6	7	8	9	10	11	12	13	14	15
						4	7	6	11	3	9	14	T	

H (below 7)

(b)

1	2	3	4	5	6	7	8	9	10	11	12	13	14	15
						4	7	6	11	3	9	14	10	T

H (below 7)

(c)

1	2	3	4	5	6	7	8	9	10	11	12	13	14	15
T						4	7	6	11	3	9	14	10	5

H (below 7)

(d)

1	2	3	4	5	6	7	8	9	10	11	12	13	14	15
12	T					4	7	6	11	3	9	14	10	5

H (below 7)

(e)

1	2	3	4	5	6	7	8	9	10	11	12	13	14	15
12	18	T				4	7	6	11	3	9	14	10	5

H (below 7)

(f)

1	2	3	4	5	6	7	8	9	10	11	12	13	14	15
12	18	16	T			4	7	6	11	3	9	14	10	5

H (below 7)

(g)

1	2	3	4	5	6	7	8	9	10	11	12	13	14	15
12	18	16	T				7	6	11	3	9	14	10	5

H (below 8)

(h)

1	2	3	4	5	6	7	8	9	10	11	12	13	14	15
12	18	16	T					6	11	3	9	14	10	5

H (below 9)

Fig. 7.3

In Fig. 7.3(a), observe that Head is indicated by writing H below the box occupied by the first entry, i.e., 4 whereas tail is indicated by writing T in the box immediately after the box occupied by the last entry namely 14. When the two enqueue operations $Enqueue(Q, 10)$ and $Enqueue(Q, 5)$ in that order have been implemented, the entry 5 occupies the last position in the array $Q[1 \ldots 15]$. The tail has to be moved to the next available position in the array and goes to box 1 of the array and this is because the array were in the form of a circular ring rather than a straight line. This has been shown in Fig. 7.3(c). Then on implementing the next three enqueue operations $Enqueue(Q, 12)$, $Enqueue(Q, 18)$, $Enqueue(Q, 16)$ in order, the tail moves as it should. The head moves

1	2	3	4	5	6	7	8	9	10	11	12	13	14	15
12	18	16	8	2	T	4	7	6	11	3	9	14	10	5

(a)   H

1	2	3	4	5	6	7	8	9	10	11	12	13	14	15
4	7	6	11	3	9	14	10	5	12	18	16	8	2	T

(b)   H

Fig. 7.4

only when a dequeue operation is implemented. On the implementation of first dequeue operation, the head moves to the position occupied by the next element in the queue, namely 7. On the implementation of the second operation $Dequeue(Q)$, the head moves to the position occupied by the next element, namely 6. Movement of the head is shown in Figs. 7.3(g) and 7.3(h).

When it so happens that $head[Q] = tail[Q]$, it indicates that there are no customers to be served and we say that the **queue is empty**. On the other hand, when $head[Q] = tail[Q] + 1$, the number of customers waiting to be served is the maximum allowed for the queue and we say that the **queue is full**. The full position for the queue $Q[1 \ldots 15]$ is indicated in Fig. 7.4.

Observe that in the case of Fig. 7.4(b), the condition $head[Q] = tail[Q] + 1$ is satisfied in view of circular movement of the queue. When the queue is empty, operation $Dequeue(Q)$ is not possible. In this case $head(Q) = tail(Q)$ and $tail(Q)$ points to an empty space in the array. Since in the operation $Dequeue(Q)$, the element in the position $head(Q)$ is to be ignored or deleted and this box being already empty, there is nothing to be ignored. Similarly, when the queue is full and $Enqueue(Q, x)$ is performed, then $tail(Q)$ shifts to the next spot in the array but that is already the position of $head(Q)$ i.e., after the operation $Enqueue(Q, x)$ is implemented, the queue becomes empty which cannot happen. This operation will become possible only when an additional space is created in the array, i.e., the length of the queue is increased.

The operations $Enqueue(Q, x)$ and $Dequeue(Q)$ when the queue is not full and empty, respectively, can be implemented using the following procedures.

Algorithm 7.1. Procedure $Enqueue(Q, x)$

---

1. replace $tail[Q]$ position by x i.e., $Q[tail\{q\}] \to x$
2. if the original $tail[Q] = length[Q]$, new $tail[Q] = 1$ or if $tail[Q] = length[Q]$, then $tail[Q] \to 1$
3. otherwise $tail[Q] \to tail[Q] + 1$
4. there is no change in the position $head[Q]$

Algorithm 7.2. Procedure $Dequeue(Q)$

---

1. let $head[Q] = x$ i.e., $head[Q] \to x$
2. if $head[Q] = length[Q]$
3. then new $head[Q] = 1$ i.e., $head[Q] \to 1$
4. otherwise $head[Q] = head[Q] + 1$
5. there is no change in the position $tail[Q]$

Exercise 7.1. Using an $array Q[1 \ldots 8]$ of length 8 and using Fig. 7.3 as a model, illustrate the result of the operations $Enqueue(Q, 3)$, $Enqueue(Q, 6)$, $Enqueue(Q, 2)$, $Enqueue(Q, 5)$, $Dequeue(Q)$, $Enqueue(Q, 7)$, $Enqueue(Q, 1)$, and $Dequeue(Q)$ in order, when the queue is initially assumed to be empty.

7.3. Shortest Paths

Let $G = (V, E)$ be a graph or a directed graph with a weight function $w : E \to R$ (the real numbers). For an edge $e \in E$, $w(e)$ is called the weight or also the length of the edge e. A walk in G is a sequence $W = (e_1, \ldots, e_n)$ of edges $e_1, e_2, \ldots, e_n$ no two of which are equal and terminal vertex of e_i is the initial vertex of e_{i+1} for every i, $1 \le i \le n - 1$. For any walk $W = (e_1, \ldots, e_n)$ we define the **length** of W to be the number $w(W) = w(e_1) + \cdots + w(e_n)$. In the case of an undirected graph we need to say that e_i and e_{i+1} have one vertex in common. The **distance** (or **minimum distance**) $\delta(a, b)$ between two vertices a and b in V is the minimum of all possible lengths of walks starting at a and ending at b. In case there is no walk starting at a and ending at b, we define $\delta(a, b) = \infty$. Also to avoid the existence of arbitrarily large negative lengths $\delta(a, b)$, we assume that the graph has no cycles of negative length. One way to achieve this is by taking $w(e) > 0$ for every $e \in E$. A walk between vertices a and b not involving any cycle is called a **path** (or a **simple path**) from a to b or b to a and a path of length $\delta(a, b)$ is called a **path of minimum length** or a **minimal path** or a **shortest path**. In case G is not a weighted graph, we may define $w(e) = 1$ for every edge e in E. Then the length of a path from vertex a to vertex b is simply the number of edges in the path from a to b.

7.4. Breadth-First Search

Given a graph $G = (V, E)$, one of the procedures or algorithms for searching G is **breadth-first search** (BFS) suggested by Moore (cf. [15]). In the breadth-first search algorithm we systematically explore the edges of the graph. Starting from a given (specified) vertex s (say), the algorithm first visits/explores all vertices that are adjacent to s. Then for every such vertex we visit all vertices that are adjacent to this vertex. Vertices adjacent to s may be called vertices at level one. Once we have visited vertices that are adjacent to every one of the vertices at level one (the vertices that are adjacent to vertices at level one are called vertices at level two), we continue the process till there are no vertices adjacent to the ones last arrive at. The given specified vertex s is called a **source**. The breadth-first search algorithm is quite useful for finding shortest path between pairs of vertices.

Let G be a graph given by adjacency list representation, s be a given source vertex in G and Q be a queue of vertices which is initially empty. With every vertex u in Q, when Q is non-empty, we associate a number $d(u)$. Initially set $d(s) = 0$ and $Enqueue(Q, s)$. For each $u \in adj[s]$, set $d(u) = 1$ and $Enqueue(Q, u)$ unless $d(u)$ is already defined in which case u is ignored. At an intermediary step when Q is non-empty, let $u = head(Q)$. For each $v \in adj[u]$, set $d(v) = d(u) + 1$ and $Enqueue(Q, v)$ unless $d(v)$ is already defined in which case v is ignored. $Dequeue(Q)$. Continue the process of assigning a label $d(u)$ to vertices and enqueueing and dequeueing till Q again becomes empty. Then we have visited all the vertices in G that are reachable from s and the process stops.

Algorithm 7.3. *BFS* (G, s, d)

---

1. given a graph G, a source vertex s, a label d and a queue Q
2. initially $Q = \phi$, set $d(s) = 0$
3. *Enqueue*(Q, s)
4. while $Q \neq \phi$ and *head*$(Q) = u$, for $v \in adj[u]$ do, if $d(v)$ is undefined, then $d(v) = d(u) + 1$. *Enqueue*(Q, v). Otherwise ignore v
5. *Dequeue*(Q)
6. repeat steps 4 and 5 till Q is non-empty

Time complexity of BFS algorithm

In the breadth-first search algorithm/procedure run on a graph G, we have to go over every edge of the graph. If G is undirected and e is an edge connecting vertices u and v, then $u \in adj[v]$ and $v \in adj[u]$ while if G is directed, either $u \in adj[v]$ or $v \in adj[u]$ but not both. For each vertex u, $adj[u]$ is scanned only when we are to dequeue u and enqueue some of the vertices from $adj[u]$, each adjacency list is scanned at most once. Therefore, the total time of scanning the adjacency lists is a constant multiple of the sum of the orders of the adjacency lists. Hence the running time or the time complexity of BFS is a constant multiple of a number $\leq |E|$ or $2|E|$ or that the time complexity is $O(|E|)$.

Lemma 7.1. *Given a graph $G = (V, E)$ with a source vertex s, a vertex v is reachable or accessible from s if and only if v is assigned a label $d(v)$ on running BFS on G.*

Proof. Suppose that a vertex v is assigned a label $d(v)$. If $d(v) = 1$, then $v \in adj[s]$, i.e., there is an edge between s and v and, so, v is reachable from s. Now suppose that $d(v) > 1$ and suppose that every vertex u with $d(u) < d(v)$ is accessible from s. Then $d(v) = d(u) + 1$ for some vertex u of G. Therefore $d(u) < d(v)$ and, by induction hypothesis, u is accessible from s. Also $d(v) = d(u) + 1$ implies that $v \in adj[u]$ so that there is an edge from u to v. Composing a path from s to u with this edge from u to

v, we get a path from s to v and v is accessible from s. This completes induction and, therefore, every vertex v which is assigned a label $d(v)$ by BFS is accessible from s.

Conversely, suppose that v is a vertex that is accessible from s. Then there is a path $s = v_1, v_2, \ldots, v_r = v$ from s to v. There being an edge from $s = v_1$ to v_2, v_2 is assigned the label $d(v_2) = 1$. Now, suppose that $i \geq 2$ and that v_i is already assigned a label $d(v_i)$. There being an edge from v_i to v_{i+1}, when the adjacency list $adj[v_i]$ is scanned, either v_{i+1} is already assigned a label $d(v_{i+1})$ or $d(v_{i+1}) = d(v_i) + 1$. In either case v_{i+1} is assigned a label $d(v_{i+1})$. Hence every v_i, $1 \leq i \leq r$, is assigned a label $d(v_i)$. In particular $v_r = v$ is assigned a label $d(v)$. $\square$

Corollary 7.2. *Let G be an undirected graph with source vertex s. Then G is connected if and only if every vertex v is assigned a label $d(v)$ on running BFS.*

Exercise 7.2. Give an example to show that the result of the Corollary 7.2 is false for directed graphs.

Corollary 7.3. *Let G be a graph with source vertex s. Then for any vertex of G, $\delta(s, v) = \infty$ if and only if v is not assigned a label $d(v)$ when BFS is run on G. (If a vertex v is not assigned a label, we say that $d(v)$ is not defined for v).*

Theorem 7.4. *When algorithm BFS is run on a graph G with source vertex s, at the end of the algorithm, for any vertex v of G,*

$$\delta(s, v) = \begin{cases} d(v) & \text{if } d(v) \text{ is defined,} \\ \infty & \text{if } d(v) \text{ is not defined.} \end{cases}$$

Proof. If v is a vertex in G for which $d(v)$ is not defined, then $\delta(s, v) = \infty$ follows from Corollary 7.3. So, suppose that v is a vertex for which $d(v)$ is defined. Then $\delta(s, v)$ is finite. We prove that $\delta(s, v) = d(v)$ by induction on $\delta(s, v)$. If $\delta(s, v) = 0$, then $v = s$ and, so, $d(v) = d(s) = 0$, i.e., $\delta(s, v) = d(v)$. If $\delta(s, v) = 1$, then there is an edge from s to v and $v \in adj[s]$ and, therefore, the way vertices are assigned a label then implies that $d(v) = 1 = \delta(s, v)$. Now suppose that $\delta(s, v) > 1$ and that for any vertex u for which $\delta(s, u) < \delta(s, v), \delta(s, u) = d(u)$. Let $(s, v_1, \ldots, v_n, v)$ be a shortest path of length $n + 1 = \delta(s, v)$. Then $(s, v_1, \ldots, v_n)$ is a shortest path from s to v_n and its length is n. Therefore, by induction hypothesis, $\delta(s, v_n) = n = d(v_n)$. Since for any vertex u which is assigned a label $d(u)$ there is a path of length $d(u)$ from s to u, $\delta(s, v) \leq d(v)$, i.e., $n + 1 \leq d(v)$. Therefore $d(v_n) < d(v)$. Hence, when BFS is run on G, the vertex v_n is treated before v. Since there is an edge from v_n to v, $v \in adj[v_n]$. Therefore v is assigned a label $d(v) = d(v_n) + 1 = n + 1$ if v is not already assigned a label, i.e., either $d(v) \leq d(v_n)$ or $d(v) = d(v_n) + 1 = n + 1$. But $n + 1 \leq d(v)$. Therefore $d(v) = n + 1 = \delta(s, v)$. This completes induction. Hence $\delta(s, v) = d(v)$ for every v for which $d(v)$ is defined. $\qquad \square$

7.4.1. Breadth-first tree

When BFS procedure is run on a graph $G = (V, E)$ with a given source vertex s and G is represented through adjacency lists, there results a tree with root s. To see how this tree is constructed, we first consider the source vertex s. We then adjoin all the vertices in $adj[s]$ along with edges from s to the vertices in $adj[s]$. Suppose that at an intermediary step, a vertex u has been assigned a label $d(u)$, is adjoined to the vertices already obtained and the adjacency list $adj[u]$ is being scanned. If $v \in adj[u]$ and v is not already assigned a label $d(v)$, then v is adjoined to the vertices already obtained and the edge (u, v) is adjoined to the edges chosen. The process continues till the algorithm has been fully run. In this fashion we get a subset of vertices of V and a subset of edges of E. At any moment when $adj[u]$ is being scanned, and $v \in adj[u]$, then v is ignored if a label has already been assigned to v and is not added to the vertices being chosen. So no vertex v is put more than once in the subset of vertices being chosen. Hence there is no cycle in the subgraph considered. To make the things more formal, we introduce a function $\pi : V \to V$ called the **predecessor function**. Suppose that a vertex u is assigned a label $d(u)$ and $v \in adj[u]$. If v is not already assigned a label, v is assigned the label $d(u) + 1$. Then we call u a **predecessor** of v and set $\pi(v) = u$. It is clear from the BFS procedure that every $v \in V$ has at most one predecessor. If $v \in V$ has no predecessor, we set $\pi(v) = NIL$. Observe that the source vertex has no predecessor. Also a vertex v which is not reachable from s when BFS procedure is run, then v has no predecessor. Such a v is a vertex which is not assigned a label.

Let

$$V_\pi = \{v \in V \mid \pi(v) \neq NIL\} \cup \{s\},$$

$$E_\pi = \{(\pi(v), v) \mid v \in V_\pi - \{s\}\}.$$

Then $(V_\pi, E_\pi) = G_\pi$ is a subgraph of G and since, every $v \in V$ has at most one predecessor, for every $v \in V_\pi, v \neq s$, there is a unique edge $(\pi(v), v)$. Therefore, $|E_\pi| = |V_\pi| - 1$. As already observed earlier, this subgraph has no cycles (i.e., it is acyclic). Therefore (V_π, E_π) is a tree with s as its root. We thus have proved the following theorem.

Theorem 7.5. *The subgraph* $(V_\pi, E_\pi) = G_\pi$ *is a tree with root s.*

*The tree G_π is called a **BFS-tree**.*

For different elements chosen as source vertices in G, we get different BFS trees. Also BFS-tree depends on the order in which the elements in adjacency lists are listed.

We have proved that for a vertex v in G, the label $d(v)$ is the shortest path length from s to v. Using the predecessor function π, we also determine a corresponding shortest path for a BFS procedure from s to v. If $\pi(v) = u$, then $d(v) = d(u) + 1$, i.e., u is assigned a label $d(u)$. Therefore u is reachable from s on BFS traversal. Therefore $\pi(u) \neq NIL$. Using this observation, if $\pi(v) = v_n, \pi(v_n) = v_{n-1}, \ldots, \pi(v_1) = s$, then $(s, v_1, \ldots, v_n, v)$ is a corresponding shortest path.

Algorithm 7.4. Shortest-path (G, s, v) Procedure

---

1. G is a graph with source vertex s and v any vertex of G
2. if $v = s$
3. then (s) is the shortest path (i.e., it is a path of length 0)
4. if $\pi(v) = NIL$
5. then there is no shortest path from s to v
6. if $\pi(v) = u$
7. and $(s, v_1, \ldots, v_{n-1}, u)$ is a shortest path from s to u
8. then $(s, v_1, \ldots, v_{n-1}, u, v)$ is a shortest path from s to v

---

In this procedure, at every step we choose and so add one vertex and, therefore, the number of choices is at most $|V|$. Hence the running time of this algorithm is $O(|V|)$.

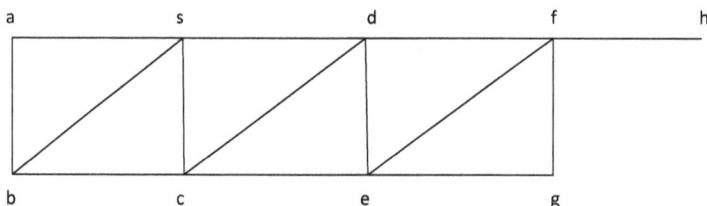

Fig. 7.5

Example 7.2. Construct a BFS- tree for the graph G of Fig. 7.5 with s as a source vertex.

Solution. The adjacency lists of G are

$adj[s] = \{a, b, c, d\}, \quad adj[a] = \{s, b\}, \quad\quad adj[c] = \{s, b, d, e\},$
$adj[b] = \{s, a, c\}, \quad\quad adj[d] = \{s, c, e, f\}, \quad adj[e] = \{c, d, f, g\},$
$adj[f] = \{d, e, g, h\}, \quad adj[g] = \{e, f\}, \quad\quad adj[h] = \{f\}.$

Initially $Q = \varphi$. As a first step $d(s) = 0$. $Enqueue(Q, s)$, so $Q = \{s\}$.

$a \in adj[s], \ d(a) = 1, \quad$ and $\quad Enqueue(Q, a), \ so \ Q = \{s, a\}.$
$b \in adj[s], \ d(b) = 1, \quad$ and $\quad Enqueue(Q, b), \ so \ Q = \{s, a, b\}.$
$c \in adj[s], \ d(c) = 1, \quad$ and $\quad Enqueue(Q, c), \ so \ Q = \{s, a, b, c\}.$
$d \in adj[s], \ d(d) = 1, \quad$ and $\quad Enqueue(Q, d), \ so \ Q = \{s, a, b, c, d\}.$

Since $adj[s]$ is exhausted, $Dequeue(Q)$. So $Q = \{a, b, c, d\}$.

Now $s \in adj[a]$ and $d(s)$ is already defined, so ignore s. Again $b \in adj[a]$ but $d(b)$ is already defined, so ignore b.

Now $adj[a]$ is exhausted, so $Dequeue(Q)$. Thus $Q = \{b, c, d\}$.

Now $s, a, c \in adj[b]$ and $d(s)$, $d(a)$, $d(c)$ are already defined. Thus $adj[b]$ is exhausted, so, $Dequeue(Q)$. Thus $Q = \{c, d\}$.

Now $s, b, d \in adj[c]$ but $d(s)$, $d(b)$, $d(d)$ are already defined, so ignore s, b, d. As $e \in adj[c]$ and $d(e)$ is not already defined and, so, define $d(e) = d(c) + 1 = 2$. $Enqueue(Q, e)$. Then $Q = [c, d, e]$.

As $adj[c]$ is exhausted, $Dequeue(Q)$. Thus $Q = \{d, e\}$.

Now $s, c, e \in adj[d]$ and $d(s)$, $d(c)$, $d(e)$ are already defined. So, ignore s, c, e. As $f \in adj[d]$ and $d(f)$ is not already defined. So, define $d(f) = d(d) + 1 = 1 + 1 = 2$, $Enqueue(Q, f)$. So $Q = \{d, e, f\}$.

Since $adj[d]$ is exhausted, $Dequeue(Q)$. Then $Q = \{e, f\}$.

Now $c, d, f \in adj[e]$ but $d(c), d(d), d(f)$ are already defined, ignore c, d, f. Next, $g \in adj[e]$ and $d(g)$ is not already defined. So, define $d(g) = d(e) + 1 = 2 + 1 = 3$ and $Enqueue(Q, g)$. So $Q = \{e, f, g\}$.

Since $adj[e]$ is exhausted, $Dequeue(Q)$. $Q = \{f, g\}$.

Now $d, e, g \in adj[f]$ but $d(d), d(e), d(g)$ are already defined, so, ignore d, e, g. Now $h \in adj[f]$ and $d(h)$ is not defined. So, define $d(h) = d(f) + 1 = 2 + 1 = 3$. $Enqueue(Q, h)$ so that $Q = \{f, g, h\}$. Since $adj[f]$ has been exhausted, $Dequeue(Q)$. Then $Q = \{g, h\}$.

Now $e, f \in adj[g]$ and $d(e), d(f)$ are already defined. So, ignore e, f. As $adj[g]$ is exhausted, $Dequeue(Q)$. Then $Q = \{h\}$.

Now $f \in adj[h]$ and $d(f)$ is already defined. Ignore f. Also $adj[h]$ is exhausted. Therefore $Dequeue(Q)$, Then $Q = \varphi$. The process stops.

Every time an Enqueue operation is executed, one vertex and an edge are enjoined to the subgraph already obtained. Therefore, considering the Enqueue operations in order, the BFS-tree for the graph G is as given in Fig. 7.6.

Observe that the number of Enqueue operations while constructing BFS-tree is equal to the number of vertices of the graph reachable from the source vertex. Thus the number of Enqueue operations for a connected graph is equal to the number of vertices in the graph.

Example 7.3. Construct a **BFS**-tree for the directed graph G of Fig. 7.7 with s as a source vertex.

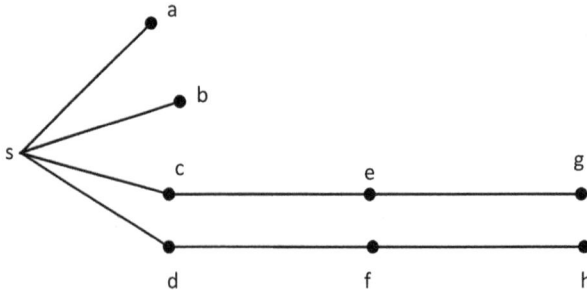

Fig. 7.6

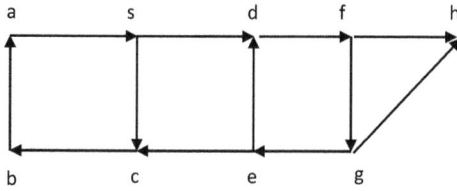

Fig. 7.7

Solution. The adjacency lists for G are

$$adj[s] = \{c, d\}, \quad adj[a] = \{s\}, \quad adj[b] = \{a\}, \quad adj[c] = \{b\},$$
$$adj[d] = \{f\}, \quad adj[e] = \{c, d\}, \quad adj[f] = \{g, h\}, \quad adj[g] = \{e, h\},$$
$$adj[h] = \varphi.$$

Initially $Q = \varphi$. Set $d(s) = 0$, $Enqueue(Q, s)$. Then $Q = \{s\}$.

$c \in adj[s]$, $d(c)$ is not already defined. Set $d(c) = 1$ and $Enqueue(Q, c)$. Then $Q = \{s, c\}$.

Now $d \in adj[s]$, $d(d)$ is not already defined. Set $d(d) = 1$ and $Enqueue(Q, d)$. Then $Q = \{s, c, d\}$.

Since $adj[s]$ is exhausted, $Dequeue(Q)$. Then $Q = \{c, d\}$.

$b \in adj[c]$ and $d(b)$ is not already defined. Set $d(b) = d(c) + 1 = 1 + 1 = 2$ and $Enqueue(Q, b)$. Then $Q = \{c, d, b\}$.

Since $adj[c]$ is exhausted, $Dequeue(Q)$. Then $Q = \{d, b\}$.

$f \in adj[d]$, $d(f)$ is not already defined, Set $d(f) = d(d) + 1 = 1 + 1 = 2$ and $Enqueue(Q, f)$. Then $Q = \{d, b, f\}$.

As $adj[d]$ is exhausted, $Dequeue(Q)$. Then $Q = \{b, f\}$.

$a \in adj[b]$, $d(a)$ is not already defined. Set $d(a) = d(b) + 1 = 2 + 1 = 3$ and $Enqueue(Q, a)$. Then $Q = \{b, f, a\}$.

Since $adj[b]$ is exhausted, $Dequeue(Q)$. Then $Q = \{f, a\}$.

$g \in adj[f]$ and $d(g)$ is not already defined. Define $d(g) = d(f) + 1 = 2 + 1 = 3$ and $Enqueue(Q, g)$. Then $Q = \{f, a, g\}$.

Again $h \in adj[f]$ and $d(h)$ is not already defined. Set $d(h) = d(f) + 1 = 2 + 1 = 3$ and $Enqueue(Q, h)$. Then $Q = \{f, a, g, h\}$. Since $adj[f]$ is exhausted, $Dequeue(Q)$. Then $Q = \{a, g, h\}$.

$s \in adj[a]$ and $d(s)$ is already defined. Ignore s. Since $adj[a]$ is exhausted, $Dequeue(Q)$. Then $Q = \{g, h\}$.

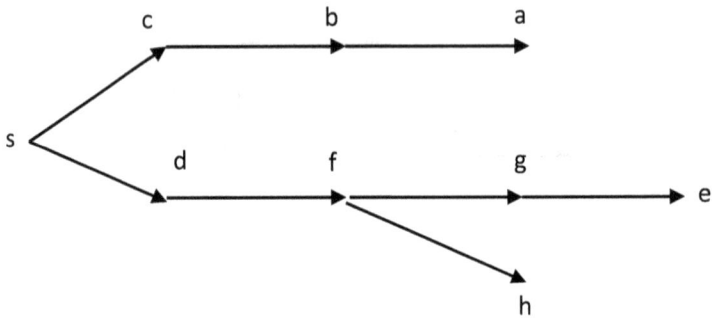

Fig. 7.8

$e \in adj[g]$ and $d(e)$ is not already defined. Set $d(e) = d(g) + 1 = 3 + 1 = 4$ and $Enqueue(Q, e)$. Then $Q = \{g, h, e\}$.

$h \in adj[g]$ and $d(h)$ has already been defined. Ignore h. As $adj[g]$ is exhausted, $Dequeue(Q)$. Then $Q = \{h, e\}$. Since $adj[h] = \varphi$, $adj[h]$ is exhausted. $Dequeue(Q)$. Then $Q = \{e\}$.

Since $c, d \in adj[e]$ and $d(c), d(d)$ are already defined, ignore c, d. As $adj[e]$ is exhausted, $Dequeue(Q)$. Then $Q = \varphi$. Therefore, the process stops here and the BFS-tree for the graph is as given in Fig. 7.8.

Exercise 7.3.

1. Construct a BFS-tree for the graph G as in Fig. 7.5 with source vertex (a) c, (b) b, (c) e.
2. Construct a BFS-tree for the graph G as in Fig. 7.7 with source vertex (a) c, (b) d, (c) f, (d) h.
3. Find a shortest path from source vertex s to each of the vertices e, g, h on running BFS on the graph G as in Fig. 7.5.
4. Find a shortest path from source vertex s to each of the vertices e, g, h for the diagraph G as in Fig. 7.7 on running BFS.
5. Prove that in a breadth-first search, the label $d(u)$ assigned to a vertex u is independent of the order in which the vertices in each adjacency list are listed.
6. Give an example to show that a BFS-tree for a graph depends on the order in which the vertices in the adjacency lists of the graph are given.

7.5. Depth-First Search

Depth-first search is such an important example of backtracking that it is also called **backtracking**. As the very name suggests, we search a given graph as deep as possible. We arbitrarily choose a vertex as a source vertex and then form a path starting at this vertex by successively adding vertices and edges, where each new edge is incident with the last vertex in the path and a vertex which is not already in the path. Given a source vertex s, we choose a vertex $u \in adj[s]$. We say that u is then searched. In BFS each vertex u visited or searched is assigned a label $d(u)$ but in the depth-first search (DFS), each vertex u is assigned a pair $(d(u), f(u))$ as a label, where as before $d(u)$ is the number given when u is first reached, $f(u)$ called the **finish time** is the number allotted when the search for u is complete. Once having chosen/discovered u in $adj[s]$, we set $d(u) = 2$ where we have already defined $d(s) = 1$. Observe that in BFS we assign $d(v) = 1$ for every $v \in adj[s]$ while $d(s) = 0$. But in this case we move on from u to some v in $adj[u]$ which has not already been reached and set $d(v) = 3$. Then we find (or choose) a w in $adj[v]$ which has not already been discovered and set $d(w) = 4$. We continue the process till we arrive at a vertex t(say), assign a value $d(t)$ to it but we cannot find a vertex z (say) in $adj[t]$ which has not been assigned a value $d(z)$. We then say that the search finished at t and assign the finish time $f(t)$ to t by setting $f(t) = d(t) + 1$. Thus t is assigned a pair $(d(t), f(t))$ which is also called **a pair of time stamps**. Suppose that in the above process we had reached t from a vertex p (say), i.e., $p = \pi(t)$ is a parent or a predecessor of t. Then we search for a vertex q (say) in $adj[p]$ which has not already been reached, i.e., for which a d-value has not been given. We then set $d(q) = f(t) + 1$. We continue the onward search and come to a vertex a (say) to which we assign a d-value $d(a)$ but cannot find a vertex in $adj[a]$ which has not already been assigned a d-value. We then assign the finish time $f(a) = d(a) + 1$ and backtrack to $\pi(a)$. Then look for a vertex in $adj[\pi(a)]$ which has not already been reached. Continue the process till we finally backtrack to s and cannot continue the search from s. The finish time $f(s)$ is taken as the largest finish time so far $+1$. The scanning with s is then complete. If there is a vertex x (say) which has not been reached and scanned, we begin the search with x as a source vertex. The process stops when all the vertices in the graph are searched.

To fix up our ideas about DFS, we consider a couple of examples.

Example 7.4. Consider the graph as in Fig. 7.9 with source vertex s.

The adjacency lists of the vertices are

$$adj[s] = \{a, c, d, e\}, \quad adj[a] = \{s, b, c\}, \quad adj[b] = \{a, c\},$$
$$adj[c] = \{s, a, b, e\}, \quad adj[d] = \{s, e, f, g\}, \quad adj[e] = \{s, c, d, g\},$$
$$adj[f] = \{d, g, h\}, \quad adj[g] = \{d, e, f\}, \quad adj[h] = \{f\}.$$

Set $d(s) = 1$. Now $a \in adj[s]$, set $d(a) = 2$. Both b, c are in $adj[a]$ which have not been reached earlier. We choose b and set $d(b) = 3$. The only vertex in $adj[b]$ which has not been reached is c and we set $d(c) = 4$. The only element in $adj[c]$ which has not been visited earlier is e and we set $d(e) = 5$. Now d, g are two vertices in $adj[e]$ which have not been visited earlier. We choose d and set $d(d) = 6$. Out of the two vertices f, g in $adj[d]$ which have not been visited, we choose f and set $d(f) = 7$. Again, out of the two vertices g, h in $adj[f]$ which have not been visited, choose g and set $d(g) = 8$. Since all the vertices in $adj[g]$ are already visited, we cannot go further and, so, we set $f(g) = 9$.

In the above search, we reached g from f. So, we backtrack to f. The only vertex in $adj[f]$ that has not been visited is h. So, we set $d(h) = 10$. The only vertex in $adj[h]$ has already been visited and, so, the search for h is complete. Therefore, set $f(h) = 11$.

We reached h from f, i.e., $\pi(h) = f$, so, we backtrack to f. All the three vertices in $adj[f]$ have already been visited. So, search at f is complete (or finished). Set $f(f) = 12$. Now $\pi(f) = d$. So, backtrack to d. All the vertices in $adj[d]$ have already been visited. So, the search at d is complete. Set $f(d) = 13$. As $\pi(d) = e$, we backtrack to e. All the vertices in $adj[e]$ are already visited. Thus, the search at e is complete. Set $f(e) = 14$.

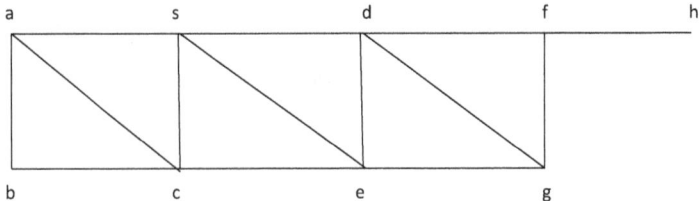

Fig. 7.9

Now $\pi(e) = c$ and all the vertices in $adj[c]$ are visited. So, the search at c is complete. Set $f(c) = 15$. Again, $\pi(c) = b$ and the vertices in $adj[b]$ have already been visited. Thus, the search at b is complete. Set $f(b) = 16$. Backtrack to $\pi(b) = a$. The vertices in $adj[a]$ have already been visited. Set $f(a) = 17$. Now $\pi(a) = s$ and all the vertices in $adj[s]$ have been visited. So, the search at s is complete. Set $f(s) = 18$.

Since all the vertices in G are scanned, DFS terminates. The graph arrived at the end is then given in Fig. 7.10.

When there are more than one choices for a vertex in $adj[u]$, different choices for this new vertex will, in general, lead to different trees at the end of DFS. A different tree that results on running DFS on the graph of Fig. 7.9 is as given in Fig. 7.11.

Observe that in the DFS, although all the vertices are visited, there may be some edges which are not traversed. Edges which are not traversed in DFS are called **back edges** or **backward edges** and those that are traced are called **forward edges**.

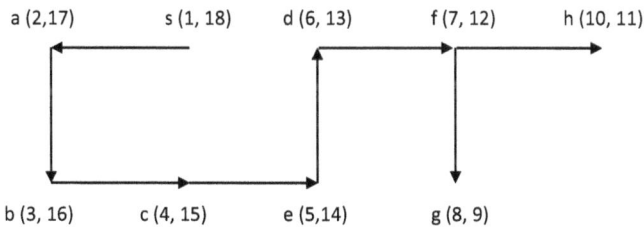

Fig. 7.10

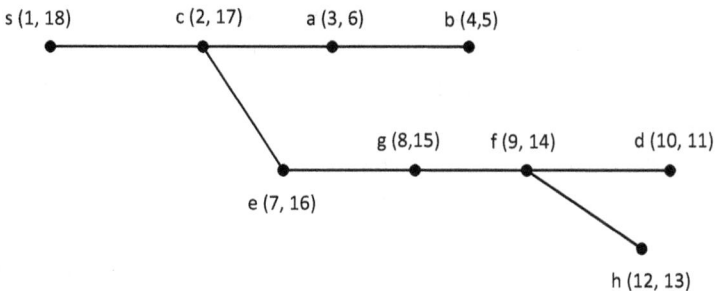

Fig. 7.11

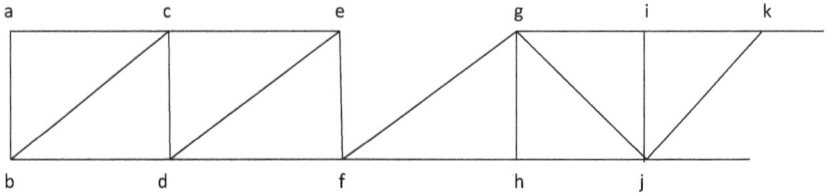

Fig. 7.12

In Example 7.4, while for tree of Fig. 7.10, the backward edges are $(s, c), (a, c), (s, e), (s, d), (e, g), (d, g)$ but in the case of tree of Fig. 7.11, the backward edges are $(s, a), (s, e), (s, d), (s, c), (d, g), (d, e)$.

Example 7.5. Consider the graph of Fig. 7.12 to which we apply DFS with f as source vertex.

The adjacency lists of the vertices are

$$adj[a] = \{b, c\}, \qquad adj[b] = \{a, c, d\}, \quad adj[c] = \{a, b, d, e\},$$
$$adj[d] = \{b, c, e, f\}, \quad adj[e] = \{c, d, f\}, \quad adj[f] = \{d, e, g, h\},$$
$$adj[g] = \{f, h, i, j\}, \quad adj[h] = \{f, g, j\}, \quad adj[i] = \{g, j, k\},$$
$$adj[j] = \{g, h, i, k\}, \quad adj[k] = \{i, j\}.$$

For the source vertex f set $d(f) = 1$. Now $d \in adj[f]$ and it has not been visited earlier, so set $d(d) = 2$. Now $b \in adj[d]$ and it has not been visited earlier, so set $d(b) = 3$. Now $a \in adj[b]$ and it has not been visited earlier, so set $d(a) = 4$. The only vertex in $adj[a]$ which has not been visited earlier is c. Therefore take $d(c) = 5$. As e is the only vertex in $adj[c]$ which has not been visited earlier, we set $d(e) = 6$.

Since all the vertices in $adj[e]$ have already been visited, search is complete at the vertex e. So, we set $f(e) = 7$ and backtrack to $\pi(e) = c$. All the vertices in $adj[c]$ have already been visited, search at the vertex c is complete. Set $f(c) = 8$ and backtrack to $\pi(c) = a$. Both the vertices in $adj[a]$ have been visited earlier and the search at the vertex a is complete. Therefore set $f(a) = 9$ and backtrack to $\pi(a) = b$. Again, all the vertices in $adj[b]$ have been visited earlier and the search at the vertex b is complete. Therefore set $f(b) = 10$ and backtrack to $\pi(b) = d$. All the vertices in $adj[d]$ have been visited earlier and the search at the vertex d is complete. Set $f(d) = 11$ and backtrack to $\pi(d) = f$.

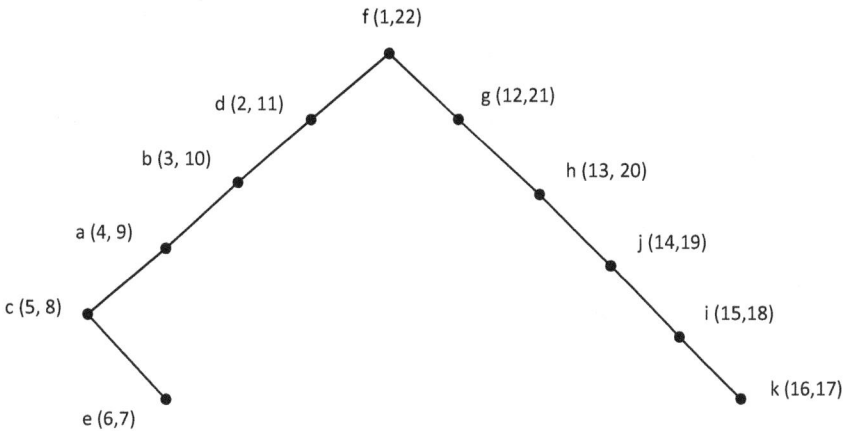

Fig. 7.13

Now $g \in adj[f]$ and it has not been visited earlier. Set $d(g) = 12$. As $h \in adj[g]$ is not already visited, set $d(h) = 13$. The only vertex in $adj[h]$ that is not visited earlier is j. So, set $d(j) = 14$. Vertex $i \in adj[j]$ is not visited earlier. So, set $d(i) = 15$. The only vertex in $adj[i]$ that has not been visited earlier is k. So, set $d(k) = 16$.

Both the vertices in $adj[k]$ have already been visited and, so, the search at the vertex k is complete. Set $f(k) = 17$ and backtrack to $\pi(k) = i$. As all the vertices in $adj[i]$ have already been assigned a d-value, the search at the vertex i is complete. Set $f(i) = 18$ and backtrack to $\pi(i) = j$. All the vertices in $adj[j]$ have been visited already. Therefore search at j is complete. Set $f(j) = 19$ and backtrack to $\pi(j) = h$. Again all the vertices in $adj[h]$ have been assigned a d-value already, the search at the vertex h is complete. Set $f(h) = 20$ and backtrack to $\pi(h) = g$. All the vertices in $adj[g]$ have already been visited and, so, the search at the vertex g is complete. Set $f(g) = 21$ and backtrack to $\pi(g) = f$. Since all the vertices in $adj[f]$ have already been visited, search at the vertex f is complete. Set $f(f) = 22$.

We have reached the source vertex. Also all the vertices in the graph have been reached. Therefore, the process of DFS is complete and the final tree obtained in this way is given in Fig. 7.13.

In this case (f, e), (d, e), (d, c), (b, c), (f, h), (g, i), (g, j), (j, k) are all backward edges.

While going along edges emanating from f on the left side there are two choices and two choices for going on the right. Similarly, for other vertices there are in general more than one choices. Making some different choices for edges emanating from a given vertex, the final tree that is obtained when DFS is complete will be different from the one obtained in Fig. 7.13.

Example 7.6. Consider the graph as in Fig. 7.14 with source vertex f.

The adjacency lists of the vertices are

$adj[a] = \{c\}$, $adj[b] = \{c\}$, $adj[c] = \{a, b, d\}$,

$adj[d] = \{c, e, f\}$, $adj[e] = \{d\}$, $adj[f] = \{d, g, h\}$,

$adj[g] = \{f\}$, $adj[h] = \{f, i, j\}$, $adj[i] = \{h\}$,

$adj[j] = \{h, k\}$, $adj[k] = \{j\}$.

We are to run DFS with source vertex f. Set $d(f) = 1$. Now $d \in adj[f]$ and d is not visited yet. Set $d(d) = 2$. Next, $c \in adj[d]$ which is not yet assigned a d-value time stamp, so we define $d(c) = 3$. As $a \in adj[c]$ and is not yet visited, we set $d(a) = 4$. As c is the only vertex in $adj[a]$ and it has already been visited, the search at the vertex a is complete. So, we set $f(a) = 5$. We then backtrack to $\pi(a) = c$. The only vertex in $adj[c]$ which has not been visited earlier is b, so we take $d(b) = 6$. The only vertex in $adj[b]$ is c and it has already been visited. Thus the search at the vertex b is complete. We set $f(b) = 7$ and backtrack to $\pi(b) = c$. Since

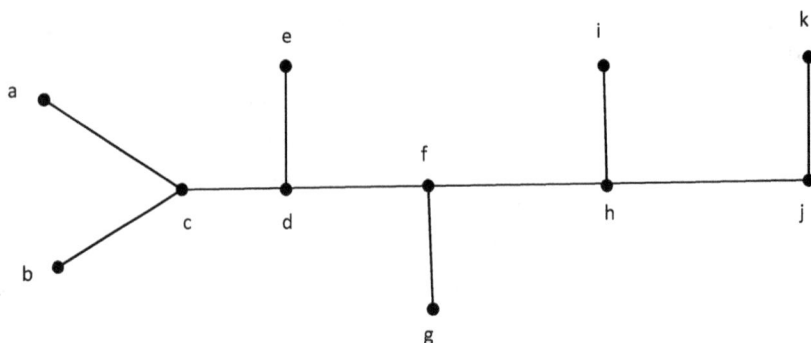

Fig. 7.14

all the vertices in $adj[c]$ have already been visited, the search at the vertex c is complete. So, we set $f(c) = 8$ and backtrack to $\pi(c) = d$. The vertex $e \in adj[d]$ is not visited earlier. We set $d(e) = 9$. The only vertex in $adj[e]$ is already visited, so the search at e is complete. We set $f(e) = 10$ and backtrack to $\pi(e) = d$. All the vertices in $adj[d]$ have already been assigned a d-value. So the search at d is complete. We set $f(d) = 11$ and backtrack to $\pi(d) = f$. Now g is in $adj[f]$ and has not been visited earlier. We set $d(g) = 12$. The only vertex in $adj[g]$ is already visited. Therefore, the search at g is complete. Set $f(g) = 13$ and backtrack to $\pi(g) = f$. The only vertex in $adj[f]$ which has not been visited is h, so we set $d(h) = 14$. The vertex i in $adj[h]$ is not already visited, so we set $d(i) = 15$. The only vertex in $adj[i]$ is already visited and the search at i is complete. We set $f(i) = 16$ and backtrack to $\pi(i) = h$. The only vertex in $adj[h]$ which is not yet visited is j. We set $d(j) = 17$. The only vertex in $adj[j]$ which is not yet visited is k and we set $d(k) = 18$. The only vertex in $adj[k]$ is already visited, so the search at k is complete. We set $f(k) = 19$ and backtrack to $\pi(k) = j$. Since both the vertices in $adj[j]$ are already visited, the search at j is complete. We set $f(j) = 20$ and backtrack to $\pi(j) = h$. All the vertices in $adj[h]$ are already visited. Thus the search at h is complete. We set $f(h) = 21$ and backtrack to $\pi(h) = f$. All the three vertices in $adj[f]$ have already been visited so that the search at f is complete. We set $f(f) = 22$. We are back to the source vertex f. Since all the vertices in the graph of Fig. 7.14 have been searched, therefore the DFS is complete. The resulting graph when DFS is complete is given in Fig. 7.15.

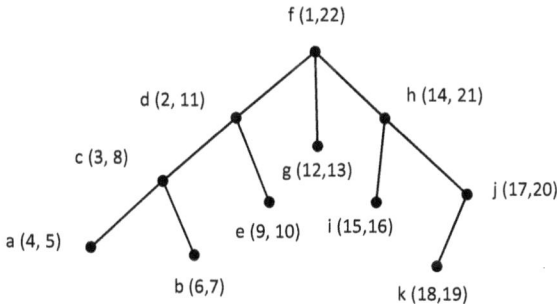

Fig. 7.15

Observe that there are no backward edges in this graph.

At each of the vertices c, d, f, h, j there are more than one choices for going along an edge emanating from the vertex. Therefore, making a different choice of an edge for going from such a vertex results in a different final tree than the one obtained in Fig. 7.15 when DFS is complete.

Example 7.7. Consider the graph as given in Fig.7.16.

We apply DFS on this graph with source vertex a. The adjacency lists of the vertices are

$$adj[a] = \{b, d, e, f, g\}, \quad adj[b] = \{a, c, d\}, \quad adj[c] = \{b, d\},$$
$$adj[d] = \{a, b, c\}, \quad adj[e] = \{a, f, g\}, \quad adj[f] = \{a, e\},$$
$$adj[g] = \{a, e\}.$$

For the source vertex a, set $d(a) = 1$. Vertex b is in $adj[a]$ and it has not been visited earlier. Set $d(b) = 2$. c is a vertex in $adj[b]$ which has not been visited earlier. Set $d(c) = 3$. The only vertex in $adj[c]$ that has not been visited earlier is d. Set $d(d) = 4$. Since all the vertices in $adj[d]$ have already been visited, search at the vertex d is complete. Set $f(d) = 5$ and backtrack to $\pi(d) = c$.

Both the vertices in $adj[c]$ have already been assigned a d-value so that the search at the vertex c is complete. Set $f(c) = 6$ and backtrack to $\pi(c) = b$. Again, all the vertices in $adj[b]$ have already been visited so that the search at the vertex b is complete. Set $f(b) = 7$ and backtrack to $\pi(b) = a$. Now $e \in adj[a]$ and it has not been visited earlier. Set $d(e) = 8$. Next, f in $adj[e]$ has not been visited earlier. So, set $d(f) = 9$. Both

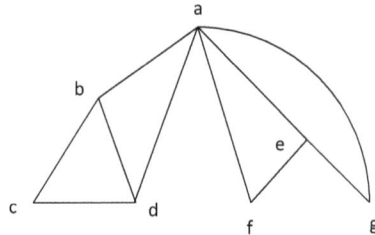

Fig. 7.16

the vertices in $adj[f]$ have already been assigned a time stamp d-value. Therefore, the search at the vertex f is complete. Set $f(f) = 10$ and backtrack to $\pi(f) = e$.

Now g is the only vertex in $adj[e]$ which has not been visited earlier. Set $d(g) = 11$. Both the vertices in $adj[g]$ have already been visited. So the search at the vertex g is complete. Set $f(g) = 12$ and backtrack to $\pi(g) = e$. All the vertices in $adj[e]$ have already been visited. Thus search at the vertex e is complete. Set $f(e) = 13$ and backtrack to $\pi(e) = a$. All the vertices in $adj[a]$ have already been visited and, so, the search at the vertex a is complete. Set $f(a) = 14$.

Since we have arrived at the source vertex and all the vertices in the graph have been visited, running DFS is complete. The final tree traced by this search is as given in Fig. 7.17.

The tree edges are (a, b), (b, c), (c, d), (a, e), (e, f), (e, g) and the backward edges are (b, d), (a, d), (a, f), (a, g).

So far the examples we have considered are examples of graphs which are undirected. But for running DFS on a graph we could equally well consider directed graphs. However, before considering an example of a directed graph, we may mention that edges in a directed graph on running DFS are divided into four categories as against two categories for undirected graphs. Let G be a directed graph and suppose DFS is run on G. We define

1. **Tree edges:** Those edges of G which are traced while running DFS on G.
2. **Forward edges:** Edges (u, v) which are not tree edges and u is an ancestor of v for the DFS run.

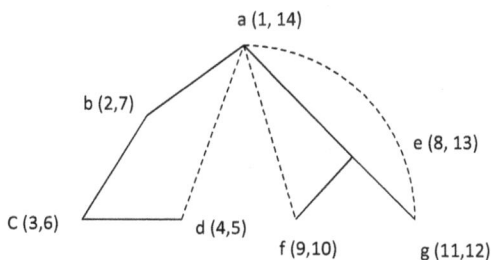

Fig. 7.17

3. **Backward or Back edges:** These are those edges (u, v) of G which are not tree edges and v is an ancestor of u in the DFS traversal.
4. **Cross edges:** Edges (u, v) which are not tree edges and neither u is an ancestor of v nor v is an ancestor of u in the DFS traversal.

Example 7.8. Perform DFS on the directed graph of Fig. 7.18 with a as a source vertex, find DFS forest and also give its forward, backward and cross edges.

Solution. The adjacency list of the vertices of the graph are

$$adj[a] = \{b, d, f\}, \ adj[b] = \{c, f\}, \ adj[c] = \{d\},$$

$$adj[d] = \{b\}, \qquad adj[e] = \{d, f\}, \ adj[f] = \{d\}.$$

Now $b \in adj[a]$. Set $d(a) = 1, d(b) = 2$. Now $c \in adj[b]$ and c has not been visited earlier. Set $d(c) = 3$. As d is in $adj[c]$ and d has not been visited earlier, we set $d(d) = 4$. The only vertex in $adj[d]$ is already visited, we set $f(d) = 5$ and backtrack to $\pi(d) = c$. The only vertex in $adj[c]$ is already visited. So, we set $f(c) = 6$ and backtrack to $\pi(c) = b$. Now f in $adj[b]$ is not visited earlier, so we choose f and set $d(f) = 7$. Since the only vertex in $adj[f]$ is already discovered, we set $f(f) = 8$ and backtrack to $\pi(f) = b$. Both the vertices in $adj[b]$ are already discovered, so we set $f(b) = 9$ and backtrack to $\pi(b) = a$. All the vertices in $adj[a]$ are already visited, so we set $f(a) = 10$.

We have come back to the source vertex. Thus search for a is complete. However, the vertex e is not yet visited. So, we set $d(e) = 11$. As both the vertices in $adj[e]$ have already been discovered, we set $f(e) = 12$.

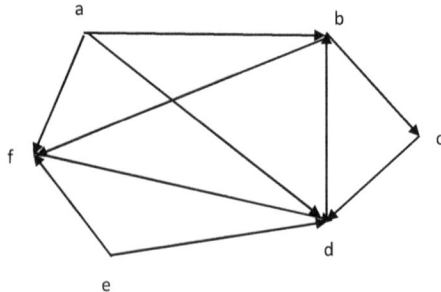

Fig. 7.18

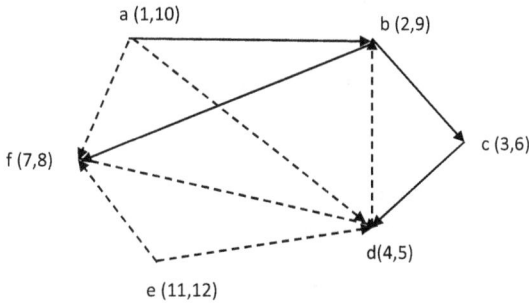

Fig. 7.19

The corresponding DFS forest is given in Fig. 7.19.

The tree edges are $(a, b), (b, c), (c, d), (b, f)$; forward edges are $(a, d), (a, f)$; the back edge is (d, b) and the cross edges are $(e, d), (e, f), (f, d)$.

Suppose we reverse the direction of the edge (d, b), i.e., the edge becomes (b, d). Then, the pairs of time stamps for the vertices remain the same but there is no back edge in this case.

Theorem 7.6 (Proper descendent theorem). *Let G be a directed or undirected graph and u, v be vertices of G. Then v is a proper descendent of u in the depth-first forest if and only if $d(u) < d(v) < f(v) < f(u)$.*

Proof. Suppose that $d(u) < d(v) < f(v) < f(u)$. The first inequality in this relation shows that v is unvisited at the time $d(u)$ of discovery of u. The relation $f(v) < f(u)$ shows that the search for v finishes, i.e., all the edges incident with or outgoing from v are examined when the depth-first search for u is still on. Therefore v occurs on the path being traced from u in the depth-first search and we backtrack to u and finish the search for u. Hence v is a proper descendent of u in the depth-first search.

Conversely, suppose that v is a proper descendent of u in the depth-first search forest. Then v is discovered after u is discovered/visited and also the search backtracks ultimately to u only after search is finished for v. Therefore $d(u) < d(v)$ and $f(v) < f(u)$. Since $d(w) < f(w)$ always, we have $d(u) < d(v) < f(v) < f(u)$. $\square$

Corollary 7.7. *Let G be a directed graph and u, v be vertices in G. In the depth-first forest, a non-tree edge (u, v) is a back edge if and only if*

$d(v) < d(u) < f(u) < f(v)$, *and it is a forward edge if and only if*
$d(u) < d(v) < f(v) < f(u)$.

For cross edges in the depth-first forest of a directed graph, we have
the following theorem.

Theorem 7.8 (Cross edge theorem). *Let G be a directed graph and (u, v)
be an edge of G. Then (u, v) is a cross edge if and only if either $f(u) <
d(v)$ or $f(v) < d(u)$.*

Proof. Exercise.   □

Theorem 7.9. *In a depth-first search of a directed or undirected graph
$G = (V, E)$, a vertex v is a descendent of a vertex u if and only if at the
time $d(u)$ the vertex u is discovered, the vertex v can be reached from u by
a path consisting entirely of not yet discovered vertices.*

Proof. Suppose that v is a descendent of u in a depth-first search of the
graph G. Let w be a vertex on the path between u and v in the depth-
first search. Then $d(w) > d(u)$ and, therefore, w is discovered after u is
discovered. Therefore v can be reached by a path consisting of vertices not
yet discovered.

Conversely, suppose that vertex v is reachable from u along a path
consisting of vertices not yet discovered at the time $d(u)$. Let w be the
closest vertex to u along the path from u to v. Then $d(u) < d(w) < f(w)$.
Now $w \in adj[u]$ and, so, the search for w must be finished before it is
finished for u. Therefore $f(w) < f(u)$, i.e., $d(u) < d(w) < f(w) <
f(u)$ so that w is a descendent of u. Working with w in place of u, we can
prove that the next vertex on the path is a descendent of w and, therefore,
of u. Continuing this argument (the number of elements on the path being
finite), let us suppose that all the vertices on this path except possibly v is
a descendent of u. Let t be the vertex immediately before v on this path.
Working with t for u and v for w, we can prove that v is a descendent of t.
Therefore v is a descendent of u.

In the procedure $DFS(G, graph)$, DFS-visit(v) is executed for each
$v \in V$. Also for each $v \in V$, DFS-visit(v) examines each edge (v, w) and
exactly once (we do not count its traversal when backtracking) and, so, the
time taken by DFS-visit is $\sum adj(v) = O(|E|)$. Therefore, the running
time of $DFS(G, graph)$ is $O(|V| + |E|)$.   □

Algorithm 7.5. Procedure *DFS* (*G*, graph)

$G = (V, E)$ is a directed or undirected graph.

1. initially for every $v \in V$, make v unvisited
2. $\pi(v) = nil$ and $time = 0$
3. for each vertex v in V, if v is unvisited
4. then *DFS*-visit (v)

Algorithm 7.6. Procedure *DFS*-Visit (v)

---

1. $G = (V, E)$ is a given graph with source vertex s
2. if the vertex v is just visited
3. then set $time = time + 1$
4. label v with $d(v)$ (i.e., the discovery time)
5. for each $w \in adj[v]$, explore $edge(v, w)$
6. if w is unvisited, set $\pi(w) = v$ and *DFS*-visit(w)
7. backtrack to v i.e., the search at v is finished
8. set $f(v) =$ the finish time $= 1+$ largest{discovery time or finish time so far}

7.6. Directed Acyclic Graphs

In this section, we consider directed acyclic graphs and then consider topological sort of the vertices of a directed acyclic graph. We also obtain a necessary and sufficient condition for acyclicity of a directed graph.

A directed graph is called **acyclic** if the graph does not have any cycles. A directed acyclic graph is also called a **dag**. Observe that every tree is a dag but every dag need not be a tree. Moreover, every directed graph need not be a dag. For example, Figs. 7.20 (a)–7.20 (c) are, respectively, a tree, a dag but not a tree and a directed graph but not a dag.

As an application of depth-first search we get the following characterization of a directed acyclic graph.

Theorem 7.10. *A directed graph G is acyclic if and only if a depth-first search of G does not encounter a back edge.*

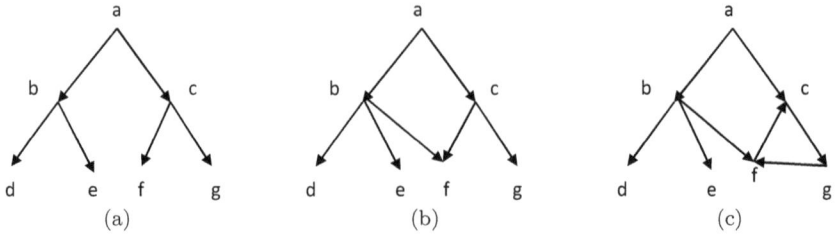

Fig. 7.20

Proof. Suppose that (u, v) is a back edge when *DFS* is run on G. Then u is a descendent of v in the *DFS* tree with some source vertex x (say). Thus there is a path from v to u in the graph. Composing this path with the edge (u, v) we get a cycle in G. Therefore G is not a dag.

Conversely, suppose that the graph G has a cycle, say $u = v_1, v_2, \ldots, v_k = v, v_{k+1} = u$. When *DFS* is performed on G, every vertex of G is discovered and, so, has a discovery time. Among all the vertices on the cycle, let u be a vertex which has the smallest discovery time $d(u)$. At the discovery time $d(u)$ of u, the other vertices in the cycle are not yet visited or discovered. We can then run *DFS* on G in such a fashion that $d(v_{i+1}) = d(v_i)+1$ for $i = 1, 2, \ldots, k-1$. Thus in this way of performing *DFS* on G, $v = v_k$ is a descendent of u. Therefore, the edge (v, u) which goes from a descendent v of u to u is a back edge in the corresponding *DFS* forest. ☐

7.6.1. Topological sort

The vertices of a dag $G = (V, E)$ are said to be in **topological order** if the vertices can be arranged as a sequence such that if (u, v) is an edge in G, then the vertex u appears on the left of v in this sequence. The process of assigning such a linear ordering to the vertices of a dag is called **topological sort**. The following simple procedure topologically sorts the vertices of a dag.

Theorem 7.11. *Let* $G = (V, E)$ *be a directed acyclic graph. Then the procedure Topological-Sort produces a topological sort of the vertices of* G.

Algorithm 7.7. Topological-Sort (G)

1. $G = (V, E)$ is a dag with n vertices
2. perform $DFS(G)$ assigning a pair of time stamps $(d(u), f(u))$ for every vertex u of G
3. as soon as search finishes at a vertex u, place u at the beginning of the sequence already arrived at
4. return the sequence of vertices

Alternatively

1. $G = (V, E)$ is a dag with n vertices
2. perform $DFS(G)$ assigning a pair of time stamps $(d(u), f(u))$ for every vertex u of G
3. if u is a first vertex for which search finishes, set $u = v_n$
4. if $v_{i+1}, \ldots, v_n$ are already obtained, and u is the next vertex at which search finishes, call $u = v_i$
5. apply 4 till all the vertices are exhausted
6. return the sequence $v_1, v_2, \ldots, v_n$

Proof. Suppose that $DFS(G)$ is performed on G and for every vertex u, the pair of time stamps $(d(u), f(u))$ is obtained. As a matter of fact, we only need the finishing time for every vertex u. Let (u, v) be an edge in the dag G which is being explored by $DFS(G)$. This means that u has just been discovered and v cannot be already visited but is not yet finished. For otherwise, v will be an ancestor of u and (u, v) will be a back edge. But G has no back edge. Therefore v is unvisited at the time $d(u)$ or search has already been finished for v. Suppose that search has already been finished for v at time $d(u)$. Then $f(v) < d(u) < f(u)$ and, therefore, v has already occurred in the sequence and u is placed in the sequence later. Thus u is placed somewhere on the left side of v in the sequence. Next, suppose that v is not yet visited at the time $d(u)$. Then v becomes a descendent of u and, therefore, by Theorem 7.6 $d(u) < d(v) < f(v) < f(u)$. Again, it follows that u appears somewhere to the left of v in the sequence of vertices being constructed. Since every edge of G is explored during DFS, the proof is complete. $\square$

Examples on the application of Theorem 7.9 and the topological sort will be considered at the end of the next section where we consider strongly connected components and reduced graph of a directed graph.

7.7. Strongly Connected Components

Let $G = (V, E)$ be a directed graph. For $u, v \in V$, say that u and v are **equivalent** if there is a path from u to v and there is a path from v to u. This relation is clearly symmetric and transitive. For every vertex u in V, we can suppose that there is a path of length 0 from u to u. Then every $u \in V$ is equivalent to itself. Hence we get an equivalence relation defined on V. Let V_i, $1 \leq i \leq k$, be the equivalence classes determined by this equivalence relation. Then every V_i is a non-empty subset of V, $V_i \cap V_j = \varphi$ for every $i \neq j$, $1 \leq i, j \leq k$ and $V = \bigcup_{1 \leq i \leq k} V_i$. For i, $1 \leq i \leq k$, let

$$E_i = \{e \in E | \text{ both the initial and terminal vertices of } e \text{ are in } V_i\}.$$

The E_i are subsets of E and $E_i = \varphi$ if and only if V_i consists of a single element (unless there is a loop around the single element of V_i). For $i \neq j$, $E_i \cap E_j = \varphi$ but we need not have $E = \bigcup E_i$ as there may be edges from a u in V_i to a v in V_j for some $i \neq j$.

For every i, $1 \leq i \leq k$, $G_i = (V_i, E_i)$ is a subgraph of G. The subgraph $G_i = (V_i, E_i)$ is called a **strongly connected component** or simply a **strong component** of G. Observe that every V_i is a maximal subset of V in which there is a path from any vertex in V_i to any other vertex in V_i. Also observe that every vertex in V is in a strong component of G.

For every strong component $G_i = (V_i, E_i)$ of G consider a vertex C_i (not of the graph G) or regard every strong component G_i as a vertex C_i and for vertices C_i, C_j say that there is an edge from C_i to C_j if there is an edge in G from some vertex in V_i to some vertex in V_j. The resulting graph is called the **reduced graph** of G.

Suppose that in the reduced graph of G there is an edge from C_i to C_j. Then there exists u_1 in V_i and v_1 in V_j with an edge $e = (u_1, v_1)$. Consider any u in V_i and v in V_j. Then $u, u_1 \in V_i$ and, so, there is a path from u to u_1. Similarly there is a path from v_1 to v. Composing the path from u to

u_1, the edge $e = (u_1, v_1)$ and the path from v_1 to v, we get a path from u to v in G.

Theorem 7.12. *The reduced graph of a directed graph is a directed acyclic graph or a dag.*

Proof. Let $G = (V, E)$ be a directed graph, $G_i = (V_i, E_i)$, $i = 1, 2, \ldots, k$ be strong components of G and $C_1, C_2, \ldots, C_k$ be vertices of the reduced graph of G. Suppose that there is a cycle $(C_{i_1}, C_{i_2}, \ldots, C_{i_j})$ in the reduced graph of G. Since there is a path from every element of V_{i_r} to every vertex of $V_{i_{r+1}}$ for $r = 1, 2, \ldots, j-1$ and also a path from every element of V_{i_j} to every element of V_{i_1}, if $u \in V_{i_1}$ and $v \in V_{i_j}$, there is a path from u to v and also a path from v to u. Thus $u \in V_{i_1} \cap V_{i_j}$, but $V_{i_1} \cap V_{i_j} = \varphi$ if $i_1 \neq i_j$ and we have a contradiction. Hence there is no cycle in the reduced graph of G. $\qquad \square$

The following procedure gives a method to compute the strong components of a given directed graph.

Algorithm 7.8. Strongly Connected-Components (G)

---

1. $G = (V, E)$ is a given directed graph
2. perform a *DFS* on G and find the finishing time $f(u)$ of every vertex u of G
3. compute the reverse graph G^t of G, i.e., G^t is obtained from G by reversing the direction of every edge of G
4. perform *DFS* on G^t starting the search on the vertex u with highest f-value $f(u)$ as source. When the *DFS* finishes on a vertex, perform *DFS* now with a source vertex v where $f(v)$ is the largest of all the f-values of vertices of G^t not visited by *DFS*-visit(u). Repeat the process of running *DFS* on the remaining vertices choosing source vertex with largest f-value for the vertices unvisited
5. each tree in the resulting *DFS* forest is a strongly connected component of G

Theorem 7.13. *Let G be a directed graph with G^t as its reverse graph. Then every strongly connected component of G is a DFS tree in the DFS forest of the reverse graph G^t as obtained in its traversal as in number 4* *of the Algorithm 7.8 and every tree in the DFS forest of G^t in its traversal as in 4 of Algorithm 7.8 is a strongly component of G.*

Proof. Suppose that u, v are vertices in the same strong component C of G. Then u, v are reachable from each other in G^t as well.

Suppose that in the depth-first search of G^t as in number 4 of Algorithm 7.8 the search with source vertex x reaches one of these two vertices, say u. When search reaches u, the vertices on a path from u to v are yet unvisited. It then follows from Theorem 7.8 that v is a descendent of u in this search. Therefore both u, v are on this tree with root x. Since every vertex of G (and, so, of G^t) is on some tree in the *DFS* forest of G^t, therefore, all vertices in the strong component C are on this tree with root x.

Conversely, suppose that u, v are in the same spanning tree of the *DFS* forest of G^t. Let x be a root of the tree containing u and v. Then u is a descendent of x and, therefore, there is a path from x to u in G^t. But then there is a path from u to x in G.

Since x is a root of the tree in *DFS* forest of G^t, it follows from condition 4 of Algorithm 7.8 that finishing time $f(x)$ of x in the *DFS* of G is greater than the finishing time $f(u)$ of u. Therefore, in the *DFS* of G, search at u is finished before it terminates at x. If in $DFS(G)$, the search at u initiated before it begins at x, then there being a path from u to x in G it follows from Theorem 7.6 that x is a descendent of u. Therefore $f(x) < f(u)$ which is a contradiction. Therefore, in $DFS(G)$, search at u occurs after it is initiated at x. Thus $d(x) < d(u) < f(u) < f(x)$ and it follows from Theorem 7.6 that u is a descendent of x in $DFS(G)$. There is, therefore, a path from x to u.

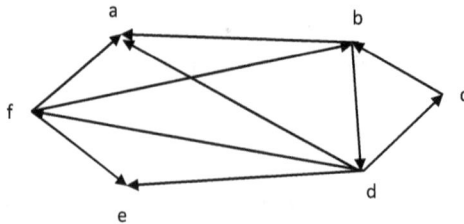

Fig. 7.21

Hence x and u are in the same strong component C of G. The same reasoning also shows that x and v are in the same strong component C' of G. Since x is a vertex in both the strong components, $C = C'$ and u, v are in the strongly connected component C of G. Therefore, all the vertices on the tree with vertex x in the *DFS* forest of G^t are contained in a strongly connected component C of G. Consider any w in C. Then x and w are in C and, therefore, x, w are on a spanning tree in the *DFS* forest of G^t. Since every vertex of G^t is on one and only one tree in the *DFS* forest of G^t and x is already on the tree containing u and v, w is also on this tree. This proves that every tree in the *DFS* forest of G^t is a strongly connected component of G and conversely. □

We now consider some examples of Topological sort and strongly connected component algorithms.

Example 7.9. Consider the graph of Fig. 7.18. As already seen in Example 7.8, (a, b) is a back edge in the graph. Therefore, the graph is not acyclic. As already computed in that example, the finishing times of the vertices are $f(a) = 10$, $f(b) = 9$, $f(c) = 6$, $f(d) = 5$, $f(e) = 12$ and $f(f) = 8$. If we order the vertices by the decreasing order of their finish times, we get the listing

$$e, a, b, f, c, d.$$

Since (d, b) is an edge in the graph and d does not appear on the left of b in this ordering, the ordering does not provide a topological ordering of the vertices of the graph.

Thus, though ordering the vertices of a dag provided by the decreasing order of their finishing times does provide a topological sort of the vertices, it fails for directed graphs which are not acyclic.

Reverse graph of the graph of Fig. 7.18 is given in Fig. 7.21.

The adjacency lists of the reverse graph G^t of G are as follows:

$$adj[a] = \varphi, \quad adj[b] = \{a, d\}, \quad adj[c] = \{b\},$$

$$adj[d] = \{a, c, e, f\}, \quad adj[e] = \varphi, \quad adj[f] = \{a, b, e\}.$$

Finishing times on running *DFS* on G as computed in Example 7.8 in decreasing order are $f(e) = 12$, $f(a) = 10$, $f(b) = 9$, $f(f) = 8$, $f(c) = 6$, $f(d) = 5$.

Perform *DFS* on G^t starting with e as a source vertex. Since $adj[e] = \varphi$, the search begins and terminates at e. Next a is the vertex with next largest f-value in the *DFS* of G. We now perform *DFS*-Visit(a). Again $adj[a] = \varphi$ and the search begins and finishes at a. The next vertex in G^t with next largest f-value is b. Then perform *DFS*-Visit(b) on G^t. We first trace the edge (b, a) but the same is not allowed and we backtrack to b. Then we trace the edges $(b, d), (d, c)$. Now we backtrack to d and can trace the edge (d, f). Although there are edges $(d, e), (d, a)$ but since a, e are already visited, the edges cannot be added to the *DFS* forest of G^t. Since all the three vertices in $adj[f]$ have already been visited, we cannot trace any of the edges with initial vertex f. We backtrack to d. Since all vertices in $adj[d]$ have been explored, the search finishes at d and we backtrack to b. Since there are no more vertices to which we can go from b, the search finishes at b. The *DFS* forest of G^t is then as given in Fig. 7.22.

Thus there are three strongly connected components of G. Let us call the component containing the only vertex e as C_1, containing the only vertex a as C_2 and the third component as C_3. Since there is an edge (e, d) as also (e, f) in G, where e is in C_1 and d as well as f are in C_3, there is an edge from C_1 to C_3. Similarly, because of the presence of an edge (a, b) or (a, d) or (a, f) in G, there is an edge from C_2 to C_3. Therefore,

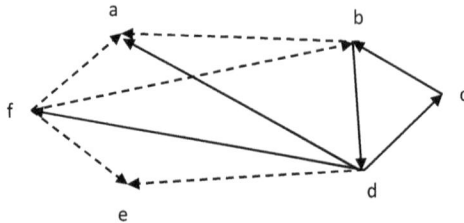

Fig. 7.22

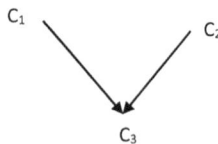

Fig. 7.23

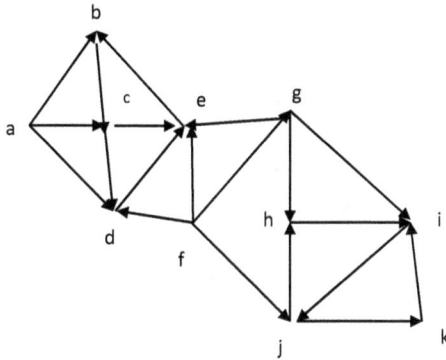

Fig. 7.24

the reduced graph of G is as given in Fig. 7.23. Since there is no edge from e to a or a to e, in G, there is no edge from C_1 to C_2 or C_2 to C_1. Hence the reduced graph of G is a dag.

Example 7.10. Consider the directed graph as given in Fig. 7.24. Perform *DFS* on G and find the corresponding tree edges, forward edges, back edges and cross edges. Is this graph a dag? Find also the reverse graph of G, strongly connected components and the reduced graph of G.

Solution. The adjacency lists of the vertices are

$$adj[a] = \{b, c, d\}, \ adj[b] = \{c\}, \ adj[c] = \{d, e\},$$

$$adj[d] = \{e\}, \qquad adj[e] = \{b\}, \ adj[f] = \{d, e, g, j\},$$

$$adj[g] = \{e, i, h\}, \ adj[h] = \{i\}, \ adj[i] = \{j\},$$

$$adj[j] = \{h, k\}, \quad adj[k] = \{i\}.$$

We perform *DFS* on G with source vertex a. From a we go to b, then to c, then to d and finally to e. The search finishes at e and we backtrack to $\pi(e) = d$. The search then finishes at d and backtrack to $\pi(d) = c$. The search finishes at c and backtrack to $\pi(c) = b$ and the search finishes at b and then at a. Thus the discovery and finishing times of the vertices visited are $a(1, 10), b(2, 9), c(3, 8), d(4, 7), e(5, 6)$.

We next start the search at f. From f we go to g, then from g to h, from h to i, from i to j and finally from j to k. The search finishes at k, we backtrack to $\pi(k) = j$, the search finishes at j and we backtrack to

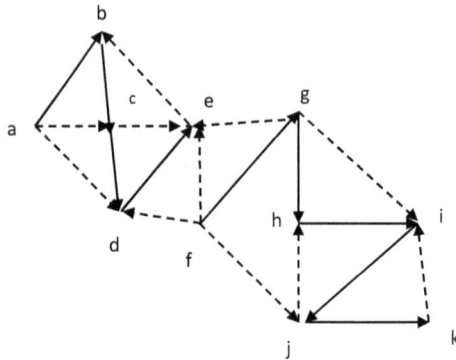

Fig. 7.25

$\pi(j) = i$, the search finishes at i and we backtrack to $\pi(i) = h$, the search finishes at h and we backtrack to $\pi(h) = g$. The search finishes at g and we backtrack to $\pi(g) = f$ and the search finishes at f as well. Since we have reached the vertex where the search was initiated, *DFS* of the graph G is complete. The discovery and finishing times of these vertices are $f(11, 22)$, $g(12, 21)$, $h(13, 20)$, $i(14, 19)$, $j(15, 18)$, $k(16, 17)$.

On the *DFS* traversal, the graph G with non-tree edges with dotted lines is given in Fig. 7.25.

Thus the

Tree edges are: (a, b), (b, c), (c, d), (d, e), (f, g), (g, h), (h, i), (i, j), (j, k);

Forward edges are: (a, c), (a, d), (c, e), (g, i), (f, j);

Back edges are: (e, b), (j, h), (k, i);

Cross edges are: (f, d), (f, e), (g, e).

In view of the presence of back edges, it follows that G is not a dag. The reverse graph G^t of G is as given in Fig. 7.26.

Linear ordering of the vertices of G in the reverse order of finishing time is $f, g, h, i, j, k, a, b, c, d, e$.

Now run *DFS* on G^t with the ordering of the vertices as above. The search originating at f terminates there itself. Then the search beginning at g terminates there. Beginning the search with vertex h, we go to j and from there to i and then finally to k. The search finishes at k, we backtrack to $\pi(k) = i$, and then backtrack to h terminating the search at h.

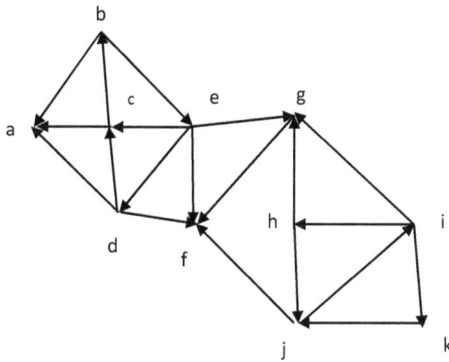

Fig. 7.26

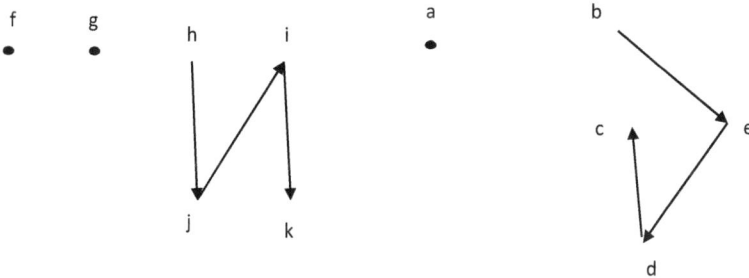

Fig. 7.27

Next, the search at a begins and ends there. Starting the search at b, we go to e, then to d and then to c. Backtrack to $\pi(c) = d$, then to $\pi(d) = e$ and then to $\pi(e) = b$ terminating the search at b. Since all the vertices in G^t have been visited, *DFS* of G^t is complete. *DFS* forest of G^t is given in Fig. 7.27.

Therefore, there are five strongly connected components of G as given by the trees in the *DFS* forest of G^t in Fig. 7.27. Let us call these strong components of G as C_1, C_2, C_3, C_4, C_5 in the order in which the trees appear in Fig. 7.27. The reduced graph of G is then as given in Fig. 7.28 the edges from C_1 to C_2, C_2 to C_3, C_1 to C_5, C_4 to C_5, C_1 to C_3 and C_2 to C_5 being provided by the edges: (f, g), (g, h), (f, d), (a, b), (f, j) and (g, e), respectively.

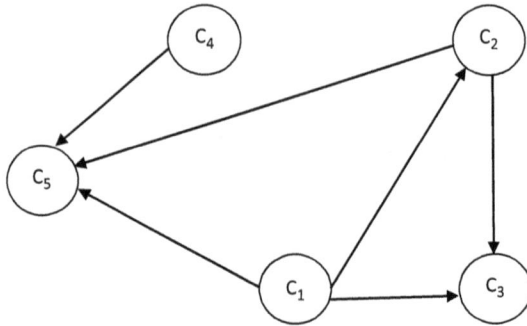

Fig. 7.28

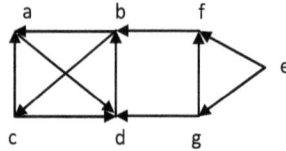

Fig. 7.29

Example 7.11. Consider the directed graph G given in Fig. 7.29. Run *DFS* on G and find all the corresponding tree edges, forward edges, back edges and cross edges and also the *DFS* forest. Also find strongly connected components and the reduced graph of G.

Solution. We run *DFS* on G with source vertex a. When the search traces an edge (x, y), we indicate it by $x \rightarrow y$ and when we backtrack from y to $\pi(y) = x$, we indicate it by $x \leftarrow{-} y$. Now, running *DFS* on G, we have $a \rightarrow d \rightarrow b \rightarrow c$, $b \leftarrow{-} c$, $d \leftarrow{-} b$, $a \leftarrow{-} d$ and the search at a terminates. Next we commence the search at e. We have $e \rightarrow f$, $e \leftarrow{-} f$, $e \rightarrow g$ and $e \leftarrow{-} g$ and search terminates at e. Since all the vertices of G have already been visited, $DFS(G)$ is complete. *DFS* forest of G is given in Fig. 7.30. Tree edges are indicated by bold lines in Fig. 7.30. There are no forward edges while back edges are (b, a), (c, a), (c, d) and cross edges are (f, b), (g, d), (g, f).

To find strong components of G, we need to first find the reverse graph G^t of G and it is given in Fig. 7.31.

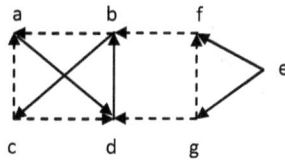

Fig. 7.30

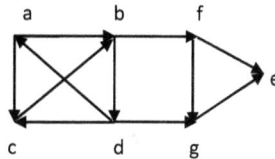

Fig. 7.31

The d- and f-values of the vertices of G when DFS is run on G are $a(1, 8), d(2, 7), b(3, 6), c(4, 5), e(9, 14), f(10, 11), g(12, 13)$.

Therefore, the vertices of G listed in the decreasing order of their f-values are: e, g, f, a, d, b, c. Now run DFS on G^t in the order of the vertices as given above. We have

$e \to e \leftarrow -- e, g \to g \leftarrow -- g, f \to f \leftarrow -- f, a \to b \to d \to c,$
$d \leftarrow -- c, b \leftarrow -- d, a \leftarrow -- b.$

Since we have arrived at the vertex where the search initiated and also since all the vertices of G^t have been visited, $DFS(G^t)$ is compete. If search begins at x and ends there itself without going to another vertex, we have indicated this by $x \to x \leftarrow -- x$.

The DFS forest of G^t consists of four trees three of which are one vertex trees, there are four strong components of G. If we take vertices C_1, C_2, C_3, C_4 corresponding to the trees in the DFS forest of G^t with C_1, C_2, C_3 corresponding to the one vertex trees e, g, f respectively, the reduced graph of G is as given in Fig. 7.32. The edges C_1 to C_2, C_1 to C_3, C_2 to C_4 and C_3 to C_4 are given by the edges $(e, g), (e, f), (g, d)$ and (f, b), respectively, of the graph.

Example 7.12. Consider the directed graph G as given in Fig. 7.33. Perform DFS on G and find the corresponding tree, forward, back and cross edges of G. Is this graph a dag? If yes give a topological sorting

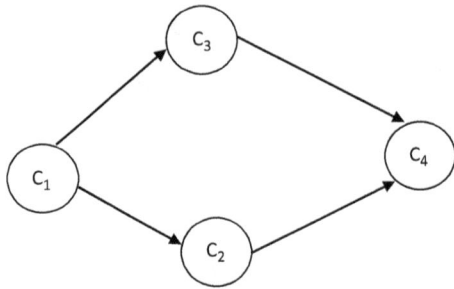

Fig. 7.32

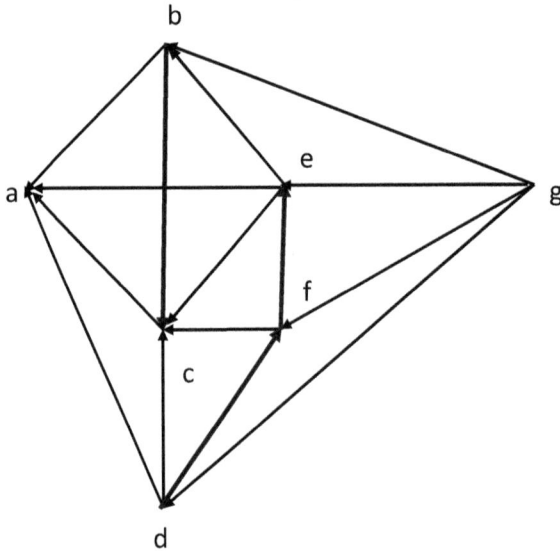

Fig. 7.33

of the vertices of G. Also find strong components and the reduced graph of G.

Solution. Adjacency lists of the vertices of G are

$adj[a] = \varphi,$ $adj[b] = \{a, c\},\ adj[c] = \{a\},\ adj[d] = \{a, c, f\},$

$adj[e] = \{a, b, c\},\ adj[f] = \{c, e\},\ adj[g] = \{b, d, e, f\}.$

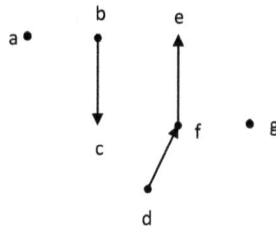

Fig. 7.34

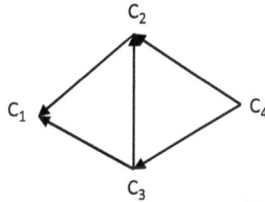

Fig. 7.35

Perform *DFS* on *G* commencing with the vertex *a*.

$$a \to a \leftarrow\!\!-- a, b \to c, b \leftarrow\!\!-- c, d \to f \to e, f \leftarrow\!\!-- e, d \leftarrow\!\!-- f, g \to g \leftarrow\!\!-- g$$

The discovery and finishing times of the vertices are given by

$$a(1, 2), b(3, 6), c(4, 5), d(7, 12), f(8, 11), e(9, 10), g(13, 14).$$

Arranging these vertices in the decreasing order of their finishing times, we get

$$g, d, f, e, b, c, a. \tag{7.1}$$

The tree edges are indicated by bold lines in Fig. 7.33 and there are no forward and back edges. The cross edges are

$(b, a), (c, a), (d, a), (d, c), (e, c), (e, b), (e, a), (f, c), (g, b), (g, e),$
$(g, d), (g, f).$

There being no back edges, *G* is a dag and topological sorting of the vertices of *G* is as given in (7.1) above. The *DFS* forest of *G* is given in Fig. 7.34 (with the nontree edges deleted).

Observe that if *DFS* is run on *G* with *g* as starting source vertex, we have $g \to d \to f \to e \to b \to c \to a$ and since all the vertices are visited, the search is complete. The discovery and finishing times of the vertices are

$$g(1, 14), d(2, 13), f(3, 12), e(4, 11), b(5, 10), c(6, 9), a(7, 8).$$

The topological sorting of the vertices is g, d, f, e, b, c, a which is the same as obtained earlier in (7.1).

If the strong components in the order in which these appear in Fig. 7.34 are called C_1, C_2, C_3, C_4 (which are regarded as vertices of the reduced graph of G), there is an edge from C_2 to C_1, from C_3 to C_2, from C_3 to C_1, from C_4 to C_2 and from C_4 to C_3 in view of the edges b to a, e to b, e to a, g to b and g to e of G, respectively, and the reduced graph of G is given in Fig. 7.35.

Exercise 7.4.

1. Is topological sorting of the vertices of a dag independent of the choice of source vertices for performing DFS on the dag? Justify.
2. Is the reverse graph of a dag again a dag? Justify.
3. If the reverse graph is again a dag, is it possible to obtain linear ordering of the vertices of one dag from that of the other?
4. Prove that every dag G has at least one vertex v with degree of v zero.
5. Let G be a connected dag and v be a vertex with degree 0. Delete v and all edges with initial vertex v. Prove that the resulting graph is again acyclic. Repeat the process till all vertices of G as well as edges are removed. Prove that the process of deletion provides a topological sorting of the vertices of G. Write an algorithm for this process and give its complexity.
6. Let $G = (V, E)$ and $G' = (V, E')$ be two directed graphs which have the same strongly connected components and the same reduced graphs. How are E and E' related?
7. Let G be a connected undirected graph and (u, v) be a back edge when DFS is run on G. Then u is an ancestor or descendent of v. What shape this result takes when G is a directed connected graph?
8. Let G be a graph and u, v be vertices in G. If $v \in adj[u]$ and v is not yet discovered when u is discovered, then v is a descendent of u in the depth-first search of G.

Chapter 8

Minimal Spanning Tree

Let G be a connected weighted graph where the weight function w associates a (positive) real number $w(e)$ to every edge e of G. For any subset S of edges of G, the weight $w(S)$ is the sum of the weights of the edges in S. If $G = (V, E)$, a subgraph $G' = (V, E')$ of G is called a **spanning subgraph** of G. If the subgraph $G' = (V, E')$ of G is a tree, we call G' a **spanning tree** of G. If T is a spanning tree of G, the weight of T is the sum of the weights of all the edges in T. The graph G being a finite graph, G has only a finite number of spanning trees and, therefore, has at least one spanning tree T (say) the weight of which is the least among the weights of all spanning trees. This T is then called a **minimal/minimum/shortest spanning tree** of G. Minimal spanning trees (MST) have applications in a wide variety of areas. A typical area of application being in designing a communication network. As mentioned by D. Jungnickel [15] minimal spanning trees were considered by Boruvka as early as in 1926 but more popular among the algorithms for constructing spanning trees are those of Prim and Kruskel. Both these algorithms are good examples of greedy strategy. In this chapter, we study both the Prim's algorithm and Kruskel's algorithm. Since Prim's algorithm for MST was developed by M. Jarnik much earlier than used by Prim, the same is also called the **Prim–Jarnik algorithm** (cf. Jungnickel [15]).

Throughout this chapter G is a connected weighted graph with the set of vertices V and the weight function w that assigns to each edge e in the set E of edges of G a positive real number $w(e)$.

8.1. Prim's Algorithm

By a **cut** of G we mean a partition (X, Y) of V so that X, Y are non-empty subsets of V with $V = X \cup Y$ and $X \cap Y = \varphi$. Prim's algorithm constructs a minimal spanning tree (MST) in a step-by-step manner. At every step we find a cut $V = X \cup Y$ where X contains all the vertices already included in the MST being constructed and $Y = V - X$, then we pick a minimum weight edge from among the edges with one vertex in X and the other in Y, include this minimum weight edge to the set of edges of the tree under construction and continue the process till the final tree obtained contains all the vertices of G.

A vertex u in a subset U of V is called a **nearest neighbor** of a vertex v in $V - U$ if u, v are adjacent and $w(u, v) \leq w(x, v)$ for all vertices x in U. We write $u = \text{nnr}(v)$. Also $w(v, \text{nnr}(v))$ is called the **key of** v and is denoted by $\text{key}(v)$ relative to the subset U of V.

Algorithm 8.1. Prim's Algorithm (G, MST)

---

1. initially choose a vertex $u \in V$ and set $U = \{u\}$, $S = \varphi$, ($T = (U, S)$ is a subgraph of G with one vertex and no edge)
2. choose a $v \in V - U$ such that $w(u, v) \leq w(u, x)$ for every $x \in V - U$. Set $U = U \cup \{v\} = \{u, v\}$, $S = S \cup \{(u, v)\} = \{(u, v)\}$
3. if $U \neq V$, choose $v \in V - U$ with $\text{key}(v) \leq \text{key}(x)$ for all $x \in V - U$ i.e., $w(v, \text{nnr}(v)) \leq w(x, \text{nnr}(x))$ for all $x \in V - U$. Set $U = U \cup \{v\}$, $S = S \cup \{(v, \text{nnr}(v))\}$
4. repeat step 3 till $U = V$

To prove that Prim's algorithm yields an MST, we have a couple of auxiliary results.

Lemma 8.1. *Let $G = (V, E)$ be a connected weighted undirected graph and $T = (V, S)$ be a spanning tree of G. Then, if any edge in $E - S$ is added to S, a unique cycle exists.*

Proof. Add the edge $e = (u, v)$ to S where e is not already in S. Since u, v are vertices in T and any two vertices in T are connected, there exists a path from u to v in T. Combining this path with the edge e there exists a cycle in $T' = (V, S \cup \{e\})$. Suppose that there exist two cycles

$(u, v_1, v_2, \ldots, v_k = v, u)$ and $(u, u_1, \ldots, u_l = v, u)$. Then there exists two paths $(u, v_1, v_2, \ldots, v_k = v)$ and $(u, u_1, \ldots, u_l = v)$ from u to v in T. Combining these two paths we get a cycle $(u, v_1, v_2, \ldots, v_k = v = u_l, u_{l-1}, \ldots, u_1, u)$ in T which is a contradiction (as a tree has no cycle). $\square$

Lemma 8.2. *Let* $G = (V, E)$ *be a connected weighted undirected graph. Among all edges incident with a vertex one with smallest weight is contained in a minimal spanning tree.*

Proof. Let $u \in V$ and $e = (u, v)$ be an edge the weight of which is less than or equal to the weight of all edges $(u, x), x \in V$. Let $T = (V, S)$ be a minimal spanning tree of G. Let $G' = (V, S \cup \{e\})$ be the subgraph of G which is obtained from T by adjoining e to S. Then G' has a unique cycle and this cycle is made up of the edge $e = (u, v)$ and the unique path from u to v in T. Let the unique path from u to v in T be $(v_1, v_2, \ldots, v_k)$ with $v_1 = u$, $v_k = v$. Removing the edge (u, v_2) from the graph G' we get an acyclic connected subgraph $T' = (V, S \cup \{e\} - \{(u, v_2)\})$ of G which is a spanning tree of G. Now $w(T') = w(S \cup \{e\} - \{(u, v_2)\}) = w(S) + w(e) - w((u, v_2)) \leq w(S) = w(T)$ as e is an edge of smallest possible weight incident with u. Since T is a minimal spanning tree, $w(T) \leq w(T')$. Therefore, $w(T) = w(T')$ and T' is a minimal spanning tree of G. $\square$

Corollary 8.3. *If e is an edge of smallest possible weight in G, then there exists a minimal spanning tree of G containing e.*

Lemma 8.4 (MST property). *Let* $G = (V, E)$ *be a connected weighted undirected graph with the weight function w associating a (positive) real number to every edge in E. Let U be a proper subset of V so that $(U, V - U)$ is a cut set of G. If $e = (u, v)$, $u \in U$, $v \in V - U$, is an edge of lowest possible weight among all possible edges incident with one vertex in U and the other in $V - U$, then there is a minimal spanning tree $T = (V, E')$ of G that includes $e = (u, v)$ as an edge.*

Proof. Let $T = (V, E')$ be a minimal spanning tree of G. If T contains e as an edge, we have nothing to prove. Suppose that e is not an edge of T. Then $T' = (V, E' \cup \{e\})$ is a subgraph of G which contains a unique cycle with edge e included in this cycle. Let this cycle be

$(u, v_1, v_2, \ldots, v_k, v, u)$. Some of the vertices $v_1, v_2, \ldots, v_k$ may be in U and the rest in $V - U$ or all of these are in U or all of these are in $V - U$. In any case, there is an edge $e' = (u', v') \neq e$ with u' in U and v' in $V - U$ in this cycle. Delete the edge e' from T' to get a subgraph $T'' = (V, E' \cup \{e\} - \{e'\})$. Let x, y be two vertices in V. If the unique path from x to y involves the edge $e' = (u', v')$, replacing this edge in the path by the path $(u' = v_i, \ldots, v_1, u, v, v_k, \ldots, v_{i+1} = v')$, we get a path from x to y in T''. Therefore, T'' is a connected subgraph of G. Since $|E' \cup \{e\} - \{e'\}| = |E'| = |V| - 1$, T'' is a spanning tree of G. By the choice of e, $w(e) \leq w(e')$ and $w(T'') = w(E' \cup \{e\} - \{e'\}) = w(E') + w(e) - w(e') \leq w(E') = w(T)$.

Since T is a minimal spanning tree, $w(T) \leq w(T'')$. Therefore $w(T) = w(T'')$ and T'' is a minimal spanning tree which contains e as an edge. $\qquad\square$

Theorem 8.5. *Prim's algorithm determines a minimal spanning tree of G.*

Proof. Let $T_i = (U_i, S_i)$ be the subgraph of G obtained on the ith iteration of Prim's algorithm. Then $T_1 = (U_1, S_1)$, where $U_1 = \{u\}$, $S_1 = \varphi$ so that T_1 is a tree with a single vertex and no edge. Also $|U_2| = 2$ and $S_2 = 1$ and T_2 is a graph with two vertices and a single edge connecting the two vertices. Thus T_2 is also a tree. Now suppose that $T_i = (U_i, S_i)$ — the subgraph of G obtained on ith iteration is a tree for $i \geq 1$ with $U_i \neq V$. Then $T_{i+1} = (U_{i+1}, S_{i+1})$ with $S_{i+1} = S_i \cup \{(u, v)\}$ with u in U_i and v in $V - U_i$ such that $w(u, v) \leq w(x, y)$ for all edges (x, y) with $x \in U_i$ and $y \in V - U_i$. Since every two vertices in T_i are connected and v is connected to u through the edge (u, v) in S_{i+1}, every two vertices in T_{i+1} are connected. Also there is no cycle in T_i. If there is a cycle in T_{i+1}, it must involve the edge (u, v). Let there be a cycle $(v_1 = u, v_2 = v, v_3, \ldots, v_k = v_1 = u)$. Then there is an edge (v_3, v) with v_3 in U_i also in T_{i+1} with $v_3 \neq u$. This is not the case and, so, T_{i+1} has no cycle and hence T_{i+1} is a tree. This completes induction and, therefore, every T_i is a tree.

Again we use induction on i and prove that every T_i is contained in a minimal spanning tree of G. That T_1 is contained in a minimal spanning tree of G is trivially true and that T_2 is contained in a minimal spanning tree follows from Lemma 8.2. Suppose that $i \geq 2$ and that T_i is contained in a minimal spanning tree $T = (V, S)$ of G. Now $T_{i+1} = (U_{i+1}, S_{i+1})$ where $U_{i+1} = U_i \cup \{v\}$ and $S_{i+1} = S_i \cup \{e\}$, where $e = (u, v)$ with $u \in U_i$, $v \in V - U_i$ such that $w(e) \leq w(x, y)$ for all $x \in U_i$, $y \in V - U_i$. If the edge e is in S, then $S_{i+1} = S_i \cup \{e\}$ is contained in S and T_{i+1} is contained in T. Otherwise, the argument as in the proof of Lemma 8.3 shows that $T' = (V, S \cup \{e\} - \{e'\})$ for some edge $e' = (u', v')$, $u' \in U_i$, $v' \in V - U_i$ is a minimal spanning tree of G. Since $S_i \subseteq S$, $e' \notin S_i$, $S_{i+1} = S_i \cup \{e\} \subseteq S \cup \{e\} - \{e'\}$. Therefore T_{i+1} is contained in the minimal spanning tree T' of G. This completes induction. Then T_n, where $n = |V|$, is contained in a minimal spanning tree $T = (V, S)$ of G. Therefore $w(T_n) \leq w(T)$. But $T_n = (V, S_n)$ is itself a spanning tree of G. Therefore $w(T) \leq w(T_n)$ and, therefore, $w(T_n) = w(T)$. Hence T_n is a minimal spanning tree of G. $\square$

Complexity of Prim's algorithm

In steps 2 and 3 of the algorithm, we have to examine $n - |U|$ vertices of G and to find a nearest neighbor of $a \in U$ (the vertex we have already arrived at). For finding a nearest neighbor, we need to compare weights $w(a, x)$ of $n - |U| < n$ vertices and the number of comparisons required goes on decreasing by 1 at each iteration. Hence, the total number of comparisons is $\leq 1 + 2 + \cdots + (n - 1) = \frac{n(n-1)}{2}$ and, therefore, the complexity of the algorithm is $O(n^2)$.

8.2. Kruskal's Algorithm

We next consider another algorithm for obtaining a minimal spanning tree of an undirected weighted connected graph $G = (V, E)$, which is a bit faster than Prim's algorithm. Here the edges of G are taken in the non-decreasing order of their weights and in every iterative step an edge is taken in order. Once an edge is considered and is either merged with one of the subgraphs obtained or is ignored, it is deleted from the list of edges. At every iteration, the number of subgraphs constructed is reduced by one and the process stops when the number of subgraphs constructed becomes one, which is then a minimal spanning tree.

Theorem 8.6. *Kruskal's algorithm determines a minimal spanning tree for the undirected, weighted, connected graph $G = (V, E, w)$.*

Algorithm 8.2. Kruskal's Algorithm

$G = (V, E)$ is a weighted undirected graph with weight function w defined on the set E of edges. To construct an MST $(V, S) = T$ with $|V| = n$. We also write $E(T)$ for the edge set S of T

1. initially $T = \varphi$, V is partitioned into subsets V_i each V_i having only one vertex
2. the edge set E of G is put in a priority queue Q (say)
3. consider the lowest weight edge $e = (u, v)$ in Q
4. find V_i, V_j such that $u \in V_i$, $v \in V_j$
5. merge V_i, V_j: take $V_i \cup V_j$ and $E(T) = E(T) \cup \{e\}$ (i.e., $V_i \cup V_j = \{u, v\}$ and $E(T) = \{e\}$)
6. delete e from Q
7. if $E(T)| < n - 1$, consider the next e in Q, $e = (u, v)$
8. find i, j so that $u \in V_i$, $v \in V_j$
9. if $V_i = V_j$, ignore e, otherwise merge V_i, V_j to get $V_i \cup V_j = V_k$ and $E(T) = E(T) \cup \{e\}$
10. delete e from Q
11. repeat steps 7 to 9 till $E(T) = n - 1$

Proof. Let $T = (V, S)$, where S denotes the edge set of the graph T, be the subgraph of G obtained by Kruskal's algorithm. Then $|S| = |V| - 1 = n - 1$, where n is the number of vertices of G. In order to prove that T is a tree and, therefore, a spanning tree of G, we need to prove that T is a connected graph.

Initially $T_i = (V_i, S_i)$, with $|V_i| = 1$, $S_i = \varphi$ for every i, $1 \le i \le n$ and each one of these is a tree with one vertex and no edges. In the first iteration, we consider $e = (u, v)$ where $w(e)$ is the least among the weights of all edges of G. Then $u \in V_i$, $v \in V_j$ for some $i \ne j$. We merge the subtrees T_i, T_j together with edge e and get the subtree $(V_i \cup V_j, \varphi \cup \varphi \cup \{e\}) = (\{u, v\}, \{e\})$ having two vertices and one edge connecting these two vertices. Therefore, this subgraph is a subtree of G. Suppose that when some edges less than $n - 1$ in number have already been considered by Krusal's algorithm, the subgraphs of G obtained are trees. Observe that no two subtrees obtained have any vertices or edges in common. Next, we consider an edge $e = (u, v)$, where the vertices u, v are not both in the same subgraph so that e is not to be ignored. Let $u \in T_u = (V_u, S_u)$, $v \in T_v = (V_v, S_v)$ so that $V_u \cap V_v = \varphi$. By induction hypothesis, both T_u, T_v are trees so that $|V_u| - 1 = |S_u|$, $|S_v| = |V_v| - 1$. Adjoining the edge e to the forest of subgraphs of G (i.e., merging T_u, T_v and adjoining the edge e), we get a new subgraph $T_{uv} = (V_u \cup V_v, S_u \cup S_v \cup \{e\}) = (V_{uv}, S_{uv})$ (say) in place of the subgraphs T_u, T_v and

$$|V_u \cup V_v| = |V_u| + |V_v|,$$

$$|S_u \cup S_v \cup \{e\}| = |S_u| + |S_v| + 1$$

$$= |V_u| - 1 + |V_v| - 1 + 1 = |V_u \cup V_v| - 1. \quad (8.1)$$

Let $x, y \in V_u \cup V_v$. If both x, y are in V_u or in V_v, then x, y are connected in T_u or connected in T_v and, so, connected in T_{uv}. Suppose that $x \in V_u$ and $y \in V_v$. Since any two elements of V_u and of V_v are connected in T_u and in T_v, respectively, there exist paths in T_u from x to u and from v to y in T_v, respectively. Combining these paths with the edge e in the middle, we get a path from x to y in T_{uv}. Thus the subgraph T_{uv} obtained is connected. It then follows from (8.1) that T_{uv} is a tree.

Now, initially we started with n trees, at each iteration, two subtrees already obtained are joined together to get a new subtree, the number of subtrees is reduced by one. The process stops when $n - 1$ edges of G have been adjoined with the subtrees and we get a single tree (U, S) with $|S| = n - 1$, so that $|U| = |S| + 1 = n$ and $U = V$. Thus, Kruskal's algorithm finally obtains a spanning tree $T = (V, S)$.

Let $T' = (V, S')$ be a minimal spanning tree of G. If $w(T) = w(T')$, then T is a minimal spanning tree of G. If $w(T) \neq w(T')$, then $w(T') < w(T)$. Let e be an edge of least possible weight in T such that $e \notin T'$, i.e., all the edges in T of weight less than $w(e)$ are in T'. Then $T'' = (V, S' \cup \{e\})$ contains a unique cycle, say $e, e_1, \ldots, e_k$, where $e_1, \ldots, e_k$ are in T'. If all the e_i are in T, then as e is also in T, T contains the cycle $e, e_1, \ldots, e_k$ which is not possible. Therefore, there exists an i, $1 \leq i \leq k$, such that $e_i \notin T$. If $w(e_i) < w(e)$, there are two possibilities. (i) Either e_i has both its vertices in the same subset V_i of V and e_i together with the edges already selected by Kruskal's algorithm forms a cycle in T. Since the edges selected by Kruskal's algorithm before e_i are of weight $\leq w(e_i) < w(e)$, these edges are in T' (by the choice of e). Therefore, we get a cycle in T' which, T' being a tree, is not possible. (ii) Otherwise e_i will be selected by Kruskal's algorithm before e and e_i will be in T which is a contradiction. Therefore $w(e_i) \geq w(e)$. Removing the edge e_i from T'', we get a spanning tree $T''' = (V, S' \cup \{e\} - \{e_i\})$ of G and $w(T''') = w(S' \cup \{e\} - \{e_i\}) = w(S') + w(e) - w(e_i) \leq w(S') = w(T')$ and T''' is a minimal spanning tree of G which contains one more edge namely e than the edges of weight $\leq w(e)$ which were already in T'. Repeating the argument a finite number of times, we find that all the edges in T are included in a minimal spanning tree of G and, hence, T itself is a minimal spanning tree of G. □

Running time of Kruskal's algorithm

The running time to sort the edges according to their non-decreasing weight using quicksort is $O(|E| \log |E|)$. At each iteration, we have to look for the subsets V_i, V_j of V such that one end vertex of an edge e under consideration is in V_i and the other is in V_j and merge V_i and V_j, if $i \neq j$, to get a new V_i, if $i < j$. Time taken for this find and merge operation is $O(n \log n)$. The graph being connected and simple, $n - 1 \leq |E| \leq n^2$. Therefore, the running time of Kruskal's algorithm is $O(n \log n + |E| \log |E|\} = O(|E| \log |E|)$. But $\log |E| \leq 2 \log n$. Hence the running time becomes $O(|E| \log n)$ (cf. Aho, Hopcroft and Ullman [2] and/or Corman, Leiserson and Rivest [5] for a detailed discussion about it).

8.3. Some Examples

Example 8.1. Construct the minimal spanning tree of the graph G as given in Fig. 8.1 using (a) Prim's algorithm with (i) a, (ii) f as root; (b) Kruskal's algorithm.

Solution. Label the edges of G as follows:

$$e_1 = (a, e), e_2 = (f, i), e_3 = (a, c), e_4 = (d, e), e_5 = (d, i),$$

$$e_6 = (h, i), e_7 = (b, e), e_8 = (d, h), e_9 = (f, h), e_{10} = (b, d),$$

$$e_{11} = (b, f), e_{12} = (f, g), e_{13} = (c, g), e_{14} = (c, f), e_{15} = (e, f),$$

$$e_{16} = (g, i), e_{17} = (a, b).$$

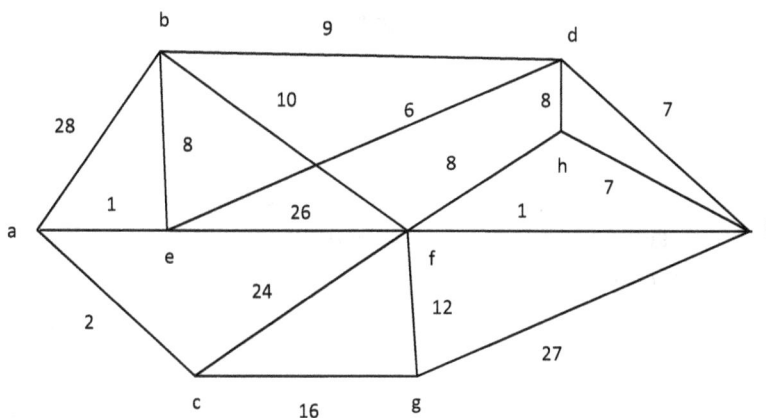

Fig. 8.1

We have labeled the edges in the non-decreasing order of their weights although that is not necessary for using Prim's algorithm. However, such an ordering is needed for using Kruskal's algorithm. Let V be the set of vertices of G. Also, let T denote the set of edges that we have to find to get a MST of G.

(a) (i) Choose a as root so that initially $U = \{a\}$.

Iteration 1. The elements of $V - U$ to be considered are b, c, e. Now nnr$(b) = a$, key$(b) = 28$; nnr$(c) = a$, key$(c) = 2$: nnr$(e) = a$, key$(e) = 1$.

The least among the keys being 1, we have $U = \{a, e\}$, $T = \{e_1\}$.

Iteration 2. The elements of $V - U$ to be considered are b, c, d, f. Now nnr$(b) = e$, key$(b) = 8$; nnr$(c) = a$, key$(c) = 2$, nnr$(d) = e$, key$(d) = 6$, nnr$(f) = e$, key$(f) = 26$.

Therefore, $U = \{a, e, c\}$, $T = \{e_1, e_3\}$, $w(T) = w(e_1) + w(e_3) = 1 + 2 = 3$.

Iteration 3. The elements of $V - U$ to be considered are b, d, f, g. Now nnr$(b) = e$, key$(b) = 8$, nnr$(d) = e$, key$(d) = 6$, nnr$(f) = c$, key$(f) = 24$, nnr$(g) = c$, key$(g) = 16$.

Therefore, $U = \{a, e, c, d\}$, $T = \{e_1, e_3, e_4\}$, $w(T) = 3 + 6 = 9$.

Iteration 4. The elements of $V - U$ to be considered are $b, f, g.h, i$. Now $\text{nnr}(b) = e, \text{key}(b) = 8, \text{nnr}(f) = c, \text{key}(f) = 24, \text{nnr}(g) = c, \text{key}(g) = 16, \text{nnr}(h) = d, \text{key}(h) = 8, \text{nnr}(i) = d, \text{key}(i) = 7$.

Therefore, $U = \{a, e, c, d, i\}, T = \{e_1, e_3, e_4, e_5\}, w(T) = 9 + 7 = 16$.

Iteration 5. The elements of $V - U$ to be considered are b, f, g, h. Now $\text{nnr}(b) = e, \text{key}(b) = 8, \text{nnr}(f) = i, \text{key}(f) = 1, \text{nnr}(g) = c, \text{key}(g) = 16, \text{nnr}(h) = i, \text{key}(h) = 7$.

Therefore, $U = \{a, e, c, d, i, f\}, T = \{e_1, e_3, e_4, e_5, e_2\}, w(T) = 16 + 1 = 17$.

Iteration 6. The elements of $V - U$ to be considered are b, g, h. Now $\text{nnr}(b) = e, \text{key}(b) = 8, \text{nnr}(g) = f, \text{key}(g) = 12, \text{nnr}(h) = i, \text{key}(h) = 7$.

Therefore, $U = \{a, e, c, d, i, f, h\}, T = \{e_1, e_3, e_4, e_5, e_2, e_6\}, w(T) = 17 + 7 = 24$.

Iteration 7. The elements of $V - U$ to be considered are b, g. Now $\text{nnr}(b) = e, \text{key}(b) = 8, \text{nnr}(g) = f, \text{key}(g) = 12$.

Therefore, $U = \{a, e, c, d, i, f, h, b\}, T = \{e_1, e_3, e_4, e_5, e_2, e_6, e_7\}, w(T) = 24 + 8 = 32$.

Iteration 8. The only element in $V - U$ is g. Now $\text{nnr}(g) = f, \text{key}(g) = 12$.

Therefore, $U = \{a, e, c, d, i, f, h, b, g\} = V, T = \{e_1, e_3, e_4, e_5, e_2, e_6, e_7, e_{12}\}, w(T) = 32 + 12 = 44$.

Since we have arrived at the set of all vertices of G, the process terminates and the minimal spanning tree with a as root is as given in Fig. 8.2.

(ii) Choose f as root so that initially $U = \{f\}$.

Iteration 1. The elements of $V - U$ to be considered are b, c, e, g, h, i. Now $\text{nnr}(b) = f, \text{key}(b) = 10, \text{nnr}(c) = f, \text{key}(c) = 24, \text{nnr}(e) = f, \text{key}(e) = 26, \text{nnr}(g) = f, \text{key}(g) = 12, \text{nnr}(h) = f, \text{key}(h) = 8, \text{nnr}(i) = f, \text{key}(i) = 1$.

The least among the keys being 1, we have $U = \{f, i\}, T = \{e_2\}, w(T) = 1$.

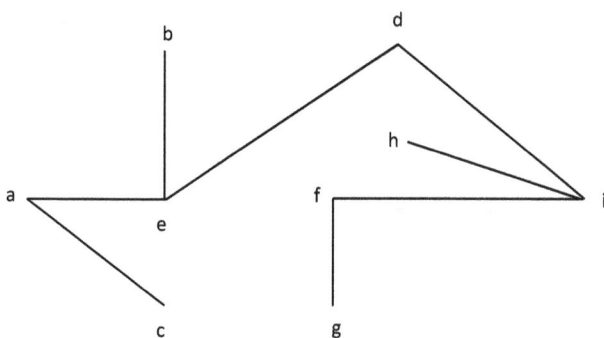

Fig. 8.2

Iteration 2. The elements of $V - U$ to be considered are b, c, d, e, g, h. Now nnr(b) $= f$, key(b) $= 10$, nnr(c) $= f$, key(c) $= 24$, nnr(d) $= i$, key(d) $= 7$; nnr(e) $= f$, key(e) $= 26$, nnr(g) $= f$, key(g) $= 12$, nnr(h) $= i$, key(h) $= 7$.

There being a tie, let us take $U = \{f, i, d\}$, $T = \{e_2, e_5\}$, $w(T) = 1 + 7 = 8$.

Iteration 3. The elements of $V - U$ to be considered are b, c, e, g, h. Now nnr(b) $= d$, key(b) $= 9$, nnr(c) $= f$, key(c) $= 24$, nnr(e) $= d$, key(e) $= 6$, nnr(g) $= f$, key(g) $= 12$, nnr(h) $= i$, key(h) $= 7$.

Therefore, $U = \{f, i, d, e\}$, $T = \{e_2, e_5, e_4\}$, $w(T) = 8 + 6 = 14$.

Iteration 4. The elements of $V - U$ to be considered are a, b, c, g, h. Now nnr(a) $= e$, key(a) $= 1$, nnr(b) $= e$, key(b) $= 8$, nnr(c) $= f$, key(c) $= 24$, nnr(g) $= f$, key(g) $= 12$, nnr(h) $= i$, key(h) $= 7$.

Therefore, $U = \{f, i, d, e, a\}$, $T = \{e_2, e_5, e_4, e_1\}$, $w(T) = 14 + 1 = 15$.

Iteration 5. The elements of $V - U$ to be considered are c, b, g, h. Now nnr(c) $= a$, key(c) $= 2$, nnr(b) $= e$, key(b) $= 8$, nnr(g) $= f$, key(g) $= 12$, nnr(h) $= i$, key(i) $= 7$.

Therefore, $U = \{f, i, d, e, a, c\}$, $T = \{e_2, e_5, e_4, e_1, e_3\}$, $w(T) = 15 + 2 = 17$.

Iteration 6. The elements of $V - U$ to be considered are b, g, h. Now nnr(b) $= e$, key(b) $= 8$; nnr(g) $= f$, key(g) $= 12$: nnr(h) $= i$, key(h) $= 7$.

Therefore, $U = \{f, i, d, e, a, c, h\}$, $T = \{e_2, e_5, e_4, e_1, e_3, e_6\}$, $w(T) = 17 + 7 = 24$.

Iteration 7. The only elements of $V - U$ are b, g. Now $\text{nnr}(b) = e$, $\text{key}(b) = 8$, $\text{nnr}(g) = f$, $\text{key}(g) = 12$.

Therefore, $U = \{f, i, d, e, a, c, h, b\}$, $T = \{e_2, e_5, e_4, e_1, e_3, e_6, e_7\}$, $w(T) = 24 + 8 = 32$.

Iteration 8. The only vertex in $V - U$ is g. Now $\text{nnr}(g) = f$, $\text{key}(g) = 12$.

Therefore, $U = \{f, i, d, e, a, c, h, b, g\} = V$, $T = \{e_2, e_5, e_4, e_1, e_3, e_6, e_7, e_{12}\}$, $w(T) = 32 + 12 = 44$.

Since T obtained here is the same as obtained in case (i), the minimal spanning tree obtained using Prim's algorithm with f as root is the same as obtained with root a and is given in Fig. 8.2.

(b) Now, we obtain minimal spanning tree of G using Kruskal's algorithm. Initially

$$V_1 = \{a\}, \ V_2 = \{b\}, \ V_3 = \{c\}, \ V_4 = \{d\}, \ V_5 = \{e\},$$
$$V_6 = \{f\}, \ V_7 = \{g\}, \ V_8 = \{h\}, \ V_9 = \{i\}.$$

Iteration 1. $e_1 = (a, e)$ and $a \in V_1$, $e \in V_5$. Merge V_1, V_5 to get new $V_1 = \{a, e\}$ and the set of edges chosen is $T = \{e_1\}$.

Iteration 2. $e_2 = (f, i)$ and $f \in V_6$, $i \in V_9$. Merge V_6, V_9 to get new $V_6 = \{f, i\}$.

Iteration 3. $e_3 = (a, c)$ and $a \in V_1$, $c \in V_3$. Merge V_1, V_3 to get new $V_1 = \{a, e, c\}$.

Iteration 4. $e_4 = (d, e)$ and $d \in V_4$, $e \in V_1$. Merge V_1, V_4 to get new $V_1 = \{a, e, c, d\}$.

Iteration 5. $e_5 = (d, i)$ and $d \in V_1$, $i \in V_6$. Merge V_1, V_6 to get new $V_1 = \{a, e, c, d, i, f\}$.

Iteration 6. $e_6 = (h, i)$ and $i \in V_1$, $h \in V_8$. Merge V_1, V_8 to get new $V_1 = \{a, e, c, d, i, f, h\}$.

Iteration 7. $e_7 = (b, e)$ and $e \in V_1$, $b \in V_2$. Merge V_1, V_2 to get new $V_1 = \{a, e, c, d, i, f, h, b\}$.

Iteration 8. Since both the vertices of each of $e_8 = (d, h)$, $e_9 = (f, h)$, $e_{10} = (b, d)$, $e_{11} = (b, f)$ are in V_1, ignore e_8, e_9, e_{10} and e_{11}.

Iteration 9. $e_{12} = (f, g)$ and $f \in V_1$, $g \in V_7$. Merge V_1, V_7 to get new $V_1 = \{a, e, c, d, i, f, h, b, g\} = V$.

Since we have arrived at V, the process terminates here and the minimal spanning tree obtained is $T = \{e_1, e_2, e_3, e_4, e_5, e_6, e_7, e_{12}\}$ which is the same as given in Fig. 8.2.

Diagrammatically, Kruskal's algorithm for this example works as shown in Fig. 8.3.

Example 8.2. Find the minimal spanning tree of the graph G given in Fig. 8.4 using (a) Prim's algorithm with (i) a, (ii) c as root; (b) Kruskal's algorithm.

Solution. Label the edges of the graph as follows:

$$e_1 = (a, h), e_2 = (d, i), e_3 = (b, c), e_4 = (b, h), e_5 = (e, g),$$

$$e_6 = (c, d), e_7 = (b, f), e_8 = (c, i), e_9 = (d, e), e_{10} = (c, h),$$

$$e_{11} = s(a, b), e_{12} = (d, g), e_{13} = (d, f), e_{14} = (c, f), e_{15} = (h, i).$$

(a) (i) Let us begin with a as root. Then $U = \{a\}$.

Iteration 1. The elements of $V - U$ to be considered are b, h. $\text{nnr}(b) = a$, $\text{key}(b) = 8$, $\text{nnr}(h) = a$, $\text{key}(h) = 1$.
Therefore, $U = \{a, h\}$, $T = \{e_1\}$, $w(T) = 1$.

Iteration 2. The elements of $V - U$ to be considered are b, c, i. $\text{nnr}(b) = h$, $\text{key}(b) = 4$; $\text{nnr}(c) = h$, $\text{key}(c) = 7$; $\text{nnr}(i) = h$, $\text{key}(i) = 12$.
Therefore, $U = \{a, h, b\}$, $T = \{e_1, e_4\}$, $w(T) = 1 + 4 = 5$.

Iteration 3. The elements of $V - U$ to be considered are c, f, i. $\text{nnr}(c) = b$, $\text{key}(c) = 3$; $\text{nnr}(f) = b$, $\text{key}(f) = 6$; $\text{nnr}(i) = h$, $\text{key}(i) = 12$.
Therefore, $U = \{a, h, b, c\}$, $T = \{e_1, e_4, e_3\}$, $w(T) = 5 + 3 = 8$.

Iteration 4. The elements of $V - U$ to be considered are d, f, i. $\text{nnr}(d) = c$, $\text{key}(d) = 5$; $\text{nnr}(f) = b$, $\text{key}(f) = 6$; $\text{nnr}(i) = c$, $\text{key}(i) = 6$.
Therefore, $U = \{a, h, b, c, d\}$, $T = \{e_1, e_4, e_3, e_6\}$, $w(T) = 8 + 5 = 13$.

Iteration 5. The elements of $V - U$ to be considered are f, i, e, g. $\text{nnr}(f) = b$, $\text{key}(f) = 6$; $\text{nnr}(i) = d$, $\text{key}(i) = 2$; $\text{nnr}(e) = d$, $\text{key}(e) = 6$; $\text{nnr}(g) = d$, $\text{key}(g) = 8$.

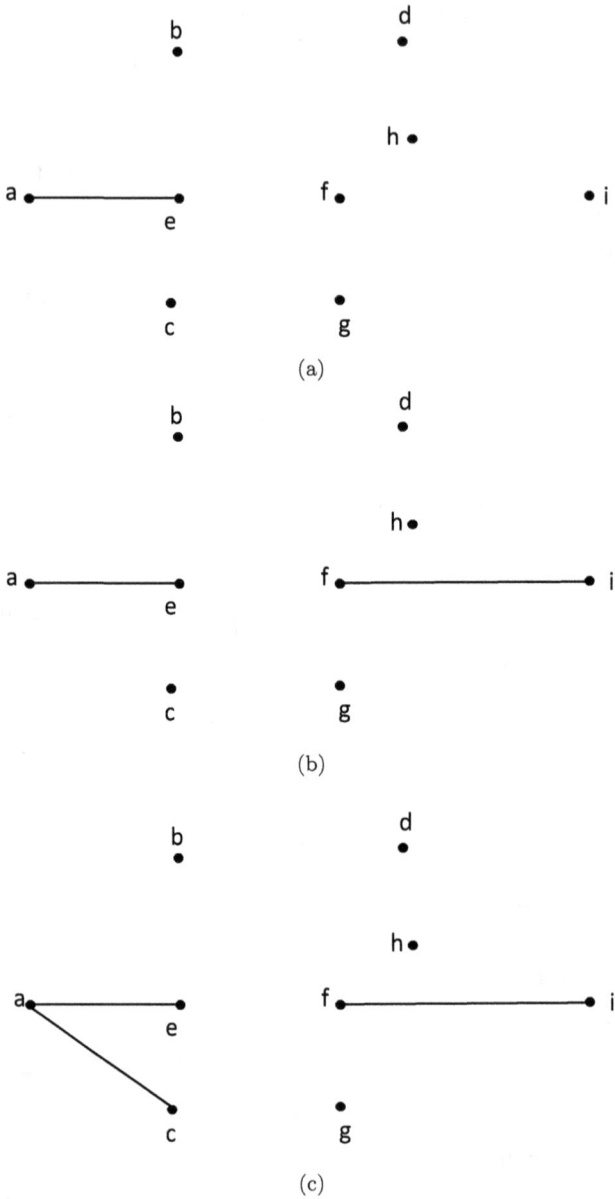

(a)

(b)

(c)

Fig. 8.3

(d)

(e)

(f)

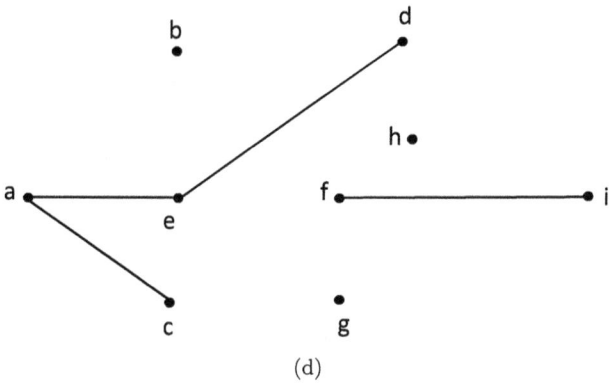

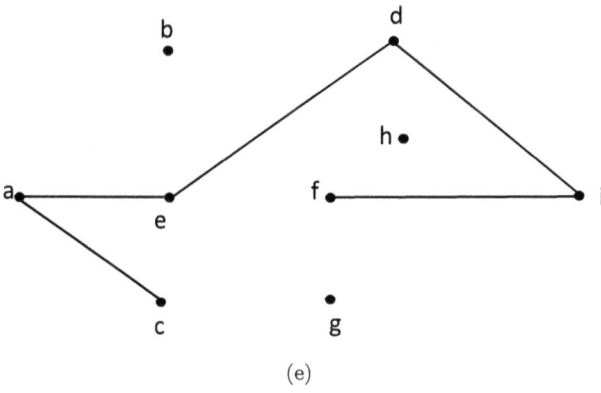

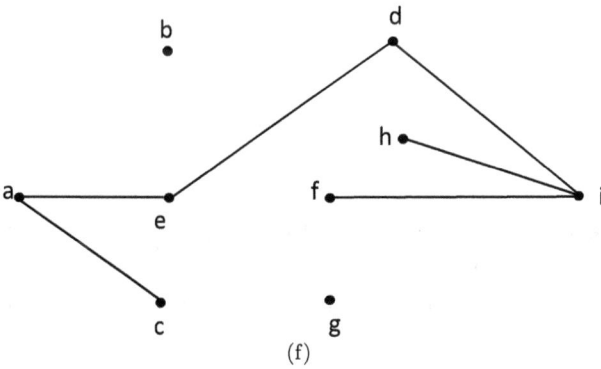

Fig. 8.3 (*Continued*)

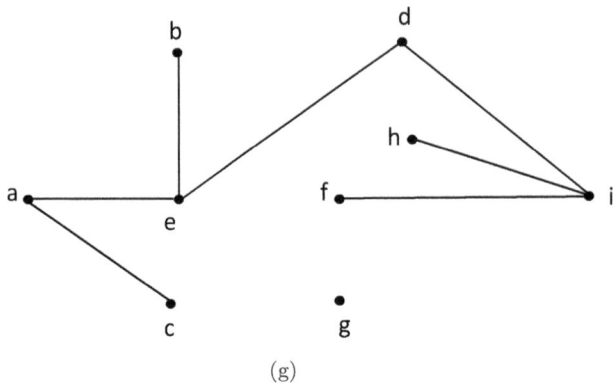

(g)

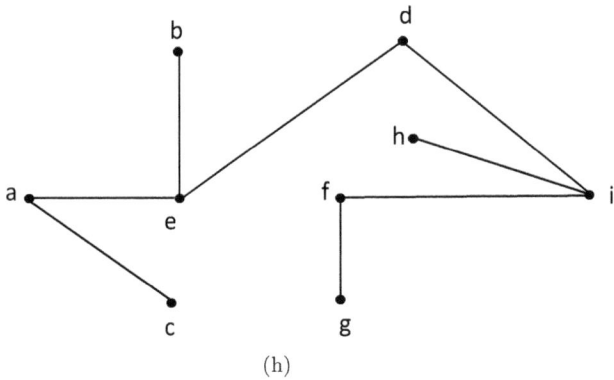

(h)

Fig. 8.3 (*Continued*)

Therefore, $U = \{a, h, b, c, d, i\}$, $T = \{e_1, e_4, e_3, e_6, e_2\}$, $w(T) = 13 + 2 = 15$.

Iteration 6. The elements of $V - U$ to be considered are f, e, g. $\text{nnr}(f) = b$, $\text{key}(f) = 6$; $\text{nnr}(e) = d$, $\text{key}(e) = 6$; $\text{nnr}(g) = d$, $\text{key}(g) = 8$.

There being a tie, let us choose the next vertex to be added to U as f. Therefore, $U = \{a, h, b, c, d, i, f\}$, $T = \{e_1, e_4, e_3, e_6, e_2, e_7\}$, $w(T) = 15 + 6 = 21$.

Iteration 7. The elements of $V - U$ to be considered are g, e. $\text{nnr}(e) = d$, $\text{key}(e) = 6$; $\text{nnr}(g) = d$, $\text{key}(g) = 8$. Therefore, $U = \{a, h, b, c, d, i, f, e\}$, $T = \{e_1, e_4, e_3, e_6, e_2, e_7, e_9\}$, $w(T) = 21 + 6 = 27$.

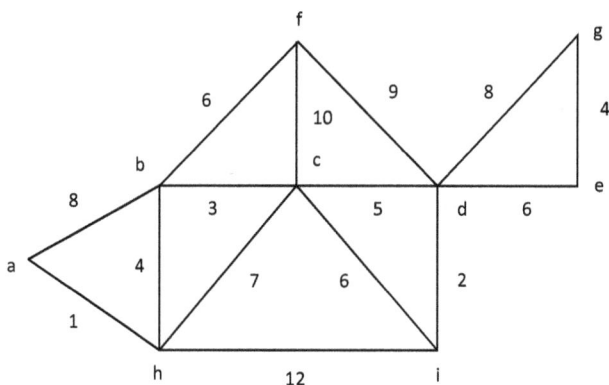

Fig. 8.4

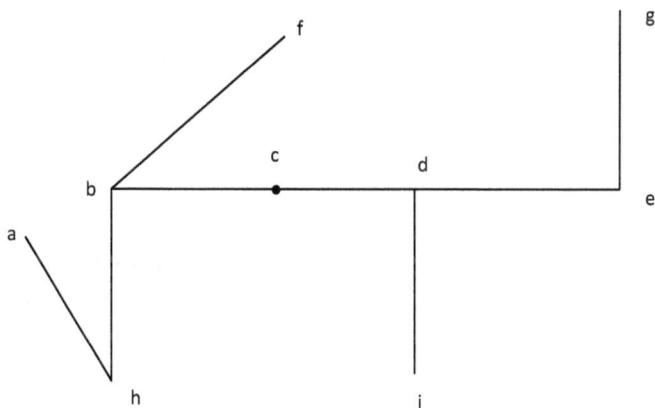

Fig. 8.5

Iteration 8. The only element of $V - U$ is g, nnr$(g) = e$, key$(g) = 4$. Therefore, $U = \{a, h, b, c, d, i, f, e, g\} = V$, $T = \{e_1, e_4, e_3, e_6, e_2, e_7, e_9, e_5\}$, $w(T) = 27 + 4 = 31$.

Since, now $U = V$, we have obtained minimal spanning tree as obtained by using Prim's algorithm and is given in Fig. 8.5.

(ii) Let us now begin with c as root so that initially $U = \{c\}$.

Iteration 1. The elements of $V - U$ to be considered are b, d, f, h, i nnr$(b) = $ nnr$(d) = $ nnr$(f) = $ nnr$(h) = $ nnr$(i) = c$, key$(b) = 3$, key$(d) = 5$, key$(f) = 10$, key$(h) = 7$, key$(i) = 6$.

Therefore, $U = \{c, b\}$, $T = \{e_3\}$, $w(T) = 3$.

Iteration 2. The elements of $V - U$ to be considered are a, f, d, h, i. $\mathrm{nnr}(a) = \mathrm{nnr}(h) = \mathrm{nnr}(f) = b$, $\mathrm{nnr}(d) = \mathrm{nnr}(i) = c$, $\mathrm{key}(a) = 8$, $\mathrm{key}(h) = 4$, $\mathrm{key}(f) = 6$, $\mathrm{key}(d) = 5$, $\mathrm{key}(i) = 6$.

Therefore, $U = \{c, b, h\}$, $T = \{e_3, e_4\}$, $w(T) = 3 + 4 = 7$.

Iteration 3. The elements of $V - U$ to be considered are a, d, f, i. $\mathrm{nnr}(a) = h$, $\mathrm{nnr}(d) = c$, $\mathrm{nnr}(f) = b$, $\mathrm{nnr}(i) = c$, $\mathrm{key}(a) = 1$, $\mathrm{key}(d) = 5$, $\mathrm{key}(f) = 6$, $\mathrm{key}(i) = 6$.

Therefore, $U = \{c, b, h, a\}$, $T = \{e_3, e_4, e_1\}$, $w(T) = 7 + 1 = 8$.

Iteration 4. The elements of $V - U$ to be considered are d, f, i. $\mathrm{nnr}(d) = c$, $\mathrm{nnr}(f) = b$, $\mathrm{nnr}(i) = c$, $\mathrm{key}(d) = 5$, $\mathrm{key}(f) = 6$, $\mathrm{key}(i) = 6$.

Therefore, $U = \{c, b, h, a, d\}$, $T = \{e_3, e_4, e_1, e_6\}$, $w(T) = 8 + 5 = 13$.

Iteration 5. The elements of $V - U$ to be considered are e, f, g, i. $\mathrm{nnr}(e) = d$, $\mathrm{nnr}(f) = b$, $\mathrm{nnr}(g) = d$, $\mathrm{nnr}(i) = d$, $\mathrm{key}(e) = 6$, $\mathrm{key}(f) = 6$, $\mathrm{key}(g) = 8$, $\mathrm{key}(i) = 2$.

Therefore, $U = \{c, b, h, a, d, i\}$, $T = \{e_3, e_4, e_1, e_6, e_2\}$, $w(T) = 13 + 2 = 15$.

Iteration 6. The elements of $V - U$ to be considered are f, e, g. $\mathrm{nnr}(f) = b$, $\mathrm{nnr}(e) = d$, $\mathrm{nnr}(g) = d$, $\mathrm{key}(f) = 6$, $\mathrm{key}(e) = 6$, $\mathrm{key}(g) = 8$.

There being a tie, we take $U = \{c, b, h, a, d, i, e\}$, $T = \{e_3, e_4, e_1, e_6, e_2, e_9\}$, $w(T) = 15 + 6 = 21$.

Iteration 7. The only elements of $V - U$ are f, g. $\mathrm{nnr}(f) = b$, $\mathrm{nnr}(g) = e$, $\mathrm{key}(f) = 6$, $\mathrm{key}(g) = 4$.

Therefore, $U = \{c, b, h, a, d, i, e, g\}$, $T = \{e_3, e_4, e_1, e_6, e_2, e_9, e_5\}$, $w(T) = 21 + 4 = 25$.

Iteration 8. The only element of $V - U$ is f, $\mathrm{nnr}(f) = b$, $\mathrm{key}(f) = 6$.

Therefore, $U = \{c, b, h, a, d, i, e, g, f\} = V$, $T = \{e_3, e_4, e_1, e_6, e_2, e_9, e_5, e_7\}$, $w(T) = 25 + 6 = 31$.

Observe that the edges in T here are the same as those in T in the case (i) although in a different order. Therefore, the minimal spanning tree obtained by using Prim's algorithm with c as root is the same as the one obtained with a as root and is given in Fig. 8.5.

(b) Now we find minimal spanning tree of G using Kruskal's algorithm.

Initially we consider the subsets of the vertex set V of G as follows:

$$V_1 = \{a\}, V_2 = \{b\}, V_3 = \{c\}, V_4 = \{d\}, V_5 = \{e\},$$
$$V_6 = \{f\}, V_7 = \{g\}, V_8 = \{h\}, V_9 = \{i\}.$$

Iteration 1. $e_1 = (a, h)$ and $a \in V_1$, $h \in V_8$. Merge V_1, V_8 to get new $V_1 = \{a, h\}$.

Iteration 2. $e_2 = (d, i)$ and $d \in V_4$, $i \in V_9$. Merge V_4, V_9 to get new $V_4 = \{d, i\}$.

Iteration 3. $e_3 = (b, c)$ and $b \in V_2$, $c \in V_3$. Merge V_2, V_3 to get new $V_2 = \{b, c\}$.

Iteration 4. $e_4 = (b, h)$ and $b \in V_2$, $h \in V_1$. Merge V_2, V_1 to get new $V_1 = \{a, h, b, c\}$.

Iteration 5. $e_5 = (e, g)$ and $e \in V_5$, $g \in V_7$. Merge V_5, V_7 to get new $V_5 = \{e, g\}$.

Iteration 6. $e_6 = (c, d)$ and $c \in V_1$, $d \in V_4$. Merge V_1, V_4 to get new $V_1 = \{a, h, b, c, d, i\}$.

Iteration 7. $e_7 = (b, f)$ and $b \in V_1$, $f \in V_6$. Merge V_1, V_6 to get new $V_1 = \{a, h, b, c, d, i, f\}$.

Iteration 8. $e_8 = (c, i)$ and both c, i belong to the same subset v_1. So ignore e_8.

Iteration 9. $e_9 = (d, e)$ and $d \in V_1$, $e \in V_5$. Merge V_1, V_5 to get new $V_1 = \{a, h, b, c, d, i, f.e, g\} = V$.

Since $V_1 = V$, iteration stops here and the minimal spanning tree obtained has edge set $T = \{e_1, e_2, e_3, e_4, e_5, e_6, e_7, e_9\}$ and is given in Fig. 8.6.

Example 8.3. Find minimal spanning tree for the graph G as given in Fig. 8.7 using (a) Prim's algorithm with root a; (b) Kruskal's algorithm.

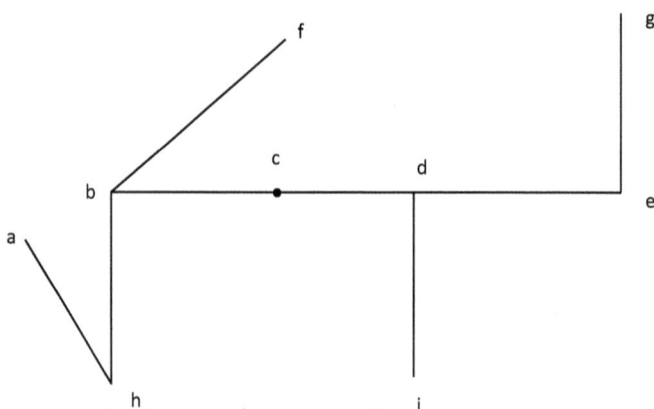

Fig. 8.6

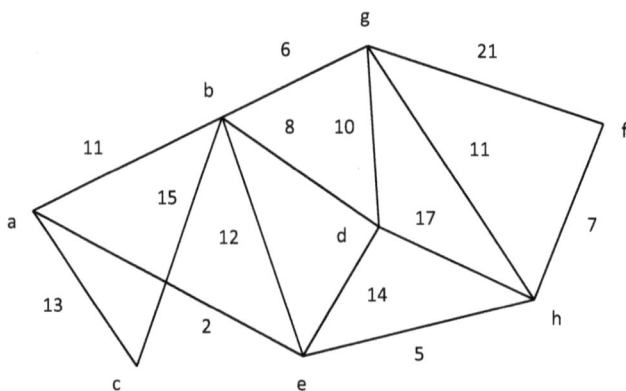

Fig. 8.7

Solution. Label the edges of G as follows:

$$e_1 = (a, e), e_2 = (e, h), e_3 = (b, g), e_4 = (f, h), e_5 = (b, d),$$
$$e_6 = (d, g), e_7 = (a, b), e_8 = (g, h), e_9 = (b, e), e_{10} = (a, c),$$
$$e_{11} = (d, e), e_{12} = (b, c), e_{13} = (d, h), e_{14} = (f, g).$$

(a) Although it is not necessary, we have labeled the edges in non-decreasing order. Such an ordering is useful for using Kruskal's algorithm.

Let also V denote the set of vertices of G and T denote the set of edges that we have to find to get MST of G.

(b) Initially we set

$$V_1 = \{a\}, \ V_2 = \{b\}, \ V_3 = \{c\}, \ V_4 = \{d\},$$
$$V_5 = \{e\}, \ V_6 = \{f\}, \ V_7 = \{g\}, \ V_8 = \{h\}.$$

Iteration 1. $e_1 = (a, e)$ and $a \in V_1$, $e \in V_5$. Merge V_1, V_5 to get new $V_1 = \{a, e\}$ and the set of edges chosen is $T = \{e_1\}$.

Iteration 2. $e_2 = (e, h)$ and $e \in V_1$, $h \in V_8$. Merge V_1, V_8 to get new $V_1 = \{a, e, h\}$ and the set of edges chosen is $T = \{e_1, e_2\}$.

Iteration 3. $e_3 = (b, g)$ and $b \in V_2$, $g \in V_7$. Merge V_2, V_7 to get new $V_2 = \{b, g\}$ and the set of edges chosen is $T = \{e_1, e_2, e_3\}$.

Iteration 4. $e_4 = (f, h)$ and $f \in V_6$, $h \in V_1$ (as in iteration 2). Merge V_1, V_6 to get new $V_1 = \{a, e, h, f\}$ and the set of edges chosen is $T = \{e_1, e_2, e_3, e_4\}$.

Iteration 5. $e_5 = (b, d)$ and $d \in V_4$, $b \in V_2$ (as in iteration 3). Merge V_2, V_4 to get new $V_2 = \{b, g, d\}$ and the set of edges chosen is $T = \{e_1, e_2, e_3, e_4, e_5\}$.

Iteration 6. $e_6 = (d, g)$ and both d, $g \in V_2$, (as in iteration 5). Therefore ignore the edge e_6.

Iteration 7. $e_7 = (a, b)$ and $a \in V_1$, $b \in V_2$ (as in iterations 4 and 5). Merge V_2, V_1 to get new $V_1 = \{a, e, h, f, b, d, g\}$ and the set of edges chosen is $T = \{e_1, e_2, e_3, e_4, e_5, e_7\}$.

Iteration 8. $e_8 = (g, h)$, $e_9 = (b, e)$ and the vertices incident with both these edges are all in V_1, (as in iteration 7). Therefore ignore these two edges.

Iteration 9. $e_{10} = (a, c)$ and $c \in V_3$, $a \in V_1$ (as in iteration 7). Merge V_3, V_1 to get new $V_1 = \{a, e, h, f, b, d, g, c\} = V$ and the set of edges chosen is $T = \{e_1, e_2, e_3, e_4, e_5, e_7, e_{10}\}$.

Since all the vertices of G have been taken care of in the above iteration and the number of edges chosen is $7 = |V| - 1$, we have arrived at a minimal spanning tree.

The edges e_7 and e_8 both are of weight 11. So, we could interchange the order of these edges. We may take

$$e_7' = (g, h)(= e_8) \quad \text{and} \quad e_8' = (a, b)(= e_7)$$

With the new ordering of the edges of G, we find that there is no change up to iteration 6. Now we consider the next iterations.

Iteration 7′. $e_7' = (g, h)$ and $g \in V_2$, $h \in V_1$ (as in iterations 4 and 5). Merge V_2, V_1 to get new $V_1 = \{a, e, h, f, b, d, g\}$ and the set of edges chosen is $T = \{e_1, e_2, e_3, e_4, e_5, e_7'\}$.

Iteration 8′. $e_8' = (a, b)$, $e_9 = (b, e)$ and the vertices with which these edges are incident are all in V_1 (as in iterations 7′). So we ignore these two edges.

Iteration 9′. $e_{10} = (a, c)$ and $c \in V_3$, $a \in V_1$ (as in iteration 7′). Merge V_3, V_1 to get new $V_1 = \{a, e, h, f, b, d, g, c\} = V$ and the set of edges chosen is $T = \{e_1, e_2, e_3, e_4, e_5, e_7', e_{10}\}$.

Since all the vertices of G have been taken care of in the above iteration, we arrive at another minimal spanning tree of G. The two minimal spanning trees of G are given in Figs. 8.8(a) and 8.8(b).

(a) We next apply Prim's algorithm to find a minimal spanning tree of G with root a. Then initially $U = \{a\}$.

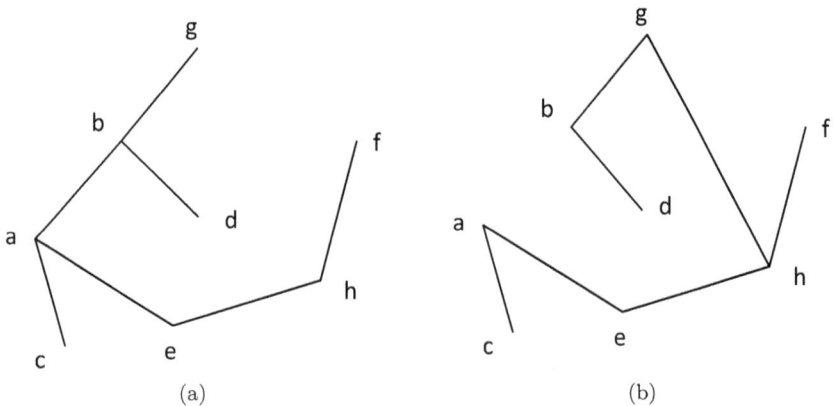

(a)   (b)

Fig. 8.8

Iteration 1. The elements of $V - U$ to be considered are b, c, e. $\text{nnr}(b) = a$, $\text{nnr}(c) = a$, $\text{nnr}(e) = a$ $\text{key}(b) = 11$, $\text{key}(c) = 13$, $\text{key}(e) = 2$. The least among the keys being 2, we have $U = \{a, e\}$, $T = \{e_1\}$.

Iteration 2. The elements of $V - U$ to be considered are b, c, d, h. $\text{nnr}(b) = a$, $\text{nnr}(c) = a$, $\text{nnr}(d) = e$, $\text{nnr}(h) = e$, $\text{key}(b) = 11$, $\text{key}(c) = 13$, $\text{key}(d) = 14$, $\text{key}(h) = 5$. Therefore, $U = \{a, e, h\}$, $T = \{e_1, e_2\}$ and $w(T) = 2 + 5 = 7$.

Iteration 3. The elements of $V - U$ to be considered are b, c, d, f, g. $\text{nnr}(b) = a$, $\text{nnr}(c) = a$, $\text{nnr}(d) = e$, $\text{nnr}(f) = h$, $\text{nnr}(g) = h$, $\text{key}(b) = 11$, $\text{key}(c) = 13$, $\text{key}(d) = 14$, $\text{key}(f) = 7$, $\text{key}(g) = 11$.
 Therefore, $U = \{a, e, h, f\}$, $T = \{e_1, e_2, e_4\}$ and $w(T) = 7 + 7 = 14$.

Iteration 4. The elements of $V - U$ to be considered are b, c, d, g. $\text{nnr}(b) = a$, $\text{nnr}(c) = a$, $\text{nnr}(d) = e$, $\text{nnr}(g) = h$, $\text{key}(b) = 11$, $\text{key}(c) = 13$, $\text{key}(d) = 14$, $\text{key}(g) = 11$. There are now two choices for the vertex to be included in U. Therefore, (i) $U = \{a, e, h, f, b\}$ or (ii) $U = \{a, e, h, f, g\}$, and (i) $T = \{e_1, e_2, e_4, e_7\}$ or (ii) $T = \{e_1, e_2, e_4, e_8\}$ and $w(T) = 14 + 11 = 25$.

Iteration 5. The elements of $V - U$ to be considered are (i) c, d, g or (ii) b, c, d.
 Case (i)

$$\text{nnr}(c) = a, \text{nnr}(d) = b, \text{nnr}(g) = b,$$

$$\text{key}(c) = 13, \text{key}(d) = 8, \text{key}(g) = 6.$$

Therefore, $U = \{a, e, h, f, g, b\}$ and $T = (e_1, e_2, e_4, e_7, e_3)$ and $w(T) = 25 + 6 = 31$.
 Case (ii)

$$\text{nnr}(b) = g, \text{nnr}(c) = a, \text{nnr}(d) = g,$$

$$\text{key}(b) = 6, \text{key}(c) = 13, \text{key}(d) = 10.$$

Therefore, $U = \{a, e, h, f, g, b\}$ and $T = \{e_1, e_2, e_4, e_8, e_3\}$ and $w(T) = 25 + 6 = 31$.
 Observe that U in both the cases is the same but the subsets T in the two cases are different.

Iteration 6. The elements of $V - U$ to be considered are c, d.

$$\text{nnr}(c) = a, \text{nnr}(d) = b,$$

$$\text{key}(c) = 13, \text{key}(d) = 8.$$

Therefore, $U = \{a, e, h, f, b, g, d\}$ and $T = \{e_1, e_2, e_4, e_7, e_3, e_5\}$ in case (i) and $T = \{e_1, e_2, e_4, e_8, e_3, e_5\}$ in case (ii) and $w(T) = 31 + 8 = 39$.

Iteration 7. The element of $V - U$ to be considered is c.

$$\text{nnr}(c) = a, \text{key}(c) = 13.$$

Therefore, $U = \{a, e, h, f, b, g, d, c\} = V$ and $T = \{e_1, e_2, e_4, e_7, e_3, e_5, e_{10}\}$ in case (i) and $T = \{e_1, e_2, e_4, e_8, e_3, e_5, e_{10}\}$ in case (ii) and $w(T) = 39 + 13 = 52$ in both the cases.

Comparing the edge sets of the two spanning trees with the edge sets of the spanning trees obtained by Kruskal's algorithm, we find that the two spanning trees obtained by Prim's algorithm are the ones obtained by Kruskal's algorithm.

Exercise 8.1.

1. Find the minimal spanning tree of the graph G as in Fig. 8.1 using Prim's algorithm with (a) e, (b) g, (c) d as root.
2. Find the minimal spanning tree of the graph G as in Fig. 8.4 using Prim's algorithm with (a) f, (b) i, (c) d as root.

Chapter 9

Single-Source Shortest Paths

A common problem in graphs is to find shortest paths to all reachable vertices from a given source. We have earlier considered this problem when the edges of the graph are not assigned any weight so that path length in that case is simply the number of edges in the path. Now we consider the case in which the graph is a directed weighted graph in which each edge in the graph is assigned a real number as its weight. In this case, the weight of a path is the sum of the weights of all the edges in the path.

Let $G = (V, E, w)$ be a directed graph where V is the set of vertices, E the set of edges of G and $w : E \rightarrow R$ (R the set of all real numbers) is a function that associates a real number $w(e)$ to every edge e in E. Given a path $p : e_1, \ldots, e_k$ or $(v_0, v_1, \ldots, v_k)$ in G, then the **weight of the path** p is defined by

$$w(p) = \sum_{i=1}^{k} w(e_i) = \sum_{i=0}^{k-1} w(v_i, v_{i+1}).$$

By a **shortest path** from a vertex u to a vertex v we mean a path p (say) from u to v such that $w(p) \leq w(q)$ for every path q from u to v. The **shortest-path-length** from u to v is the weight of a shortest path from u to v and is denoted by $\delta(u, v)$. The problem that we consider in this chapter is to determine the cost (length/weight) of a shortest path from a given vertex called source vertex to every other vertex in V. This problem is called the **single-source shortest-path problem**. We consider two procedures/algorithms to solve this problem. One of the two algorithms for shortest path problem is for graphs in which the weights of all the edges are positive (or non-negative) while in the other algorithm some of the edges may have negative weights but the graph has no cycle of negative length. If G is a directed weighted graph in which a vertex u reachable from a source s has a path from s to u containing a cycle of negative length $-l$ (say), then a path from s to u going round the cycle k times has length $a - kl$ where a is the length of a path from s to u not going over the cycle. The distance $a - kl$ goes on decreasing when k increases. Therefore, there is a path from s to u of length $-M$, where M is an arbitrarily chosen positive number i.e., the length of the path is $-\infty$. Thus, the shortest path length problem has no solution when the graph has cycles of negative weight (we use length or weight of a path interchangeably).

The single-source-shortest-path problem arose from a real-life problem and many more problems of this nature appeared over the years. As one such example, we may consider the construction of a road map of a city transport, all buses leaving a central depot and reaching points in the city such that the distance covered from the central depot to the points connected is the least possible. Another problem may be to minimize the amount of pipe line needed to connect the main gas supply depot to the various users.

9.1. Preliminaries

Throughout this chapter $G = (V, E, w)$ is a directed weighted graph with the weight function w associating a weight $w(e)$ to every edge e in E and s is a source vertex. Also for vertices $u, v \in V$, $\delta(u, v)$ denotes the length of a shortest path from u to v. For an edge (u, v) on a path from s, u is the predecessor of v and we indicate this by $\pi(v) = u$. We need the following preliminary lemmas.

Lemma 9.1 (Triangle inequality). *For all edges $(u, v) \in E$, we have* $\delta(s, v) \leq \delta(s, u) + w(u, v)$.

Proof. Let $p = (s, u_1, \ldots, u_k = u)$ be a shortest path from s to u. Then $q = (s, u_1, \ldots, u_k = u, v)$ is a path from s to v. Therefore

$$\delta(s, v) \leq w(q) = w(p) + w(u, v) = \delta(s, u) + w(u, v).$$

$\square$

Lemma 9.2 (Sub-paths of shortest paths). *If* $p = (s, u_1, \ldots, u_k = u)$ *is a shortest path from s to u, then for every i, j, $0 \leq i < j \leq k$, $q = (u_i, u_{i+1}, \ldots, u_j)$ is a shortest path from u_i to u_j.*

Proof. Let $p' = (s = u_0, \ldots, u_{i-1})$ and $p'' = (u_{j+1}, \ldots, u_k)$ be subpaths of p. Then $p = p' \rightsquigarrow q \rightsquigarrow p''$ and, therefore, $w(p) = w(p') + w(q) + w(p'')$. Let q' be a shortest from u_i to u_j. Then $w(q') \leq w(q)$ and $p' \rightsquigarrow q' \rightsquigarrow p''$ is a path from s to u_k. Therefore $w(p' \rightsquigarrow q' \rightsquigarrow p'') \geq \delta(s, u_k) = w(p)$ and $w(p) \leq w(p' \rightsquigarrow q' \rightsquigarrow p'') = w(p') + w(q') + w(p'') \leq w(p') + w(q) + w(p'') = w(p)$. Therefore $w(p') + w(q') + w(p'') = w(p') + w(q) + w(p'')$ which implies that $w(q) = w(q')$ and q is a shortest path from u_i to u_j. $\square$

Here, for paths p, q with terminal vertex of p equal to initial vertex of q, $p \rightsquigarrow q$ means the concatenation of the paths p and q.

In the algorithm that we consider, we assign to every vertex v in V an **estimated shortest path length** $d(v)$ (which is an upper bound on $\delta(s, v)$) from s to v and as the algorithm proceeds, we improve upon the value $d(v)$ till we finally arrive at $d(v) = \delta(s, v)$. Initially, we set $d(s) = 0$, $d(v) = \infty$ for every $v \in V$, $v \neq s$ and $\pi(v) = \text{nil}$. We may formalize this as the following procedure:

Algorithm 9.1. Initialize-Single-Source (G, s)

---

1. for each $v \in V$
2. do $d(v) = \infty$ for $v \neq s$
3. $d(s) = 0$
4. $\pi(v) = \text{nil}$

The initialize-single-source procedure could equally well have been taken as: $d(s) = 0$, $\pi(s) = \text{nil}$, for each $v \in V - \{s\}$, $d(v) = w(s, v)$ and $\pi(v) = s$. The alternative form of initialize-single-source procedure results in reducing the number of iterative steps in the final algorithm by one. In the algorithm that we consider, we build a subset S of V by initially taking $S = \varphi$ (or taking $S = \{s\}$ in the alternative initializing procedure) and adding one vertex at a time to S. When a vertex u is added to S at an iterative step, we need to improve upon the estimated shortest path length

$d(v)$ for every v in $V - S$. Indeed we only need to do this for every v in adj(u) (= the set of vertices that are adjacent to u).

As already mentioned, single-source-shortest-path problem is meaningful only in the case when the graph G has no negative weight cycles. We then have the following result.

Theorem 9.3. *If the graph G has no cycles of negative weight, then a shortest weight path from s to a vertex v, if it exists, is a path having at most $|V| - 1$ edges.*

Proof. Suppose that $v \in V$ is a vertex for which a shortest weight path from s to v exists. Let $p = (s = v_0, v_1, \ldots, v_k = v)$ be a shortest weight path. Suppose that p contains a cycle $q = (v_i, v_{i+1}, \ldots, v_j, v_{j+1} = v_i)$ for some $i \geq 0$. Let $p' = (s = v_0, v_1, \ldots, v_{i-1})$ and $p'' = (v_{j+1}, \ldots, v_k)$ be sub-paths of p. Then p is $p' \rightsquigarrow q \rightsquigarrow p''$ and, therefore, $w(p) = w(p') + w(q) + w(p'')$ (observe that p' or p'' may be empty).

Since G has no cycles of negative weight, $w(q) \geq 0$. Therefore $w(p) \geq w(p') + w(p'')$ and $p' \rightsquigarrow p''$ is a path from s to v. Since p is a shortest path from s to v, it follows that $p' \rightsquigarrow p''$ is a shortest weight path from s to v. Repeating this argument a finite number of times, we get a shortest path $p''' = (s = v_0, v_1, \ldots, v_r = v)$ which does not contain a cycle. Therefore, $v_0, v_1, \ldots, v_r$ are all distinct vertices in V. Therefore, the number of edges in the shortest path p''' from s to v is at most $|V| - 1$. □

So far we talked of shortest path length from the source s to the other vertices of G. However, we may also be interested in finding shortest paths from s to the vertices of G. These may be obtained by using the concept of predecessor $\pi(v)$ of $v \in V$. Recall that given $v \in V$, $\pi(v)$ is another vertex in V such that $(\pi(v), v)$ is an edge in E or $\pi(v) = $ nil. Then the chain of predecessors of a vertex v in the construction of shortest path length terminates at the source vertex s in a finite number of steps and we get a shortest length path from s to v. We will come back to this problem after we have considered at least one of the shortest-path algorithm.

9.2. Dijkstra's Algorithm

In this section, we consider single-source-shortest-path problem when the weight $w(e)$ for every edge e in E is non-negative. Therefore, the restriction that the graph G has no cycles of negative length is automatically satisfied.

Algorithm 9.2. Dijkstra's Algorithm (G, w, s)

---

1. initialize-single-source (G, s)
2. $S = \varphi$
3. while $V - S \neq \varphi$
4. choose a vertex u in $V - S$ such that
5. $d(u)$ is minimum
6. add u to S
7. for each v in $V - S$ (S when u has been added)
8. if $d(v) \leq d(u) + w(u, v)$
9. keep $d(v)$ and $\pi(v)$ unchanged
10. if $d(v) > d(u) + w(u, v)$
11. set $d(v) = d(u) + w(u, v)$
12. $\pi(v) = u$

Items 8–12 are taken for each $v \in V - S$ when u has been added to S and the steps are taken relative to u. Therefore v is, as a matter of fact, to be chosen from adj$[u]$ — the set of vertices that are adjacent to u.

Theorem 9.4. *When Dijkstra's algorithm is run on $G = (V, E, w)$ with source vertex s and with $w(e)$ non-negative for every edge $e \in E$, then*

(a) *$d(v)$ is finite for $v \in V$ if and only if v is reachable from the source vertex s;*
(b) *$\delta(s, v) \leq d(v)$ for every $v \in V$;*
(c) *at the end of the algorithm $\delta(s, v) = d(v)$ for every $v \in V$.*

Proof. (a) Suppose that $v \in V$ is reachable from the source vertex s. Let $p = (s = v_0, v_1, \ldots, v_k = v)$ be a path from s to v. We prove by induction on i that $d(v_i) < \infty$ for every $i \geq 0$.

Now $d(v_0) = d(s) = 0$, so that the assertion is true for $i = 0$. Suppose that $i \geq 0$ and that $d(v_i) < \infty$. Suppose that v_i is added to S just before u is added to S i.e., u is chosen in adj$[v_i]$. Then, by items 7–11 of Dijkstra;s algorithm, $d(u) = \min\{d(u), d(v_i) + w(v_i, u)\} < \infty$, as both $d(v_i)$ and $w(v_i, u)$ are finite.

If v_{i+1} is added to S before v_i is added, then $d(v_{i+1}) \leq d(v_i) < \infty$ (because of items 4–6 of the algorithm). On the other hand, suppose that v_{i+1} is added to S after v_i is added. Then by the above observation and by induction hypothesis it follows that $d(v_{i+1}) < \infty$. This completes induction and, therefore, $d(v_i) < \infty$ for every $i \geq 0$. In particular, this is true for $i = k$ and $d(v) < \infty$.

Conversely, suppose that $d(v) < \infty$. Since initially $d(v) = \infty$ for every $v \neq s$, $d(v)$ becomes finite only if after a finite number of iterations, we arrive at a vertex $u \in V$ with $d(u) < \infty$, $d(v) = d(u) + w(u, v)$ and $u = \pi(v)$, i.e., u is a predecessor of v so that $(\pi(v), v) = (u, v)$ is an edge in E. Let us call $u = u_1$. Repeating this argument, we arrive at a sequence $v, u_1, u_2, \ldots, u_t$ such that $u_1 = \pi(v)$ and $u_{i+1} = \pi(u_i)$ for $i \geq 1$. The process continues till we arrive at a u_k such that $d(u_k) = 0$ and $\pi(u_k) = $ nil. But $\pi(u) = $ nil for only those u for which $d(u) = \infty$ or $u = s$. Therefore $u_k = s$ and $(s = u_k, u_{k-1}, \ldots, u_1 = u, v)$ is a path from s to v. Hence v is reachable from s.

(b) Initially $d(s) = 0 = \delta(s, s)$. If $d(v) = \infty$, then also $\delta(s, v) \leq d(v)$ is trivially true. Suppose that $d(v) < \infty$. Then v is reachable from s and there exists a path $p = (s, v_1, \ldots, v_k = v)$ with $d(v) = d(v_k) = d(v_{k-1}) + w(v_{k-1}, v_k) = d(v_{k-2}) + w(v_{k-2}, v_{k-1}) + w(v_{k-1}, v_k) = \cdots = \sum_{0 \leq i \leq k-1} w(v_i, v_{i+1})$, with $v_0 = s, v_k = v$. Thus $d(v) = w(p) \geq \delta(s, v)$.

(c) In view of (b) above, we need only prove that $\delta(s, v) \geq d(v)$ for every v.

Initially $d(s) = 0 = \delta(s, s)$ and if $d(v) = \infty$, then v is not reachable from s so that $\delta(s, v) = \infty$ as well. Therefore, we need only prove that $d(v) \leq \delta(s, v)$ if $d(v) < \infty$.

Suppose that $\delta(s, x) \geq d(x)$ for all vertices x that are added to S before u is added. We may suppose that $d(u) < \infty$. Let $(s, v_1, \ldots, v_k = u)$ be a shortest path from s to u and let e_i be the edge v_{i-1}, v_i, $1 \leq i \leq k$. In view of Lemma 9.2, for every $r = 0, 1, \ldots, k$, $(v_0, v_1, \ldots, v_r)$ is a shortest path from s to v_r. Therefore

$$\delta(s, v_r) = \sum_{1 \leq i \leq r} w(e_i). \tag{9.1}$$

Choose j to be the maximal index such that v_j is added to S before u is added. Then, by induction hypothesis,

$$\delta(s, v_j) \geq d(v_j) \quad \text{or} \quad \sum_{1 \leq i \leq j} w(e_i) \geq d(v_j). \tag{9.2}$$

Now suppose that $v_{j+1} \neq u$ (i.e., $j \neq k - 1$). Then after v_j is added to S in the while loop (steps 7–11)

$$d(v_{j+1}) = \min\{d(v_{j+1}), d(v_j) + w(v_j, v_{j+1})\}$$
$$\leq d(v_j) + w(v_j, v_{j+1}) \leq \sum_{1 \leq i \leq j+1} w(e_i)$$

(by (9.2)). As $d(v_{j+1})$ can, if at all, only be decreased (it never increases) in the following iterations, the above inequality is still valid when u is added to S. Thus

$$d(v_{j+1}) \leq \sum_{1 \leq i \leq j+1} w(e_i) \leq \sum_{1 \leq i \leq k} w(e_i) = \delta(s, u). \tag{9.3}$$

Therefore, if $\delta(s, u) < d(u)$, then $d(v_{j+1}) < d(u)$ showing that v_{j+1} is added to S before u is added. This contradicts the choice of j and, therefore

$$d(u) \leq \delta(s, u). \tag{9.4}$$

When $v_{j+1} = u$, i.e., when $j = k - 1$, then relation (9.3) implies that $d(u) \leq \delta(s, u)$. Therefore the relation (9.4) holds for u in all cases. This completes induction and, therefore, $d(x) \leq \delta(s, x)$ for all $x \in V$. Combining this result with that of Theorem 9.4(a) and (b), we have $d(x) = \delta(s, x)$ for all $x \in V$ and Dijkstra's algorithm is correct. $\qquad\square$

Theorem 9.5. *The running time or complexity of Dijkstra's algorithm is*
$O(|V|^2)$.

Proof. In step 4 of the algorithm, the minimum of $d(u)$ can be computed
for $v \in V - S$ which requires $|V|-|S|-1$ comparisons. Initially $V-S = V$
as $S = \varphi$, and at each iteration one element is added to S, so this
involves $|V|$ choices for u. Therefore, the total number of comparisons
needed for this step is $1 + 2 + \cdots + (|V| - 1) = \frac{|V|(|V|-1)}{2}$. Thus,
this process involves $O(|V|^2)$ steps. Initialization in step 1 involves $2|V|$
steps. Steps 8 and 10 need one comparison for each $v \in V - S$ and, so,
require $|V| - |S|$ comparisons, i.e., again the total number of comparisons
needed is $O(|V|^2)$. In steps 11 and 12, the number of steps needed is
$2|V| = O(|V|)$. Therefore, the total running time of Dijkstra's algorithm
is $O(|V|^2) + O(|V|^2) + O(|V|) = O(|V|^2)$.   □

Remark. Observe that results of Theorem 9.4(a) and (b) are true whether
weights of edges of G are all non-negative or not. However, in the proof
of Theorem 9.4(c), the inequality (9.3) may not be valid if G has some
edges of negative weight. Therefore, Theorem 9.4(c) is valid only when
the graph has no edges of negative weight.

Consider the directed weighted graph as given in Fig. 9.1.

We consider the use of Dijkstra's algorithm to find a possible solution
to single-source-shortest-path problem with 1 as source vertex.

Initially $S = \varphi$, $d(1) = 0$, $d(i) = \infty$ for $2 \le i \le 5$ and $\pi(i) = $ nil for
all i.

Iteration 1. $u = 1$, $S = \{1\}$, $\overline{d(2) = -1}$, $\pi(2) = 1$, $d(3) = 4$, $\pi(3) = 1$.

Iteration 2. $u = 2$, $S = \{1, 2\}$, $d(3) = 2$, $\pi(3) = 2$, $d(4) = 1$, $\pi(4) = 2$,
$\overline{d(5) = 1, \pi(5) = 2.}$

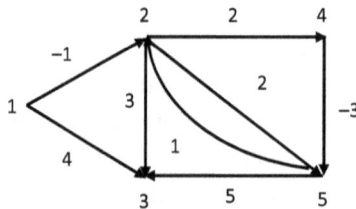

Fig. 9.1

Iteration 3. $u = 5$, $S = \{1, 2, 5\}$, $d(3) = 2$, $\pi(3) = 2$, $\overline{d(4) = 1}, \pi(4) = 2$.

Iteration 4. $u = 4$, $S = \{1, 2, 5, 4\}$, $\overline{d(3) = 2, \pi(3) = 2}$.

Iteration 5. $u = 3$, $S = \{1, 2, 5, 4, 3\}$

and the process of iteration stops here. As a result of the iterative steps, we find that

$\delta(1, 2) = -1$ and the corresponding shortest path is $(1, 2)$,
$\delta(1, 5) = 1$ and the corresponding shortest path is $(1, 2, 5)$,
$\delta(1, 4) = 1$ and the corresponding shortest path is $(1, 2, 4)$,
$\delta(1, 3) = 2$ and the corresponding shortest path is $(1, 2, 3)$.

Observe that $(1, 2, 4, 5)$ is a path from the source vertex to the vertex 5 and its weight is $-1 + 2 - 3 = -2$ which is less than $\delta(1, 5)$ as calculated above by Dijkstra's algorithm. This proves that Dijkstra's algorithm is not valid for graphs having edges of negative weight.

We now consider a couple of examples on the use of Dijkstra's algorithm. In the above example and other examples to follow, we indicate u with least value $d(u)$ in the iteration as $\overline{d(u)}$.

Example 9.1. Use Dijkstra's algorithm and solve single-source-shortest-path problem with source vertex 1 for the graph of Fig. 9.2.

Solution. Initially $S = \varphi$, $d(1) = 0$, $d(i) = \infty$ for $2 \le i \le 9$ and $\pi(i) =$ nil for all i.

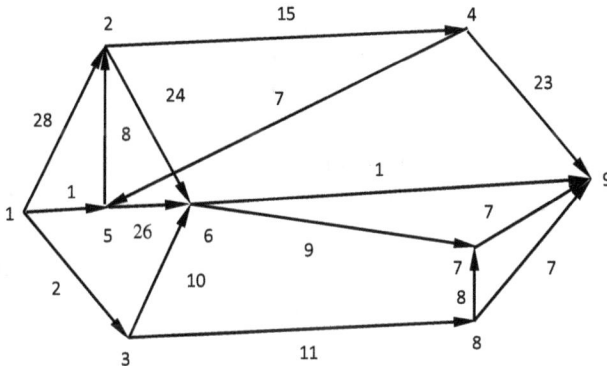

Fig. 9.2

Iteration 1. $u = 1$, $S = \{1\}$, $d(2) = 28$, $\pi(2) = 1$, $\overline{d(5) = 1, \pi(5) = 1}$, $d(3) = 2$, $\pi(3) = 1$.

Iteration 2. $u = 5$, $S = \{1, 5\}$, $d(2) = 9$, $\pi(2) = 5$, $\overline{d(3) = 2, \pi(3) = 1}$, $d(6) = 27$, $\pi(6) = 5$.

Iteration 3. $u = 3$, $S = \{1, 5, 3\}$, $\overline{d(2) = 9, \pi(2) = 5}$, $d(6) = 12$, $\pi(6) = 3$, $d(8) = 13$, $\pi(8) = 3$.

Iteration 4. $u = 2$, $S = \{1, 5, 3, 2\}$, $\overline{d(6) = 12, \pi(6) = 3}$, $d(8) = 13$, $\pi(8) = 3$, $d(4) = 24$, $\pi(4) = 2$.

Iteration 5. $u = 6$, $S = \{1, 5, 3, 2, 6\}$, $\overline{d(8) = 13, \pi(8) = 3}$, $d(4) = 24$, $\pi(4) = 2$, $d(7) = 21$, $\pi(7) = 6$, $d(9) = 13$, $\pi(9) = 6$.

(In this case, there is a tie between vertices 8 and 9 and we may choose one of these two arbitrarily.)

Iteration 6. $u = 8$, $S = \{1, 5, 3, 2, 6, 8\}$, $d(4) = 24$, $\pi(4) = 2$, $d(7) = 21$, $\pi(7) = 6$, $\overline{d(9) = 13, \pi(9) = 6}$.

Iteration 7. $u = 9$, $S = \{1, 5, 3, 2, 6, 8, 9\}$, $d(4) = 24$, $\pi(4) = 2$, $\overline{d(7) = 21, \pi(7) = 6}$.

Iteration 8. $u = 7$, $S = \{1, 5, 3, 2, 6, 8, 9, 7\}$, $\overline{d(4) = 24, \pi(4) = 2}$.

Iteration 9. $u = 4$, $S = V$, and the process stops here. Observe that

    $\delta(1, 5) = 1$ and a shortest path is $(1, 5)$;
    $\delta(1, 3) = 2$ and a shortest path is $(1, 3)$;
    $\delta(1, 2) = 9$ and a shortest path is $(1, 5, 2)$;
    $\delta(1, 6) = 12$ and a shortest path is $(1, 3, 6)$;
    $\delta(1, 8) = 13$ and a shortest path is $(1, 3, 8)$;
    $\delta(1, 9) = 13$ and a shortest path is $(1, 3, 6, 9)$;
    $\delta(1, 7) = 21$ and a shortest path is $(1, 3, 6, 7)$;
    $\delta(1, 4) = 24$ and a shortest path is $(1, 5, 2, 4)$.

Example 9.2. Use Dijkstra's algorithm and solve single-source-shortest-paths problem with source vertex 1 for the graph of Fig. 9.3.

Solution. Initially $S = \varphi$, $d(1) = 0$, $d(i) = \infty$ for $2 \leq i \leq 6$ and $\pi(i) =$ nil for all i.

Iteration 1. $u = 1$, $S = \{1\}$, $\overline{d(2) = 5, \pi(2) = 1}$, $d(3) = 10$, $\pi(3) = 1$.

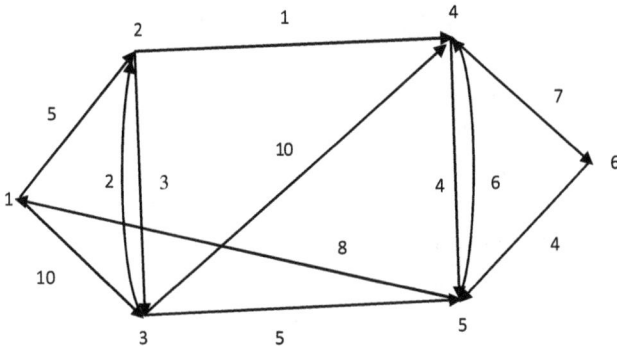

Fig. 9.3

Iteration 2. $u = 2$, $S = \{1, 2\}$, $d(3) = 8$, $\pi(3) = 2$, $\overline{d(4) = 6, \pi(4) = 2}$.

Iteration 3. $u = 4$, $S = \{1, 2, 4\}$, $\overline{d(3) = 8, \pi(3) = 2}$, $d(5) = 10$, $\pi(5) = 4$, $d(6) = 13$, $\pi(6) = 4$.

Iteration 4. $u = 3$, $S = \{1, 2, 4, 3\}$, $\overline{d(5) = 10, \pi(5) = 4}$, $d(6) = 13$, $\pi(6) = 4$.

Iteration 5. $u = 5$, $S = \{1, 2, 4, 3, 5\}$, $\overline{d(6) = 13, \pi(6) = 4}$.

Iteration 6. $u = 6$, $S = V$, and the process stops here. Also

$\delta(1, 2) = 5$ and the corresponding shortest path is $(1, 2)$;
$\delta(1, 4) = 6$ and the corresponding shortest path is $(1, 2, 4)$;
$\delta(1, 3) = 8$ and the corresponding shortest path is $(1, 2, 3)$;
$\delta(1, 5) = 10$ and the corresponding shortest path is $(1, 2, 4, 5)$;
$\delta(1, 6) = 13$ and the corresponding shortest path is $(1, 2, 4, 6)$.

Remark. After the choice of u in $V - S$ as at step 4, the process as at steps 10–12 for the choice of v as at step 7 may be abbreviated as a process of relaxing over the edge (u, v). We may describe this as follows:

Algorithm 9.3. Relax (u, v, w)

---

1. if $d(v) > d(u) + w(u, v)$
2. then $d(v) = d(u) + w(u, v)$
3. $\pi(v) = u$

The proofs of Theorems 9.4(a) and (b) as given seem to depend on Dijkstra's algorithm. However, this is not so and the two results only depend on the property of "relaxing" an edge and the initializing procedure.

Let us consider proving that $\delta(s, v) \leq d(v)$ for every $v \in V$.

Initially $d(s) = 0$ and $d(v) = \infty$ for $v \neq s$. Since $\delta(s, s) = 0$ and $\delta(s, v)$, $v \neq s$, may be finite or infinity, we have $\delta(s, v) \leq d(v)$ for every $v \in V$ initially. We now use induction on the number of steps using Relax(u, v, w). Suppose that $\delta(s, x) \leq d(x)$ for every $x \in V$ prior to using Relax(u, v, w) for an edge (u, v) in E. When this relaxation is used, the only $d(x)$ that changes is $d(v)$. Relaxing the edge (u, v), we have if $d(v) > d(u) + w(u, v)$, then $d(v) = d(u) + w(u, v) \geq \delta(s, u) + w(u, v) \geq \delta(s, v)$. This completes induction and $\delta(s, v) \leq d(v)$ for every $v \in V$.

For a given $v \in V$, if $d(v) < \infty$, then $\delta(s, v)$ is finite and there exists a path of finite length from s to v. Therefore v is reachable from s. If, on the other hand, v is not reachable from s, then there is no path from s to v. Therefore $\delta(s, v) = \infty$. Since $\delta(s, v) \leq d(v)$, we also have $d(v) = \infty$. We have thus proved.

Theorem 9.6. *Theorem 9.4(a) and (b) are true independent of Dijkstra's algorithm and independent of the assumption that $w(e)$ is non-negative for every e in E.*

Moreover, once $d(v)$ attains its minimum (finite) value $\delta(s, v)$, it can never decrease beyond that whatever relaxing steps over any edges we may use.

We consider Dijkstra's algorithm for directed, weighted graph with all edge weights being non-negative. However, there is one exception as provided by the following result.

Remark. Let $G = (V, E, s)$ be a directed weighted graph with source vertex s in which the edges that leave the source may have negative weights while all other edge weights are non-negative. If G has no negative weight cycles, then Dijkstra's algorithm correctly finds shortest paths from s to the other vertices in G.

Proof. As observed above, Theorem 9.4(a) and (b) are true in this case as well. In order to prove our contention, we now need to prove the assertion in Theorem 9.4(c).

Outgoing edges from the source vertex s being the only possible edges with negative weight, negative weight cycles, if any, will have to be cycles from s to s. It being given that there are no negative weight cycles, there are no negative weight cycles from s to s. Hence for every vertex v which is adjacent to s, the shortest path from s to v is the weight of the edge from s to v. With the notations as in the proof of Theorem 9.4(c), it follows that $w(e_i) > 0$ for every $i \geq 2$. It follows from these observations that the relations as in (9.3) hold in the present case as well. Thus the proof of Theorem 9.4(c) carries over in the present case and the proof of the above assertion is true. □

9.3. Bellman–Ford Algorithm

So far we have restricted our attention to graphs G with non-negative edge weights. There may, however, exist problems in which some of the edges of G may have negative weight. Dijkstra's algorithm is a good example of "greedy algorithm". It is a greedy algorithm because in steps 4–5 of the algorithm, we make a choice of u in $V - S$ with $d(u)$ minimum. The algorithm that we intend to study in this section is a good example of "dynamic programming". Unlike in Dijkstra's algorithm, there is no subset S of V to be constructed till $S = V$. In Dijkstra's algorithm, once u is chosen in $V - S$ with $d(u)$ minimum and u is added to S, we have to relax every edge (u, v) with v in $V - S$. In contrast, in the Bellman–Ford algorithm that we intend to study, at every iterative step we have to relax all the edges in the graph.

Algorithm 9.4. Bellman–Ford Algorithm (G, w, s)

---

1. initialize–single–source (G, s) for each $v \in V$, $v \neq s$, $d(v) = \infty$ for the source vertex s, $d(s) = 0$
   for each $v \in V$ $\pi(v) = \text{nil}$
2. for $i = 1$ to $|V| - 1$
   for each edge (u, v) in E relax (u, v)
   if $d(v) > d(u) + w(u, v)$ set $d(v) = d(u) + w(u, v)$, $\pi(v) = u$
   otherwise keep $d(v)$ and $\pi(v)$ unchanged
3. check for negative weight cycles
   for $i = |V|$
   if there is an edge (u, v) with $d(v) > d(u) + w(u, v)$
   then the graph contains a negative weight cycle

Running time of the Bellman–Ford algorithm

Since the Bellman–Ford algorithm relaxes all the edges in E at every iteration and $|V|-1$ iterative steps are taken, the total number of relaxation operations is $(|V|-1)|E|$. Also the initialization process involves $2|V|$ steps. Therefore, if the graph contains no cycles of negative weight the running time of the algorithm is $O((|V|-1)|E|) + O(2|V|) = O((|V|-1)|E|)$. On the other hand, if an indication of the existence of a cycle of negative weight or otherwise is also required, the number of relaxation steps is $|V||E|$ and, so, the running time is $O(|V||E|) + O(|V|) = O(|V||E|)$.

Before proving the correctness of the Bellman–Ford algorithm, we need to look at the effect of relaxations on a shortest path from s to a vertex v.

Lemma 9.7. *Suppose that the graph G is initialized by the procedure Initialize–single–source (G, s).*

(a) Let $s \leadsto u \to v$ be a shortest path from s to v and a sequence of relaxation calls be executed on the edges of G including the call Relax(u, v, w). If $d(u) = \delta(s, u)$ prior to the call to Relax(u, v, w), then $d(v) = \delta(s, v)$.

(b) Let $p = (v_0, v_1, \ldots, v_k)$, with $v_0 = s$, be a shortest path from s to v_k. Suppose that a sequence of relaxation calls occurs that includes, in order, relation of edges $(v_0, v_1), (v_1, v_2), \ldots, (v_{k-1}, v_k)$. Then $d(v_k) = \delta(s, v_k)$ after the relaxations.

Proof. (a) On relaxing the edge (u, v), we have

$$d(v) \leq d(u) + w(u, v) = \delta(s, u) + w(u, v) = \delta(s, v).$$

But $\delta(s, v) \leq d(v)$ always. Therefore $d(v) = \delta(s, v)$.

(b) Since $(v_0, v_1, \ldots, v_k)$ is a shortest path from s to v_k, $(v_0, v_1, \ldots, v_i)$ is a shortest path from s to v_i for $1 \leq i \leq k$. We prove by induction on i that $d(v_i) = \delta(s, v_i)$ for $i \geq 0$. For $i = 0$, $d(v_0) = d(s) = 0 = \delta(s, s)$ and, so, the claim holds. For $i = 1$, $d(v_1) = d(v_0) + w(v_0, v_1) = w(s, v_1) = \delta(s, v_1)$ because (v_0, v_1) is a shortest path. Now, suppose that $i \geq 0$ and that $d(v_i) = \delta(s, v_i)$. Now use the relaxation call (v_i, v_{i+1}). By part (a) above, $d(v_{i+1}) = d(v_i) + w(v_i, v_{i+1}) = \delta(s, v_i) + w(v_i, v_{i+1}) = \delta(s, v_{i+1})$ again using the fact that $(v_0, v_1, \ldots, v_i, v_{i+1})$ is a shortest path from s to v_{i+1}. Thus induction is complete. In particular $d(v_k) = \delta(s, v_k)$. $\square$

Lemma 9.8. *Suppose that Bellman–Ford algorithm is run on G. After i operations of the for loop:*

(a) *if $d(u) < \infty$, then $d(u) = $ length of some path from s to u;*
(b) *if there is a path from s to u with at most i edges, then $d(u) \leq \delta^i(s, u)$, where $\delta^i(s, u)$ denotes the length of a shortest path from s to u with at most i edges.*

Proof. We need to prove the lemma for all vertices u that are reachable from s so that $d(u) < \infty$. We prove the result by induction on i.

(a) For $i = 1, d(u) = d(s) + w(s, u) = w(s, u) = $ length of the path (s, u) from s to u. Suppose that $i \geq 1$ and that after i iterations of the for loop, $d(u) = $ length of a path p from s to u. Let $p = (s, u_1, \ldots, u_k, u)$ be a path from s to u with $d(u) = w(p)$. Then $d(u_k)$ is also length of a path from s to u_k. Relax the edge (u_k, u). Then either $d(u) < d(u_k) + w(u_k, u)$ or $d(u) = d(u_k) + w(u_k, u)$.

If $d(u) < d(u_k) + w(u_k, u)$, then $d(u)$ is the same as obtained after i iterations and, so, $d(u)$ is the length of some path from s to u. On the other hand, $d(u) = d(u_k) + w(u_k, u) = $ (length of a path from s to u_k) $+ w(u_k, u) = $ length of a path from s to u. This completes induction. Hence after i operations, $d(u) = $ length of a path from s to u.

(b) If there is a path from s to u with at most one edge, then $d(u) = w(s, u) = \delta^1(s, u)$ as (s, u) is a shortest path from s to u with one edge.

We denote by $\delta^k(s, u)$ the length of a shortest path from s to u with at most k edges. Suppose that for every u for which there is a path with at most i edges, then $d(u) \leq \delta^i(s, u)$. Now, let u be a vertex with a path from s to u with at most $i + 1$ edges. Among all the paths with at most $i + 1$ edges, let $p = (s, u_1, \ldots, u_k, u)$ be a shortest path from s to u with $\leq i + 1$ edges. Then $(k \leq i)$ $q = (s, u_1, \ldots, u_k)$ is a shortest path from s to u_k with at most i edges. Therefore, $d(u_k) \leq \delta^i(s, u_k)$. Now relax the edge (u_k, u). Then

$$d(u) \leq d(u_k) + w(u_k, u) \leq \delta^i(s, u_k) + w(u_k, u) \leq \delta^{i+1}(s, u).$$

This completes induction. Hence (b) holds for all $i \geq 1$. $\square$

Theorem 9.9. *The Bellman–Ford algorithm returns the shortest path length from s to every vertex reachable from s or indicates that there is a negative weight cycle in G.*

Proof. Let $|V| = n$. At each iteration i that the edges are scanned and relaxed, the algorithm finds all shortest paths having at most i edges. The longest possible path not containing a cycle can have $n - 1$ edges. Therefore, the edges must be scanned $n - 1$ times (there being $n - 1$ vertices besides the source vertex) to make sure that the shortest path has been found for all vertices. Hence, if the graph has no cycles of negative weight, then $\delta^{n-1}(s, u) = \delta(s, u)$. After $n - 1$ iterations of the for loop, we have

$$d(u) \leq \delta^{n-1}(s, u) = \delta(s, u).$$

Since $d(u) \geq \delta(s, u)$ always, we have $d(u) = \delta(s, u)$ and the algorithm leads to the shortest path lengths from the source vertex s to all the vertices of the graph.

In case, it is not given that the graph has no negative weight cycles, the edges are scanned one more time. If it is found that $\delta^n(s, u) = \delta^{n-1}(s, u)$ for every $u \in V$, then as above $d(u) = \delta^{n-1}(s, u)$ and we get the shortest path lengths from s to all vertices. On the other hand, if there exists even one vertex u for which $d(u)$ on the nth iteration is less than $d(u)$ on the $(n - 1)$th iteration, then it follows that the graph has a negative weight cycle(s) and the algorithm does not yield shortest path lengths from s to the vertices of G. □

Example 9.3. Apply Bellman–Ford algorithm to the directed weighted graph as given in Fig. 9.4 with source vertex s and find, if possible, shortest path lengths from s to all other vertices of G. If these exist, also find corresponding shortest paths.

Solution.

Iteration 1. Only the edges (s, x), (s, y), (s, z) are effectively relaxed and we get $d(x) = 6$, $d(y) = 5$, $d(z) = 5$ and $d(-) = \infty$ for the other vertices.

Iteration 2. Edges effectively relaxed are (x, y), (y, z), (x, t), (y, t), (z, u) and $d(x) = 6$, $d(y) = 4$, $d(z) = 3$, $d(t) = 5$, $d(u) = 4$ and $d(v) = \infty$.

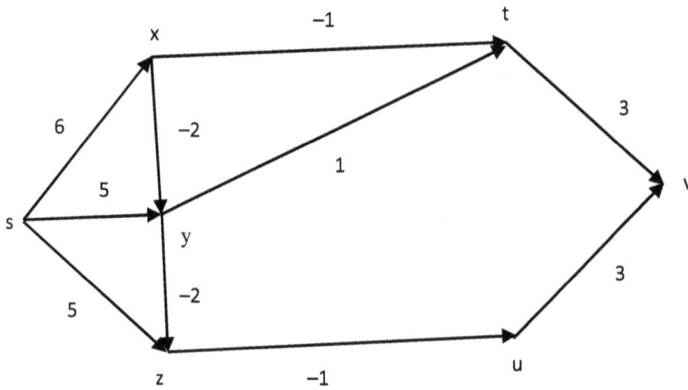

Fig. 9.4

Iteration 3. Edges effectively relaxed are (y, z), (t, v), (u, v) and $d(x) = 6$, $d(y) = 4$, $d(z) = 2$, $d(t) = 5$, $d(u) = 2$, $d(v) = 7$.

Iteration 4. Edges (y, z), (z, u), (u, v) are relaxed and $d(x) = 6$, $d(y) = 4$, $d(z) = 2$, $d(t) = 5$, $d(u) = 1$, $d(v) = 5$.

Iteration 5. Edges (z, u), (u, v) are effectively relaxed and $d(x) = 6$, $d(y) = 4$, $d(z) = 2$, $d(t) = 5$, $d(u) = 1$, $d(v) = 4$.

Iteration 6. Edge (u, v) is the only one effectively relaxed and $d(x) = 6$, $d(y) = 4$, $d(z) = 2$, $d(t) = 5$, $d(u) = 1$, $d(v) = 4$.

Since no changes take place in the estimated shortest lengths from iteration 5 to iteration 6, these will remain the same after every future iteration. Hence the shortest path lengths are $\delta(s, x) = 6$, $\delta(s, y) = 4$, $\delta(s, z) = 2$, $\delta(s, t) = 5$, $\delta(s, u) = 1$ and $\delta(s, v) = 4$.

Since no change takes place in $d(x)$ after first iteration, in $d(y)$, $d(t)$ after second iteration and in $d(z)$ after the third iteration, the corresponding shortest paths from s to x, y, z and t are, respectively, (s, x), (s, x, y), (s, x, y, z), (s, x, t). No change occurs in $d(u)$ after iteration 4 and in $d(v)$ after iteration 5. Therefore, the corresponding shortest paths to u, v are (s, x, y, z, u) and (s, x, y, z, u, v), respectively.

Example 9.4. Apply Bellman–Ford algorithm to find shortest path lengths, if possible, from the source vertex s to the other vertices in the graph G as given in Fig. 9.5.

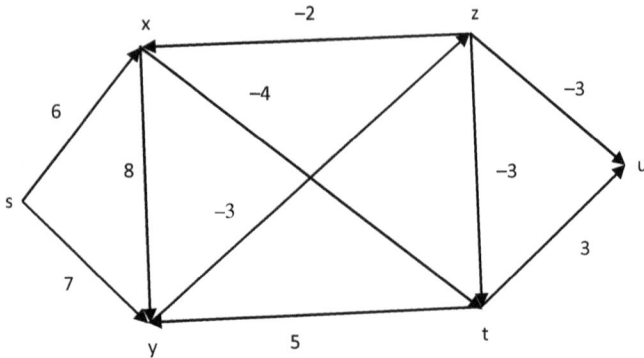

Fig. 9.5

Solution.

Iteration 1. In view of the initialization conditions, the only edges effectively relaxed are (s, x), (s, y) and we get $d(x) = 6$, $d(y) = 7$ and $d(v) = \infty$ for all other vertices $v \neq s$.

Iteration 2. The only edges effectively relaxed are (x, y), (x, t), (y, z) and $d(x) = 6$, $d(y) = 7$, $d(t) = 2$, $d(z) = 4$, $d(u) = \infty$.

Iteration 3. The edges effectively relaxed are (z, x), (z, t), (z, u), (t, u), (y, z) and

$$d(x) = 2, d(y) = 7, d(t) = 1, d(z) = 4, d(u) = 1.$$

Iteration 4. Relaxing the edges (x, t) and (t, y), we get $d(x) = 2, d(y) = 6, d(t) = -2, d(z) = 4, d(u) = 1$.

Observe that once we have obtained $d(v)$ for every v in iteration i, we need compute $d(v)$ in the $(i + 1)$th iteration and this is done by

$$d(v) = \min\{d(v), d(u) + w(u, v)| \text{ for all } u \text{ with } (u, v) \in E\}.$$

Iteration 5. $d(x) = 2, d(y) = 3, d(t) = -2, d(z) = 3, d(u) = 1.$

Iteration 6. $d(x) = 1, d(y) = 3, d(t) = -2, d(z) = 0, d(u) = 0.$

Iteration 7. $d(x) = -2, d(y) = 3, d(t) = -3, d(z) = 0, d(u) = -3.$

There are six vertices in the graph and the estimated shortest distance values have not stabilized up to the sixth iteration. This proves that the

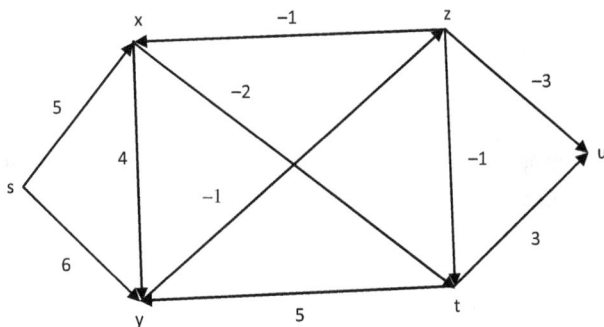

Fig. 9.6

graph has a cycle of negative weight. Observe that $x \xrightarrow{-4} t \xrightarrow{5} y \xrightarrow{-3} z \xrightarrow{-2} x$ is a cycle of weight -4. Thus the single-source-shortest-path problem for this graph has no solution.

Example 9.5. Apply Bellman–Ford algorithm to find shortest path lengths, if possible, from the source vertex s to the other vertices in the graph G as given in Fig. 9.6.

Solution.

Iteration 1. $d(x) = 5, d(y) = 6, d(v) = \infty$ for all other vertices $v \neq s$.

Iteration 2. $d(x) = 5, \pi(x) = s, d(y) = 6, \pi(y) = s, d(t) = 3, \pi(t) = x,$ $d(z) = 5, \pi(z) = y, d(u) = \infty, \pi(u) =$ nil.

Iteration 3. $d(x) = 4, \pi(x) = z, d(y) = 6, \pi(y) = s, d(t) = 3, \pi(t) = x,$ $d(z) = 5, \pi(z) = y, d(u) = 2, \pi(u) = z.$

Iteration 4. $d(x) = 4, \pi(x) = z, d(y) = 6, \pi(y) = s, d(t) = 2, \pi(t) = x,$ $d(z) = 5, \pi(z) = y, d(u) = 2, \pi(u) = z.$

Iteration 5. $d(x) = 4, \pi(x) = z, d(y) = 6, \pi(y) = s, d(t) = 2, \pi(t) = x,$ $d(z) = 5, \pi(z) = y, d(u) = 2, \pi(u) = z.$

Since the estimated shortest distances from s to the other vertices remain the same in iteration 5 as these are in iteration 4, these remain the same in every future iteration. Therefore

$$\delta(s, x) = d(x) = 4, \delta(s, y) = d(y) = 6, \delta(s, z) = d(z) = 5,$$

$$\delta(s, t) = d(t) = 2, \delta(s, u) = d(u) = 2$$

and the corresponding shortest paths are $(s, y, z, x), (s, y), (s, y, z),$ (s, y, z, x, t) and (s, y, z, u), respectively.

Exercise 9.1.

1. Give an example, if possible, of a directed, weighted graph with no cycle of negative weight in which some vertex has more than one shortest path from the source vertex.
2. In a weighted directed graph with weights of edges being positive but not all 1, shortest path to any vertex from the source vertex is unique. Comment.
3. Consider the graphs as in Examples 9.3–9.5 with all the arrows except the arrows from s to x, s to y and s to z reversed. Apply Bellman–Ford algorithm to the resulting graphs and find, if possible, shortest distances from the source vertex s to the other vertices in the graphs. Also indicate, if any of these graphs has a negative weight cycle.

9.4. Predecessor Subgraph

Let $G = (V, E)$ be a directed graph. For every $v \in V$, define $\pi(v) = u$ if $(u, v) \in E$ and if there is no $u \in V$ with $(u, v) \in E$, define $\pi(v) = $ nil (i.e., $\pi(x)$ is predecessor of x as defined earlier). Let

$$V_\pi = \{v \in V \mid \pi(v) \neq \text{nil}\} \cup \{\text{nil}\} \quad \text{and}$$
$$E_\pi = \{(\pi(v), v) \mid v \in V_\pi - \{\text{nil}\}\}.$$

Then $G_\pi = (V_\pi, E_\pi)$ is a subgraph of G and is called a **predecessor subgraph** of G.

Now suppose that $G = (V, E)$ is a weighted directed graph with a source vertex s and having no negative weight cycle reachable from s. Recall that a cycle is reachable from a vertex s(say) if any and, therefore, every vertex in the cycle is reachable from s. Then as seen already shortest path weights from the source s to the other vertices are well defined.

A shortest-paths tree rooted at s is a subgraph $G' = (V', E')$ (also called a predecessor subgraph) such that

(i) V' is a set of vertices reachable from s in G;
(ii) G' forms a rooted tree with root s;

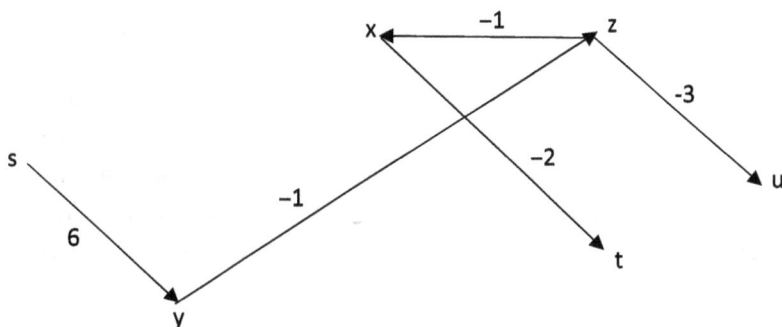

Fig. 9.7

(iii) for every $v \in V'$ the unique simple path from s to v in G' is a shortest path from s to v in G.

Consider the graph G as Fig. 9.6. Since iterations stabilize at the fourth iteration, we consider the predecessor subgraph for iteration 4. The same is given in Fig. 9.7.

The shortest paths from the source vertex s to the various vertices are:

$$s \rightarrow y, s \rightarrow y \rightarrow z, s \rightarrow y \rightarrow z \rightarrow x,$$

$$s \rightarrow y \rightarrow z \rightarrow u, s \rightarrow y \rightarrow z \rightarrow x \rightarrow t.$$

9.5. Difference Constraints and Shortest Paths

In linear programming problem we are concerned with optimization (maximization or minimization) of a linear function $c(x) = \sum_{i=1}^{n} c_i x_i$, called optimization/objective function, where $c_1, \ldots, c_n$ are given constants and $x_1, \ldots, x_n$ are the variables (called decision variables). The variables are to satisfy certain constraints given in the form of linear inequalities and or equalities and also satisfy the non-negativity conditions. Formally, let $A = (a_{ij})$ be an $m \times n$ matrix, c a $1 \times n$ vector and b an $m \times 1$ vector. We are required to find a solution $x = (x_1, x_2, \ldots, x_n)^t$ (i.e., x is an $n \times 1$ vector) which optimizes the objective function $c(x) = c\,x$ subject to the constraints $Ax \leq b$ and the non-negativity restrictions $x_i \geq 0$ for all i, $1 \leq i \leq n$. A solution x which may or may not optimize the objective function but satisfies the constraints is called a **feasible**

solution. Optimization problems arise naturally in many areas including production, distribution of goods, economics, to name just a few. Many times it is just required to know whether a feasible solution exists or not without considering the objective function. We do not consider solving the general linear programming problem here but are concerned only with the existence of a feasible solution and that too in a very special case. Although the non-negativity restrictions $x_i \geq 0$ are an essential part of the general linear programming problem, we do not insist on these here.

Let A be a matrix in which every row has one entry equal to 1, another entry equal to -1 and all other entries equal to zero. Then $Ax \leq b$ is called a **system of difference constraints**. This system of constraints may be written as $x_i - x_j \leq b_k, 1 \leq i, j \leq n, 1 \leq k \leq m$.

Consider, for example, finding a vector x of length 5 that satisfies $Ax \leq b$, where

$$
A = \begin{pmatrix}
1 & -1 & 0 & 0 & 0 \\
-1 & 0 & 1 & 0 & 0 \\
0 & -1 & 0 & 0 & 1 \\
0 & 1 & 0 & -1 & 0 \\
1 & 0 & 0 & 0 & -1 \\
0 & 0 & -1 & 1 & 0 \\
0 & 1 & -1 & 0 & 0
\end{pmatrix}, \quad
b = \begin{pmatrix}
1 \\
-1 \\
-3 \\
3 \\
4 \\
-3 \\
2
\end{pmatrix}
$$

Writing the inequalities $Ax \leq b$ in full, we have

$$x_1 - x_2 \leq 1, \quad -x_1 + x_3 \leq -1, \quad -x_2 + x_5 \leq -3,$$
$$x_2 - x_4 \leq 3, \quad x_1 - x_5 \leq 4, \quad -x_3 + x_4 \leq -3, \quad x_2 - x_3 \leq 2 \quad (9.5)$$

as the difference constraints. A simple computation shows that $x_1 = 0$, $x_2 = -1$, $x_3 = -1$, $x_4 = -4$, $x_5 = -4$ satisfy the above inequalities. A simple computation also shows that $(1, 0, 0, -3, -3)$ and $(-1, -2, -2, -5, -5)$ also satisfy the difference constraints. This is so because if for any i, j, $x_i - x_j \leq b_k$, then $(x_i + d) - (x_j + d) \leq b_k$ where d is an arbitrarily chosen constant.

Our concern in this section is to find a feasible solution of the difference constraints or to indicate that no such solution exists. We use the Bellman–Ford algorithm for this purpose. But to do that, we first need to associate a directed weighted graph to the given set of difference constraints. The associated graph is called the **constraints graph** and is defined as follows.

Corresponding to every decision variable x_i consider a vertex v_i, $1 \leq i \leq n$ and take an additional vertex v_0. Thus $V = \{v_0, v_1, \ldots, v_n\}$. Corresponding to every constraint $x_j - x_i \leq b_k$, introduce an edge (v_i, v_j) and set $w(v_i, v_j) = b_k$. Also consider additional edges (v_0, v_i) with weight 0 for every i, $1 \leq i \leq n$. Then $G = (V, E, v_0)$ is a directed weighted graph with source vertex v_0. In view of $x_j - x_i \leq b_k$, x_i, x_j corresponding to v_i, v_j, respectively, it follows that $x_j - x_i \leq w(v_i, v_j)$.

Observe that the directed weighted graph corresponding to the difference constraints (9.5) is as given in Fig. 9.8.

A feasible solution to the given constraints is provided by the following result.

Theorem 9.10. *Let $G = (V, E, v_0)$ be the constraints graph corresponding to given system $Ax \leq b$ of difference constraints. Then $x_i = \delta(v_0, v_i)$,*

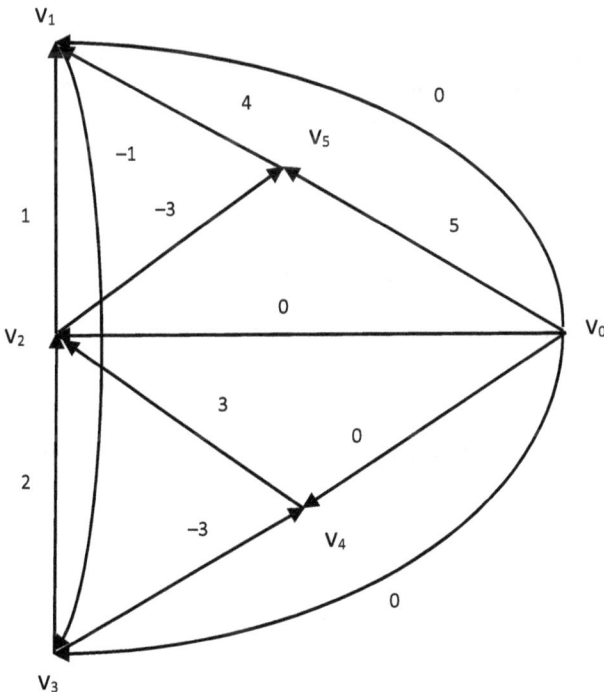

Fig. 9.8

$1 \leq i \leq n$, *is a feasible solution of the given difference constraints provided the graph G has no negative weight cycle. If the graph G has a negative weight cycle, then the system of difference constraints has no solution.*

Proof. Suppose that G has no negative weight cycle. Then it follows from Theorem 9.9 that shortest path distances from the source vertex v_0 to the other vertices exist and are well defined. Let $x_i = \delta(v_0, v_i)$, $1 \leq i \leq n$. By the triangle law of inequality for the shortest path distances, we have for all i, j,

$$\delta(v_0, v_j) \leq \delta(v_0, v_i) + \delta(v_i, v_j) \text{ or}$$
$$x_j \leq x_i + w(v_i, v_j) \text{ or } x_j - x_i \leq b_k$$

showing that $x_i = \delta(v_0, v_i)$ satisfy the given system of difference constraints.

Now suppose that G contains a negative weight cycle. Although as already observed that single-source-shortest-path problem has no solution, it is not enough to say that then the given system of difference constraints has no solution. Suppose that $c = (v_1, v_2, \ldots, v_k)$ with $v_k = v_1$ is a negative weight cycle. From the definition of G, we find that G has no edge with v_0 as a terminal vertex. Therefore, no cycle in G can involve v_0 as one of its vertices. Now

$$0 > w(c) = w(v_1, v_2) + w(v_2, v_3) + \cdots + w(v_{k-1}, v_k)$$
$$\geq (x_2 - x_1) + (x_3 - x_2) + \cdots + (x_k - x_{k-1})$$
$$= (x_2 - x_1) + (x_3 - x_2) + \cdots + (x_{k-1} - x_{k-2}) + (x_1 - x_{k-1})$$
$$= (x_2 + x_3 + \ldots + x_{k-1} + x_1) - (x_1 + x_2 + \cdots + x_{k-1}) = 0$$

which is a contradiction. Hence the given system of difference constraints has no solution. □

Observe that in the proof of the second part we have not talked of shortest path lengths of the vertices from the source vertex but have only used the definition of weight of edges in G.

Example 9.6. Find a feasible solution of the system (9.5) of constraints differences.

Solution. As already observed, a solution of this system of constraint differences exists. Now we use Theorem 9.9 to find a solution. The constraints graph is given in Fig. 9.8. We run Bellman–Ford algorithm on this graph with v_0 as a source vertex.

Initially $d(v_0) = 0, d(v_i) = \infty$ for $1 \leq i \leq 5$.

Iteration 1. Since every edge (v_0, v_i) has weight $0, d(v_0) = 0, d(v_i) = 0,$ $1 \leq i \leq 5.$

Iteration 2. $d(v_1) = 0, \quad d(v_2) = 0, \quad d(v_3) = -1, \quad d(v_4) = -3,$ $d(v_5) = -3.$

Iteration 3. $d(v_1) = 0, d(v_2) = 0, d(v_3) = -1, d(v_4) = -4,$ $d(v_5) = -3.$

Iteration 4. $d(v_1) = 0, d(v_2) = -1, d(v_3) = -1, d(v_4) = -4,$ $d(v_5) = -3.$

Iteration 5. $d(v_1) = 0, d(v_2) = -1, d(v_3) = -1, d(v_4) = -4,$ $d(v_5) = -4.$

Iteration 6. $d(v_1) = 0, d(v_2) = -1, d(v_3) = -1, d(v_4) = -4,$ $d(v_5) = -4.$

There being no change in the estimated shortest lengths from iteration 5 to iteration 6, the estimated values have stabilized and are, therefore, the shortest path lengths. Hence $\delta(v_0, v_1) = 0, \delta(v_0, v_2) = -1, \delta(v_0, v_3) = -1, \delta(v_0, v_4) = -4, \delta(v_0, v_5) = -4$ and $(0, -1, -1, -4, -4)$ is a feasible solution of the given system of difference constraints. A general feasible solution will be $(d, -1+d, -1+d, -4+d, -4+d)$ for any constant d.

Example 9.7. Find a feasible solution or determine that no such solution exists for the following system of constraints.

$$x_1 - x_2 \leq 2, \quad x_1 - x_4 \leq 7, \quad x_1 - x_5 \leq 5, \quad x_2 - x_5 \leq 10,$$

$$x_3 - x_1 \leq 2, \quad x_4 - x_6 \leq -1, \quad x_4 - x_3 \leq 3, \quad x_5 - x_2 \leq -8,$$

$$x_6 - x_1 \leq 10, \quad x_6 - x_3 \leq -4.$$

Solution. There being 6 variables in the given system of ten constraints, the graph representing the constraints has 7 vertices and $10 + 6 = 16$ edges.

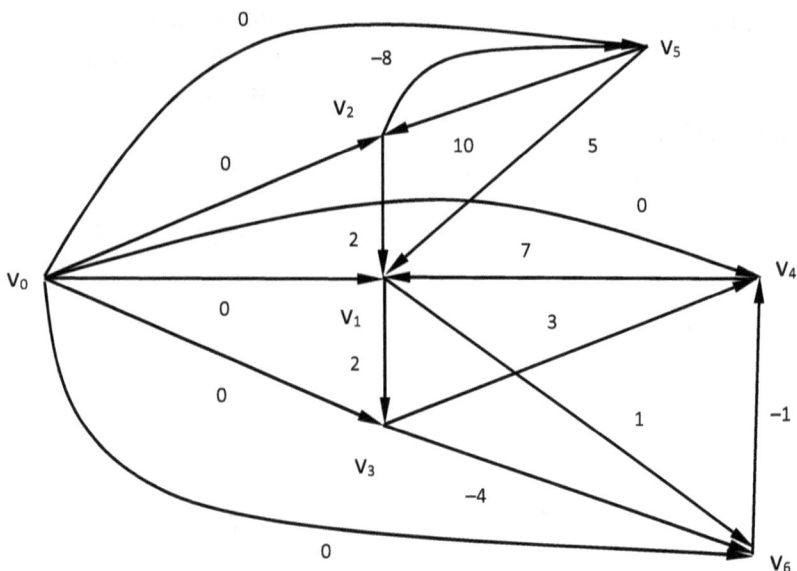

Fig. 9.9

The difference constraints graph is given in Fig. 9.9. Now run Bellman–Ford algorithm on this graph.

Initially $d(v_0) = 0, d(v_i) = \infty$ for every i, $1 \leq i \leq 6$.

Iteration 1. The weight of every edge (v_0, v_i) for $i = 1$ to 6 being 0, $d(v_i) = 0, 1 \leq i \leq 6$.

Iteration 2. We only need to relax the edges (v_2, v_5), (v_3, v_6) and (v_6, v_4). $d(v_1) = 0, d(v_2) = 0, d(v_3) = 0, d(v_4) = -1, d(v_5) = -8, d(v_6) = -4$.

Iteration 3. $d(v_1) = -3, d(v_2) = 0, d(v_3) = 0, d(v_4) = -5, d(v_5) = -8, d(v_6) = -4$.

Iteration 4. $d(v_1) = -3, d(v_2) = 0, d(v_3) = -1, d(v_4) = -5, d(v_5) = -8, d(v_6) = -4$.

Iteration 5. $d(v_1) = -3, d(v_2) = 0, d(v_3) = -1, d(v_4) = -5, d(v_5) = -8, d(v_6) = -5$.

Iteration 6. $d(v_1) = -3, d(v_2) = 0, d(v_3) = -1, d(v_4) = -6, d(v_5) = -8, d(v_6) = -5$.

Iteration 7. $d(v_1) = -3$, $d(v_2) = 0$, $d(v_3) = -1$, $d(v_4) = -6$, $d(v_5) = -8$, $d(v_6) = -5$.

There being no change in the estimated shortest path lengths from iteration 6 to iteration 7, these stabilize here and are the actual shortest path lengths from v_0 to the vertices. Therefore $\delta(v_0, v_1) = -3$, $\delta(v_0, v_2) = 0$, $\delta(v_0, v_3) = -1$, $\delta(v_0, v_4) = -6$, $\delta(v_0, v_5) = -8$, $\delta(v_0, v_6) = -5$ and $(-3, 0, -1, -6, -8, -5)$ is a feasible solution for the given difference constraints.

Example 9.8. Find a feasible solution or determine that no feasible solution exists for the following system of difference constraints:

$$x_1 - x_2 \leq -7, \quad x_2 - x_4 \leq 2, \quad x_2 - x_5 \leq 4, \quad x_2 - x_3 \leq 9,$$

$$x_3 - x_1 \leq -5, \quad x_3 - x_5 \leq -9, \quad x_4 - x_1 \leq 3, \quad x_4 - x_3 \leq 4.$$

Solution. There being 5 decision variables and 8 constraints, the graph representing the given system of difference constraints has 6 vertices and $8 + 5 = 13$ edges. The difference constraints graph is then given in Fig. 9.10. Now we run Bellman–Ford algorithm on this graph. We show

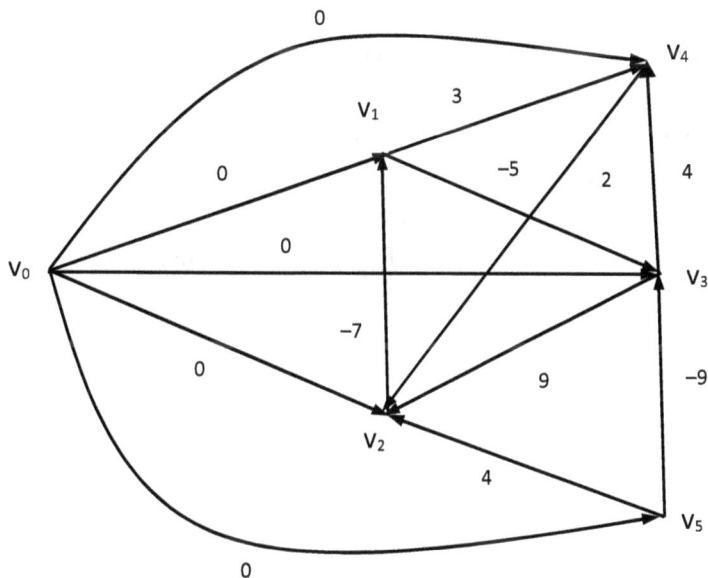

Fig. 9.10

Table 9.1

Iteration	$d(v_0)$	$d(v_1)$	$d(v_2)$	$d(v_3)$	$d(v_4)$	$d(v_5)$
0	0	∞	∞	∞	∞	∞
1	0	0	0	0	0	0
2	0	-7	0	-9	0	0
3	0	-7	0	-12	-5	0
4	0	-7	-3	-12	-8	0
5	0	-10	-6	-12	-8	0
6	0	-13	-6	-15	-8	0

the estimated shortest path distance values for the iterations in the tabular form (see Table 9.1).

Since the estimated shortest path lengths change from iteration 5 to iteration 6, these values do not stabilize and, so, there are no shortest path lengths from v_0 to the other vertices. Hence there is a negative weight cycle and the constraints have no solution.

Observe that $v_3 \xrightarrow{9} v_2 \xrightarrow{-7} v_1 \xrightarrow{-5} v_3$ is a cycle of weight -3.

Exercise 9.2.

1. Find a feasible solution or determine that no solution exists for the system of difference constraints as in Example 9.8 with
    (a) the constraint $x_1 - x_2 \leq -7$ replaced by $x_1 - x_2 \leq -3$;
    (b) the constraints $x_1 - x_2 \leq -7$, $x_3 - x_1 \leq -5$, $x_2 - x_3 \leq 9$ replaced by $x_1 - x_2 \leq -4$, $x_3 - x_1 \leq -2$, $x_2 - x_3 \leq 7$.

2. Can any shortest-path-weight from the new vertex v_0 in a constraints graph be positive? Explain.

3. Give some realistic problems in which the difference constraints arise.

Chapter 10

All Pairs Shortest Paths

Given a directed weighted graph with no cycles of negative weight, we considered in the last chapter the problem of finding shortest path lengths of all vertices from a given vertex as a source vertex. We mainly discussed two algorithms namely Dijkstra's algorithm and Bellman–Ford algorithm for this problem. We prove that while the running time of Bellman–Ford algorithm is $O(|V||E|)$ that of Dijkstra's algorithm is $O(|V|^2)$. Of course Dijkstra's algorithm considers only the case when the weights of the edges of the graph are non-negative. Regarding each vertex of the given graph as a source vertex, we can find the shortest path lengths for every pair of vertices. Therefore, the running time for this problem will be $O(|V|^2|E|)$ or $O(|V|^3)$ depending upon whether Bellman–Ford or Dijkstra's algorithm is used.

In this chapter, we use matrix representation of the given directed graph and consider two algorithms for solving the all pairs shortest path problem. Yet another algorithm due to Johnson is also considered for the all pairs shortest path length problem for sparse graphs.

10.1. A Recursive Algorithm for Shortest Path Lengths

Let $G = (V, E)$ be a directed weighted graph with the weight function w associating a real number with every edge of G. We also suppose that G has no cycles of negative weight. For the sake of convenience, we take the vertex set $V = \{1, 2, \ldots, n\}$. The graph G is represented by the adjacency matrix $W = (w(i, j))$ where W is a square matrix of order n and the (i, j)th element of W is $w(i, j)$ which is the weight of the edge (i, j).

Then

$$w(i, j) = \begin{cases} 0 & \text{if } i = j, \\ \text{weight of the edge}(i, j) & \text{if } i \neq j \quad \text{and} \quad (i, j) \in E, \\ \infty & \text{if } i \neq j \quad \text{and} \quad (i, j) \notin E. \end{cases}$$

(10.1)

Observe that $w(i, j)$ also represents the weight of a shortest path from i to j among all paths having at most one edge. For an $m \geq 1$, let $l_{ij}^{(m)}$ denote the length of a shortest path from i to j among all paths having at most m edges. Let $L^{(m)} = (l_{ij}^{(m)})$. Then the square matrix $L^{(m)}$ represents the shortest path lengths of all pairs of vertices of G among all paths having at most m edges. Consider a path p from vertex i to vertex k having at most m edges with $w(p) = l_{ik}^{(m)}$. Composing p with edge (k, j) we get a path p' from i to j having at most $m + 1$ edges. Therefore

$$l_{ij}^{(m+1)} \leq w(p') = w(p) + w(k, j) = l_{ik}^{(m)} + w(k, j).$$

Since this inequality holds for all k,

$$l_{ij}^{(m+1)} \leq \min_{\substack{1 \leq k \leq n \\ k \neq j}} (l_{ik}^{(m)} + w(k, j)).$$

Since $w(j, j) = 0$, the above relation also holds for $k = j$, Therefore, we have

$$l_{ij}^{(m+1)} \leq \min_{1 \leq k \leq n} (l_{ik}^{(m)} + w(k, j)).$$

Also a path of minimal weight having at most $(m + 1)$ edges can be obtained from a path having at most m edges followed by one edge, we have

$$l_{ij}^{(m+1)} = \min_{1 \leq k \leq n} (l_{ik}^{(m)} + w(k, j)).$$

(10.2)

Given two "distance" square matrices $A = (a_{ij})$, $B = (b_{ij})$ of order n, define a "product" $A \odot B$ by $A \odot B = C = (c_{ij})$, where

$$c_{ij} = \min_{1 \leq k \leq n} \{a_{ik} + b_{kj}\}.$$

Then, in view of (10.2), it follows that

$$L^{(m+1)} = L^{(m)} \odot L^{(1)} = L^{(m)} \odot W.$$

Theorem 10.1. *All pairs shortest path lengths for the graph G are given by the entries of the matrix $L^{(n-1)}$, i.e., for every i, j, $1 \leq i, j \leq n$, the shortest path length from i to j is $l_{ij}^{(n-1)}$.*

Proof. It is clear from the definition of $l_{ij}^{(m+1)}$ that for all $m \geq 1$, $1 \leq i$, $j \leq n$, $l_{ij}^{(m+1)} \leq l_{ij}^{(m)}$. In particular, $l_{ij}^{(n)} \leq l_{ij}^{(n-1)}$ for all i, j.

Let p be a shortest length path from i to j having at most n edges so that $w(p) = l_{ij}^{(n)}$. If p is a path having at most $n-1$ edges, then $l_{ij}^{(n-1)} \geq w(p) \geq l_{ij}^{(n-1)}$ so that $l_{ij}^{(n-1)} = w(p)$. If, on the other hand, p has n edges, p contains a cycle (there being only n vertices, one of the vertices is repeated in p). Since G does not have any cycle of negative weight, removing this cycle, we get a path p' having at most $n-1$ edges with $w(p') \leq w(p)$. Then $l_{ij}^{(n-1)} \leq w(p') \leq w(p) = l_{ij}^{(n)} \leq l_{ij}^{(n-1)}$ and $l_{ij}^{(n)} = l_{ij}^{(n-1)}$. Thus, in either case, $l_{ij}^{(n)} = l_{ij}^{(n-1)}$. $\square$

Corollary 10.2. $L^{(m)} = L^{(n-1)}$ for all $m \geq n-1$.

Theorem 10.1 immediately suggests the following recursive algorithm for finding all pairs shortest path lengths.

Algorithm 10.1. Recursive Algorithm-all Pairs Shortest Path Lengths

---

1. given $G = (V, E)$ with weight function w associating a real number to every edge and $V = \{1, 2, \ldots, n\}$. W the weight adjacency matrix representing G
2. set $L^{(1)} = (l_{ij}^{(1)}) = W$
3. for $m = 1$ to $n - 1$, $L^{(m)} = (l_{ij}^{(m)})$
4. for $i = 1$ to n, for $j = 1$ to n
5. set $l_{ij}^{(m+1)} = \min_{1 \leq k \leq n}\{l_{ik}^{(m)} + w(k, j)\}$
6. return $L^{(n-1)}$

Step 5 computes $l_{ij}^{(m+1)}$ by making n computations. Since this is done for $i = 1$ to n and $j = 1$ to n (step 4), the running time for obtaining $L^{(m+1)}$ from $L^{(m)}$ is $O(n^3)$. Since m varies from 1 to $n - 1$, the total running time of the algorithm is $O(n^4) = O(|V|^4)$.

We can reduce the running time of this algorithm by using a variation of this algorithm and using Corollary 10.2. For this we also need the following lemma.

Lemma 10.3. *The matrix multiplication $\odot$ satisfies the associative law.*

Proof. Let $A = (a_{ij})$, $B = (b_{ij})$, $C = (c_{ij})$ be three "distance" matrices of order n. Let
$$A \odot B = (d_{ij}), (A \odot B) \odot C = (e_{ij}), B \odot C = (x_{ij}), A \odot (B \odot C) = (y_{ij}).$$
Then for $1 \leq i, j \leq n$,

$$d_{ij} = \min_{1 \leq k \leq n} \{a_{ik} + b_{kj}\},$$

$$e_{ij} = \min_{1 \leq k \leq n} \{d_{ik} + c_{kj}\}$$

$$= \min_{1 \leq k \leq n} \{\min_{1 \leq t \leq n} \{a_{it} + b_{tk}\} + c_{kj}\}$$

$$= \min_{1 \leq k \leq n} \{\min_{1 \leq t \leq n} \{a_{it} + b_{tk} + c_{kj}\}\}$$

$$= \min_{1 \leq t \leq n} \{\min_{1 \leq k \leq n} \{a_{it} + b_{tk} + c_{kj}\}\}$$

$$= \min_{1 \leq t \leq n} \{a_{it} + \min_{1 \leq k \leq n} \{b_{tk} + c_{kj}\}\}$$

$$= \min_{1 \leq t \leq n} \{a_{it} + x_{tj}\} = y_{ij}.$$

Hence $(A \odot B) \odot C = A \odot (B \odot C)$.   □

Corollary 10.4. *For any "distance" matrix A and positive integers m, t, $A^m \odot A^t = A^t \odot A^m$.*
   The $\odot$ product being associative, we have

$$L^{(4)} = L^{(3)} \odot L^{(1)} = (L^{(2)} \odot L^{(1)}) \odot L^{(1)}$$
$$= L^{(2)} \odot (L^{(1)} \odot L^{(1)}) = L^{(2)} \odot L^{(2)},$$

and, in general, $L^{(2m)} = L^{(m)}) \odot L^{(m)}$ (which can be proved by an easy induction on m the base case $m = 1$ being trivially true and the case $m = 2$ proved above).

Let k be a least positive integer with $2^{k-1} < n - 1 \leq 2^k$. Then

$$L^{(n-1)} = L^{(2^k)} = \left(L^{(2^{(k-1)})}\right)^{(2)} = \left(\left(L^{(2^{(k-2)})}\right)^{(2)}\right)^{(2)}$$

and, therefore, $L^{(n-1)}$ can be obtained in k steps instead of $n - 1$ steps. In view of this the recursive algorithm-all pairs shortest path lengths may be replaced by the following algorithm.

Algorithm 10.1. Recursive Algorithm (Revised)

---

1. given $G = (V, E)$ with weight function w associating a real number to every edge, $V = \{1, 2, \ldots, n\}$ and k a positive integer with $2^{k-1} < n - 1 \leq 2^k$ so that $k - 1 < \lg(n - 1) \leq k$. W is the weight adjacency matrix representation of G
2. set $L^{(1)} = (l_{ij}^{(1)}) = W$
3. for $m = 1$ to k, set $L^{(2^m)} = \left(L^{(2^{(m-1)})}\right)^2$
4. return $L^{(2^k)} = L^{(n-1)}$

The running time for the computation of $L^{(2^m)} = \left(L^{(2^{(m-1)})}\right)^{(2)}$ for a given m is $O(n^3)$ and then the running time of the revised recursive algorithm is $O(n^3 k) = O(n^3 \lg(n - 1)) = O(n^3 \lg n)$.

Example 10.1. Using recursive algorithm and revised recursive algorithm, find all pairs shortest path lengths for the graph G as given in Fig. 10.1.

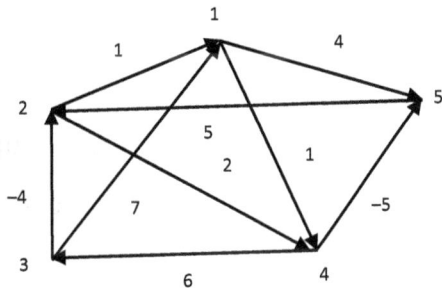

Fig. 10.1

Solution. The weight adjacency matrix for the graph G is

$$W = \begin{pmatrix} 0 & \infty & \infty & 1 & 4 \\ 1 & 0 & \infty & 2 & \infty \\ 7 & -4 & 0 & \infty & \infty \\ \infty & \infty & 6 & 0 & -5 \\ \infty & 5 & \infty & \infty & 0 \end{pmatrix},$$

which we denote by $L^{(1)}$. There being five vertices in G, the shortest path weights for the pairs of vertices of G are given by the entries of the matrix $L^{(4)}$ provided there are no negative weight cycles in G. For this purpose, we will also compute $L^{(5)}$ and compare $L^{(4)}$ and $L^{(5)}$. In case $L^{(4)} = L^{(5)}$, we will get $L^{(m)} = L^{(4)}$ for all $m \geq 5$ and the shortest path weights are given by $L^{(4)}$. In case $L^{(5)} \neq L^{(4)}$, it will follow that there is at least one cycle of negative weight in G.

(a) Recall that $l_{ij}^{(m)} = \min_{1 \leq k \leq n}\{l_{ik}^{(m-1)} + w(k, j)\}$. Therefore $l_{11}^{(2)} = l_{11}^{(1)} = 0$,

$$l_{12}^{(2)} = \min\{0 + \infty, \infty + 0, \infty - 4, 1 + \infty, 4 + 5\} = 9,$$

$$l_{13}^{(2)} = \min\{0 + \infty, \infty + \infty, \infty + 0, 1 + 6, 4 + \infty\} = 7,$$

$$l_{14}^{(2)} = \min\{0 + 1, \infty + 2, \infty + \infty, 1 + 0, 4 + \infty\} = 1,$$

$$l_{15}^{(2)} = \min\{0 + 4, \infty + \infty, \infty + \infty, 1 - 5, 4 + 0\} = -4,$$

$$l_{21}^{(2)} = \min\{1 + 0, 0 + 1, \infty + 7, 2 + \infty, \infty + \infty\} = 1,$$

$$l_{22}^{(2)} = l_{22}^{(1)} = 0,$$

$$l_{23}^{(2)} = \min\{1 + \infty, 0 + \infty, \infty + 0, 2 + 6, \infty + \infty\} = 8,$$

$$l_{24}^{(2)} = \min\{1 + 1, 0 + 2, \infty + \infty, 2 + 0, \infty + \infty\} = 2,$$

$$l_{25}^{(2)} = \min\{1 + 4, 0 + \infty, \infty + \infty, 2 - 5, \infty + 0\} = -3,$$

$$l_{31}^{(2)} = \min\{7 + 0, -4 + 1, 0 + 7, \infty + \infty, \infty + \infty\} = -3,$$

$$l_{32}^{(2)} = \min\{7 + \infty, -4 + 0, 0 - 4, \infty + \infty, \infty + 5\} = -4,$$

$$l_{33}^{(2)} = l_{33}^{(1)} = 0,$$

$$l_{34}^{(2)} = \min\{7 + 1, -4 + 2, 0 + \infty, \infty + 0, \infty + \infty\} = -2,$$

$$l_{35}^{(2)} = \min\{7 + 4, -4 + \infty, 0 + \infty, \infty - 5, \infty + 0\} = 11,$$

$$l_{41}^{(2)} = \min\{\infty + 0, \infty + 1, 6 + 7, 0 + \infty, -5 + \infty\} = 13,$$

$$l_{42}^{(2)} = \min\{\infty + \infty, \infty + 0, 6 - 4, 0 + \infty, -5 + 5\} = 0,$$

$$l_{43}^{(2)} = \min\{\infty + \infty, \infty + \infty, 6 + 0, 0 + 6, -5 + \infty\} = 6,$$

$$l_{44}^{(2)} = l_{44}^{(1)} = 0,$$

$$l_{45}^{(2)} = \min\{\infty + 4, \infty + \infty, 6 + \infty, 0 - 5, -5 + 0\} = -5,$$

$$l_{51}^{(2)} = \min\{\infty + 0, 5 + 1, \infty + 7, \infty + \infty, 0 + \infty\} = 6,$$

$$l_{52}^{(2)} = \min\{\infty + \infty, 5 + 0, \infty - 4, \infty + \infty, 0 + 5\} = 5,$$

$$l_{53}^{(2)} = \min\{\infty + \infty, 5 + \infty, \infty + 0, \infty + 6, 0 + \infty\} = \infty,$$

$$l_{54}^{(2)} = \min\{\infty + 1, 5 + 2, \infty + \infty, \infty + 0, 0 + \infty\} = 7,$$

$$l_{55}^{(2)} = 0.$$

Therefore

$$L^{(2)} = \begin{pmatrix} 0 & 9 & 7 & 1 & -4 \\ 1 & 0 & 8 & 2 & -3 \\ -3 & -4 & 0 & -2 & 11 \\ 13 & 0 & 6 & 0 & -5 \\ 6 & 5 & \infty & 7 & 0 \end{pmatrix},$$

$$l_{11}^{(3)} = l_{11}^{(2)} = 0,$$

$$l_{12}^{(3)} = \min\{0 + \infty, 9 + 0, 7 - 4, 1 + \infty, -4 + 5\} = 1,$$

$$l_{13}^{(3)} = \min\{0 + \infty, 9 + \infty, 7 + 0, 1 + 6, -4 + \infty\} = 7,$$

$$l_{14}^{(3)} = \min\{0 + 1, 9 + 2, 7 + \infty, 1 + 0, -4 + \infty\} = 1,$$

$$l_{15}^{(3)} = \min\{0 + 4, 9 + \infty, 7 + \infty, 1 - 5, -4 + 0\} = -4,$$

$$l_{21}^{(3)} = \min\{1 + 0, 0 + 1, 8 + 7, 2 + \infty, -3 + \infty\} = 1,$$

$$l_{22}^{(3)} = l_{22}^{(2)} = 0,$$

$l_{23}^{(3)} = \min\{1 + \infty, 0 + \infty, 8 + 0, 2 + 6, -3 + \infty\} = 8,$

$l_{24}^{(3)} = \min\{1 + 1, 0 + 2, 8 + \infty, 2 + 0, -3 + \infty\} = 2,$

$l_{25}^{(3)} = \min\{1 + 4, 0 + \infty, 8 + \infty, 2 - 5, -3 + 0\} = -3,$

$l_{31}^{(3)} = \min\{-3 + 0, -4 + 1, 0 + 7, -2 + \infty, 11 + \infty\} = -3,$

$l_{32}^{(3)} = \min\{-3 + \infty, -4 + 0, 0 - 4, -2 + \infty, 11 + 5\} = -4,$

$l_{33}^{(3)} = l_{33}^{(2)} = 0,$

$l_{34}^{(3)} = \min\{-3 + 1, -4 + 2, 0 + \infty, -2 + 0, 11 + \infty\} = -2,$

$l_{35}^{(3)} = \min\{-3 + 4, -4 + \infty, 0 + \infty, -2 - 5, 11 + 0\} = -7,$

$l_{41}^{(3)} = \min\{13 + 0, 0 + 1, 6 + 7, 0 + \infty, -5 + \infty\} = 1,$

$l_{42}^{(3)} = \min\{13 + \infty, 0 + 0, 6 - 4, 0 + \infty, -5 + 5\} = 0,$

$l_{43}^{(3)} = \min\{13 + \infty, 0 + \infty, 6 + 0, 0 + 6, -5 + \infty\} = 6,$

$l_{44}^{(3)} = l_{44}^{(2)} = 0,$

$l_{45}^{(3)} = \min\{13 + 4, 0 + \infty, 6 + \infty, 0 - 5, -5 + 0\} = -5,$

$l_{51}^{(3)} = \min\{6 + 0, 5 + 1, \infty + 7, 7 + \infty, 0 + \infty\} = 6,$

$l_{52}^{(3)} = \min\{6 + \infty, 5 + 0, \infty - 4, 7 + \infty, 0 + 5\} = 5,$

$l_{53}^{(3)} = \min\{6 + \infty, 5 + \infty, \infty + 0, 7 + 6, 0 + \infty\} = 13,$

$l_{54}^{(3)} = \min\{6 + 1, 5 + 2, \infty + \infty, 7 + 0, 0 + \infty\} = 7,$

$l_{55}^{(3)} = l_{55}^{(2)} = 0.$

Therefore

$$L^{(3)} = \begin{pmatrix} 0 & 1 & 7 & 1 & -4 \\ 1 & 0 & 8 & 2 & -3 \\ -3 & -4 & 0 & -2 & -7 \\ 1 & 0 & 6 & 0 & -5 \\ 6 & 5 & 13 & 7 & 0 \end{pmatrix}.$$

Next, we compute $L^{(4)} = L^{(3)} \odot L^{(1)}$. Now

$$l_{11}^{(4)} = l_{11}^{(3)} = 0,$$

$$l_{12}^{(4)} = \min\{0 + \infty, 1 + 0, 7 - 4, 1 + \infty, -4 + 5\} = 1,$$

$$l_{13}^{(4)} = \min\{0 + \infty, 1 + \infty, 7 + 0, 1 + 6, -4 + \infty\} = 7,$$

$$l_{14}^{(4)} = \min\{0 + 1, 1 + 2, 7 + \infty, 1 + 0, -4 + \infty\} = 1,$$

$$l_{15}^{(4)} = \min\{0 + 4, 1 + \infty, 7 + \infty, 1 - 5, -4 + 0\} = -4,$$

$$l_{21}^{(4)} = \min\{1 + 0, 0 + 1, 8 + 7, 2 + \infty, -3 + \infty\} = 1,$$

$$l_{22}^{(4)} = l_{22}^{(3)} = 0,$$

$$l_{23}^{(4)} = \min\{1 + \infty, 0 + \infty, 8 + 0, 2 + 6, -3 + \infty\} = 8,$$

$$l_{24}^{(4)} = \min\{1 + 1, 0 + 2, 8 + \infty, 2 + 0, -3 + \infty\} = 2,$$

$$l_{25}^{(4)} = \min\{1 + 4, 0 + \infty, 8 + \infty, 2 - 5, -3 + 0\} = -3,$$

$$l_{31}^{(4)} = \min\{-3 + 0, -4 + 1, 0 + 7, -2 + \infty, -7 + \infty\} = -3,$$

$$l_{32}^{(4)} = \min\{-3 + \infty, -4 + 0, 0 - 4, -2 + \infty, -7 + 5\} = -4,$$

$$l_{33}^{(4)} = l_{33}^{(3)} = 0,$$

$$l_{34}^{(4)} = \min\{-3 + 1, -4 + 2, 0 + \infty, -2 + 0, -7 + \infty\} = -2,$$

$$l_{35}^{(4)} = \min\{-3 + 4, -4 + \infty, 0 + \infty, -2 - 5, -7 + 0\} = -7,$$

$$l_{41}^{(4)} = \min\{1 + 0, 0 + 1, 6 + 7, 0 + \infty, -5 + \infty\} = 1,$$

$$l_{42}^{(4)} = \min\{1 + \infty, 0 + 0, 6 - 4, 0 + \infty, -5 + 5\} = 0,$$

$$l_{43}^{(4)} = \min\{1 + \infty, 0 + \infty, 6 + 0, 0 + 6, -5 + \infty\} = 6,$$

$$l_{44}^{(4)} = l_{44}^{(3)} = 0,$$

$$l_{45}^{(4)} = \min\{1 + 4, 0 + \infty, 6 + \infty, 0 - 5, -5 + 0\} = -5,$$

$$l_{51}^{(4)} = \min\{6 + 0, 5 + 1, 13 + 7, 7 + \infty, 0 + \infty\} = 6,$$

$$l_{52}^{(4)} = \min\{6 + \infty, 5 + 0, 13 - 4, 7 + \infty, 0 + 5\} = 5,$$

$$l_{53}^{(4)} = \min\{6 + \infty, 5 + \infty, 13 + 0, 7 + 6, 0 + \infty\} = 13,$$

$$l_{54}^{(4)} = \min\{6 + 1, 5 + 2, 13 + \infty, 7 + 0, 0 + \infty\} = 7,$$

$$l_{55}^{(4)} = l_{55}^{(3)} = 0.$$

Therefore

$$L^{(4)} = \begin{pmatrix} 0 & 1 & 7 & 1 & -4 \\ 1 & 0 & 8 & 2 & -3 \\ -3 & -4 & 0 & -2 & -7 \\ 1 & 0 & 6 & 0 & -5 \\ 6 & 5 & 13 & 7 & 0 \end{pmatrix}.$$

In order to see whether G contains a cycle of negative weight or not, we also need to compute $L^{(5)}$.

$$l_{11}^{(5)} = l_{11}^{(4)} = 0,$$

$$l_{12}^{(5)} = \min\{0 + \infty, 1 + 0, 7 - 4, 1 + \infty, -4 + 5\} = 1,$$

$$l_{13}^{(5)} = \min\{0 + \infty, 1 + \infty, 7 + 0, 1 + 6, -4 + \infty\} = 7,$$

$$l_{14}^{(5)} = \min\{0 + 1, 1 + 2, 7 + \infty, 1 + 0, -4 + \infty\} = 1,$$

$$l_{15}^{(5)} = \min\{0 + 4, 1 + \infty, 7 + \infty, 1 - 5, -4 + 0\} = -4,$$

$$l_{21}^{(5)} = \min\{1 + 0, 0 + 1, 8 + 7, 2 + \infty, -3 + \infty\} = 1,$$

$$l_{22}^{(5)} = l_{22}^{(4)} = 0,$$

$$l_{23}^{(5)} = \min\{1 + \infty, 0 + \infty, 8 + 0, 2 + 6, -3 + \infty\} = 8,$$

$$l_{24}^{(5)} = \min\{1 + 1, 0 + 2, 8 + \infty, 2 + 0, -3 + \infty\} = 2,$$

$$l_{25}^{(5)} = \min\{1 + 4, 0 + \infty, 8 + \infty, 2 - 5, -3 + 0\} = -3,$$

$$l_{31}^{(5)} = \min\{-3 + 0, -4 + 1, 0 + 7, -2 + \infty, -7 + \infty\} = -3,$$

$$l_{32}^{(5)} = \min\{-3 + \infty, -4 + 0, 0 - 4, -2 + \infty, -7 + 5\} = -4,$$

$$l_{33}^{(5)} = l_{33}^{(4)} = 0,$$

$$l_{34}^{(5)} = \min\{-3 + 1, -4 + 2, 0 + \infty, -2 + 0, -7 + \infty\} = -2,$$

$l_{35}^{(5)} = \min\{-3+4, -4+\infty, 0+\infty, -2-5, -7+0\} = -7,$

$l_{41}^{(5)} = \min\{1+0, 0+1, 6+7, 0+\infty, -5+\infty\} = 1,$

$l_{42}^{(5)} = \min\{1+\infty, 0+0, 6-4, 0+\infty, -5+5\} = 0,$

$l_{43}^{(5)} = \min\{1+\infty, 0+\infty, 6+0, 0+6, -5+\infty\} = 6,$

$l_{44}^{(5)} = l_{44}^{(4)} = 0,$

$l_{45}^{(5)} = \min\{1+4, 0+\infty, 6+\infty, 0-5, -5+0\} = -5,$

$l_{51}^{(5)} = \min\{6+0, 5+1, 13+7, 7+\infty, 0+\infty\} = 6,$

$l_{52}^{(5)} = \min\{6+\infty, 5+0, 13-4, 7+\infty, 0+5\} = 5,$

$l_{53}^{(5)} = \min\{6+\infty, 5+\infty, 13+0, 7+6, 0+\infty\} = 13,$

$l_{54}^{(5)} = \min\{6+1, 5+2, 13+\infty, 7+0, 0+\infty\} = 7,$

$l_{55}^{(5)} = l_{55}^{(4)} = 0.$

Therefore

$$L^{(5)} = \begin{pmatrix} 0 & 1 & 7 & 1 & -4 \\ 1 & 0 & 8 & 2 & -3 \\ -3 & -4 & 0 & -2 & -7 \\ 1 & 0 & 6 & 0 & -5 \\ 6 & 5 & 13 & 7 & 0 \end{pmatrix},$$

which equals $L^{(4)}$. Therefore, there is no negative weight cycle in G and the shortest path weights for the pairs of vertices of G are given by the matrix $L^{(4)}$.

(b) For using the revised recursive algorithm, observe that we have already computed $L^{(2)}$ and then we need to compute $L^{(2)} \odot L^{(2)}$. For this, we have

$l_{11}^{(4)} = l_{11}^{(2)} = 0,$

$l_{12}^{(4)} = \min\{0+9, 9+0, 7-4, 1+0, -4+5\} = 1,$

$l_{13}^{(4)} = \min\{0+7, 9+8, 7+0, 1+6, -4+\infty\} = 7,$

$$l_{14}^{(4)} = \min\{0 + 1, 9 + 2, 7 - 2, 1 + 0, -4 + 7\} = 1,$$

$$l_{15}^{(4)} = \min\{0 - 4, 9 - 3, 7 + 11, 1 - 5, -4 + 0\} = -4,$$

$$l_{21}^{(4)} = \min\{1 + 0, 0 + 1, 8 - 3, 2 + 13, -3 + 6\} = 1,$$

$$l_{22}^{(4)} = l_{22}^{(2)} = 0,$$

$$l_{23}^{(4)} = \min\{1 + 7, 0 + 8, 8 + 0, 2 + 6, -3 + \infty\} = 8,$$

$$l_{24}^{(4)} = \min\{1 + 1, 0 + 2, 8 - 2, 2 + 0, -3 + 7\} = 2,$$

$$l_{25}^{(4)} = \min\{1 - 4, 0 - 3, 8 + 11, 2 - 5, -3 + 0\} = -3,$$

$$l_{31}^{(4)} = \min\{-3 + 0, -4 + 1, 0 - 3, -2 + 13, 11 + 6\} = -3,$$

$$l_{32}^{(4)} = \min\{-3 + 9, -4 + 0, 0 - 4, -2 + 0, 11 + 5\} = -4,$$

$$l_{33}^{(4)} = l_{33}^{(2)} = 0,$$

$$l_{34}^{(4)} = \min\{-3 + 1, -4 + 2, 0 - 2, -2 + 0, 11 + 7\} = -2,$$

$$l_{35}^{(4)} = \min\{-3 - 4, -4 - 3, 0 + 11, -2 - 5, 11 + 0\} = -7,$$

$$l_{41}^{(4)} = \min\{13 + 0, 0 + 1, 6 - 3, 0 + 13, -5 + 6\} = 1,$$

$$l_{42}^{(4)} = \min\{13 + 9, 0 + 0, 6 - 4, 0 + 0, -5 + 5\} = 0,$$

$$l_{43}^{(4)} = \min\{13 + 7, 0 + 8, 6 + 0, 0 + 6, -5 + \infty\} = 6,$$

$$l_{44}^{(4)} = l_{44}^{(2)} = 0,$$

$$l_{45}^{(4)} = \min\{13 - 4, 0 - 3, 6 + 11, 0 - 5, -5 + 0\} = -5,$$

$$l_{51}^{(4)} = \min\{6 + 0, 5 + 1, \infty - 3, 7 + 13, 0 + 8\} = 6,$$

$$l_{52}^{(4)} = \min\{6 + 9, 5 + 0, \infty - 4, 7 + 0, 0 + 5\} = 5,$$

$$l_{53}^{(4)} = \min\{6 + 7, 5 + 8, \infty + 0, 7 + 6, 0 + \infty\} = 13,$$

$$l_{54}^{(4)} = \min\{6 + 1, 5 + 2, \infty - 2, 7 + 0, 0 + 7\} = 7,$$

$$l_{55}^{(4)} = l_{55}^{(2)} = 0.$$

The above computations confirm that $L^{(4)}$ obtained in this fashion is the same as $L^{(4)}$ computed as in (a).

Although using the recursive method computes the shortest path lengths between all pairs of vertices of the given directed graph, this method does not suggest a procedure for constructing the corresponding shortest paths. Of course finding the shortest path lengths uses a method very similar to the method of matrix multiplication.

10.2. The Floyd–Warshall Algorithm

In the last section, we considered a procedure for finding shortest path lengths between all pairs of vertices of a directed weighted graph by considering the number of edges in a path. In this section, we consider a procedure which considers the set of vertices that occur on a path from one vertex to another. If p is a path from a vertex i to a vertex j in the given directed weighted graph $G = (V, E)$ with weight function w associating a real number to every edge in E, the vertices that occur on the path p except the vertices i, j are called **intermediate vertices** of the path p. We assume that there are no negative weight cycles in G. Therefore, all the vertices on a shortest path are distinct and, so, if p is a shortest path from i to j, none of the intermediate vertices on p is i or j. Let d_{ij}^k denote the length of a shortest path from i to j with intermediate vertices lying in the set $\{1, 2, \ldots, k\}$. If the vertex k is not an intermediate vertex of p, then the intermediate vertices of p are in the set $\{1, 2, \ldots, k - 1\}$ and, so, $w(p) = d_{ij}^{k-1}$, i.e., $d_{ij}^k = w(p) = d_{ij}^{k-1}$. On the other hand, suppose that k is indeed an intermediate vertex of p. Then we can consider p as a composition of two paths p' from i to k and p'' from k to j. The path p being a shortest path, both p' and p'' are shortest paths from i to k and from k to j, respectively. Therefore

$$d_{ij}^k = w(p) = w(p') + w(p'').$$

Since the intermediate vertices of p are in $\{1, 2, \ldots, k\}$, the intermediate vertices of p' are in $\{1, 2, \ldots, k - 1\}$ and those of p'' are also in $\{1, 2, \ldots, k - 1\}$. Therefore $w(p') = d_{ik}^{k-1}$, $w(p'') = d_{kj}^{k-1}$. Hence

$$d_{ij}^k = w(p) = w(p') + w(p'') = d_{ik}^{k-1} + d_{kj}^{k-1}.$$

Therefore, we finally have, for $k \geq 1$,

$$d_{ij}^k = \min\{d_{ij}^{k-1}, d_{ik}^{k-1} + d_{kj}^{k-1}\}. \tag{10.3}$$

When $k = 0$,

d_{ij}^0 = the length of a shortest path from i to j with no intermediate vertex

= weight of the edge $(i, j) = w(i, j)$ which is ∞ if there is no path from i to j.

For $k = 0, 1, 2, \ldots, n$, we define

$$D^0 = W = (w(i, j)), \quad D^k = (d_{ij}^k) \text{ for } k \geq 1.$$

Since no intermediate vertex from i to j can be larger than n, the entries of D^n give the shortest path lengths between every pair of vertices of G.

The procedure described is known as Floyd–Warshall algorithm. We formalize this in the following algorithm.

Algorithm 10.2. Floyd–Warshall Algorithm for Shortest Path Lengths $G(V, E), w$

---

1. $G = (V, E)$ is a weighted directed graph, w the weight function associating a real number to every edge of G, G has no negative weight cycle, $V = \{1, 2, \ldots, n\}$
2. initially $d_{ij}^0 = w(i, j)$ if there is an edge from i to j, $d_{ij}^0 = \infty$ otherwise
3. $d_{ii}^0 = 0$
4. for $k = 1$ to n
5.   for $i = 1$ to n
6.   for $j = 1$ to n
7. $d_{ij}^k = \min\{d_{ij}^{k-1}, d_{ik}^{k-1} + d_{kj}^{k-1}\}$
8. return D^n

The for loop in steps 4 to 7 computes d_{ij}^k in $2n^3$ computations as for a given k, d_{ij}^k is computed from d_{ij}^{k-1} by one addition and one comparison. As i, j both take values from 1 to n, the number of computations involved is $2n^2$. As k takes values from 1 to n, the total number of computations is $2n^2 \times n = 2n^3$. Hence the running time of this algorithm is $O(n^3)$ which is far better than the running time $O(n^4)$ of the recursive algorithm and the running time $O(n^3 \lg n)$ of the revised recursive algorithm considered in the last section. For this reason the recursive and revised recursive algorithms considered there are called **slow algorithms**.

We now use Floyd–Warshall algorithm to find the shortest path lengths for the pair of vertices in the graph G of Example 10.1.

Example 10.2. Find the shortest path lengths for all the pairs of vertices of the graph G given in Fig. 10.1.

Solution. Here

$$D^0 = W = \begin{pmatrix} 0 & \infty & \infty & 1 & 4 \\ 1 & 0 & \infty & 2 & \infty \\ 7 & -4 & 0 & \infty & \infty \\ \infty & \infty & 6 & 0 & -5 \\ \infty & 5 & \infty & \infty & 0 \end{pmatrix}.$$

Now

$$d_{1j}^1 = \min\{d_{1j}^0, d_{11}^0 + d_{1j}^0\} = d_{1j}^0,$$

$$d_{2j}^1 = \min\{d_{2j}^0, d_{21}^0 + d_{1j}^0\} = \min\{d_{2j}^0, 1 + d_{1j}^0\} = (1, 0, \infty, 2, 5),$$

$$d_{3j}^1 = \min\{d_{3j}^0, d_{31}^0 + d_{1j}^0\} = \min\{d_{3j}^0, 7 + d_{1j}^0\} = (7, -4, 0, 8, 11),$$

$$d_{4j}^1 = \min\{d_{4j}^0, d_{41}^0 + d_{1j}^0\} = d_{4j}^0$$

$$d_{5j}^1 = \min\{d_{5j}^0, d_{51}^0 + d_{1j}^0\} = d_{5j}^0,$$

$$d_{1j}^2 = \min\{d_{1j}^1, d_{12}^1 + d_{2j}^1\} = d_{1j}^1,$$

$$d_{2j}^2 = \min\{d_{2j}^1, d_{22}^1 + d_{2j}^1\} = d_{2j}^1,$$

$$d_{3j}^2 = \min\{d_{3j}^1, d_{32}^1 + d_{2j}^1\} = \min\{d_{3j}^1, -4 + d_{2j}^1\} = (-3, -4, 0, -2, 1),$$

$$d_{4j}^2 = \min\{d_{4j}^1, d_{42}^1 + d_{2j}^1\} = d_{4j}^1,$$

$$d_{5j}^2 = \min\{d_{5j}^1, d_{52}^1 + d_{2j}^1\} = \min\{d_{5j}^1, 5 + d_{2j}^1\} = (6, 5, \infty, 7, 0),$$

$$d_{1j}^3 = \min\{d_{1j}^2, d_{13}^2 + d_{3j}^2\} = d_{1j}^2,$$

$$d_{2j}^3 = \min\{d_{2j}^2, d_{23}^2 + d_{3j}^2\} = d_{2j}^2,$$

$$d_{3j}^3 = \min\{d_{3j}^2, d_{33}^2 + d_{3j}^2\} = d_{3j}^2,$$

$$d_{4j}^3 = \min\{d_{4j}^2, d_{43}^2 + d_{3j}^2\} = \min\{d_{4j}^2, 6 + d_{3j}^2\} = (3, 2, 6, 0, -5),$$

$$d_{5j}^3 = \min\{d_{5j}^2, d_{53}^2 + d_{3j}^2\} = d_{5j}^2,$$

$$d_{1j}^4 = \min\{d_{1j}^3, d_{14}^3 + d_{4j}^3\} = \min\{d_{1j}^3, 1 + d_{4j}^3\} = (0, 3, 7, 1, -4),$$

$$d_{2j}^4 = \min\{d_{2j}^3, d_{24}^3 + d_{4j}^3\} = \min\{d_{2j}^3, 2 + d_{4j}^3\} = (1, 0, 8, 2, -3),$$

$$d_{3j}^4 = \min\{d_{3j}^3, d_{34}^3 + d_{4j}^3\} = \min\{d_{3j}^3, -2 + d_{4j}^3\} = (-3, -4, 0, -2, -7),$$

$$d_{4j}^4 = \min\{d_{4j}^3, d_{44}^3 + d_{4j}^3\} = d_{4j}^3,$$

$$d_{5j}^4 = \min\{d_{5j}^3, d_{54}^3 + d_{4j}^3\} = \min\{d_{5j}^3, 7 + d_{4j}^3\} = (6, 5, 13, 7, 0),$$

$$d_{1j}^5 = \min\{d_{1j}^4, d_{15}^4 + d_{5j}^4\} = \min\{d_{1j}^4, -4 + d_{5j}^4\} = (0, 1, 7, 1, -4),$$

$$d_{2j}^5 = \min\{d_{2j}^4, d_{25}^4 + d_{5j}^4\} = \min\{d_{2j}^4, -3 + d_{5j}^4\} = (1, 0, 8, 2, -3),$$

$$d_{3j}^5 = \min\{d_{3j}^4, d_{35}^4 + d_{5j}^4\} = \min\{d_{3j}^4, -7 + d_{5j}^4\} = (-3, -4, 0, -2, -7),$$

$$d_{4j}^5 = \min\{d_{4j}^4, d_{45}^4 + d_{5j}^4\} = \min\{d_{4j}^4, -5 + d_{5j}^4\} = (1, 0, 6, 0, -5),$$

$$d_{5j}^5 = \min\{d_{5j}^4, d_{55}^4 + d_{5j}^4\} = d_{5j}^4.$$

Therefore, the matrices D^k are as given below:

$$D^1 = \begin{pmatrix} 0 & \infty & \infty & 1 & 4 \\ 1 & 0 & \infty & 2 & 5 \\ 7 & -4 & 0 & 8 & 1 \\ \infty & \infty & 6 & 0 & -5 \\ \infty & 5 & \infty & \infty & 0 \end{pmatrix}, \quad D^2 = \begin{pmatrix} 0 & \infty & \infty & 1 & 4 \\ 1 & 0 & \infty & 2 & 5 \\ -3 & -4 & 0 & -2 & 1 \\ \infty & \infty & 6 & 0 & -5 \\ 6 & 5 & \infty & 7 & 0 \end{pmatrix},$$

$$D^3 = \begin{pmatrix} 0 & \infty & \infty & 1 & 4 \\ 1 & 0 & \infty & 2 & 5 \\ -3 & -4 & 0 & -2 & 1 \\ 3 & 2 & 6 & 0 & -5 \\ 6 & 5 & \infty & 7 & 0 \end{pmatrix}, \quad D^4 = \begin{pmatrix} 0 & 3 & 7 & 1 & -4 \\ 1 & 0 & 8 & 2 & -3 \\ -3 & -4 & 0 & -2 & -7 \\ 3 & 2 & 6 & 0 & -5 \\ 6 & 5 & 13 & 7 & 0 \end{pmatrix},$$

$$D^5 = \begin{pmatrix} 0 & 1 & 7 & 1 & -4 \\ 1 & 0 & 8 & 2 & -3 \\ -3 & -4 & 0 & -2 & -7 \\ 1 & 0 & 6 & 0 & -5 \\ 6 & 5 & 13 & 7 & 0 \end{pmatrix}.$$

The shortest path lengths between the pairs of vertices of G are then given by the entries of the matrix D^5. Observe that D^5 is the same as the matrix $L^{(4)}$ obtained in the last section (as it should be).

10.2.1. Construction of a shortest path

In addition to the saving in the running time for finding shortest path lengths for pairs of vertices of G using Floyd–Warshall algorithm, there is an advantage that it also helps us in actual construction of shortest paths between every pair of vertices of G. We discuss two procedures for this construction. One of these uses the idea of predecessor of every vertex on a shortest path starting with the end vertex while the other uses the idea of largest intermediate vertex on a shortest path from vertex i to vertex j.

For a shortest path from vertex i to vertex j for which the intermediate vertices are in $\{1, 2, \ldots, k\}$, let π_{ij}^k denote the predecessor of j on this path. We write π^k for the square matrix of order n in which the (i, j)th entry is π_{ij}^k. Let p be a shortest path from i to j for a given pair of vertices i, j. All the intermediate vertices of p are in $\{1, 2, \ldots, n\}$. Suppose that we have already computed the matrix $\pi = \pi^n$. Ignoring the superscript n from the entries of π, we first obtain π_{ij}. Then we consider $\pi_{i\pi_{ij}}$ and continue the process of finding the predecessor till we finally arrive at i itself. There being only a finite number of vertices in G, the process must terminate with i in a finite number of steps and in this way we have actually constructed the path p. Observe that if p is just an edge (i, j), then $\pi_{ij}^k = i$ if $i \neq j$ while $\pi_{ij}^k = nil$ if $j = i$, i.e., there is no predecessor of i on a path from i to i. We write 0 instead of nil for simplification of notation.

We now define π_{ij}^k recursively as follows.

For any $k \geq 1$, and for vertices i, j set

$$
\pi_{ij}^k = \begin{cases} \pi_{ij}^{k-1} & \text{if } d_{ij}^{k-1} \leq d_{ik}^{k-1} + d_{kj}^{k-1}, \\ \pi_{kj}^{k-1} & \text{if } d_{ij}^{k-1} > d_{ik}^{k-1} + d_{kj}^{k-1}, \end{cases}
$$

where the d_{ij}^k are the distances as already used for Floyd–Warshall algorithm. Thus the definition of π_{ij}^k is directly related to the definition of d_{ij}^k and justification for this is the same as that given for the definition of d_{ij}^k. Formally, we may define this in the following procedure.

Algorithm 10.3. Procedure-Construction of Predecessor Matrix

---

1. given a directed graph $G = (V, E)$ with weight function w, d_{ij}^k the length of a shortest path from i to j with the intermediate vertices in $\{1, 2, \ldots, k\}$, π^k the square matrix (π_{ij}^k) of order n
2. initially $\pi_{ij}^0 = i$ for $i \neq j$ and (i, j) an edge in E
3. $\pi_{ii}^k = nil$ or 0
4. for $k = 1$ to n
5. for $i = 1$ to n
6. for $j = 1$ to n
7. $\pi_{ij}^k = \pi_{ij}^{k-1}$ if $d_{ij}^{k-1} \leq d_{ik}^{k-1} + d_{kj}^{k-1}$
   $= \pi_{kj}^{k-1}$ if $d_{ij}^{k-1} > d_{ik}^{k-1} + d_{kj}^{k-1}$
8. return π^n

We next consider the other procedure for the construction of a shortest path from i to j. For $k \geq 1$, let p_{ij}^k denote the index of a largest intermediate vertex (or just the largest intermediate vertex) on a shortest path from i to j with intermediate vertices in $\{1, 2, \ldots, k\}$. If k is not an intermediate vertex on this path, then $p_{ij}^k = p_{ij}^{k-1}$ while if k actually occurs as an intermediate vertex on this path, then $p_{ij}^k = k$. To relate this problem with Floyd–Warshall algorithm, we define

$$p_{ij}^k = \begin{cases} p_{ij}^{k-1} & \text{if } d_{ij}^{k-1} \leq d_{ik}^{k-1} + d_{kj}^{k-1}, \\ k & \text{if } d_{ij}^{k-1} > d_{ik}^{k-1} + d_{kj}^{k-1}. \end{cases}$$

Define $P^k =$ the square matrix (p_{ij}^k) of order n. We write $P^n = P$. Again d_{ij}^k is shortest distance from i to j considered in the Floyd–Warshall algorithm. For a path consisting of a single edge (i, j), there are no intermediate vertices and conversely, we have $p_{ij}^0 = 0$. We may also set $p_{ij}^0 = 0$ if $i = j$ or (i, j) is not in E. Then $P^0 = 0$ the zero matrix.

Once P is obtained, to find a shortest path from i to j, we have p_{ij} as the largest intermediate vertex on this path. Then we find the largest vertex on a shortest path from i to p_{ij} and a largest vertex on a path from p_{ij} to j. In fact the shortest path from i to j can be decomposed as a shortest path i to p_{ij} ($i \rightsquigarrow p_{ij}$) and a shortest path from p_{ij} to j ($p_{ij} \rightsquigarrow j$). We continue this process of subdivision till we arrive at paths none of which can be further subdivided-always keeping a track of the order in which the sub-paths are being obtained. Finally we will have constructed shortest path we started with to find. We may call this procedure as the largest intermediate vertex procedure for construction of shortest paths. Finally, we have the following procedure.

Algorithm 10.4. Procedure-largest Intermediate Vertex Matrix

---

1. given a directed graph $G = (V, E)$ with weight function w, d_{ij}^k the length of a shortest path from i to j with the intermediate vertices in $\{1, 2, \ldots, k\}$. To construct the square matrix P^k of order n
2. initially $p_{ij}^0 = 0$ for all i, j
3. for $k = 1$ to n
4. for $i = 1$ to n
5. for $j = 1$ to n
6. $p_{ij}^k = p_{ij}^{k-1}$ if $d_{ij}^{k-1} \leq d_{ik}^{k-1} + d_{kj}^{k-1}$
7. $p_{ij}^k = k$ if $d_{ij}^{k-1} > d_{ik}^{k-1} + d_{kj}^{k-1}$
8. return P^n

Observe that the running time of both the procedures for construction of shortest paths is $O(n^3)$. Also observe that for finding π_{ij}^k and p_{ij}^k we do not have to compare the d_{ij}^k's De novo but only have to compare the consecutive entries of the matrices D^k constructed in the Floyd–Warshall procedure. We illustrate this point through a couple of examples.

Example 10.3. Construct shortest paths

(a) for all pairs of vertices using the predecessor procedure;
(b) from the vertices 1 to 2, 3 to 5, 4 to 1 and 5 to 3 using the intermediate vertex procedure for the graph G as given in Fig. 10.1.

Solution. (a) Observe that, by first writing the matrix π^0 and then comparing corresponding rows of D^{k-1} and D^k for $k = 1$ to 5, we get

$$\pi^0 = \begin{pmatrix} 0\,0\,0\,1\,1 \\ 2\,0\,0\,2\,0 \\ 3\,3\,0\,0\,0 \\ 0\,0\,4\,0\,4 \\ 0\,5\,0\,0\,0 \end{pmatrix}, \quad \pi^1 = \begin{pmatrix} 0\,0\,0\,1\,1 \\ 2\,0\,0\,2\,1 \\ 3\,3\,0\,1\,1 \\ 0\,0\,4\,0\,4 \\ 0\,5\,0\,0\,0 \end{pmatrix},$$

$$\pi^2 = \begin{pmatrix} 0\,0\,0\,1\,1 \\ 2\,0\,0\,2\,1 \\ 2\,3\,0\,2\,1 \\ 0\,0\,4\,0\,4 \\ 2\,5\,0\,2\,0 \end{pmatrix}, \quad \pi^3 = \begin{pmatrix} 0\,0\,0\,1\,1 \\ 2\,0\,0\,2\,1 \\ 2\,3\,0\,2\,1 \\ 2\,3\,4\,0\,4 \\ 2\,5\,0\,2\,0 \end{pmatrix},$$

$$\pi^4 = \begin{pmatrix} 0\,3\,4\,1\,4 \\ 2\,0\,4\,2\,4 \\ 2\,3\,0\,2\,4 \\ 2\,3\,4\,0\,4 \\ 2\,5\,4\,2\,0 \end{pmatrix}, \quad \pi^5 = \begin{pmatrix} 0\,5\,4\,1\,4 \\ 2\,0\,4\,2\,4 \\ 2\,3\,0\,2\,4 \\ 2\,5\,4\,0\,4 \\ 2\,5\,4\,2\,0 \end{pmatrix}.$$

The matrix π^5 then leads to the following shortest paths:

$1 \to 4 \to 5 \to 2$	$2 \to 1$	$3 \to 2 \to 1$	$4 \to 5 \to 2 \to 1$
$5 \to 2 \to 1$	$1 \to 4 \to 3$	$2 \to 4 \to 3$	$3 \to 2$
$4 \to 5 \to 2$	$5 \to 2$	$1 \to 4$	$2 \to 4$
$3 \to 2 \to 4$	$4 \to 3$	$5 \to 2 \to 4 \to 3$	$1 \to 4 \to 5$
$2 \to 4 \to 5$	$3 \to 2 \to 4 \to 5$	$4 \to 5$	$5 \to 2 \to 4.$

(b) With $P^0 = 0$ and comparing corresponding rows of D^{k-1} and D^k for $k = 1$ to 5, we get

$$P^1 = \begin{pmatrix} 0\,0\,0\,0\,0 \\ 0\,0\,0\,0\,1 \\ 0\,0\,0\,1\,1 \\ 0\,0\,0\,0\,0 \\ 0\,0\,0\,0\,0 \end{pmatrix}, \quad P^2 = \begin{pmatrix} 0\,0\,0\,0\,0 \\ 0\,0\,0\,0\,1 \\ 2\,0\,0\,2\,1 \\ 0\,0\,0\,0\,0 \\ 2\,0\,0\,2\,0 \end{pmatrix},$$

$$P^3 = \begin{pmatrix} 0\,0\,0\,0\,0 \\ 0\,0\,0\,0\,1 \\ 2\,0\,0\,2\,1 \\ 3\,3\,0\,0\,0 \\ 2\,0\,0\,2\,0 \end{pmatrix}, \quad P^4 = \begin{pmatrix} 0\,4\,4\,0\,4 \\ 0\,0\,4\,0\,4 \\ 2\,0\,0\,2\,4 \\ 3\,3\,0\,0\,0 \\ 2\,0\,4\,2\,0 \end{pmatrix},$$

$$P^5 = \begin{pmatrix} 0\,5\,4\,0\,4 \\ 0\,0\,4\,0\,4 \\ 2\,0\,0\,2\,4 \\ 5\,5\,0\,0\,0 \\ 2\,0\,4\,2\,0 \end{pmatrix} = P \text{ (say)}.$$

We now use the matrix P to calculate the required shortest paths.

For the shortest path from 1 to 2, the largest intermediate vertex on this path is 5, p_{52} being 0, the shortest path from 5 to 2 just consists of the edge $(5, 2)$. The shortest path from 1 to 5 has 4 as the largest intermediate vertex so $1 \rightsquigarrow 5$ consists of paths $1 \rightsquigarrow 4$ followed by $4 \rightsquigarrow 5$. Now $p_{45} = 0$, $p_{14} = 0$ show that the path $4 \rightsquigarrow 5$ is just the edge $(4, 5)$ and the path $1 \rightsquigarrow 4$ is the edge $(1, 4)$. Therefore, the path from 1 to 2 is $1 \rightarrow 4 \rightarrow 5 \rightarrow 2$.

For the shortest path from 3 to 5, we have $p_{35} = 4$, $p_{34} = 2$, $p_{45} = 0$ and $p_{32} = 0$. Therefore, the shortest path is $3 \rightarrow 2 \rightarrow 4 \rightarrow 5$.

For the shortest path from 4 to 1, we have $p_{41} = 5$, $p_{45} = 0$, $p_{51} = 2$, $p_{21} = 0$ and $p_{52} = 0$. The required shortest path is $4 \rightarrow 5 \rightarrow 2 \rightarrow 1$.

For the shortest path from 5 to 3, we have $p_{53} = 4$, $p_{54} = 2$, $p_{43} = 0$, $p_{52} = 0$, $p_{24} = 0$. Then the required path is $5 \rightarrow 2 \rightarrow 4 \rightarrow 3$.

Example 10.4. Find the shortest paths and their lengths for all pairs of vertices of the directed graph G as given in Fig. 10.2.

Solution. Let D^i be the matrices as computed by Floyd–Warshall procedure for the graph G. These are

$$D^0 = W = \begin{pmatrix} 0 & 1 & \infty & 4 \\ \infty & 0 & 2 & \infty \\ 6 & \infty & 0 & \infty \\ \infty & 5 & 3 & 0 \end{pmatrix}, \quad D^1 = \begin{pmatrix} 0 & 1 & \infty & 4 \\ \infty & 0 & 2 & \infty \\ 6 & 7 & 0 & 10 \\ \infty & 5 & 3 & 0 \end{pmatrix},$$

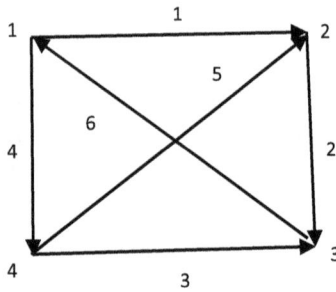

Fig. 10.2

$$D^2 = \begin{pmatrix} 0 & 1 & 3 & 4 \\ \infty & 0 & 2 & \infty \\ 6 & 7 & 0 & 10 \\ \infty & 5 & 3 & 0 \end{pmatrix}, \quad D^3 = \begin{pmatrix} 0 & 1 & 3 & 4 \\ 8 & 0 & 2 & 12 \\ 6 & 7 & 0 & 10 \\ 9 & 5 & 3 & 0 \end{pmatrix} = D^4.$$

The above matrices result because of the following calculations:

$$d^1_{1j} = \min\{d^0_{1j}, d^0_{11} + d^0_{1j}\} = d^0_{1j}, \ d^1_{2j} = \min\{d^0_{2j}, d^0_{21} + d^0_{1j}\} = d^0_{2j},$$

$$d^1_{3j} = \min\{d^0_{3j}, d^0_{31} + d^0_{1j}\} = \min\{d^0_{3j}, 6 + d^0_{1j}\} = (6, 7, 0, 10),$$

$$d^1_{4j} = \min\{d^0_{4j}, d^0_{41} + d^0_{1j}\} = d^0_{4j},$$

$$d^2_{1j} = \min\{d^1_{1j}, d^1_{12} + d^1_{2j}\} = \min\{d^1_{1j}, 1 + d^1_{2j}\} = (0, 1, 3, 4),$$

$$d^2_{2j} = \min\{d^1_{2j}, d^1_{22} + d^1_{2j}\} = d^1_{2j},$$

$$d^2_{3j} = \min\{d^1_{3j}, d^1_{32} + d^1_{2j}\} = \min\{d^1_{3j}, 7 + d^1_{2j}\} = (6, 7, 0, 10),$$

$$d^2_{4j} = \min\{d^1_{4j}, d^1_{42} + d^1_{2j}\} = \min\{d^1_{4j}, 5 + d^1_{2j}\} = (\infty, 5, 3, 0),$$

$$d^3_{1j} = \min\{d^2_{1j}, d^2_{13} + d^2_{3j}\} = \min\{d^2_{1j}, 3 + d^2_{3j}\} = (0, 1, 3, 4),$$

$$d^3_{2j} = \min\{d^2_{2j}, d^2_{23} + d^2_{3j}\} = \min\{d^2_{2j}, 2 + d^2_{3j}\} = (8, 0, 2, 12),$$

$$d^3_{3j} = \min\{d^2_{3j}, d^2_{33} + d^2_{3j}\} = d^2_{3j},$$

$$d^3_{4j} = \min\{d^2_{4j}, d^2_{43} + d^2_{3j}\} = \min\{d^2_{4j}, 3 + d^2_{3j}\} = (9, 5, 3, 0),$$

$$d^4_{1j} = \min\{d^3_{1j}, d^3_{14} + d^3_{4j}\} = \min\{d^3_{1j}, 4 + d^3_{4j}\} = (0, 1, 3, 4) = d^3_{1j},$$

$$d_{2j}^4 = \min\{d_{2j}^3, d_{24}^3 + d_{4j}^3\} = \min\{d_{2j}^3, 12 + d_{4j}^3\} = (8, 0, 2, 12) = d_{2j}^3,$$

$$d_{3j}^4 = \min\{d_{3j}^3, d_{34}^3 + d_{4j}^3\} = \min\{d_{3j}^3, 10 + d_{4j}^3\} = (6, 7, 0, 10) = d_{3j}^3,$$

$$d_{4j}^4 = \min\{d_{4j}^3, d_{44}^3 + d_{4j}^3\} = d_{4j}^3.$$

For the construction of shortest paths, we construct the predecessor matrices.

$$\pi^0 = \begin{pmatrix} 0\ 1\ 0\ 1 \\ 0\ 0\ 2\ 0 \\ 3\ 0\ 0\ 0 \\ 0\ 4\ 4\ 0 \end{pmatrix}, \quad \pi^1 = \begin{pmatrix} 0\ 1\ 0\ 1 \\ 0\ 0\ 2\ 0 \\ 3\ 1\ 0\ 1 \\ 0\ 4\ 4\ 0 \end{pmatrix},$$

$$\pi^2 = \begin{pmatrix} 0\ 1\ 2\ 1 \\ 0\ 0\ 2\ 0 \\ 3\ 1\ 0\ 1 \\ 0\ 4\ 4\ 0 \end{pmatrix}, \quad \pi^3 = \begin{pmatrix} 0\ 1\ 2\ 1 \\ 3\ 0\ 2\ 1 \\ 3\ 1\ 0\ 1 \\ 3\ 4\ 4\ 0 \end{pmatrix} = \pi^4, \text{ as } D^3 = D^4.$$

The shortest paths are:

$1 \to 2$	$2 \to 3 \to 1$	$3 \to 1$	$4 \to 3 \to 1$
$1 \to 2 \to 3$	$2 \to 3$	$3 \to 1 \to 2$	$4 \to 2$
$1 \to 4$	$2 \to 3 \to 1 \to 4$	$3 \to 1 \to 4$	$4 \to 3.$

10.3. Transitive Closure

We now consider a couple of applications of the Floyd–Warshall algorithm or a procedure based on these ideas. In many problems, we are only interested in knowing whether a vertex j is reachable from a vertex i. As such we consider just a directed graph $G = (V, E)$ with weight of every edge being 1. As before we consider the vertex set V to be $\{1, 2, \ldots, n\}$. Consider a square matrix $A = (a_{ij})$ of order n in which $a_{ij} = 1$ if and only if there is a path of length 1 or more from i to j in G. Then A is the adjacency matrix of a graph G^* on the vertex set V which is obtained from G by introducing an edge (i, j) whenever j is reachable from i in G.

Formally, we have $G^* = (V, E^*)$ where

$$E^* = \{(i, j) | 1 \leq i, j \leq n \text{ and } j \text{ is reachable from } i \text{ in } G\}$$
$$= \{(i, j) | 1 \leq i, j \leq n \text{ and there is a path of length 1}$$
$$\text{or more from } i \text{ to } j \text{ in } G\}$$

The graph G^* is called the **transitive closure** of G. Also the matrix A is called the **transitive closure of the adjacency matrix** of G. Thus, to construct the transitive closure of the graph G it is enough to construct the matrix A. We construct A by a recursive procedure based on the ideas of Floyd–Warshall.

Initially we set $a_{ij}^0 = 0$ for $i \neq j$ if (i, j) is not an edge in E and $a_{ij}^0 = 1$ for $i \neq j$ if (i, j) is an edge in E or $i = j$. For $k = 1$ to n, to construct a_{ij}^k, we consider a path from vertex i to vertex j with intermediate vertices in $\{1, 2, \ldots, k\}$. There are two possibilities: either k is not an intermediate vertex on this path so that intermediate vertices for this path are in $\{1, 2, \ldots, k-1\}$ or k is indeed an intermediate vertex for this path. In the second case, the path may be considered as a path from i to k composed with a path from k to j. The graph being assumed to be acyclic (or at least the path under consideration may be assumed to be simple), the two sub-paths from i to k and from k to j have their intermediate vertices in the set $\{1, 2, \ldots, k-1\}$. We thus have either a_{ij}^{k-1} or a_{ij}^k is such that both a_{ik}^{k-1} and a_{kj}^{k-1} are available simultaneously. As every a_{ij}^l is either 0 or 1, composition of the sub-paths from i to k and from k to j may be taken as the product $a_{ik}^{k-1} a_{kj}^{k-1}$. In the sense of logic, we need to have the logical conjunction of the occurrence of the sub-paths. We may thus consider a_{ij}^l elements of the set $\{0, 1\}$ with Boolean compositions $+$ and $.$ which are defined as follows:

$+$	0	1
0	0	1
1	1	1

$.$	0	1
0	0	0
1	0	1

Consequently, we have that for the Boolean numbers $a_1 + a_2 + \cdots + a_m = 1$ when at least one of the a_i's is 1 and $a_1.a_2 \cdots \cdots a_m = 1$ only when $a_i = 1$ for every i, $1 \leq i \leq m$. Using the Boolean compositions, we then have

$$a_{ij}^k = a_{ij}^{k-1} + a_{ik}^{k-1} a_{kj}^{k-1}.$$

Since every path from i to j has to have intermediate vertices in $\{1, 2, \ldots, n\}$, we need only define a_{ij}^k for k, $1 \leq k \leq n$. We may summarize the recursive definition of the a_{ij}^m and so, of the transitive closure in the following procedure.

Algorithm 10.5. Procedure Transitive Closure $G = (V, E)$

---

1. G is a directed graph with vertex set $V = \{1, 2, \ldots, n\}$ and weight of every edge is 1
2. initially, for $i = 1$ to n, $j = 1$ to n $a_{ij}^0 = 0$ for $i \neq j$ but $(i, j) \notin E$, and $a_{ij}^0 = 1$ if $(i, j) \in E$ or $i = j$
3. for $k = 1$ to n
4. for $i = 1$ to n
5. for $j = 1$ to n
6. $a_{ij}^k = a_{ij}^{k-1} + a_{ik}^{k-1} a_{kj}^{k-1}$
7. return $A = (a_{ij}^k)$

Running time and correctness of the procedure transitive closure.

As is the case for Floyd–Warshall algorithm, the running time of the procedure transitive closure is $O(n^3)$.

We prove by induction on k, $k = 0$ to n, *that for all i, j, $1 \leq i, j \leq n$, $a_{ij}^k = 1$ if and only if either $j = i$ or there is a path from i to j with intermediate vertices in $\{1, 2, \ldots, k\}$.*

The result is trivially true for $k = 0$. Now, suppose that $k \geq 1$ and that the result holds for $k - 1$. Since by definition $a_{ii}^k = a_{ii}^{k-1} + a_{ik}^{k-1} a_{ki}^{k-1}$ and $a_{ii}^{k-1} = 1$ by induction hypothesis, $a_{ii}^k = 1$. Now suppose that $i \neq j$. If $a_{ij}^k = 0$, then $a_{ij}^{k-1} = 0$ and $a_{ik}^{k-1} a_{kj}^{k-1} = 0$. In view of the Boolean composition (product) we have either $a_{ik}^{k-1} = 0$ or $a_{kj}^{k-1} = 0$. Therefore, there is no path from i to j with intermediate vertices in $\{1, 2, \ldots, k - 1\}$ and either there is no path from i to k or from k to j with intermediate vertices in $\{1, 2, \ldots, k - 1\}$. Hence there is no path from i to j with intermediate vertices in $\{1, 2, \ldots, k\}$. If $a_{ij}^k = 1$, then one at least of a_{ij}^{k-1}

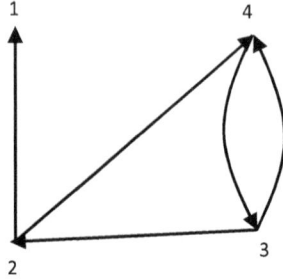

Fig. 10.3

and $a_{ik}^{k-1} a_{kj}^{k-1}$ is 1. If $a_{ij}^{k-1} = 1$, then, by induction hypothesis, there is a path from i to j with intermediate vertices in $\{1, 2, \ldots, k-1\}$ and hence in $\{1, 2, \ldots, k\}$. On the other hand, if $a_{ik}^{k-1} a_{kj}^{k-1} = 1$, then both $a_{ik}^{k-1} = 1$ and $a_{kj}^{k-1} = 1$. Again, by induction hypothesis, there is a path from i to k and a path from k to j with intermediate vertices in $\{1, 2, \ldots, k-1\}$. Composing these two paths we get a path from i to j with intermediate vertices in $\{1, 2, \ldots, k\}$. This completes induction and, hence, correctness of the procedure.

Example 10.5. Find the transitive closure of the directed graph as given in Fig. 10.3.

Solution. Now

$$A^0 = \begin{pmatrix} 1 & 0 & 0 & 0 \\ 1 & 1 & 0 & 1 \\ 0 & 1 & 1 & 1 \\ 0 & 0 & 1 & 1 \end{pmatrix}.$$

Now,

$$a_{1j}^1 = a_{1j}^0 + a_{11}^0 a_{1j}^0 = a_{1j}^0,$$

$$a_{2j}^1 = a_{2j}^0 + a_{21}^0 a_{1j}^0 = a_{2j}^0 + a_{1j}^0,$$

$$a_{3j}^1 = a_{3j}^0 + a_{31}^0 a_{1j}^0 = a_{3j}^0,$$

$$a_{4j}^1 = a_{4j}^0 + a_{41}^0 a_{1j}^0 = a_{4j}^0,$$

$$a_{1j}^2 = a_{1j}^1 + a_{12}^1 a_{2j}^1 = a_{1j}^1,$$

$$a_{2j}^2 = a_{2j}^1 + a_{22}^1 a_{2j}^1 = a_{2j}^1 + a_{2j}^1 = a_{2j}^1,$$

$$a_{3j}^2 = a_{3j}^1 + a_{32}^1 a_{2j}^1 = a_{3j}^1 + a_{2j}^1,$$

$$a_{4j}^2 = a_{4j}^1 + a_{42}^1 a_{2j}^1 = a_{4j}^1,$$

$$a_{1j}^3 = a_{1j}^2 + a_{13}^2 a_{3j}^2 = a_{1j}^2,$$

$$a_{2j}^3 = a_{2j}^2 + a_{23}^2 a_{3j}^2 = a_{2j}^2,$$

$$a_{3j}^3 = a_{3j}^2 + a_{33}^2 a_{3j}^2 = a_{3j}^2 + a_{3j}^2 = a_{3j}^2,$$

$$a_{4j}^3 = a_{4j}^2 + a_{43}^2 a_{3j}^2 = a_{4j}^2 + a_{3j}^2,$$

$$a_{1j}^4 = a_{1j}^3 + a_{14}^3 a_{4j}^3 = a_{1j}^3,$$

$$a_{2j}^4 = a_{2j}^3 + a_{24}^3 a_{4j}^3 = a_{2j}^3 + a_{4j}^3,$$

$$a_{3j}^4 = a_{3j}^3 + a_{34}^3 a_{4j}^3 = a_{3j}^3 + a_{4j}^3,$$

$$a_{4j}^4 = a_{4j}^3 + a_{44}^3 a_{4j}^3 = a_{4j}^3 + a_{4j}^3 = a_{4j}^3$$

so that

$$A^1 = \begin{pmatrix} 1 & 0 & 0 & 0 \\ 1 & 1 & 0 & 1 \\ 0 & 1 & 1 & 1 \\ 0 & 0 & 1 & 1 \end{pmatrix}, \quad A^2 = \begin{pmatrix} 1 & 0 & 0 & 0 \\ 1 & 1 & 0 & 1 \\ 1 & 1 & 1 & 1 \\ 0 & 0 & 1 & 1 \end{pmatrix}, \quad A^3 = \begin{pmatrix} 1 & 0 & 0 & 0 \\ 1 & 1 & 0 & 1 \\ 1 & 1 & 1 & 1 \\ 1 & 1 & 1 & 1 \end{pmatrix}.$$

Since a_{4j}^3 is 1 for every j, $a_{2j}^4 = a_{4j}^3$ and $a_{3j}^4 = a_{4j}^3 = a_{3j}^3$. The transitive closure of the graph G is given by the matrix

$$A^4 = \begin{pmatrix} 1 & 0 & 0 & 0 \\ 1 & 1 & 1 & 1 \\ 1 & 1 & 1 & 1 \\ 1 & 1 & 1 & 1 \end{pmatrix},$$

which in turn implies that the incidence matrix of the transitive closure is the matrix $C = A^4 - I$.

In many applications, it is required to find most central vertex of a directed weighted graph — a problem which we now consider.

Let $G = (V, E)$ be a directed weighted graph with weight function w associating a real number $w(e)$ to every edge e of G. For a vertex v of G, we define **eccentricity** $e(v)$ of v to be maximum of the shortest paths from v to the vertices of G i.e.,

$$\varepsilon(v) = \max_u \{\delta(v, u) | u \in V\}.$$

A vertex u of G for which $\varepsilon(u) = \min\{\varepsilon(v) \mid v \in V\}$ is called **center** of G. Center of a graph may not be unique. Finding center(s) of a graph is easy. We may adopt the following procedure:

1. Using Floyd–Warshall algorithm, find shortest distances for all pairs of vertices of G.
2. In the shortest cost (path length) matrix, in the row corresponding to v find the maximum entry (which equals the eccentricity $\varepsilon(v)$ of v).
3. In the column of eccentricities, find the least entry.
4. The vertices for which this least (as in 3 above) occurs are the centers of G.

Suppose that $|V| = n$. Then running time for finding the shortest path lengths as in number 1 above takes time $O(n^3)$. Finding maximum of each row of the cost matrix requires $n - 1$ comparisons and, so, the total number of comparisons required is $n(n-1)$. Again finding minimum in the column of eccentricities requires $n-1$ comparisons. Thus, altogether $(n+1)(n-1)$ comparisons are needed for the steps at 2 and 3. Therefore, the running time for finding the center(s) of G is $O(n^3) + O(n^2) = O(n^3)$.

Example 10.6. Find the center(s) of the graph G as in Fig. 10.2 with the vertices 1, 2, 3, 4 named as a, b, c, d, respectively.

Solution. All pairs shortest path lengths matrix already computed in Example 10.4 is

$$D = \begin{pmatrix} 0 & 1 & 3 & 4 \\ 8 & 0 & 2 & 12 \\ 6 & 7 & 0 & 10 \\ 9 & 5 & 3 & 0 \end{pmatrix}.$$

Therefore, the eccentricities of the vertices are $\varepsilon(a) = 4$, $\varepsilon(b) = 12$, $\varepsilon(c) = 10$, $\varepsilon(d) = 9$ and, therefore, a is the only center of G.

Example 10.7. Find the center(s) of the graph G as given in Fig. 10.1 with the vertices 1, 2, 3, 4, 5 named as a, b, c, d, e, respectively.

Solution. As already computed in Example 10.1 (as well as in Example 10.2) the all pairs shortest path lengths matrix is

$$D = \begin{pmatrix} 0 & 1 & 7 & 1 & -4 \\ 1 & 0 & 8 & 2 & -3 \\ -3 & -4 & 0 & -2 & -7 \\ 1 & 0 & 6 & 0 & -5 \\ 6 & 5 & 13 & 7 & 0 \end{pmatrix}.$$

The eccentricities are given by $\varepsilon(a) = 7$, $\varepsilon(b) = 8$, $\varepsilon(c) = 0$, $\varepsilon(d) = 6$, $\varepsilon(e) = 13$ and, therefore, c is the center of G.

Exercise 10.1.

1. Give an example of a graph which has more than one center.
2. Let G be a directed graph with weight of every edge equal to 1 and A be the adjacency matrix of G. Regarding A as a Boolean matrix, prove that $A^k(i, j) = 1$ if and only if there is a path from vertex i to vertex j of length exactly k, for every k, $0 \le k \le n$, where n is the number of vertices of G. Deduce that the transitive closure A^* of A is $I + A + A^2 + \cdots + A^n$, where I is the identity matrix of order n.

10.4. Johnson's Algorithm

Let $G = (V, E)$ be a directed weighted graph with weight function w associating a real number $w(e)$ to every edge e in E and with $|V| = n$. Floyd–Warshall algorithm that we discussed has $O(n^3)$ as running time. Johnson proposed a procedure which considerably improves upon the running time for the problem in many cases but not always. If the graph G were a complete graph so that $|E| = n^2$, the two running times are identical (or the same). On the other hand, if $|E|$ is much smaller than n^2, which is the case when G is a sparse graph, improvement in running time obtained by Johnson's procedure is appreciable.

Johnson's procedure is, as a matter of fact, a combination of Dijkstra and of Bellman–Ford. Since Dijkstra's algorithm for single source shortest paths works only for directed weighted graphs with all weights non-negative, we need some how to transform G into a graph G' in which the edge weights become non-negative. This process of transforming G is called **reweighting**.

Let $h : V \to R$ (real numbers) be a function and for an edge (u, v) in G, define

$$\widehat{w}(u, v) = w(u, v) + h(u) - h(v)$$

This introduces a weight function $\widehat{w}$ on the edges of G. For vertices u, v in V, let $\delta(u, v)$ and $\widehat{\delta}(u, v)$ denote the shortest path lengths from u to v with respect to the weight functions w and $\widehat{w}$, respectively.

Lemma 10.5. *(a) For $u, v \in V$, $\delta(u, v) = \widehat{\delta}(u, v) - h(u) + h(v)$.*

(b) $c = (u = u_0, \ldots, u_k = u)$ is a negative weight cycle in $G = (V, E, w)$ if and only if c is a negative weight cycle in $G = (V, E, \widehat{w})$.

Proof. (a) Let $u, v \in V$ and p be a path from u to v. Let $p = (u = u_1, \ldots, u_k = v)$. Then

$$\widehat{w}(p) = \widehat{w}(u_1, u_2) + \widehat{w}(u_2, u_3) + \cdots + \widehat{w}(u_{k-1}, u_k)$$

$$= \sum_{1 \leq i \leq k-1} \{w(u_i, u_{i+1}) + h(u_i) - h(u_{i+1})\}$$

$$= \sum_{1 \leq i \leq k-1} w(u_i, u_{i+1}) + (h(u_1)$$

$$- h(u_2)) + (h(u_2) - h(u_3)) + \cdots + (h(u_{k-1}) - h(u_k))$$

$$= w(p) + h(u_1) - h(u_k) = w(p) + h(u) - h(v). \qquad (10.4)$$

Suppose q is another path from u to v in G. Then as above

$$\widehat{w}(q) = w(q) + h(u) - h(v).$$

Thus $w(p) \leq w(q)$ if and only if $\widehat{w}(p) \leq \widehat{w}(q)$. Thus p is a shortest path from u to v in $G = (V, E, w)$ if and only if p is a shortest path from u to v in $G = (V, E, \widehat{w})$. The result then follows from (10.4).

(b) Since $u_k = u = u_1$, it follows as in (a) above that $\widehat{w}(c) = w(c)$. Hence c is a negative weight cycle in $G = (V, E, w)$ if and only if c is a negative weight cycle in $G = (V, E, \widehat{w})$. $\qquad \square$

Now the problem is to choose a function $h : V \rightarrow R$ so that $\widehat{w}(u, v) = w(u, v) + h(u) - h(v) \geq 0$ for all edges (u, v) in G. For this we use Bellman–Ford. Let G' be obtained from G by adding a new vertex s which is not already in V, i.e., we take $V' = V \cup \{s\}$ and take $E' = E \cup \{(s, v) | v \in V\}$, i.e., add edges by connecting s to every vertex v in V. Also extend the weight function w by assigning weight 0 to every new edge (s, v) introduced. Let the resulting graph be $G' = (V', E')$. Now run Bellman–Ford to find the shortest distance $\delta(s, v)$ from the source vertex s to every $v \in V$. Then, by the triangle inequality, for vertices $u, v \in V$, $\delta(s, v) \leq \delta(s, u) + w(u, v)$ so that if we set $h(u) = \delta(s, u)$, then

$$w(u, v) + h(u) - h(v) \geq 0.$$

It thus follows that if h is chosen as above then $\widehat{w}(u, v) \geq 0$ for all edges (u, v) in E.

Observe that in G', there being no edge with terminal vertex s, there is no cycle which involves s. Hence any cycle in G' is already a cycle in G. Therefore, if G has no cycle with negative weight, there is no negative weight cycle in G'. So, when Bellman–Ford is run on G' and the algorithm indicates the existence of a negative weight cycle in G', then there is a negative weight cycle in G and the function h as above cannot be found. Hence Johnson's algorithm cannot be applied here.

Once h is defined using Bellman–Ford on G', we now remove s from G' and all edges (s, v) introduced and get back to the graph G with weight function $\widehat{w}$. Run Dijkstra's algorithm on $G = (V, E, \widehat{w})$ with every vertex v in V as source vertex to get the shortest path weights $\widehat{\delta}(x, y)$ for x, y in V. Finally get shortest path weights for $G = (V, E, w)$ by setting $\delta(x, y) = \widehat{\delta}(x, y) + h(y) - h(x)$.

Step 2 of Johnson's algorithm takes a constant amount of time. The running time of Bellman–Ford as in steps 3 to 5 is $O((|V| + 1)(|E| + |V|)) = O(|V||E| + |V|^2)$. Dijkstra's algorithm with every vertex in turn regarded as source vertex has running time $O(|V||E| + |V|^2 \lg |V|)$. Steps 9 and 10 only need a constant amount of time. Therefore, the running time of Johnson's algorithm is $O(|V||E| + |V|^2) + O(|V||E| + |V|^2 \lg |V|) = O(|V||E| + |V|^2 \lg |V|)$. If G is a full (complete) matrix, $|E| = |V|^2$ so that the running time of Johnson's algorithm becomes $O(|V|^3 + |V|^2 \lg |V|) = O(|V|^3)$ which is the same as that of Floyd–Warshall but if the graph is

Algorithm 10.6. Johnson's Algorithm

1. given a directed weighted graph $G = (V, E, w)$ with w as the weight function
2. add a new vertex s to G, $s \notin V$, along with edges (s, v), $v \in V$, each of weight 0. Call the new graph $G' = (V', E', w)$
3. run Bellman–Ford on G'
4. if Bellman–Ford indicates the existence of a negative weight cycle in G', then stop. The graph G has a negative weight cycle
5. else, find shortest path lengths $\delta(s, v)$ for vertices v in V
6. define $h(v) = \delta(s, v)$, $v \in V$
7. define weight function $\widehat{w}$ on G by $\widehat{w}(u, v) = w(u, v) + h(u) - h(v)$, $u, v \in V$
8. run Dijkstra's algorithm on $G = (V, E, \widehat{w})$ with every vertex v in V as a source vertex
9. get all pairs shortest path lengths $\widehat{\delta}(u, v)$ for $u, v \in V$
10. for $u, v \in V$, set $\delta(u, v) = \widehat{\delta}(u, v) - h(u) + h(v)$

sparse $|E|$ is much smaller than $|V|^2$ so that the running time of Johnson's algorithm shows a great improvement on that of Floyd–Warshall.

Example 10.8. Using Johnson's algorithm find the shortest path lengths for all pairs of vertices for the directed weighted graph as in Fig. 10.1.

Solution. Adding a new vertex s not already in V and edges (s, i), $i = 1$ to 5, all of these with weight 0, the graph G' is given in Fig. 10.4.

We now use Bellman–Ford to find the shortest distance $d(i) = \delta(s, i)$ for $i = 1$ to 5.

Initially $d(s) = 0$ and $d(i) = \infty$ for all i, $1 \leq i \leq 5$.

Iteration 1. Relax the edges (s, i) for $i = 1$ to 5 and we get $d(i) = 0$ for all i.

Iteration 2. Relax the edges $(2, 1)$, $(3, 2)$, $(4, 5)$, $(4, 3)$, $(1, 4)$ and all the remaining edges, we get $d(1) = 0$, $d(2) = -4$, $d(3) = 0$, $d(4) = 0$, $d(5) = -5$.

Iteration 3. Relax the edges $(2, 1)$, $(2, 4)$ along with the remaining edges and we get $d(1) = -3$, $d(2) = -4$, $d(3) = 0$, $d(4) = -2$, $d(5) = -5$.

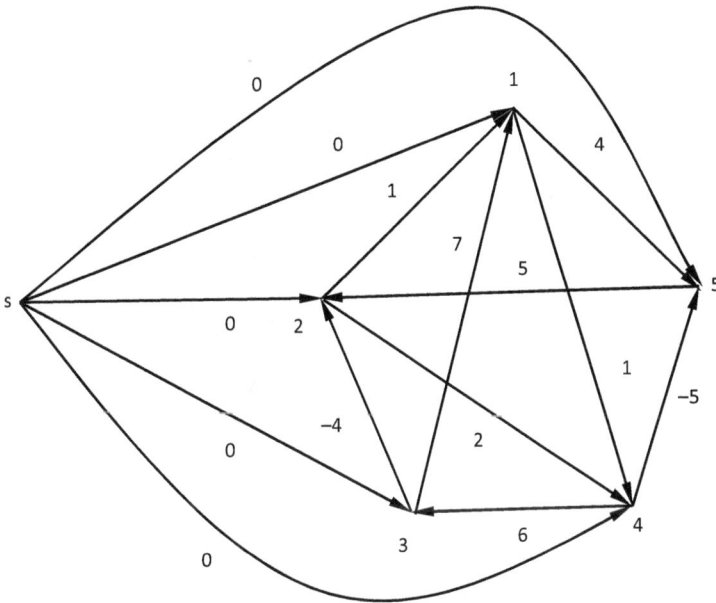

Fig. 10.4

Iteration 4. Relax the edge $(4, 5)$ along with the remaining edges, and we get $d(1) = -3, d(2) = -4, d(3) = 0, d(4) = -2, d(5) = -7$.

In all iterations only the edges that are effectively relaxed are explicitly mentioned. In the final iteration, we find that the d-values remain unchanged. Therefore, the function h is defined by

$$h(1) = -3, h(2) = -4, h(3) = 0, h(4) = -2, h(5) = -7.$$

Now reweight the edges of G using the function h. We have

$$\widehat{w}(1, 5) = w(1, 5) + h(1) - h(5) = 8,$$
$$\widehat{w}(1, 4) = w(1, 4) + h(1) - h(4) = 0,$$
$$\widehat{w}(2, 1) = w(2, 1) + h(2) - h(1) = 0,$$
$$\widehat{w}(2, 4) = w(2, 4) + h(2) - h(4) = 0,$$

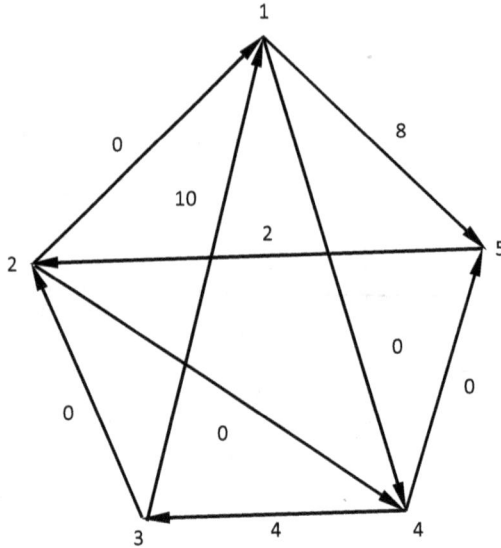

Fig. 10.5

$$\widehat{w}(3, 1) = w(3, 1) + h(3) - h(1) = 10,$$

$$\widehat{w}(3, 2) = w(3, 2) + h(3) - h(2) = 0,$$

$$\widehat{w}(4, 5) = w(4, 5) + h(4) - h(5) = 0,$$

$$\widehat{w}(4, 3) = w(4, 3) + h(4) - h(3) = 4,$$

$$\widehat{w}(5, 2) = w(5, 2) + h(5) - h(2) = 2.$$

The reweighted graph G is as given in Fig. 10.5 on which we need to run Dijkstra's algorithm with each vertex as source vertex.

(a) When the source vertex is 1.

Initially $S = \varphi$, $d(1) = 0$, $d(i) = \infty$ for $2 \leq i \leq 5$.

Iteration 1. $u = 1$, $S = \{1\}$, $d(4) = 0$, $d(5) = 8$,

Iteration 2. $u = 4$, $S = \{1, 4\}$, $d(3) = 4$, $d(5) = 0$,

Iteration 3. $u = 5$, $S = \{1, 4, 5\}$, $d(2) = 2$, $d(3) = 4$,

Iteration 4. $u = 2$, $S = \{1, 4, 5, 2\}$, $d(3) = 4$,

Iteration 5. $u = 3$, $S = \{1, 4, 5, 2, 3\}$,

and the process of iteration stops here. Therefore $\widehat{\delta}(1, i) = (0, 2, 4, 0, 0)$ and the shortest distances are

$$\delta(1, i) = (0, 2 - h(1) + h(2), 4 - h(1) + h(3),$$
$$0 - h(1) + h(4), 0 - h(1) + h(5)) = (0, 1, 7, 1, -4).$$

(b) When the source vertex is 2.
Initially $S = \varphi, d(2) = 0, d(i) = \infty$ for $i = 1, 3, 4, 5$.

Iteration 1. $u = 2$, $S = \{2\}$, $d(1) = 0$, $d(4) = 0$,

Iteration 2. $u = 1$, $S = \{2, 1\}$, $d(4) = 0$, $d(5) = 8$,

Iteration 3. $u = 4$, $S = \{2, 1, 4\}$, $d(3) = 4$, $d(5) = 0$,

Iteration 4. $u = 5$, $S = \{2, 1, 4, 5\}$, $d(3) = 4$,

Iteration 5. $u = 3$, $S = \{2, 1, 4, 5, 3\}$,

and the process of iteration stops here giving

$$\widehat{\delta}(2, i) = (d(1), 0, d(3), d(4), d(5)) = (0, 0, 4, 0, 0).$$

Then the shortest path lengths $\delta(2, i)$ for G with weight function w are

$$\delta(2, i) = (0 - h(2) + h(1), 0, 4 - h(2) + h(3),$$
$$0 - h(2) + h(4), 0 - h(2) + h(5)) = (1, 0, 8, -2, -3).$$

(c) When the source vertex is 3.
Initially $S = \varphi, d(3) = 0, d(i) = \infty$ for $i = 1, 2, 4, 5$.

Iteration 1. $u = 3$, $S = \{3\}$, $d(1) = 10$, $d(2) = 0$,

Iteration 2. $u = 2$, $S = \{3, 2\}$, $d(1) = 0$, $d(4) = 0$,

Iteration 3. $u = 1$, $S = \{3, 2, 1\}$, $d(4) = 0$, $d(5) = 8$,

Iteration 4. $u = 4$, $S = \{3, 2, 1, 4\}$, $d(5) = 0$,

Iteration 5. $u = 5$, $S = \{3, 2, 1, 4, 5\}$.

and the process of iteration stops here giving $\widehat{\delta}(3, i) = (0, 0, 0, 0, 0)$. Then the shortest path distances $\delta(3, i)$ are given by

$$\delta(3, i) = (0 - h(3) + h(1), 0 - h(3) + h(2), 0, 0 - h(3) + h(4),$$
$$0 - h(3) + h(5)) = (-3, -4, 0, -2, -7).$$

(d) When the source vertex is 4.
Initially $S = \varphi, d(4) = 0, d(i) = \infty$ for $i = 1, 2, 3, 5$.

Iteration 1. $u = 4, S = \{4\}, d(3) = 4, d(5) = 0,$

Iteration 2. $u = 5, S = \{4, 5\}, d(2) = 2, d(3) = 4,$

Iteration 3. $u = 2, S = \{4, 5, 2\}, d(1) = 2, d(3) = 4,$

Iteration 4. $u = 1, S = \{4, 5, 2, 1\}, d(3) = 4,$

Iteration 5. $u = 3, S = \{4, 5, 2, 1, 3\},$

and the process of iteration stops giving $\widehat{\delta}(4, i) = (2, 2, 4, 0, 0)$. The shortest path lengths $\delta(4, i)$ are given by

$$\delta(4, i) = (2 - h(4) + h(1), 2 - h(4) + h(2), 4 - h(4) + h(3), 0,$$
$$0 - h(4) + h(5)) = (1, 0, 6, 0, -5).$$

(e) When the source vertex is 5.
Initially $S = \varphi, d(5) = 0, d(i) = \infty$ for $i = 1, 2, 3, 4$.

Iteration 1. $u = 5, S = \{5\}, d(2) = 2,$

Iteration 2. $u = 2, S = \{5, 2\}, d(1) = 2, d(4) = 2,$

Iteration 3. $u = 1, S = \{5, 2, 1\}, d(4) = 2,$

Iteration 4. $u = 4, S = \{5, 2, 1, 4\}, d(3) = 6,$

Iteration 5. $u = 3, S = \{5, 2, 1, 4, 3\},$

and the process of iteration stops here giving $\widehat{\delta}(5, i) = (2, 2, 6, 2, 0)$. The shortest path lengths $\delta(5, i)$ for the graph G with weight function w are given by

$$\delta(5, i) = (2 - h(5) + h(1), 2 - h(5) + h(2), 6 - h(5) + h(3),$$
$$2 - h(5) + h(4), 0) = (6, 5, 13, 7, 0).$$

Collecting the information provided by the five processes of Dijkstra's algorithm, the all pairs shortest path lengths are given by the matrix

$$D = \begin{pmatrix} 0 & 1 & 7 & 1 & -4 \\ 1 & 0 & 8 & 2 & -3 \\ -3 & -4 & 0 & -2 & -7 \\ 1 & 0 & 6 & 0 & -5 \\ 6 & 5 & 13 & 7 & 0 \end{pmatrix},$$

which is the same as already obtained by the slow recursive procedure and also by Floyd–Warshall algorithm.

Exercise 10.2. Find the five shortest path trees with vertices 1, 2, 3, 4 and 5 as provided by the five processes of Dijkstra's algorithm in Example 10.8.

Chapter 11

Flow Networks

Let G be a given connected, weighted, simple directed graph with two specified vertices called a source and a sink. A function f which associates a non-negative real number to every edge e of G is called a flow in G if $f(e)$ is less than or equal to the weight of e for every edge e of G. Total flow leaving the source equals the total flow entering the sink and is called the flow value. Our aim in this chapter is not just to find a maximal flow value but also the distribution of flow, i.e., $f(e)$ for every edge e which gives the maximal flow value. If G is a bipartite (undirected) graph, a subset M of edges of G is called a matching in G if no two edges in M have a common vertex. Finding a maximal matching in G can be solved by using the procedure for maximal flow in a flow network and is also considered in this chapter. Labeling procedure for labeling the vertices due to Ford and Fulkerson is used to obtain maximal flow in a flow network.

11.1. Ford and Fulkerson Theorem

Let $G = (V, E)$ be a connected, weighted simple diagraph, where V is the set of vertices and E the set of edges with the weight function $c : E \rightarrow R_0$, where R_0 is the set of all non-negative real numbers. (See the change of notation — the weight function w used earlier is being replaced by c (the capacity function).) For $e \in E$, the weight $c(e)$ is called the capacity of the edge e. Suppose there are two distinguished vertices, s called a **source** and t called a **sink**. Then $N = (G, c, s, t)$ is called a **flow network**. For an edge e, let e^- denote the initial vertex of e and e^+ the terminal vertex of e. By a **flow** on the network N, we mean a function $f : E \to R_0$ such that

$$(F1) \quad f(e) \leq c(e) \quad \text{for every } e \in E, \tag{11.1}$$

$$(F2) \quad \text{for each vertex } v \in V, v \neq s, t, \sum_{e^-=v} f(e) = \sum_{e^+=v} f(e)$$

$$\tag{11.2}$$

Observe that if $e = (u, v)$ so that $e^- = u$ and $e^+ = v$, then $f(e)$ denotes the flow leaving u and it denotes the flow entering v. Thus the condition $(F2)$ means that whatever is flow entering v leaves it, i.e., there is no accumulation of flow at any internal vertex (i.e., a vertex other than the source and the sink). The feasibility condition $(F1)$ means that the flow in any edge is non-negative and is bounded above by the capacity of the edge. An edge e for which $f(e) = c(e)$ is called a **saturated edge**.

We now prove that the flow leaving the source equals the flow entering the sink.

Lemma 11.1. *Let* $N = (G, c, s, t)$ *be a flow network with flow* f. *Then*

$$\sum_{e^-=s} f(e) - \sum_{e^+=s} f(e) = \sum_{e^+=t} f(e) - \sum_{e^-=t} f(e).$$

Proof. Observe that

$$\sum_{e^-=s} f(e) + \sum_{e^-=t} f(e) + \sum_{e^-=v\in V-\{s,t\}} f(e) = \sum_{e^-=v\in V} f(e) = \sum_{e} f(e)$$

$$= \sum_{e^+=s} f(e) + \sum_{e^+=t} f(e) + \sum_{e^+=v\in V-\{s,t\}} f(e).$$

Therefore, using $(F2)$, we get

$$\sum_{e^-=s} f(e) + \sum_{e^-=t} f(e) = \sum_{e^+=s} f(e) + \sum_{e^+=t} f(e).$$

Or

$$\sum_{e^-=s} f(e) - \sum_{e^+=s} f(e) = \sum_{e^+=t} f(e) - \sum_{e^-=t} f(e).$$

$\square$

Definition 11.1.

1. The quantity $\sum_{e^-=s} f(e) - \sum_{e^+=s} f(e)$ which is the net flow leaving s (and, therefore, also the net flow entering t) is called the **value of the flow** f and is denoted by $w(f)$.
2. A flow f on N is called **maximal** if $w(f) \geq w(f')$ for every flow (pattern) f' on N.
   The problem that we like to study in this chapter is to find a maximal flow on a given flow network N.
3. By a cut of a flow network N we mean a partition $V = S \cup T$, $S \cap T = \varphi$ of the vertex set V of G with $s \in S, t \in T$.
4. If X, Y are disjoint subsets of V, then $c(X, Y) = \sum_{e^- \in X, e^+ \in Y} c(e)$, i.e., the sum of the capacities of all edges in E which have their initial vertex in X and terminal vertex in Y is called the capacity of the pair (X, Y). In particular, the sum $c(S, T) = \sum_{e^- \in S, e^+ \in T} c(e)$ is called the **capacity of the cut** (S, T).
5. A cut (S, T) of the flow network N is called a **minimal cut** of N if $c(S, T) \leq c(S', T')$ for every cut (S', T') of N.

   In the direction of finding a maximal flow pattern, we first have the following theorem.

Theorem 11.2. *Let* $N = (G, c, s, t)$ *be a flow network,* (S, T) *a cut of* N *and* f *a flow pattern on* N. *Then* $w(f) = \sum_{e^- \in S, e^+ \in T} c(e) - \sum_{e^+ \in S, e^- \in T} c(e)$. *In particular,* $w(f) \leq c(S, T)$.

Proof. By the conservation of flow for any intermediate vertex $v \in V - \{s, t\}$, we have

$$\sum_{e^-=v} f(e) - \sum_{e^+=v} f(e) = 0 \text{ or } \sum_{e^-=v, e^+ \in V} f(e) - \sum_{e^+=v, e^- \in V} f(e) = 0.$$

This is, in particular, true for every $v \in S - \{s\}$. Therefore

$$\sum_{v \in S-\{s\}} \sum_{e^-=v, e^+ \in V} f(e) - \sum_{v \in S-\{s\}} \sum_{e^+=v, e^- \in V} f(e) = 0. \tag{11.3}$$

Also, for $v = s$, we have

$$\sum_{e^- = s} f(e) - \sum_{e^+ = s} f(e) = w(f) \text{ or}$$

$$\sum_{e^- = s, e^+ \in V} f(e) - \sum_{e^+ = s, e^- \in V} f(e) = w(f). \tag{11.4}$$

Adding (11.3) and (11.4), we get

$$w(f) = \sum_{e^- \in S, e^+ \in V} f(e) - \sum_{e^+ \in S, e^- \in V} f(e)$$

$$= \sum_{e^- \in S, e^+ \in S} f(e) + \sum_{e^- \in S, e^+ \in T} f(e)$$

$$- \sum_{e^+ \in S, e^- \in S} f(e) - \sum_{e^+ \in S, e^- \in T} f(e)$$

$$= \sum_{e^- \in S, e^+ \in T} f(e) - \sum_{e^+ \in S, e^- \in T} f(e).$$

Since $\sum_{e^+ \in S, e^- \in T} f(e)$ is always a non-negative quantity, we have

$$w(f) \leq \sum_{e^- \in S, e^+ \in T} f(e) \leq \sum_{e^- \in S, e^+ \in T} c(e) = c(S, T).$$

□

Let $p = (s, v_1, \ldots, v_j, v_{j+1}, \ldots, v_k, t)$ be a (undirected) path in the flow network $N = (G, c, s, t)$ with flow pattern f in which for every edge $e = (v_j, v_{j+1})$, $f(e) < c(e)$ while $f(e) > 0$ when $e = (v_{j+1}, v_j)$. In the path p, an edge $e = (v_j, v_{j+1})$ directed from v_j to v_{j+1} is called a **forward edge** and an edge $e = (v_{j+1}, v_j)$ directed from v_{j+1} to v_j is called a **backward edge**. Also the path p from s to t is then called an **augmenting path** with respect to f.

Lemma 11.3. *Let* $N = (G, c, s, t)$ *be a flow network with flow pattern* f, p *be an augmenting path with respect to* f, $d = \min\{d_1, d_2\}$, *where* $d_1 = \min\{c(e) - f(e) \mid e$ *a forward edge*$\}$ *and* $d_2 = \min\{f(e) \mid e$ *a* *backward edge}. Then* $f' : E \to R_0$ *defined by*

$$f'(e) = \begin{cases} f(e) + d & e \text{ a forward edge,} \\ f(e) - d & e \text{ a backward edge,} \\ f(e) & \text{otherwise} \end{cases}$$

is also a flow pattern on the network N and the flow value $w(f') = w(f) + d$.

Proof. Let $e \in E$. If e is a

(a) forward edge, then $f'(e) = f(e) + d \le f(e) + c(e) - f(e) = c(e)$,
(b) backward edge, then $f'(e) = f(e) - d \le c(e) - d \le c(e)$.

For all other e, $f'(e) = f(e) \le c(e)$.
Hence $f'(e) \le c(e)$ for all $e \in E$.

The graph G being simple, p is a simple path. Therefore, there is only one edge in p incident with s and only one edge in p incident with t. If e in p is incident with s, then either $e^- = s$ or $e^+ = s$ but not both and then $f'(e) = f(e) + d$ or $f(e) - d$ as the case may be (but not both). Now

$$w(f') = \sum_{e^- = s} f'(e) - \sum_{e^+ = s} f'(e)$$

$$= \left(\sum_{e^- = s, e \in p} (f(e) + d) - \sum_{e^+ = s, e \in p} (f(e) - d) \right)$$

$$+ \left(\sum_{e^- = s, e \notin p} f(e) - \sum_{e^+ = s, e \notin p} f(e) \right)$$

$$= d + \left(\sum_{e^- = s, e \in p} f(e) - \sum_{e^+ = s, e \in p} f(e) \right)$$

$$+ \left(\sum_{e^- = s, e \notin p} f(e) - \sum_{e^+ = s, e \notin p} f(e) \right)$$

$$= d + w(f). \qquad \square$$

We now prove the theorem due to Ford and Fulkerson.

Theorem 11.4. *In a given transport network $N = (G, c, s, t)$, the maximum value of flow pattern is equal to the minimum value of the capacities of all the cuts of N.*

Proof. Let f be a flow pattern for N such that the value $w(f)$ of the flow is the maximum for all possible flows for N. Define a subset S of the vertex set V of G recursively as follows:

(a) $s \in S$.
(b) If a vertex $u \in S$ and either $f(e) < c(e)$ for some edge $e = (u, v)$ in E (i.e., $e^- = u$) or $f(e) > 0$ for some edge $e = (v, u)$ in E (i.e., $e^+ = u$), then $v \in S$.

Take $T = V - S = \{v \in V | v \notin S\}$. Then $V = S \cup T$, $S \cap T = \varphi$.
We claim that $t \notin S$ so that $t \in T$.
Suppose to the contrary that $t \in S$.
If $v \in S$, it follows from the construction of S that there exists $u \in S$ such that either $(u, v) = e$ is in E and $f(e) < c(e)$ or $(v, u) = e$ is in E and $f(e) > 0$. Then a repeated application of this and the fact that there are only a finite number of vertices in G shows that there exists a path (undirected) p from s to t, say $p = (s, v_1, \ldots, v_j, v_{j+1}, \ldots, v_k, t)$ with $v_1, \ldots, v_j, v_{j+1}, \ldots, v_k$ in S such that either $(v_j, v_{j+1}) = e$ is in E and $f(e) < c(e)$ or $(v_{j+1}, v_j) = e$ is in E and $f(e) > 0$. In the path p, let $d_1 = \min\{c(e) - f(e) \mid e$ a forward edge in $p\}$ and $d_2 = \min\{f(e) \mid e$ a backward edge in $p\}$ and $d = \min(d_1, d_2)$ so that $d > 0$. Define a function $f' : E \to R_0$ by

$$f'(e) = \begin{cases} f(e) + d & e \text{ a forward edge,} \\ f(e) - d & e \text{ a backward edge,} \\ f(e) & \text{otherwise.} \end{cases}$$

It follows from Lemma 11.3 that f' is also a flow in the network N and $w(f') = w(f) + d > w(f)$. This contradicts the assumption that f is a flow on N for which $w(f)$ is maximal. Hence $t \notin S$ so that $t \in T$. Therefore (S, T) is a cut of the flow network N.

By condition (b) in the construction of S, for each vertex u in S and v in T, we have (there are three possibilities: (i) either $e = (u, v)$ is in E or (ii) $e = (v, u)$ is in E or (iii) neither of (u, v), (v, u) is in E) $f(e) = c(e)$

if $e = (u, v)$ is in E and $f(e) = 0$ if $e = (v, u)$ is in E. Therefore, by Theorem 11.2, we have

$$w(f) = \sum_{e^- \in S, e^+ \in T} f(e) - \sum_{e^+ \in S, e^- \in T} f(e) \cdot$$

$$= \sum_{e^- \in S, e^+ \in T} f(e) = c(S, T).$$

This completes the proof of the theorem in view of Theorem 11.2. (In this case (S, T) is a minimal cut.) ☐

The above theorem gives a simple procedure for finding maximum value of flow in a flow network but does not give a flow pattern for the maximal flow. However, a procedure using vertex labeling is available for finding a maximal flow pattern. We first consider a couple of examples for maximum flow in a flow network.

Example 11.1. Find a maximal flow and a corresponding cut for the network N given by the diagram as given in Fig. 11.1.

Solution. We find all cuts (S, T) with the corresponding capacities $c(S, T)$. For any cut (S, T) of N, we only consider the vertex set S, the vertex set T being understood.

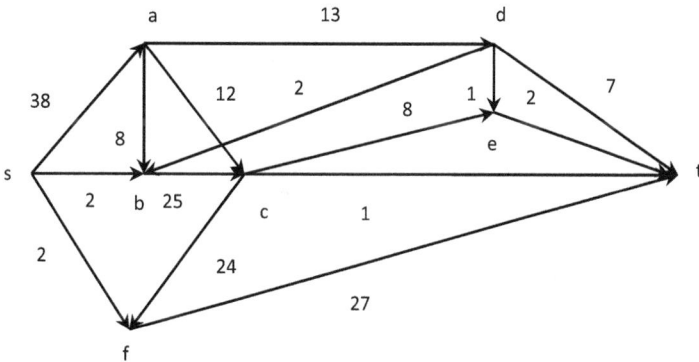

Fig. 11.1

Vertex set S	Capacity $c(S, T)$
s	$38 + 1 + 2 = 41$
s, a	$1 + 2 + 8 + 13 + 12 = 36$
s, b	$38 + 2 + 25 = 65$
s, c	$38 + 1 + 2 + 8 + 1 + 24 = 74$
s, d	$38 + 1 + 2 + 1 + 7 + 2 = 51$
s, e	$38 + 1 + 2 + 2 = 43$
s, f	$38 + 1 + 27 = 66$
s, a, b	$2 + 12 + 13 + 25 = 52$
s, a, c	$1 + 2 + 8 + 13 + 8 + 1 + 24 = 57$
$\boxed{s, a, d}$	$1 + 2 + 8 + 12 + 1 + 7 + 2 = \boxed{33}$
s, a, e	$1 + 2 + 8 + 12 + 13 + 2 = 38$
s, a, f	$1 + 8 + 12 + 13 + 27 = 61$
s, b, c	$38 + 2 + 24 + 1 + 8 = 73$
s, b, d	$38 + 2 + 25 + 1 + 7 = 73$
s, b, e	$38 + 2 + 25 + 2 = 67$
s, b, f	$38 + 25 + 27 = 90$
s, c, d	$38 + 1 + 2 + 8 + 1 + 24 + 7 + 2 + 1 = 84$
s, c, e	$38 + 1 + 2 + 1 + 24 + 2 = 68$
s, c, f	$38 + 1 + 8 + 1 + 27 = 75$
s, d, e	$38 + 1 + 2 + 2 + 7 + 2 = 52$
s, d, f	$38 + 1 + 1 + 7 + 27 + 2 = 76$
s, e, f	$38 + 1 + 2 + 27 = 68$
s, a, b, c	$2 + 13 + 1 + 8 + 24 = 48$
s, a, b, d	$2 + 12 + 25 + 1 + 7 = 47$
s, a, b, e	$2 + 13 + 12 + 25 + 2 = 54$
s, a, b, f	$12 + 13 + 25 + 27 = 77$
s, a, c, d	$1 + 2 + 8 + 1 + 24 + 1 + 7 + 2 + 8 = 54$
s, a, c, e	$1 + 2 + 8 + 13 + 1 + 24 + 2 = 51$
s, a, c, f	$1 + 8 + 13 + 8 + 1 + 27 = 58$
s, a, d, e	$1 + 2 + 8 + 12 + 2 + 7 + 2 = 34$
s, a, d, f	$1 + 8 + 12 + 2 + 1 + 7 + 27 = 58$
s, a, e, f	$1 + 8 + 12 + 13 + 2 + 27 = 63$
s, b, c, d	$38 + 2 + 8 + 1 + 24 + 1 + 7 = 81$
s, b, c, e	$38 + 2 + 1 + 24 + 2 = 67$

s, b, c, f	$38 + 8 + 1 + 27 = 74$
s, b, d, e	$38 + 2 + 25 + 2 + 7 = 74$
s, b, d, f	$38 + 25 + 1 + 7 + 27 = 98$
s, b, e, f	$38 + 25 + 2 + 27 = 92$
s, c, d, e	$38 + 1 + 2 + 1 + 24 + 7 + 2 + 2 = 77$
s, c, d, f	$38 + 1 + 8 + 1 + 2 + 1 + 7 + 27 = 85$
s, c, e, f	$38 + 1 + 1 + 27 + 2 = 69$
s, d, e, f	$38 + 1 + 2 + 7 + 27 + 2 = 77$
s, a, b, c, d	$2 + 8 + 1 + 24 + 1 + 7 = 43$
s, a, b, c, e	$2 + 13 + 1 + 24 + 2 = 42$
s, a, b, c, f	$13 + 8 + 1 + 27 = 49$
s, a, b, d, e	$2 + 12 + 7 + 25 + 2 = 48$
s, a, b, d, f	$12 + 25 + 1 + 7 + 27 = 72$
s, a, b, e, f	$12 + 13 + 25 + 2 + 27 = 79$
s, a, c, d, e	$1 + 2 + 8 + 1 + 24 + 7 + 2 + 2 = 47$
s, a, c, d, f	$1 + 8 + 8 + 1 + 2 + 1 + 7 + 27 = 55$
s, a, c, e, f	$1 + 8 + 13 + 1 + 2 + 27 = 52$
s, a, d, e, f	$1 + 8 + 12 + 2 + 7 + 2 + 27 = 59$
s, b, c, d, e	$38 + 2 + 1 + 24 + 7 + 2 = 74$
s, b, c, d, f	$38 + 8 + 1 + 1 + 7 + 27 = 82$
s, b, c, e, f	$38 + 1 + 2 + 27 = 68$
s, c, d, e, f	$38 + 1 + 1 + 2 + 7 + 2 + 27 = 78$
s, b, d, e, f	$38 + 25 + 7 + 2 + 27 = 99$
s, a, b, c, d, e	$2 + 1 + 24 + 7 + 2 = 36$
s, a, b, c, d, f	$8 + 1 + 1 + 7 + 27 = 44$
s, a, b, c, e, f	$13 + 2 + 1 + 27 = 43$
s, a, b, d, e, f	$12 + 25 + 7 + 2 + 27 = 73$
s, a, c, d, e, f	$1 + 8 + 1 + 2 + 7 + 2 + 27 = 48$
s, b, c, d, e, f	$38 + 1 + 7 + 2 + 27 = 75$
s, a, b, c, d, e, f	$1 + 2 + 7 + 27 = 37$

The minimum of the capacities of the various cuts of the network being 33, the maximum flow value of the flow network is $w = 33$ and the corresponding cut is (S, T) with $S = \{s, a, d\}$ and $T = \{b, c, e, f, t\}$.

Example 11.2. Find maximal flow and a corresponding cut for the transport network N given by the diagram as in Fig. 11.2.

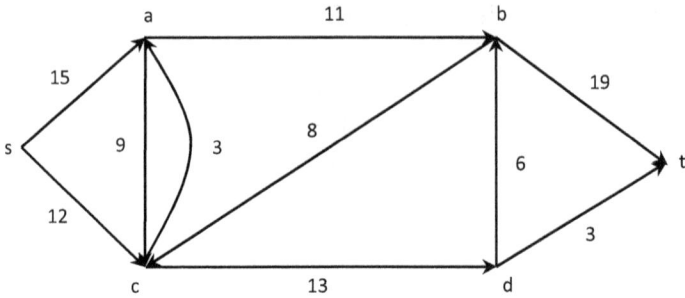

Fig. 11.2

Solution. We find all cuts (S, T) with the corresponding capacities $c(S, T)$. For any cut (S, T) of N, we only consider the vertex set S, the vertex set T being understood.

Vertex set S	Capacity $c(S, T)$
s	$15 + 12 = 27$
s, a	$12 + 9 + 11 = 32$
s, b	$15 + 12 + 8 + 19 = 54$
s, c	$15 + 3 + 13 = 31$
s, d	$15 + 12 + 6 + 3 = 36$
s, a, b	$12 + 9 + 8 + 19 = 48$
s, a, c	$11 + 13 = 24$
s, a, d	$12 + 9 + 11 + 6 + 3 = 41$
s, b, c	$15 + 19 + 3 + 13 = 50$
s, b, d	$15 + 12 + 8 + 19 + 3 = 57$
s, c, d	$15 + 3 + 6 + 3 = 27$
s, a, b, c	$19 + 13 = 32$
s, a, b, d	$12 + 9 + 19 + 8 + 3 = 51$
$\boxed{s, a, c, d}$	$11 + 6 + 3 = \boxed{20}$
s, b, c, d	$15 + 19 + 3 + 3 = 40$
s, a, b, c, d	$19 + 3 = 22$

The minimum of capacities of the various cuts of the network being 20, the maximum value of flow of the network is 20 with the corresponding cut being (S, T) where $S = \{s, a, c, d\}$ and $T = \{b, t\}$.

11.2. Labeling Procedure for Maximal Flow and Flow Pattern

Using cuts for a transport network, we can find the value of maximal flow in the network. However, this does not help us in finding an actual flow pattern for maximum flow. For finding a maximal flow pattern, we use a recursive procedure for labeling of vertices. We assume that initially the flow in every edge is zero. Also assume that using the procedure a flow pattern f has been arrived at.

We label the source vertex s with label $(-, \infty)$. The significance of this label is that the source s can supply a large amount of flow into the network. Scan all vertices that are adjacent from the source s and label these vertices. Observe that there are no vertices that are adjacent to the source. Let a be a vertex that is adjacent from s. Then label a with $(s^+, \Delta a)$, where

$$\Delta a = c(s, a) - f(s, a) \quad \text{if } c(s, a) > f(s, a)$$

and a is not labeled if $c(s, a) = f(s, a)$. Here and elsewhere we write $c(a, b)$ and $f(a, b)$ for $c(e)$ and $f(e)$, respectively, when $e = (a, b)$. Next scan and label all vertices that are adjacent from (or to) vertices that are already labeled. Let b be a vertex that is adjacent from a labeled vertex a. Then label b with $(a^+, \Delta b)$, where

$$\Delta b = \min\{\Delta a, c(a, b) - f(a, b)\} \quad \text{if } c(a, b) > f(a, b),$$

and b is not labeled if $c(a, b) = f(a, b)$.

When b is adjacent to a labeled vertex a, then label b with $(a^-, \Delta b)$, where $\Delta b = \min\{\Delta a, f(b, a)\}$ if $f(b, a) > 0$ and b is not labeled if $f(b, a) = 0$.

Labeling procedure described may not be unique. A vertex b may be adjacent to or from more than one labeled vertex a. Then label for b may be chosen arbitrarily, when there are more than one ways of labeling it. Continue the process of labeling till we can go no further.

Observe that if a is a labeled vertex and b is adjacent from a so that b is labeled $(a^+, \Delta b)$, it means that flow in the edge (a, b) can be increased by an amount Δb. Thus the incoming flow to b can be increased by Δb. There being no accumulation of flow in any vertex, by whatever amount the flow into b is increased along the edge (a, b), the outgoing flow from b to the other vertices has to increase by the same amount. On the other hand, the label $(a^-, \Delta b)$ on b means that the flow from b to a along the edge (b, a) is to be decreased by an amount Δb. Again, by the law of conservation of flow, the flow from b to the unlabeled vertices increases by Δb.

The process of labeling stops in two ways.

(a) We are not able to arrive at the sink vertex t, i.e., we cannot label the sink t. In this case, let S be the set of all labeled vertices and T be the set of vertices that cannot be labeled. Both S, T are non-empty sets and $S \cup T$ gives all the vertices of G. Thus (S, T) is a cut of the network. Then the flow in the various edges gives a maximal flow pattern and the sum of the flows from the vertices in S to the vertices in T gives the flow of the cut (S, T) and, so, the maximum flow value.

(b) We arrive at the sink vertex t which is labeled say $(z^+, \Delta t)$. We start working backwards from t. The earlier vertex is labeled either $(y^+, \Delta z)$ or $(y^-, \Delta z)$ and $\Delta t \leq \Delta z$. Continuing backwards, we get an augmenting path $p = (s, a, b, \ldots, z, t)$ from s to t and $\Delta t \leq \Delta u$ for every vertex u on the path p. Consider a function $f' : E \to R_0$ where

$$
f'(e) = \begin{cases} f(e) + d & e \text{ a forward edge on } p, \\ f(e) - d & e \text{ a backward edge on } p, \\ f(e) & \text{otherwise,} \end{cases}
$$

where $d = \Delta t > 0$. Then f' is a flow on the network N and the flow value $w(f') = w(f) + d > w(f)$. We then start the process of labeling vertices with the new flow pattern in the network. We either arrive at the case where t is not labeled proving that the flow pattern f' is maximal or that t is labeled leading to a new flow pattern f'' which is an improvement on the flow f'. In the latter case we again start the process of labeling. We continue the process till we arrive at the case as in (a) above.

We now illustrate this procedure through a couple of examples.

Example 11.3. Find a maximal flow for the transport network of Example 11.2 using the labeling procedure.

Solution. We start with the zero flow in every edge and start labeling the vertices. We label the source vertex s with $(-, \infty)$. Then label the vertices a and c with $(s^+, 15)$ and $(s^+, 12)$, respectively. Next label b with

$(a^+, 11), d$ with $(c^+, 12)$ and finally t with $(b^+, 11)$. There are two choices for labeling of b and having labeled b, there are two choices for labeling d. After having labeled b and d, there are two choices for labeling t. However, we choose the labels as shown in Fig. 11.3(a). Then the resulting flow in the edges is as shown in Fig. 11.3(b). Now we label the vertices in Fig. 11.3(b) as $s(-, \infty)$, $a(s^+, 4)$, $c(s^+, 12)$, $d(c^+, 12)$, $b(d^+, 6)$ and $t(d^+, 3)$. The resulting flow in the edges is shown in Fig. 11.3(c). We now label the vertices in Fig. 11.3(c) as $s(-, \infty)$, $a(s^+, 4)$, $c(s^+, 9)$, $d(c^+, 9)$, $b(d^+, 6)$ and $t(b^+, 6)$. The resulting flow is then given as in Fig. 11.3(d). We label the vertices in Fig. 11.3(d) as $s(-, \infty)$, $a(s^+, 4)$, $c(s^+, 3)$, $d(c^+, 3)$ and we cannot label b and t. Observe that b cannot be labeled using the labeled vertex c because of the flow in the edge (b, c) being 0. We thus get a cut with $S = \{s, a, c, d\}$ and $T = \{b, t\}$. Then the total value of the flow is $f(a, b) + f(d, b) + f(d, t) = 11 + 6 + 3 = 20$.

Example 11.4. Find maximal flow and a corresponding flow pattern for the transport network given in Fig. 11.1.

Solution. Initially, let us begin with a zero flow in every edge. We label the vertices as follows:

$$s(-, \infty), a(s^+, 38), b(s^+, 1), f(s^+, 2), c(a^+, 12) \text{ or}$$

$$(b^+, 1), d(a^+, 13), e(d^+, 1) \text{ or } (c^+, 1), t(d^+, 7) \text{ or}$$

$$(e^+, 2) \text{ or } (c^+, 1) \text{ or } (f^+, 2).$$

We choose the label $(c^+, 1)$ for t and $(b^+, 1)$ for c. Then the flow in the network on first pass is as given in Fig. 11.4(b) because of the augmenting path $s \to b \to c \to t$. We next label the vertices in Fig. 11.4(b) as follows:

$$s(-, \infty), a(s^+, 38), f(s^+, 2), b(a^+, 8), c(a^+, 12) \text{ or}$$

$$(b^+, 8), d(a^+, 13), e(d^+, 1) \text{ or } (c^+, 8), t(d^+, 7) \text{ or } (e^+, 1) \text{ or } (f^+, 2).$$

We choose the label $(e^+, 1)$ for t, $(b^+, 8)$ for c and $(d^+, 1)$ for e. Then the flow in the network on second pass is as given in Fig. 11.4(c). Now

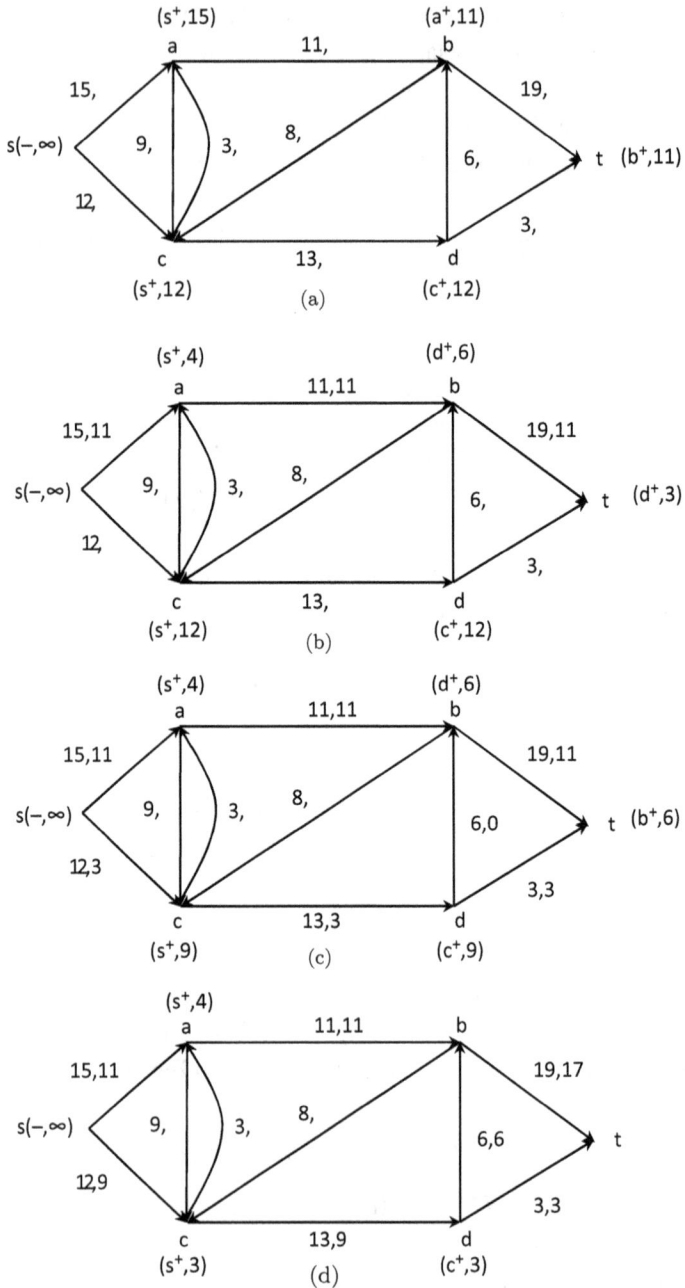

(a)

(b)

(c)

(d)

Fig. 11.3

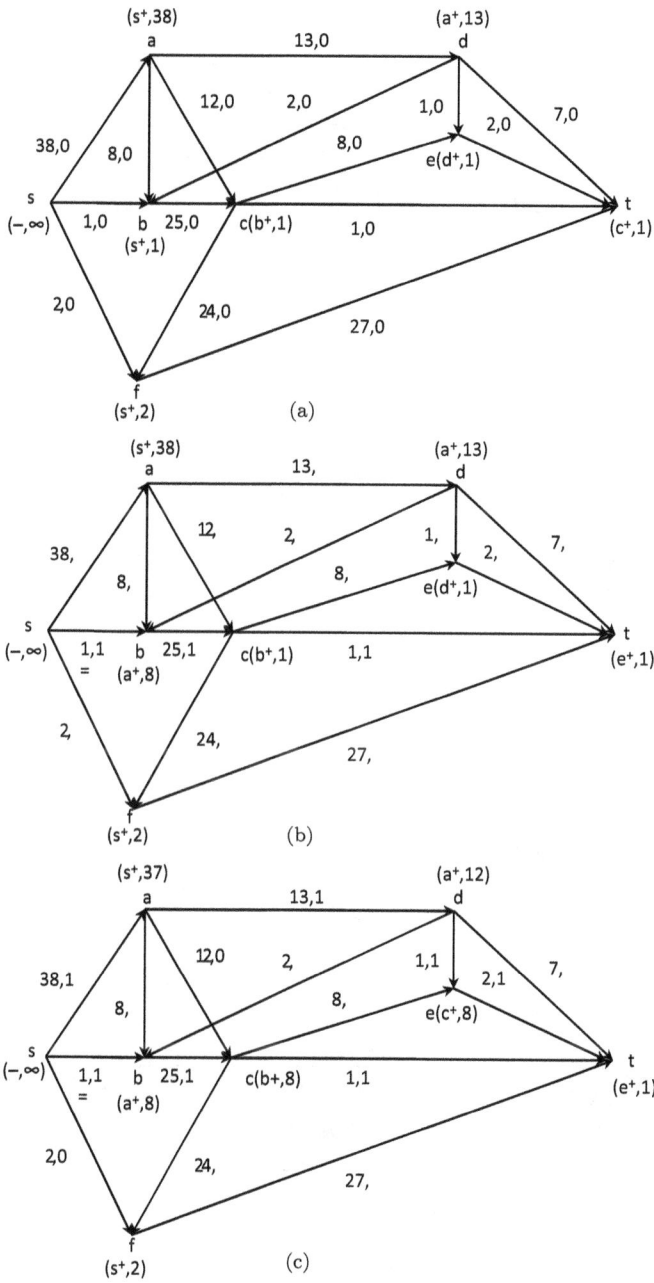

Fig. 11.4

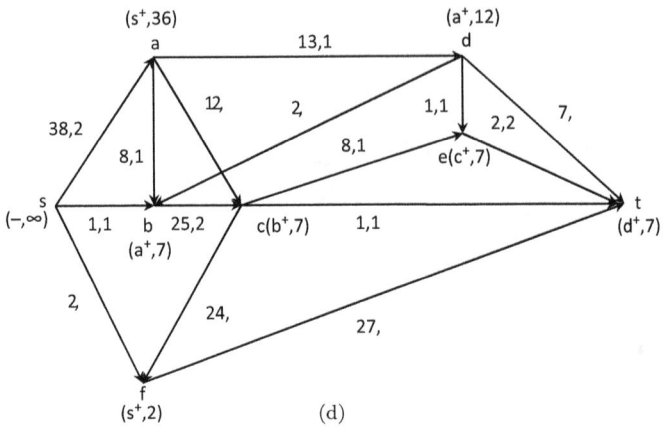

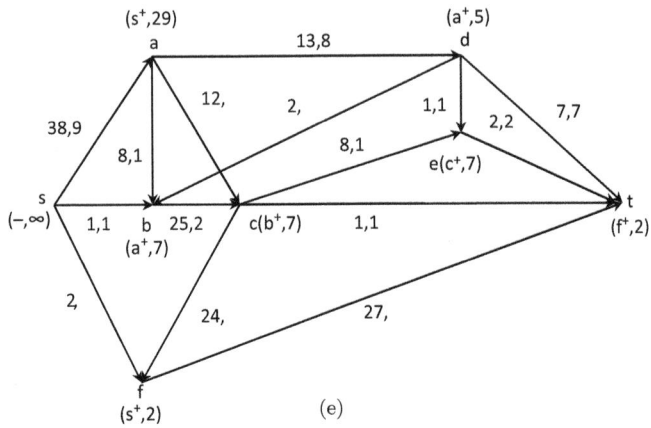

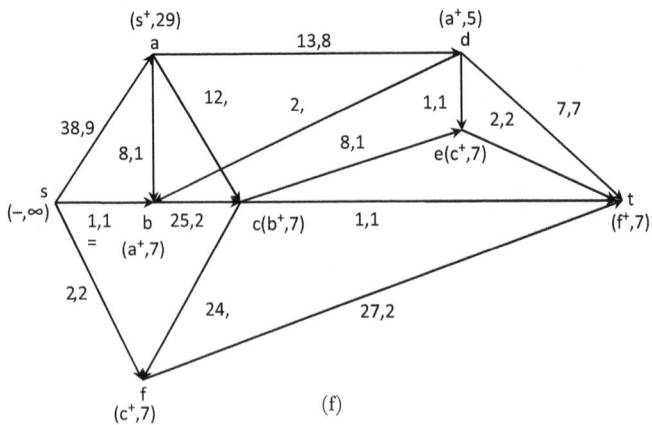

Fig. 11.4 (*Continued*)

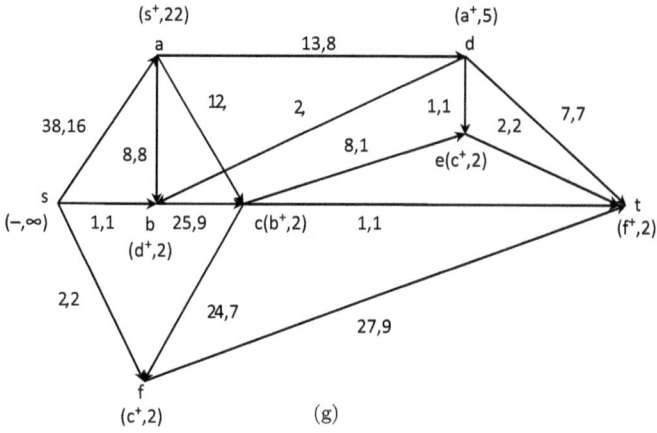

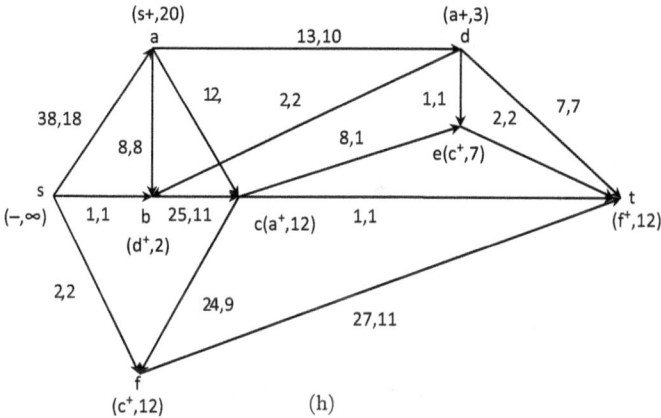

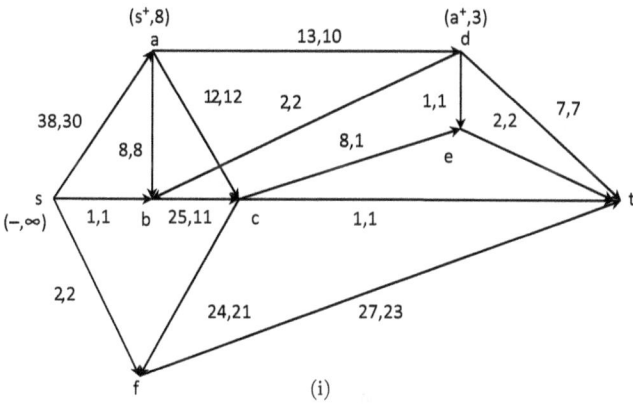

Fig. 11.4 (*Continued*)

label the vertices in Fig. 11.4(c) as given below:

$$s(-, \infty), a(s^+, 27), f(s^+, 2), b(a^+, 8), d(a^+, 12), c(a^+, 12) \text{ or}$$
$$(b^+, 8), e(c^+, 8) \text{ and } t(d^+, 7) \text{ or } (e^+, 1) \text{ or } (f^+, 2).$$

We choose the label $(b^+, 8)$ for c and $(e^+, 1)$ for t. Then the resulting flow in the network is as given in Fig. 11.4(d). We label the vertices in Fig. 11.4(d) as below:

$$s(-, \infty), a(s^+, 36), f(s^+, 2), b(a^+, 7), c(b^+, 7) \text{ or}$$
$$(a^+, 12), d(a^+, 12), e(c^+, 7) \text{ and } t(d^+, 7) \text{ or } (f^+, 2).$$

We choose the label $(b^+, 7)$ for c and $(d^+, 7)$ for t. The resulting flow is then given in Fig. 11.4(e). Next we label the vertices in Fig. 11.4(e) and the same are as given below:

$$s(-, \infty), a(s^+, 29), f(s^+, 2), b(a^+, 7), c(b^+, 7) \text{ or}$$
$$(a^+, 12), d(a^+, 5), e(c^+, 7) \text{ and } t(f^+, 2).$$

We choose label $(b^+, 7)$ for c.

The resulting flow in the edges is then given in Fig. 11.4(f). Next we label the vertices in Fig. 11.4(f) and the same are as given below:

$$s(-, \infty), a(s^+, 29), b(a^+, 7), c(b^+, 7)$$
$$d(a^+, 5), e(c^+, 7), f(c^+, 7) \text{ and } t(f^+, 7).$$

The resulting flow in the edges is then given in Fig. 11.4(g). Next label the vertices in Fig. 11.4(g). We label

$$s(-, \infty), a(s^+, 22), d(a^+, 5), b(d^+, 2), c(b^+, 2), e(c^+, 2), f(c^+, 2)$$
$$\text{and } t(f^+, 2).$$

The resulting flow is then given in Fig. 11.4(h). We label the vertices in Fig. 11.4(h) as

$$s(-, \infty), a(s^+, 20), c(a^+, 12), d(a^+, 3), e(c^+, 7), f(c^+, 12)$$
$$\text{and } t(f^+, 12).$$

The resulting flow is then given in Fig. 11.4(i). We have now arrived at the position where we can label s with $(-, \infty)$, a with $(s^+, 8)$, d with $(a^+, 3)$ and none of the remaining vertices b, c, e, f and t can be labeled. Hence no additional flow can be allowed into the network. Therefore, Fig. 11.4(i) gives the maximal flow pattern with the maximal flow being

$$f(a, b) + f(s, b) + f(s, f) + f(a, c) + f(d, b) + f(d, e) + f(d, t)$$
$$= 1 + 2 + 8 + 12 + 2 + 1 + 7 = 33.$$

Also the flow pattern is

$$f(s, a) = 30, \ f(s, b) = 1, \ f(s, f) = 2, \ f(a, b) = 8, \ f(a, c) = 12,$$
$$f(a, d) = 10, \ f(b, c) = 11, \ f(c, e) = 1, \ f(c, f) = 21, \ f(d, b) = 2,$$
$$f(d, e) = 1, \ f(d, t) = 7, \ f(e, t) = 2, \ f(c, t) = 1, \ f(f, t) = 23.$$

Example 11.5. Find a maximal flow and corresponding maximal flow pattern for the transport network given in Fig. 11.5(a).

Solution. We assume the initial flow to be zero. We simply write x, to indicate the flow $x, 0$ in any edge. Figure 11.5(f) shows that although we can label s, a, b, c, we cannot label t. Thus (S, T) is a cut of the network with $S = \{s, a, b, c\}$ and $T = \{t\}$. The maximal flow value is $f(a, t) + f(c, t) = 9 + 6 = 15$ and the maximal flow pattern is as in Fig. 11.5(f).

We next prove that the labeling procedure terminates in a finite number of steps with t not being labeled leading to a maximal flow when all the capacities $c(e)$ are integers.

Theorem 11.5. *Let $N = (G, c, s, t)$ be a flow network where the capacities $c(e)$ are integers for all edges e of G. Then there is a maximal flow on the network with $f(e)$ integers for all edges e.*

Proof. Initially suppose that the flow $f_0(e) = 0$ for all edges e. If f_0 is a maximal flow, we have nothing to prove. Suppose that f_0 is not maximal. Then the labeling procedure on N with flow f_0 leads to an augmenting

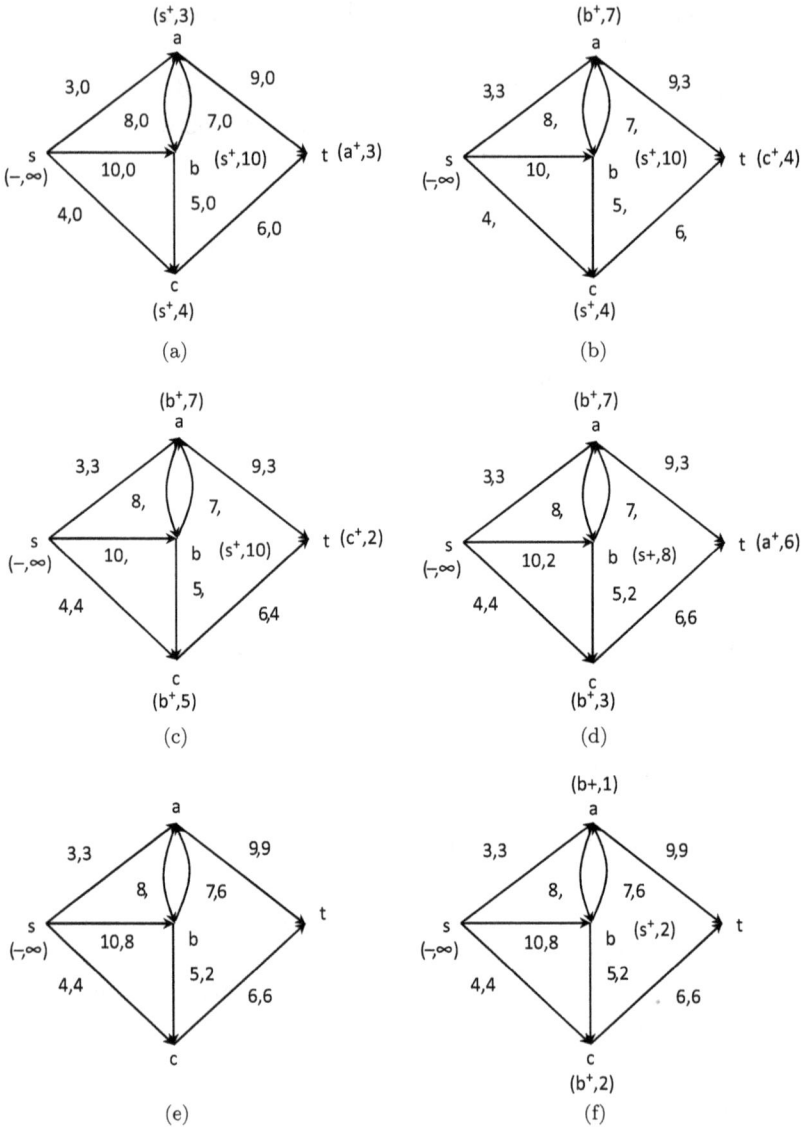

Fig. 11.5

path p_1 and a flow f_1 where

$$f_1(e) = \begin{cases} f_0(e) + d = d & \text{if } e \text{ is a forward edge on } p_1, \\ f_0(e) = 0 & \text{otherwise,} \end{cases}$$

where $d = \Delta(t)$. Since $f_0(e) = 0$ for every e, there is no backward edge on the path $p_1 = (s, a, b, \ldots, z, t)$. Also the label of a is $(s^+, \Delta a)$ where $\Delta a = \min\{c(e) - f_0(e), \infty\} = \min\{c(e), \infty\} = c(e)$, an integer, where e is the edge (s, a). Suppose that u is a vertex on the path p_1 and that Δu is an integer. If v is a vertex on p_1 adjacent from u, v is labeled $(u^+, \Delta v)$ and if e is the edge (u, v), then $\Delta v = \min\{c(e) - f_0(e), \Delta u\} = \min\{c(e), \Delta u\}$ is an integer. Therefore, Δu is an integer for every vertex u on p_1. In particular $d = \Delta t$ is an integer. Therefore $f_1(e)$ is an integer for every edge e. We repeat the above process and find that either f_1 is a maximal flow or we get an improved flow f_2 with $f_2(e)$ an integer for every edge e. Since for every e, a flow f satisfies $f(e) \leq c(e)$ which are finite integers, the above process cannot continue indefinitely, i.e., the process terminates in a finite number of steps and we get a flow f which is maximal and $f(e)$ is an integer for every edge e of G. □

Remark. The above result also holds when all the capacities $c(e)$ are rational numbers, i.e., if N is a network with all the capacities $c(e)$ rational numbers, then there exists a maximal flow f with $f(e)$ a rational number for every edge e.

Let $c = lcm$ of the denominators of all the $c(e)$. Consider the network $N' = (G, c', s, t)$ where $c'(e) = c(e)c$ for every edge e of G. Then N' is a flow network with all the capacities $c'(e)$ integers. By the above theorem, there exists a maximal flow f' and all the $f'(e)$ are integers. Consider a function $f : E \rightarrow R_0$ defined by

$$f(e) = \frac{1}{c} f'(e), \quad e \in E.$$

Then f is a flow pattern on N and $f(e)$ are rational numbers. Also the flow f' being maximal, f is a maximal flow.

The labeling procedure and Theorem 11.5 for integers as well as rational numbers as capacities is due to Ford and Fulkerson. Theorem 11.5 is also true for real number capacities but we do not prove this result here.

The argument used in the proof of Theorem 11.5 suggests the following algorithm.

Algorithm 11.1. Labeling Procedure for Vertices in G

Given a flow network $N = (G, c, s, t)$ and flow f in N. We label the vertices of G

1. if (s, a) is an edge in G label a as $(s^+, \Delta(a))$, $\Delta(a) = c(s, a) - f(s, a)$ if $c(s, a) > f(s, a)$; do not label a if $c(s, a) = f(s, a)$
2. if (a, b) is an edge in G and a is labeled, label b with $(a^+, \Delta(b))$, $\Delta(b) = \min\{\Delta(a), c(a, b) - f(a, b)\}$ if $c(a, b) > f(a, b)$; do not label b if $c(a, b) = f(a, b)$
3. if (b, a) is an edge in G and a is labeled, label b with $(a^-, \Delta(b))$, $\Delta(b) = \min\{\Delta(a), f(b, a)\}$ if $f(b, a) \neq 0$; do not label b if $f(b, a) = 0$

Algorithm 11.2. Procedure Maxflow (N)

Given a flow network $N = (G, c, s, t)$ and $G = (V, E)$, V the set of vertices and E the set of edges in G

1. initially $f(e) = 0$ for every edge e of G
2. f be a flow in N and $p = (e_1, e_2, \ldots, e_k)$ be an augmenting path with respect to flow f
3. set $d = \min\{c(e_i) - f(e_i)$ for e_i a forward edge in p, $d = f(e_i)$ for e_i a backward edge in $p\}$
4. replace $f(e_i)$ by $f(e_i) + d$ if e_i is a forward edge in p and by $f(e_i) - d$ if e_i is a backward edge in p
5. repeat steps 2 to 4 till no further improvement in the value of f is possible
6. if we finally arrive at an augmenting path, we get a maximal flow in N by working backwards from t on this path
7. otherwise we arrive at a cut (S, T) and maxflow(N)= $C(S, T)$

11.3. Maximum Bipartite Matching

Let $G = (V, E)$ be an undirected graph. A **matching in** G is a subset M of edges of G no two of which have a vertex in common. If G is a bipartite graph, the vertex set V is partitioned as $V = L \cup R$ with $L \cap R = \varphi$, $L \neq \varphi$, $R \neq \varphi$ and every edge e in E has one vertex in L and the other in R. Thus no two vertices in L are adjacent and no two vertices in R are adjacent. Although it is not essential as a part of the definition of bipartite graph, we assume that a bipartite graph G has no isolated vertices. There are several combinatorial problems which can be considered as matching problems in bipartite graphs. To cite one example, we think of job assignment problem in which there are several jobs to be done and several qualified persons to do the jobs. Each person is competent to do some jobs but not necessarily all and each job needs to be assigned to some worker. No job is assigned to more than one worker and no worker is assigned more than one job. Let L denote the set of all workers and R the set of all jobs to be done. For each $a \in L$, introduce edges (a, x) where the x runs over all the jobs a is competent to do. This leads to a bipartite graph $G = (V, E)$ with $V = L \cup R$. The restrictions of the problem ensure that there are no isolated vertices. Moreover, a solution of the problem gives a matching in G.

A vertex $v \in V$ is said to be **matched** by matching M if some edge e in M is incident with v. If there is no edge in M which is incident with v, the vertex v is said to be **unmatched**. A matching M in G is called a **maximum matching** if for every matching M' in G, order $|M'|$ of $M' \leq$ order $|M|$ of M. The problem of finding a maximum matching in a bipartite graph can be solved by the method of finding a maximal flow in a flow network.

Let $G = (V, E)$ be a bipartite graph with $V = L \cup R$ with $L \cap R = \varphi$, $L \neq \varphi$, $R \neq \varphi$ and every edge in E being incident with one vertex in L and the other vertex in R. Consider two elements (or vertices), s called a source and t called a sink which are not in V. Consider a directed graph $G' = (V', E')$ where $V' = V \cup \{s, t\}$ and

$$E' = \{(s, u) | u \in L\} \cup \{(u, v) \mid u \in L, v \in R$$

and there is an edge in E incident with u and $v\} \cup \{(v, t) \mid v \in R\}$.

Here we are writing (u, v), as usual, for an edge with initial vertex u and terminal vertex v. Assign a unit capacity, $c(e) = 1$, to each edge e in E'.

Then the network $N = (G', c, s, t)$ is called **transportation network of the or associated with the bipartite graph** G.

We next consider a connection between matching in bipartite graphs and flows in associated networks. For the purpose of this connection and otherwise, a flow f on a flow network $N = (G, c, s, t)$ is called **integer valued** if $f(e)$ is an integer for every edge e of G.

Proposition 11.6. *Let $G = (V = L \cup R, E)$ be a bipartite graph and $N = (G', c, s, t)$ with $G' = (V', E')$ be the associated flow network. If M is a matching in G, then there is an integer-valued flow f on N with value $w(f) = |M|$. Conversely, if there is an integer-valued flow f on N, then there exists a matching M in G with order $|M|$ equal to the value $w(f)$ of the flow f.*

Proof. Let M be a matching in G. For every $u \in L$ for which there is an edge in M incident with u and for every $v \in R$ for which there is an edge in M incident with v, define $f(s, u) = f(v, t) = 1$. Also for every $u \in L$ and $v \in R$ such that there is an edge in M incident with u, v, define $f(u, v) = 1$. For all other edges in E', define $f(u, v) = 0$. For every $u \in L$, either there is exactly one $v \in R$ such that (u, v) is an edge in M or there is no such v. If there is a $v \in R$ with (u, v) in M, then $f(s, u) = f(u, v) = f(v, t) = 1$ otherwise $f(s, u) = f(u, v') = 0$ for every $v' \in R$. Also $f(e) \leq c(e)$ for every e in E'. Therefore f is an integer-valued flow on N. Also $(S = L \cup \{s\}, T = R \cup \{t\})$ is a cut of N. It is clear from the definition of f that $c(S, T) = |M|$. Hence the flow value $w(f) = |M|$.

Conversely, let f be an integer-valued flow on N. Let $M = \{(u, v) \mid u \in L, v \in R \text{ and } f(u, v) > 0\}$. The flow f being integer valued, whenever $f(u, v) > 0$, we have $f(u, v) = 1$ as the capacity of each edge in E is 1. For each vertex $u \in L$, there is exactly one edge (s, u) with u as terminal vertex. Therefore, flow entering u is either 0 or 1. If the flow entering u is 1, the same flow must leave u. Non-zero flow in any edge being 1, there must be exactly one $v \in R$ such that the unit flow entering u leaves along the edge (u, v). Similarly, if $v \in R$ and a non-zero flow leaves v along the edge (v, t), there is exactly one $u \in L$ such that $f(u, v) = 1$. Hence for every $u \in L$, there exists exactly one $v \in R$ and for every $v \in R$, there exists exactly one $u \in L$ such that $f(u, v) > 0$. Therefore, the set M defined above is a matching in G. The flow value of the flow f is given by

$$w(f) = \sum_{e^- = s} f(e)$$

$$= \sum_{u \in L} f(s, u), \text{ as } s \text{ is not adjacent to any } v \in R \text{ nor to } t$$

$$= \sum_{u \in L \text{ \& there exists } v \in V \text{ with } (u,v) \in M} f(s, u),$$

$$\text{as } f(s, u) = 0 \text{ if there is no } v \in V \text{ with } (u, v) \in M$$

$$= |M|.$$

$\square$

In order to find a maximum matching M in a bipartite graph G, we consider the corresponding flow network $N = (G', c, s, t)$ as above. Construct a maximal flow pattern on N using Ford–Fulkerson method of labeling. Then the flow pattern constructed is an integer-valued flow and finally construct a matching M in G for which $|M| = w(f)$. If M is not a maximum matching in G, let M' be a matching in G with $|M| < |M'|$. By the above proposition, we can find a flow f' on N with $|M'| = w(f')$. But then $w(f) < w(f')$ which is a contradiction to f being a maximal flow. Hence the matching M constructed is a maximum matching.

In a given bipartite graph, a maximum matching may not be unique. For example, consider the bipartite graph as given in Fig. 11.6. If there is a matching in G of cardinality 4, then in this matching both b, d have to be

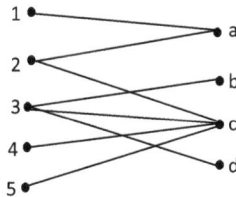

Fig. 11.6

adjacent to distinct vertices in $\{1, 2, 3, 4, 5\}$ which is not the case. Hence there is no matching in G of order 4. However

$$\{(1, a), (2, c), (3, b)\}, \{(2, a), (3, b), (4, c)\}, \{(2, a), (3, d), (5, c)\},$$

$$\{(1, a), (2, c), (3, d)\}, \{(2, a), (3, d), (4, c)\}$$

are some of the matching (but not all) of order 3. Hence each one of these is a maximum matching.

Exercise 11.1. Find all other maximum matching in the bipartite graph of Fig. 11.6.

We next consider another algorithm for finding a maximum matching in a graph $G = (V, E)$.

Let M be a matching in G. A path connecting two unmatched vertices in which alternate edges are in M is called an **augmented path** relative to M. Consider an augmented path $P = (e_1, e_2, \ldots, e_k)$ relative to M connecting unmatched vertices u as initial vertex of e_1 and v as terminal vertex of e_k. Since u, v are unmatched, e_1 and e_k are not in M. Then $e_2, e_4, \ldots,$ are in M. If k is even, then e_k is among the alternate edges $e_2, e_4, \ldots$ which are in M. This is a contradiction as e_k is not in M. Hence k is odd.

Suppose that e_3 is incident with vertices a, b and e_5 is incident with vertices c, d. Then a (say) is terminal vertex of e_2, b is initial vertex of e_4, c (say) is terminal vertex of e_4 and d is initial vertex of e_6. The edges e_2, e_4, e_6 being in M, it follows that no two of a, b, c, d are equal. In particular, e_3 and e_5 have no vertex in common. The same argument applies to every two edges in P which are not in M. Hence no two edges among $e_1, e_3, \ldots, e_k$ have a common vertex.

Observe that the set of vertices incident with $\{e_1, e_3, \ldots, e_k\} = $ the set of vertices incident with $\{e_2, e_4, \ldots, e_{k-1}\}$ together with u, v (both of which are unmatched by M).

Let $e = (a, b)$ be an edge in $M - P$. Since u, v are unmatched by M, we must have $\{a, b\} \cap \{u, v\} = \varphi$. If one of the vertices, say a, incident with e is incident with anyone of $e_1, e_3, \ldots, e_k$, then a is incident with one of $e_2, e_4, \ldots, e_k$, say e_{2l}. Then e, e_{2i} are both in M and have a common vertex. Thus a is not incident with any edge in $P - M$. Thus, e does not have a vertex common with any edge in $P - M$. It then follows that

$P \oplus M = (M - P) \cup (P - M)$ is a matching in G. Since $|M \cap P| = \frac{k-1}{2}$ we have

$$|M \oplus P| = |M| + |P| - 2|P \cap M| = |M| + k - 2\frac{k-1}{2} = |M| + 1.$$

We have thus proved the following result.

Proposition 11.7. *If M is a matching in a graph G and P is an augmenting path relative to M, then $M \oplus P$ is a matching in G and $|M \oplus P| > |M|$.*

Corollary 11.8. *A matching M in a graph G is a maximum matching if and only if there is no augmenting path relative to M.*

Observe that the empty set φ of edges in a graph $G = (V, E)$ is always a matching in G and so is any subset M of E containing exactly one edge. Using this observation and Corollary 11.8, we can now state our procedure for finding a maximum matching in a graph $G = (V, E)$.

Algorithm 11.3. Procedure for Finding a Maximum Matching

1. start with $M = \varphi$
2. find an augmenting path P relative to M and replace M by $M \oplus P$
3. repeat step 2 until no further augmenting path relative to M exists and M is a maximum matching in G

Example 11.6. Using the procedure given above, construct a maximum matching for the graph given in Fig. 11.7.

Solution. Consider a matching $M = \{e_4\}$ and unmatched vertices 3, 9. An augmented path with vertices 3, 9 relative to M is $P = \{e_7, e_4, e_5\}$ and $M \oplus P = \{e_5, e_7\} = M_1$ (say). An augmenting path relative to M_1 with $u = 4$, $v = 7$ is $P_1 = \{e_{10}, e_5, e_4, e_7, e_8\}$ and $M_2 = M_1 \oplus P_1 = \{e_4, e_8, e_{10}\}$. An augmenting path relative to M_2 with $u = 10$, $v = 1$ is $P_2 = \{e_{11}, e_{10}, e_5, e_4, e_7, e_8, e_2\}$ and $M_3 = M_2 \oplus P_2 = \{e_{11}, e_5, e_7, e_2\}$. The matching M_3 has order 4. The graph is bipartite with $V = L \cup R$, the subsets $L = \{1, 2, 3, 4, 5\}$, $R = \{6, 7, 8, 9, 10\}$ each having order 5. Therefore, there is no matching of order > 5, i.e., a maximum matching has order ≤ 5. The edges in M_3 are incident with vertices 4, 10; 2, 9; 3, 6;

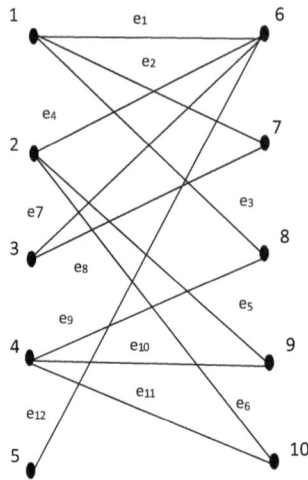

Fig. 11.7

1, 7. Therefore, any augmenting path relative to M_3 has to connect vertices 5 and 8. There being only one edge having 5 as an end vertex, we construct an augmenting path P_3 relative to M_3 by considering edge e_{12} which is incident with vertices 5, 6. Although there are three edges incident with vertex 6, there is only one edge e_7 which is in M_3. Therefore, we next take e_7 in P_3. There is only one other edge e_8 incident with the vertex 3 and e_8 is not in M_3. Therefore, we next put e_8 in P_3. The edge e_8 is incident with vertices 7, 3 and there is only one edge namely e_2 in M_3 which is incident with the vertex 7. So, we next put e_2 in P_3. The edge e_2 has vertex 1 as its other end vertex. There are two other edges e_1 and e_3 incident with the vertex 1. The edge e_1 is incident with vertices 1, 6 and e_3 is incident with vertices 1, 8. Neither of e_1, e_3 is in M_3. Since we need a path connecting vertices 5 and 8, we need to put e_3 in P_3. Thus the augmenting path P_3 relative to M_3 is $P_3 = \{e_{12}, e_7, e_8, e_2, e_3\}$. Then $M_4 = M_3 \oplus P_3 = \{e_{11}, e_5, e_{12}, e_8, e_3\}$ is a maximum matching in the graph.

Exercise 11.2. If M, N are matching in a graph G, then $M \oplus N$ has at least one augmenting path relative to M.

Proof of Proposition 11.6 is modeled on the proof of [5, Lemma 26.10].

Chapter 12

Polynomial Multiplication, FFT and DFT

The method of addition of two polynomials of degree n needs $n + 1$ additions of numbers while usual method of multiplying two such polynomials involves, in general, $(n + 1)^2$ multiplications and $2(1 + 2 + \cdots + n) = n(n + 1)$ additions of numbers. Thus multiplying two polynomials of degree n takes $\theta(n^2)$ time. However, using Finite Fourier Transforms (*FFT*) or more accurately discrete Fourier Transforms (*DFT*) reduces the time taken to $\theta(n \lg n)$. Fourier transforms and, so, the *FFT* have a wide variety of other applications such as in signal processing, image processing and host of other problems in electrical engineering. In this chapter, we discuss Fourier and Discrete Fourier Transforms.

12.1. Evaluation of a Polynomial

Consider a polynomial

$$a(x) = a_0 + a_1 x + \cdots + a_n x^n$$

of degree n in the variable x with the coefficients $a_0, a_1, \ldots, a_n$ real numbers (or integers). To find the value that the polynomial $a(x)$ takes when $x = v$ where v is a real number or an integer in the usual approach, we need to get $a_i v^i$ for $i = 0, 1, 2, \ldots, n$. Finally we add all these products to get the value $a(v)$ of the polynomial for $x = v$. This process involves $1 + 2 + \cdots + n = \frac{n(n+1)}{2}$ multiplications and n additions. We can reduce the number of multiplications considerably by adopting the following procedure. Let us evaluate the powers of v iteratively: $v^0 = 1$, $v^1 = v$, $v^2 = v.v$ and once we have obtained v^i, v^{i+1} can be evaluated by just one multiplication $v^{i+1} = v^i \cdot v$. Then the total number of multiplications involved for getting all the powers of v is n. Another n multiplications are involved in computing the products $a_i v^i$. Hence the total number of computations involved in finding $a(v)$ is $n + n = 2n$ multiplications and n additions. This process of straightforward evaluation of $a(v)$ may be given in the following algorithm:

Algorithm 12.1. Algorithm Straight Eval $(a(x), n, v)$

---

1. initially $r_0 = 1$, $s_0 = a_0$
2. for $i = 1$ to n, $r_i = r_{i-1} * v$, $s_i = s_{i-1} + a_i r_i$
3. return $s_n \ (= s)$

This algorithm needs $2n$ multiplications and n additions.

A great deal of improvement in computation on the straightforward evaluation of $a(v)$ is provided by Horner's method. For evaluating $a(v)$ by **Horner's method**, we rewrite the polynomial $a(x)$ as

$$a(x) = a_n x^n + a_{n-1} x^{n-1} + \cdots + a_1 x + a_0$$
$$= (\ldots ((a_n x + a_{n-1})x + a_{n-2})x + \cdots + a_1)x + a_0.$$

Then $a(v)$ is obtained in the following steps.

Set $s_0 = a_n$, $s_1 = a_n v + a_{n-1}$, $s_2 = s_1 v + a_{n-2}, \ldots, s_i = s_{i-1} v + a_{n-i}$.

As an algorithm, we have

Algorithm 12.2. Algorithm Horner $(a(x), n, v)$

---

1. initially set $s = a_n$
2. for $i = 1$ to n, set $s = sv + a_{n-i}$
3. return s_n

The number of computations involved for evaluation of $a(v)$ by Horner's method is n multiplications and n additions.

Given two polynomials $a(x) = a_0 + a_1 x + \cdots + a_n x^n$ and $b(x) = b_0 + b_1 x + \cdots + b_n x^n$ both of degree $\leq n$, finding the sum $a(x) + b(x)$ of these polynomials involves only $n + 1$ additions. The sum $a(x) + b(x) =$

$c(x) = c_0 + c_1 x + \cdots + c_n x^n$ where $c_i = a_i + b_i, 0 \le i \le n$ and we need only perform $n + 1$ additions.

12.2. Discrete Fourier Transforms and Polynomial Multiplication

12.2.1. Representation of polynomials

Given a polynomial $a(x) = a_0 + a_1 x + \cdots + a_n x^n$ of degree n, where one or more of the coefficients $a_0, a_1, \ldots, a_{n-1}$ could be zero, we can also represent as polynomial $(a_0, a_1, \ldots, a_n) = a$ and call such a representation as **coefficient representation** of the polynomial $a(x)$. We may also represent a polynomial in another form. For this, let $x_0, x_1, \ldots, x_n$ be constants (i.e., elements of the real or complex field) no two of which are equal. Let $y_i = a(x_i)$ be the value of the polynomial $a(x)$ evaluated for $x = x_i, 0 \le i \le n$. Thus we get $n + 1$ distinct points $(x_0, y_0), (x_1, y_1), \ldots, (x_n, y_n)$ (of the Euclidean plane when a_i and x_i are real numbers). Once $a(x)$ and distinct elements $x_0, x_1, \ldots, x_n$ are given, the points $(x_0, y_0), (x_1, y_1), \ldots, (x_n, y_n)$ with $y_i = a(x_i)$, $0 \le i \le n$, are uniquely determined and are called a **point-value representation** of the polynomial $a(x)$. Given $n + 1$ distinct points, does there always exist a unique polynomial for which the given points are a point-value representation of the polynomial? This question is answered in the affirmative by the following theorem.

Theorem 12.1. *For any set $\{(x_0, y_0), (x_1, y_1), \ldots, (x_n, y_n)\}$ of $n + 1$ points in which $x_0, x_1, \ldots, x_n$ are all distinct, there exists a unique polynomial $a(x)$ of degree $\le n$ such that $y_i = a(x_i)$ for $i = 0, 1, 2, \ldots, n$.*

Proof. Consider the square matrix

$$A = \begin{pmatrix} 1 & x_0 & x_0^2 & \cdots & x_0^n \\ 1 & x_1 & x_1^2 & \cdots & x_1^n \\ & & & \ddots & \\ 1 & x_n & x_n^2 & \cdots & x_n^n \end{pmatrix}$$

of order $n + 1$. The matrix A is a Vandermonde matrix and its determinant $\det A = \prod_{0 \le k < j \le n} (x_j - x_k)$. The x_j being all distinct, $\det A \ne 0$

and, so, A is an invertible matrix. The transpose y^t of the row vector $y = (y_0, y_1, \ldots, y_n)$ is an $(n + 1) \times 1$-matrix and, therefore, $A^{-1}y^t = a^t$ is again an $(n + 1) \times 1$-matrix where $a = (a_0, a_1, \ldots, a_n)$ is a row vector of length $n + 1$. Then $Aa^t = y^t$ shows that

$$\sum_{j=0}^{n} a_j x_i^j = y_i \quad \text{for all } i, 0 \le i \le n. \tag{12.1}$$

The relations (12.1) then show that $a(x_i) = y_i$ for all i, $0 \le i \le n$, where $a(x) = \sum_{j=0}^{n} a_j x^j$ is a polynomial of degree $\le n$. That the a_i and, so, the polynomial $a(x)$ are uniquely determined because of the matrix A being invertible. □

One of the advantages of the point-value representation of polynomials is evident in the case of multiplication of polynomials.

Let $a(x) = a_0 + a_1 x + \cdots + a_n x^n$, $b(x) = b_0 + bx + \cdots + b_n x^n$ be polynomials of degree n. Let $a(x)b(x) = c(x)$, where $c(x) = c_0 + c_1 x + \cdots + c_{2n} x^{2n}$ is a polynomial of degree $2n$. In the usual process of multiplication of polynomials, we know that

$$c_0 = a_0 b_0, \ c_1 = a_9 b_1 + a_1 b_0, \ c_2 = a_0 b_2 + a_1 b_1 + a_2 b_0, \ldots,$$

$$c_i = a_0 b_i + a_1 b_{i-1} + \cdots + a_{i-1} b_1 + a_i b_0, \ldots, c_{2n} = a_n b_n.$$

Therefore, in the computation of $a(x)b(x)$, there are

$$1 + 2 + \cdots + n + (n + 1) + n + (n - 1) + \cdots + 2 + 1$$
$$= (n + 1) + n(n + 1) = (n + 1)^2$$

multiplications and

$$1 + 2 + \cdots + n + (n - 1) + \cdots + 2 + 1 = n^2$$

additions involved.

Taking $a_{n+j} = 0 = b_{n+j}$ for $j = 1, 2, \ldots, n$, we may rewrite $a(x) = \sum_{j=0}^{2n} a_j x^j$, $b(x) = \sum_{j=9}^{2n} b_j x^j$ which are polynomials of degree $\le 2n$. Suppose that point-value representations of $a(x)$ and $b(x)$ are $(x_0, y_0), (x_1, y_1), \ldots, (x_{2n}, y_{2n})$ and $(x_0, z_0), (x_1, z_1), \ldots, (x_{2n}, z_{2n})$, respectively. Then

$$y_i = a(x_i), \ z_i = b(x_i) \quad \text{for} \quad 0 \le i \le 2n.$$

For $0 \le i \le 2n$, we have $c(x_i) = a(x_i)b(x_i) = y_i z_i$ and, therefore, the polynomial $c(x)$ has point-value representation

$$(x_0, y_0 z_0), (x_1, y_1 z_1), \ldots, (x_{2n}, y_{2n} z_{2n})$$

so that in this representation the product $a(x)b(x)$ involves only $2n + 1$ multiplications.

For obvious reasons, we define **degree-bound** of a polynomial $a(x) = \sum_{j=0}^{n} a_j x^j$, where a_n may or may not be zero, to be n. Then $a(x)$ is also a polynomial of degree bound $n + 1$ and so on.

The number of computations involved for multiplying two polynomials and many other problems is greatly reduced when the distinct numbers $x_0, x_1, \ldots, x_n$ in the point-value representation of a polynomial of degree-bound n are chosen judiciously.

12.3. Primitive Roots of Unity

Let n be an integer > 1. A complex number ω is called a **primitive nth root of unity** if $\omega^n = 1$ and $\omega^m \ne 1$ for any positive integer $m < n$. It is then immediate that $\omega^0 = 1, \omega, \omega^2, \ldots, \omega^{n-1}$ are all distinct and are nth roots of unity. Since $0 = \omega^n - 1 = (\omega - 1)(1 + \omega + \omega^2 + \cdots + \omega^{n-1})$ and $\omega \ne 1$, we have $\sum_{i=0}^{n-1} \omega^i = 0$. We can, infact, also have the following proposition.

Proposition 12.2. *For any p, $1 \le p \le n - 1$, $\sum_{i=0}^{n-1} \omega^{pi} = 0$.*

Proof. For any p, $1 \le p \le n - 1$, $\omega^p \ne 1$ so that $\omega^p - 1 \ne 0$. Then

$$\sum_{i=0}^{n-1} \omega^{pi} = \frac{\left\{ \left(\sum_{i=0}^{n-1} (\omega^p)^i \right) (\omega^p - 1) \right\}}{\omega^p - 1}$$

$$= \frac{\{(\omega^p)^n - 1\}}{\omega^p - 1} = \frac{(\{(\omega^n)^p - 1\})}{\omega^p - 1} = 0. \qquad \square$$

We also need the following simple properties of the primitive nth root ω of unity.

Suppose that $n = mk$, where m and k are integers both greater than 1. Then $1 = \omega^n = (\omega^k)^m$ showing that ω^k is an mth root of unity. If ω^k is not a primitive mth root, there exists an integer t, $1 < t < m$ such that $(\omega^k)^t = 1$, i.e., $\omega^{kt} = 1$ and $k < kt < mk = n$. This shows that ω is not a primitive nth root of unity. This contradiction proves that ω^k is a primitive mth root of unity.

When $k = 2$ so that $n = 2m$, then $1 = \omega^n = (\omega^m)^2$ and, therefore, $(\omega^m - 1)(\omega^m + 1) = 0$. Since $\omega^m \neq 1$, we have $\omega^m = -1$. Therefore, for any $j \geq 1$, $\omega^{m+j} = -\omega^j$. Also $\omega, \omega^2, \ldots, \omega^{2m-1}, \omega^{2m}$ being all distinct, $\omega^2, \omega^4, \ldots, \omega^{2(m-1)}$ are all distinct and for any j, $1 \leq j \leq m-1$, $(\omega^{2j})^m = (\omega^{2m})^j = 1$. Thus $\omega^0, \omega^2, \omega^4, \ldots, \omega^{2(m-1)}$ are all the distinct mth roots of unity.

We record these properties of the primitive nth root of unity in the following proposition.

Proposition 12.3. *If $n = mk$ with m, k integers both greater than 1, then*

(a) ω^k *is a primitive mth root of unity;*
(b) *when $k = 2$, then $\omega^{m+j} = -\omega^j$ for every $j \geq 1$;*
(c) *when $k = 2$, $\omega^0, \omega^2, \omega^4, \ldots, \omega^{2(m-1)}$ are all the distinct mth roots of unity.*

Observe that if ω is a primitive nth root of unity, then ω^j, j a positive integer, is a primitive nth root of unity if and only if $\gcd(j, n) = 1$. Observe that $e^{\frac{2\pi i}{n}} = \cos\frac{2\pi}{n} + i \sin\frac{2\pi}{n}$, where $i = \sqrt{-1}$, is a complex number and $(e^{\frac{2\pi i}{n}})^n = \cos 2\pi + i \sin 2\pi = 1$, so that $e^{\frac{2\pi i}{n}}$ is an nth root of unity. If for any integer k, $1 \leq k < n$, $(e^{\frac{2\pi i}{n}})^k = 1$, then $\cos\frac{2k\pi}{n} + i \sin\frac{2k\pi}{n} = 1$ which is not possible because $\sin\frac{2k\pi}{n} = 0$ if and only if $\frac{2k\pi}{n}$ is an integral multiple of π and $\cos\frac{2k\pi}{n} = 1$ if and only if $\frac{2k\pi}{n}$ is an integral multiple of 2π. Hence $\omega = e^{\frac{2\pi i}{n}}$ is a primitive nth root of unity and $\omega^k = e^{\frac{2k\pi i}{n}}$, $1 \leq k \leq n$ are all the nth roots of unity.

12.4. Discrete Fourier Transforms

Fourier transforms convert a continuous function, say $f(x)$, into another continuous function $g(x)$, say and the inverse Fourier transforms convert $g(x)$ into $f(x)$. However, we are not concerned with continuous functions.

Corresponding to the continuous Fourier transform there is the concept of discrete Fourier transform which converts a given set of points (numbers) into another set of the same number of points. Inverse discrete Fourier transform converts the new set of points back into the original points considered.

Let $a_0, a_1, \ldots, a_{n-1}$ be a given set of points. **Discrete Fourier Transform** for this set of points is defined by

$$A_j = \sum_{0 \le k \le n-1} a_k e^{\frac{2\pi ijk}{n}}, \; 0 \le j \le n-1 \tag{12.2}$$

and the a_j are then recovered by

$$a_k = \frac{1}{n} \sum_{0 \le j < n} A_j e^{\frac{-2\pi ijk}{n}}, 0 \le k \le n-1, \tag{12.3}$$

which is called the **Inverse Discrete Fourier Transform.**

We next prove as to how the a_j are recovered by the relation (12.3).

Let B be the square matrix (with $\omega = e^{\frac{2\pi i}{n}}$)

$$B = \begin{pmatrix} 1 & 1 & 1 & \cdots & 1 \\ 1 & \omega & \omega^2 & \cdots & \omega^{n-1} \\ 1 & \omega^2 & \omega^4 & \cdots & \omega^{2(n-1)} \\ & & & \ddots & \\ 1 & \omega^{n-1} & \omega^{2(n-1)} & \cdots & \omega^{(n-1)^2} \end{pmatrix}$$

of order n which is a Vandermonde matrix and so is invertible. The n equations in (12.2) are in the matrix form given by

$$B(a_0, a_1, \ldots, a_{n-1})^t = (A_9, A_1, \ldots, A_{n-1}). \tag{12.4}$$

The matrix B being invertible, we get

$$(a_0, a_1, \ldots, a_{n-1})^t = B^{-1}(A_0, A_1, \ldots, A_{n-1}). \tag{12.5}$$

Consider the square matrix

$$
C = \frac{1}{n}
\begin{pmatrix}
1 & 1 & \cdots & 1 & \cdots & 1 \\
1 & \omega^{-1} & \cdots & \omega^{-k} & \cdots & \omega^{-(n-1)} \\
 & & & \cdots & & \\
1 & \omega^{-j} & \cdots & \omega^{-jk} & \cdots & \omega^{-j(n-1)} \\
 & & & & \ddots & \\
1 & \omega^{-(n-1)} & \cdots & \omega^{-(n-1)k} & \cdots & \omega^{-(n-1)^2}
\end{pmatrix}.
$$

For $j, k = 0, 1, \ldots, n - 1$, the (j, k)th entry of the product BC is $\frac{1}{n} \sum_{r=0}^{n-1} \omega^{(j-k)r}$ which is 0 if $j \neq k$ (by Proposition 12.2) and is 1 if $j = k$. Thus BC is the identity matrix of order n. The matrix C also being invertible (being Vandermonde), we find that $C = B^{-1}$ or $B = C^{-1}$. Therefore, Eq. (12.5) becomes

$$
(a_0, a_1, \ldots, a_{n-1})^t = C(A_0, A_1, \ldots, A_{n-1}), \qquad (12.6)
$$

which is equivalent to the equations (12.3).

There is a close relationship between discrete Fourier transform and polynomial evaluation. If we consider a polynomial $a(x) = a_0 + a_1 x + \cdots + a_{n-1} x^{n-1}$, then observe that the discrete Fourier element A_j as in (12.2) is nothing but the evaluation of the polynomial $a(x)$ for $x = \omega^j$ and the inverse discrete Fourier element a_k as in (12.3) is the evaluation of the polynomial $\frac{1}{n} A(x)$ where $A(x) = A_0 + A_1 x + \cdots + A_{n-1} x^{n-1}$ for $x = \omega^{-k}$.

Thus, the discrete Fourier transform corresponds precisely to the evaluation of a polynomial at the n points $\omega^0, \omega^1, \omega^2, \ldots, \omega^{n-1}$, where $\omega = e^{\frac{2\pi i}{n}}$ and the inverse discrete Fourier transform corresponds to evaluation of a related polynomial at the n points $\omega^0, \omega^{-1}, \omega^{-2}, \ldots, \omega^{-(n-1)}$.

We can now derive a divide and conquer algorithm for the discrete Fourier transform.

Let n be an even positive integer, say $n = 2m$ where m is a positive integer. Let $a_0, a_1, \ldots, a_{n-1}$ be the n coefficients to be determined by discrete Fourier transform and let $a(x) = a_0 + a_1 x + \cdots + a_{n-1} x^{n-1}$. Break $a(x)$ into two parts one of which consists of the terms with even exponents and the other consists of the terms with odd exponents. Thus

$$a(x) = (a_0 + a_2x^2 + \cdots + a_{n-2}x^{n-2})$$
$$+ (a_1x + a_3x^3 + \cdots + a_{n-1}x^{n-1}).$$

Set $x^2 = y$. Then

$$a(x) = (a_0 + a_2y + \cdots + a_{n-2}y^{m-1})$$
$$+ (a_1 + a_3y + \cdots + a_{n-1}y^{m-1})x$$
$$= c(y) + xd(y), \tag{12.7}$$

where $c(y) = a_0 + a_2y + \cdots + a_{n-2}y^{m-1}$, $d(y) = a_1 + a_3y + \cdots + a_{n-1}y^{m-1}$ are polynomials in y of degree $\leq m - 1$. The values of the Fourier transforms are the values $a(\omega^j)$ for $0 \leq j \leq n - 1$ of $a(x)$ for the points ω^j, $0 \leq j \leq n - 1$. In view of (12.7), these values of $a(x)$ are expressible as

$$a(\omega^j) = c(\omega^{2j}) + d(\omega^{2j})\omega^j,$$
$$a(\omega^{j+m}) = c(\omega^{2j}) + d(\omega^{2j})\omega^{j+m}$$
$$= c(\omega^{2j}) - d(\omega^{2j})\omega^j,$$

(by Proposition 12.3(b))for $0 \leq j \leq m - 1$.

Since ω^{2j}, $0 \leq j \leq m - 1$, are the mth roots of unity (Proposition 12.3(c)), the problem reduces to finding finite Fourier transforms of two sets $\{a_0, a_2, \ldots, a_{n-2}\}$, $\{a_1, a_3, \ldots, a_{n-1}\}$ of points each having $\frac{n}{2} = m$ points. The two sub-problems we obtain are thus the evaluation of the polynomials $c(y)$ and $d(y)$ both of which are of degree $\leq m$. If m were also even, we could break each of the two sub-problems into two sub-problems. Thus, if n were a power of 2, we could continue the process of subdivision till we arrive at finding Fourier Transform of just one point b say. This then amounts to just evaluating a constant polynomial.

The above argument thus leads us to the following recursive Fast Fourier Transform (*FFT*).

Let $T(n)$ denote the running time for the *FFT* algorithm applied to an input sequence of length n. Since in step 5, the problem is reduced to

---

Algorithm 12.3. Algorithm FFT $(n, a, a(x), \omega, A)$

1. $n = 2^m$, $a = [0; n - 1]$ is the set of given n points a_i, $0 \leq i \leq n - 1$, ω is a primitive nth root of unity, $a(x) = a_0 + a_1 x + \cdots + a_{n-1} x^{n-1}$, $A[0; n - 1]$ is a set of values $a(\omega^j)$, $0 \leq j \leq n - 1$
2. c and d are polynomials; D, C and ω_p are complex arrays
3. if $n = 1$, then $a_0 = a_0$ and we are done
4. else $m = \frac{n}{2}$
5. $c(x) = a_0 + a_2 x^2 + \cdots + a_{n-2} x^{n-2}$

$$d(x) = a_1 x + a_3 x^3 + \cdots + a_{n-1} x^{n-1}$$

6. $FFT(m, c(x), c = (a_0, a_2, \ldots, a_{n-2}), \omega^2, C)$
7. $FFT(m, d(x), d = (a_1, a_3, \ldots, a_{n-1}), \omega^2, D)$
8. $\omega_p[-1] = \frac{1}{\omega}$
9. for $j = 0, 1, \ldots, m - 1$ set
10. $\omega_p[j] = \omega \times \omega_p[j - 1]$
11. $A[j] = C[j] + \omega_p[j] \times D[j]$
12. $A[j + m] = C[j] - \omega_p[j] \times D[j]$

---

solving two problems each of size $\frac{n}{2}$, we have

$$T(n) = 2T\left(\frac{n}{2}\right) + Dn \tag{12.8}$$

where D is a constant and Dn is bound on the time needed to form $c(x), d(x)$ and $A(x)$ using items 9–11. The algorithm is recursively applied till we arrive at a problem of length 1. For $n = 1$, we are given only one point a_0 so that $T(1)$ is just a constant, say d. Solving the recurrence relation (12.8), we get (for $n = 2^m$)

$$T(n) = 2T(\frac{n}{2}) + Dn = 2T(2^{m-1}) + D2^m$$

$$= 2^2 T(2^{m-2}) + 2D2^{m-1} + D2^m = 2^2 T(2^{m-2}) + 2D2^m$$

$$\vdots$$

$$= 2^m T(1) + mD2^m$$

$$= 2^m d + mD2^m = dn + Dmn = O(m2^m) = O(n \log_2 n).$$

We next apply the FFT for finding the time complexity of product of two polynomials $a(x), b(x)$ of degree $\leq n - 1$. By introducing zero coefficients, we can extend $a(x)$ and $b(x)$ to be polynomials of degree bound $2n - 1$, i.e.,

$$a(x) = a_0 + a_1 x + \cdots + a_{n-1} x^{n-1} + \cdots + a_{2n-1} x^{2n-1},$$
$$b(x) = b_0 + bx + \cdots + b_{n-1} x^{n-1} + \cdots + b_{2n-1} x^{2n-1}.$$

Using FFT, we represent the polynomials $a(x)$ and $b(x)$ in the point value representation as

$$a(x) = (x_0, y_0), (x_1, y_1), \ldots, (x_{2n-1}, y_{2n-1}), \text{ and}$$
$$b(x) = (x_0, z_0), (x_1, z_1), \ldots, (x_{2n-1}, z_{2n-1}),$$

where $x_j = \omega^{j+1}, 0 \leq j \leq 2n - 1$, with ω a primitive $2n$th root of unity. The time needed for either of the two conversions is $2n \log_2 2n$.

As already seen

$$a(x)b(x) = (x_0, y_0 z_0), (x_1, y_1 z_1), \ldots, (x_{2n-1}, y_{2n-1} z_{2n-1})$$

needing $2n$ multiplications. Now applying FFT to the $2n$ points $y_0 z_0, y_1 z_1, \ldots, y_{2n-1} z_{2n-1}$ we obtain the coefficients $c_0, c_1, \ldots, c_{2n-1}$ of the polynomial $c(x) = a(x)b(x)$ and, so, the product $c(x) = a(x)b(x)$ is obtained in time $2T(2n) + 2nd + 2T(2n) = 4T(2n) + 2nd = O(2n \log_2 2n)$. (Here $T(m) = O(m \log_2 m)$ is the complexity of FFT applied to a set of m points.)

Exercise 12.1.

1. Find the point value representation of the polynomial $a(x) = 5 - 2x + 3x^2 - 4x^3$ and of the reverse polynomial of $a(x)$, using the points $1, 2, 3$.
2. Find the product of the polynomials $a(x)$ as in exercise 1 above and $b(x) = 3 - 4x + 2x^2 - 4x^3 + x^5$ using their point-value representations with respect to the points $0, 1, 2, 3, 4$.
3. Compute the DFT of the vector $(1, 2, 3, 4)$.
4. If the DFT of a polynomial $a(x)$ is $A(x) = 1 - 2x + 3x^2 - 4x^3$, find $a(x)$.

The proof of Theorem 12.1 is based on the proof of [5, Theorem 30.1].

Chapter 13

String Matching

Formally, the string-matching or pattern-matching problem may be described as follows. Given a text T as an array $T[1 \ldots n]$ of length n and the pattern as an array $P[1 \ldots m]$ of length m, $m \leq n$, where the elements of both P and T are characters from the same alphabet $\sum$ (say), to find all s, $0 \leq s \leq n - m$ such that $T[s+1 \ldots s+m] = P[1 \ldots m]$. If for an s, $T[s+1 \ldots s+m] = P[1 \ldots m]$, we say that P **occurs** in T **with shift** s. If P occurs in T with shift s, we call s a **valid shift**, while otherwise we call s an **invalid shift**. The **string-matching problem** is thus to find all valid shifts with which the pattern P occurs in the text T.

For example, if $\sum = \{a, b, c\}$, $T = a\,b\,c\,a\,b\,b\,a\,c\,b\,a\,b\,b\,a\,c\,b$ and $P = a\,b\,b\,a$, then as shown in Fig. 13.1, the valid shifts are $s = 3$ and $s = 9$.

If T is as above and $P = a\,b\,a\,b$, then there is no valid shift for P in T.

In this chapter, we discuss four string-matching algorithms. But we first consider some notations and terminology.

13.1. Preliminaries

Let $\sum$ be an alphabet. We write $\sum^*$ for the set of all strings of finite length over $\sum$. We assume that the **empty string** denoted by ε is also in $\sum^*$. The **length of a string** x is denoted by $|x|$ so that $|\varepsilon| = 0$. If x, y are two

T	a	b	c	a	b	b	a	c	b	a	b	b	a	c	b
P				a	b	b	a			a	b	b	a		

Fig. 13.1

strings in $\sum^*$, then the concatenation of x, y denoted by xy is the string in which the elements of x are followed by the elements of y in order and $|xy| = |x| + |y|$. For example, if $x = a\,b\,a\,b$ and $y = c\,a\,b\,a\,a$, then $xy = a\,b\,a\,b\,c\,a\,b\,a\,a$.

We say that a string w is a **prefix** of a string x and express it by $w[x$ if there exists a string z in $\sum^*$ such that $x = wz$ and we say that w is a **suffix** of x denoted by $w]x$ if $x = zw$ for some z in $\sum^*$. Observe that ε is a prefix as well as a suffix of every string x.

The relations [and] on $\sum^*$ have the following properties.

Lemma 13.1. *Let* $x, y, z, w \in \sum^*$ *and* $a \in \sum$. *Then the following properties are true.*

(i) *If* $w[x,$ *or* $w]x,$ *then* $|w| \le |x|$.

(ii) $x]y$ *if and only if* $xa]ya$.

(iii) $x[y$ *if and only if* $ax[ay$.

(iv) *Both the relations* [*and*] *on* $\sum^*$ *are reflexive, antisymmetric and transitive. Neither of the two relations is symmetric.*

(v) *If* $x]z$, $y]z$ *and* $|x| \le |y|$, *then* $x]y$ *while if* $|x| = |y|$, *then* $x = y$. *The same is also true for the function* [.

Proof. (i) If $w[x$, there exists u in $\sum^*$ such that $x = wu$. Then, $|x| = |w| + |u| \ge |w|$ as $|u| \ge 0$.

The result for $w]x$ follows on similar lines.

(ii) $x]y$ if and only if there exists u in $\sum^*$ such that $y = ux$ which is true if and only if $ya = uxa$. The last relation is true if and only if $xa]ya$.

(iii) The proof is similar to that of (ii) above.

(iv) Since $x\varepsilon = x = \varepsilon x$, $x]x$ and $x[x$. Thus the relations [and] are reflexive.

Suppose that for the strings x, y in $\sum^*$ we have $x]y$ and $y]x$. Then there exist strings u, v in $\sum^*$ such that $y = ux$ and $x = vy$. Therefore

$y = ux = u(vy) = uvy$ and $|y| = |u| + |v| + |y|$ which implies that $|u| = |v| = 0$. Therefore $y = \varepsilon x = x$. Hence the relation] is antisymmetric. That the relation [is antisymmetric follows on similar lines.

Next suppose that $x]y$ and $y]z$. Then there exist u, v in $\sum^*$ such that $y = ux$ and $z = vy$. Thus $z = vy = v(ux) = (vu)x$ and it follows that $x]z$. Hence the relation] is transitive. That the relation [is transitive follows similarly.

Observe that $x]ax$ for $a \neq \varepsilon$ and, since $|ax| = |a| + |x| > |x|$, it follows from (i) above that ax is not a suffix of x. Therefore the relation] is not symmetric. That [is not symmetric can be proved similarly.

(v) Suppose that $x]z$, $y]z$ and $|x| \leq |y|$. There exist u, v in $\sum^*$ such that $z = ux$, $z = vy$. Then $|u| + |x| = |v| + |y|$ and, so, $|u| \geq |v|$. Let $u = u_1 \ldots u_k$ and $v = v_1 \ldots v_r$, where u_i, v_j are in $\sum$ and $k \geq r$. Then $u_1 \ldots u_k x = v_1 \ldots v_r y$ implies that $u_{r+1} \ldots u_k x = y$ which shows that $x]y$. If $|x| = |y|$, then $|u| = |v|$ and left cancellation in $u_1 \ldots u_k x = v_1 \ldots v_k y$ then implies that $x = y$. $\qquad \square$

Given an array $A[1 \ldots N]$, we denote the k-character prefix $A[1 \ldots k]$ of A by A_k. With this notation, observe that for the pattern $P[1 \ldots m]$ and text $T = T[1 \ldots n]$, P_k is the prefix $P[1 \ldots k]$ of P and T_k is the prefix $T[1 \ldots k]$ of T. Thus, with this notation, *the chain-matching problem requires finding all shifts* $0 \leq s \leq n - m$, *such that* $P]T_{s+m}$.

As in every algorithm, to compute the time complexity of a chain-matching algorithm we need to have a procedure or idea of how to compute the time taken for comparing characters or alphabets of strings.

We regard two equal length strings to be compared for equality as a primitive operation (i.e., an operation which cannot be described in terms of other operations like axioms in mathematics). If two strings x and y are compared from left to right and the comparison stops when a mismatch is discovered, we assume that the time taken by such a match is a constant multiple of (1+ the number of matching characters discovered). The '1' is added because if k matching characters are discovered, the comparison of the $(k + 1)$th character also takes a constant amount of time, i.e., all told $k + 1$ comparisons of characters have been made. In notational form if z is a largest string of length t (say) such that $z[x$ and $z[y$, then the test "$x = y$" is assumed to take time $\theta(t + 1)$.

13.2. The Naïve String-Matching Algorithm

The naïve string-matching procedure finds all possible valid shifts s by checking whether the equality $P = T[s+1 \ldots s+m]$ holds for all possible values of $s = 0, 1, \ldots, n - m$. For each s, $0 \le s \le n - m$, the checking of equality $P = T[s+1 \ldots s+m]$ involves at most m character comparisons. Therefore the checking of equality $P = T[s+1 \ldots s+m]$ for all s involves at most $(n - m + 1)m$ character comparisons. Since each comparison of two characters one from P and the other from T takes a constant amount of time, the running time of this algorithm is $\theta((n - m + 1)m)$. The number of character comparisons is the largest possible namely $(n - m + 1)m$ when $P = a \ldots a = a^m$ and $T = a \ldots a = a^n$ and, so, the above bound is tight in this case. The naïve method does not make use of any special characteristics of P and T. Formally, we may state the naïve string-matching algorithm as below:

Algorithm 13.1. Naïve String-Matching Algorithm (P, T)

1. n = length of text T
2. m = length of pattern P
3. for $s = 0$ to $n - m$
4. check if $P[1 \ldots m] = T[s + 1 \ldots s + m]$
5. print s i.e., s is a valid shift if the answer is yes in 4
6. otherwise move to next s or $s = s + 1$ in 4

As observed above the worst case running time of the naïve string-matching procedure is $\theta((n - m + 1)m)$. It may also be observed that the only case where the number of character comparisons is precisely $(n - m + 1)m$ is when $P = a^m$ and $T = a^n$ for some a in $\sum$.

Graphically, the naïve string-matching procedure may be represented by initially placing P above T so that $P[1]$ is above $T[1]$ and $P[m]$ is above $T[m]$. The corresponding elements of P and T are then compared.

The comparison stops when either $s = 0$ is a valid shift or a mismatch is found. In either case slide the pattern P one step to the right. Then compare the corresponding elements of P and T. Once comparison stops when either a valid shift is obtained or a mismatch occurs, slide P to the right by one step. Continue the process till P is above $T[n - m + 1 \ldots n]$ to determine whether $s = n - m$ is a valid shift or not. We now consider an example to indicate this process.

Example 13.1. Let $P = a\,b\,a$ and $T = a\,c\,a\,b\,a\,c\,a\,b\,a$. The graphical representation of the naïve string-matcher is then as given in Fig. 13.2.

Two matching characters of P and T are shown by drawing a straight vertical line between the two while a mismatch is indicated by drawing a dotted line between the two. We find that $s = 2, 6$ are the only valid shifts in this case.

Example 13.2. Show the comparisons the naïve string-matcher makes for the pattern $P = 001$ in the text $T = 000100101001$.

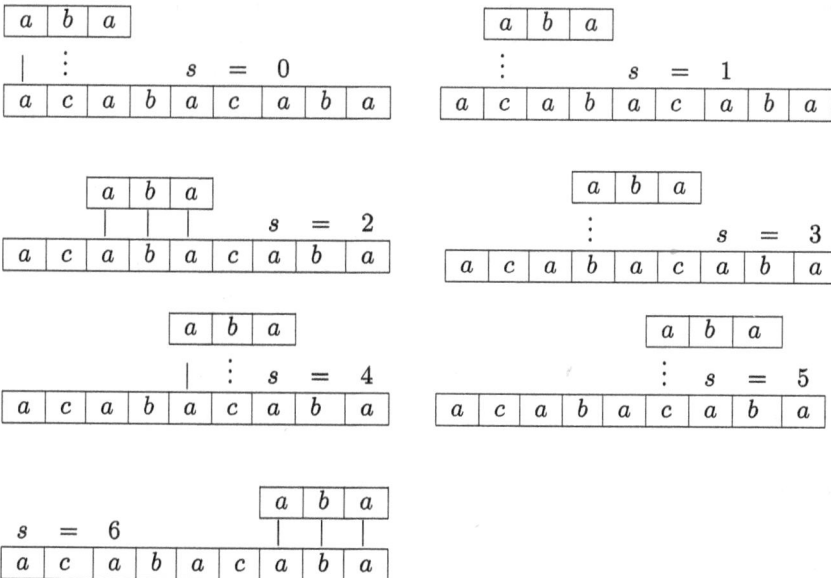

Fig. 13.2

Table 13.1

s	0	1	2	3	4	5	6	7	8	9
Number of comparisons	3	3	2	1	3	2	1	2	1	3

Solution. The comparisons made by the naïve string-matcher in this case are given in Table 13.1.

Thus the total number of comparisons made is $3 + 3 + 2 + 1 + 3 + 2 + 1 + 2 + 1 + 3 = 21$ which is less than $m(n - m + 1) = 3(12 - 3 + 1) = 30$.

Example 13.3. If all characters in the pattern P are distinct, show how to accelerate naïve string-matcher to run in time $\theta(n)$ on an n character text T.

Solution. Let T be a text string of length n and P be a pattern of length m with all characters distinct. Assume P written above T so that $P[1]$ is above $T[1]$. If $T[1] \neq P[1]$, slide P to the right one step at a time till we arrive at the first character from the left which equals $P[1]$. Let $T[r_1]$ be the first entry of T which equals $P[1]$. Let $s_1 (\leq m)$ be the largest positive integer such that $T[r_1 \ldots r_1 + s_1 - 1] = P_{s_1}$ but $T[r_1 + s_1] \neq P[s_1 + 1]$. Since all the characters in P are distinct, $T[r_1 + i] \neq P[1]$ for all i, $1 \leq i \leq s_1 - 1$. Therefore, we need to slide P to the right not just by one step but s_1 steps so that $P[1]$ now appears above $T[r_1 + s_1]$. Now using naïve string-matcher, continue sliding P one step to the right till we arrive at the first character in $T[r_1 + s_1 \ldots n]$ which equals $P[1]$. Let the first such character in $T[r_1 + s_1 \ldots n]$ be $T[r_2]$, i.e., $T[r_2] = P[1]$. Now comparing the corresponding characters in $T[r_2 \ldots n]$ with those of P we get $T[r_2 \ldots r_2 + s_2 - 1] = P_{s_2}$ for some s_2, $m \geq s_2 \geq 1$ but $T[r_2 + s_2] \neq P[s_2 + 1]$. Also, once again, $T[r_2 + i] \neq P[1]$ for all i, $1 \leq i \leq s_2 - 1$ and, therefore, we need to slide P to the right not just by one step at a time but straightaway by s_2 steps so that $P[1]$ appears above $T[r_2 + s_2]$. Now find a first r_3 such that $T[r_3] = P[1]$ in the string $T[r_2 + s_2 \ldots n]$ and continue the above process. Let r_k be the largest integer for which $T[r_k] = P[1]$. Let $s_k \geq 1$ be such that $T[r_k \ldots r_k + s_k - 1] = P_{s_k}$ but $T[r_k + s_k] \neq P[s_k + 1]$. In this fashion we find all the substrings of T namely $T[r_1 \ldots r_1 + s_1 - 1]$, $T[r_2 \ldots r_2 + s_2 - 1], \ldots, T[r_k \ldots r_k + s_k - 1]$ which equal $P_{s_1} P_{s_2}, \ldots, P_{s_k}$. Observe that these substrings of T are non-overlapping. Therefore, the total number of matching characters is $s_1 + s_2 + \cdots + s_k$ which is $\leq n$. Even accounting for the characters of T which have been compared with $P[1]$ but not found equal to it, for example, $T[1], \ldots, T[r_1 - 1], T[r_1 + s_1], \ldots, T[r_2 - 1], \ldots, T[r_{k-1} + s_{k-1}], \ldots, T[r_k - 1], T[r_k + s_k], \ldots, T[n]$, the total number of comparisons made is at most n, the length of T. Hence the running time of this improvement of the naïve string-matching procedure is $\theta(n)$ (or so to say $O(n)$).

Remark. The hypothesis that all characters in the pattern P are different in the above example is significant. For example, if $P = aab$ and $T = abaaabbaab$, then there are two valid shifts namely 3 and 7. Comparison of characters using naive string-matcher leads to the following matching strings:

$$T_1 = P_1, \; T[3 \ldots 4] = P_2, \; T[46] = P, \; T[5] = P_1, \; T[8 \ldots 10]$$
$$= P, \; T[9] = P_1$$

and the sum of the number of characters in the matching strings is $1 + 2 + 3 + 1 + 3 + 1 = 11 > 10$ — the length of T.

13.3. The Rabin–Karp Algorithm

Rabin and Karp proposed a string-matching algorithm with running time $\theta(m + n)$ although this algorithm needs a preprocessing time $\theta(m)$. However, its worst case running time is the same as that of the naïve string-matcher which is also known as the **Brute-force algorithm**. The Rabin–Karp algorithm uses quite a few results from elementary number theory. We regard elementary arithmetic operations such as multiplication, division, computing remainders as primitive operations needing, say, unit amount of time. Thus, for estimating the running time of an algorithm we need to count the number of such operations that the algorithm makes. Addition of two numbers takes much less time in comparison to their multiplication and as such the time taken by addition of two numbers is ignored. We first consider a couple of results from number theory which we need frequently.

Consider a sequence $A = a_1 a_2 \ldots a_m$ of length m where a_i are numbers from the set $[0, 1, \ldots, 9\}$ of decimal digits. Corresponding to the sequence A, we get an m digit number a (say) in the decimal system. Observe that

$$a = a_1 10^{m-1} + a_2 10^{m-2} + \cdots + a_{m-1} 10 + a_m$$
$$= a_m + 10(a_{m-1} + 10(a_{m-2} + 10(a_{m-3} + \cdots + 10(a_2 + 10a_1))\ldots)$$
$$(13.1)$$

and the number of multiplications involved is $m - 1$ in addition to $m - 1$ additions involved. Thus we can compute a from A in $\theta(m)$ time.

Again consider the sequence $A = a_1 a_2 \ldots a_m$ and let A' be the sequence $a_2 a_3 \ldots a_{m+1}$ of length m obtained from A by deleting the leftmost element and introducing a new element $a_{m+1} = a_m + 1$ on the right. Then the value a' of A' can be computed from the value a of A in constant time by using

$$a' = 10(a - 10^{m-1} a_1) + a_{m+1}. \qquad (13.2)$$

This is so because

$$a' = a_2 10^{m-1} + a_3 10^{m-2} + \cdots + a_m 10 + a_{m+1}$$
$$= 10(a_2 10^{m-2} + a_2 10^{m-3} + \cdots + a_m) + a_{m+1}$$
$$= 10(a_1 10^{m-1} + a_2 10^{m-2} + \cdots + a_{m-1} 10 + a_m - a_1 10^{m-1}) + a_{m+1}$$
$$= 10(a - a_1 10^{m-1}) + a_{m+1}.$$

Observe that if $m = 5$ and a is the number 31456, then the number $a' = 14567$ can be obtained as follows:

$$a = 3.10^4 + 1.10^3 + 4.10^2 + 5.10 + 6$$
$$a' = 1.10^4 + 4.10^3 + 5.10^2 + 6.10 + 7$$
$$= 10(3.10^4 + 1.10^3 + 4.10^2 + 5.10 + 6 - 3.10^4) + 7$$
$$= 10(a - 30000) + 7$$

which involves only one multiplication but two additions. Also, apparently, the computation of a from A involves $4 + 3 + 2 + 1 = 10$ multiplications in addition to four additions. However, we may write

$$a = 6 + 10(5 + 4.10 + 1.10^2 + 3.10^3) = 6 + 10(5 + 10(4 + 1.10 + 3.10^2))$$
$$= 6 + 10(5 + 10(4 + 10(1 + 3.10)))$$

which involves only four multiplications in addition to four additions.

Lemma 13.2. *Given integers* $a > 1$, $m > 1$, *the power* a^m *can be obtained in* $\theta(\log m)$ *time, where* $\log m$ *denotes* $\log_2 m$.

Proof. There exists an integer $k \geq 1$ such that $2^k \leq m < 2^{k+1}$. Then

$$m = 2^k + b_1 2^{k-1} + b_2 2^{k-2} + \cdots + b_{k-1}2 + b_k,$$

where for every i, $1 \leq i \leq k$, $b_i = 0$ or 1. This is equivalent to saying that $b_0 b_1 b_2 \ldots b_k$ with $b_0 = 1$ is the binary representation of the number m. We can compute the powers

$$a_1 = a^2, \; a_2 = a_1^2 = a^{2^2}, \ldots, a_k = a_{k-1}^2 = a^{2^k}$$

by squaring the number obtained last. Thus a^{2^k} is computed by using only k multiplications. Then

$$a^m = a^{2^k} \cdot (a^{2^{k-1}})^{b_1} \cdots (a^2)^{b_{k-1}} \cdot a^{b_k}.$$

This involves k multiplications in addition to the k multiplications used to obtain a^{2^k}. Thus a^m can be computed using $2k$ multiplications. Therefore a^m can be obtained in $\theta(2k)$ time. Now $2^k \leq m < 2^{k+1}$ implies that $k \leq \log m$ so that $2k \leq 2 \log m$. But $\theta(2k) = \theta(k)$, and, so, a^m can be computed in $\theta(\log m)$ time. $\qquad\qquad\square$

Before we discuss Rabin–Karp algorithm, let us observe that two sequences of the same length match if and only if they represent the same number. Also we may point out that we have considered above using decimal digits and the corresponding numbers only for simplicity. We could have discussed these results using any $d > 1$ as radix, i.e., the sequences are over the alphabet $\sum = \{0, 1, 2, \ldots, d - 1\}$.

Let $P[1 \ldots m]$ be a pattern of length m and $T[1 \cdots n]$ be a text of length n. Let p denote the value determined by P and let $t_0, t_1, \ldots, t_{n-m}$ be the values determined by the sequences T_m, $T[2 \cdots m+1]$, $T[3 \cdots m+$

2], ..., $T[s+1 \ldots s+m]$, $\cdots$, $T[n-m+1 \cdots n]$, respectively. We suppose that P and T are sequences over the alphabet $\sum = \{0, 1, 2, \ldots, d-1\}$, i.e., the characters are radix-d digits where $d > 1$. The numbers p and t_0 are pre-computed in time $\theta(m)$, by (13.1), i.e.,

$$p = P[m] + d(P[m-1] + d(P[m-2] + \cdots + d(P[2] + dP[1]) \ldots),$$

$$t_0 = T[m] + d(T[m-1] + d(T[m-2] + \cdots + d(T[2] + dT[1]) \ldots).$$

$$(13.3)$$

Observe that these are recursive computations.

We first compute $P[2]+dP[1]$, then compute $P[3]+d(P[2]+dP[1])$, then $P[4]+d(P[3]+d(P[2]+dP[1]))$, ... At each step of recursion only one multiplication and one addition are needed. Thus, as already seen the total number of arithmetic operations needed are $m-1$ multiplications and $m-1$ additions. Similarly for t_0. Once t_0 is computed, $t_1, t_2, \ldots, t_{n-m}$ are recursively obtained by using (13.2), each recursion requiring a constant number of arithmetic operations. Thus $t_1, t_2, \ldots, t_{n-m}$ are computed in $\theta(n-m)$ time. The numbers $t_0, t_1, t_2, \ldots, t_{n-m}$ are compared with p. If $t_s = p$ for an s, $0 \le s \le n-m$, then s is a valid shift because then the sequences P and $T[s+1 \ldots s+m]$ match. Otherwise s is an invalid shift. This way all the valid shifts are obtained in $\theta(n-m)$ time together with $\theta(m)$ preprocessing time. The key observation in this is that the sequences P and $T[s+1 \ldots s+m]$ match if and only if $p = t_s$.

However, there is a difficulty in this procedure because the numbers p and t_s may be very large and working with these numbers may not be convenient. To overcome such and other related difficulties a suitable number q (usually a prime) is chosen a modulus and the numbers $p, t_0, t_1, t_2, \ldots, t_{n-m}$ are all taken modulo q. The numbers p and t_s may be congruent modulo q but may not be equal. For example $2 \equiv 15 \mod 13$ but $2 \ne 15$ whereas if a and b are equal then $a \equiv b \mod q$.

Consider a radix-d sequence $A = a_1 a_2 \ldots a_m$ of length m. Then the number a given by the sequence A is given by

$$a = a_m + a_{m-1}d + a_{m-2}d^2 + \cdots + a_2 d^{m-2} + a_1 d^{m-1}.$$

Let $b_1 = a_1$ and for $i \geq 2$, $b_i \equiv a_i + db_{i-1} \bmod q$. Suppose that

$$b_i \equiv a_i + a_{i-1}d + a_{i-2}d^2 + \cdots + a_2d^{i-2} + a_1d^{i-1} \bmod q. \text{ Then}$$

$$b_{i+1} \equiv a_{i+1} + db_i \equiv a_{i+1} + d(a_i + a_{i-1}d + a_{i-2}d^2 + \cdots$$

$$+ a_2d^{i-2} + a_1d^{i-1}) \bmod q$$

$$= a_{i+1} + a_id + a_{i-1}d^2 + \cdots + a_2d^{i-1} + a_1d^i.$$

This completes induction and, therefore, $b_m \equiv a \bmod q$.

In the computation of each b_i, $i \geq 2$, three elementary arithmetic operations namely multiplication, addition and finding remainder on division by q are involved. Therefore, for computation of b_m exactly $3(m-1)$ such operations are involved. Hence $a \bmod q$ is pre-computed in $\theta(m)$ time. In particular $p \bmod q$ and $t_0 \bmod q$ are pre-computed in $\theta(m)$ time each. Since $t_{s+1} = d(t_s - T[s+1]d^{m-1}) + T[s+m+1]$, $t_{s+1} \bmod q$ equals $(d(t_s \bmod q - T[s+1]d^{m-1}) + T[s+m+1]) \bmod q$.

Therefore, for every $s \geq 0$, $t_{s+1} \bmod q$ can be obtained from $t_s \bmod q$ by using a (fixed) constant number of arithmetic operations. Hence $t_1 \bmod q, t_2 \bmod q, \ldots, t_{n-m} \bmod q$ can be obtained in $\theta(n-m)$ time. Next we compare $t_s \bmod q$ with $p \bmod q$, i.e., find all $s \geq 0$ for which $t_s \equiv p \pmod q$. Those values of s for which congruence is true are the possible valid shifts for the pattern matching but not necessarily valid shifts. If $t_s \equiv p \bmod q$, we then compare the strings P and $T[s+1 \cdots s+m]$. Only those s for which $t_s \equiv p \bmod q$ and $P = T[s+1 \cdots s+m]$ are the valid shifts and the other s for which $t_s \equiv p \bmod q$ but $P \neq T[s+1 \cdots s+m]$ are spurious hits.

There are thus two parts of the Rabin–Karp chain-matcher. One when the numbers involved are not very large and we deal with p and t_s as they are and the other when the numbers involved are large and we work with $p \bmod q$ and $t_s \bmod q$, where q is a suitably chosen modulus. We now state the Rabin–Karp algorithm only for the second case.

Example 13.4. Working *modulo* $q = 11$, how many superiors hits does the Rabin–Karp string-matcher encounter in the text $T = 3\,1\,5\,1\,4\,2\,9\,5\,6$ $3\,5\,8\,9\,7\,9\,3$ when looking for the pattern (a) $P = 2\,6\,7$; (b) $P = 2\,9$; (c) $P = 5\,1\,4$?

Algorithm 13.2. Rabin–Karp String-Matcher $(P, T, \ d, q)$

1. $n =$ length $|T|$ of the text string T
2. $m =$ length $|P|$ of the pattern P
3. $d =$ radix of number sequences
4. $q =$ chosen modulus
5. $h \equiv d^{m-1} \bmod q$
6. $p = 0$
7. $t_0 = 0$
8. for $i = 1$ to m,
9. do $p = dp + P[i] \bmod q$
10. $t_0 = dt_0 + T[i] \bmod q$ (we compute $p \bmod q$ and $t_0 \bmod q$ by using (13.3))
11. for $s = 0$ to $n - m$
12. do if $p = t_s$
13. then if $P[1 \cdots m] = T[s + 1 \cdots s + m]$
14. then print s (i.e., s is a valid shift)
15. if $s < n - m$
16. then $t_{s+1} = d(t_s - T[s + 1]d^{m-1}) + T[s + m + 1] \bmod q$ (i.e., s is checked for valid shift, then $s + 1$ is checked and so on till $s = n - m$)

Solution. (a) $p = 200 + 60 + 7 \equiv 2 - 6 + 7 = 3 \ (\bmod q)$. This is so because $q = 11$ and, so, $10^{m-1} = 10^2 \equiv 1 \ (\bmod q)$. Also $10 \equiv -1 \ (\bmod q)$. Now, *modulo q*

$$t_0 = 315 \equiv 3 - 1 + 5 = 7 \quad t_7 = 563 \equiv 5 - 6 + 3 = 2$$

$$t_1 = 151 \equiv 1 - 5 + 1 \equiv 8 \quad t_8 = 635 \equiv 6 - 3 + 5 = 8$$

$$t_2 = 514 \equiv 5 - 1 + 4 = 8 \quad t_9 = 358 \equiv 3 - 5 + 8 = 6$$

$$t_3 = 142 \equiv 1 - 4 + 2 \equiv 10 \quad t_{10} = 589 \equiv 5 - 8 + 9 = 6$$

$$t_4 = 429 \equiv 4 - 2 + 9 \equiv 0 \quad t_{11} = 897 \equiv 8 - 9 + 7 = 6$$

$$t_5 = 295 \equiv 2 - 9 + 5 \equiv 9 \quad t_{12} = 979 \equiv 9 - 7 + 9 \equiv 0$$

$$t_6 = 956 \equiv 9 - 5 + 6 = 10 \quad t_{13} = 793 \equiv 7 - 9 + 3 = 1$$

Since none of the $t_s \equiv 3 \pmod q$, there is no hit, spurious or real.

(c) $p = 514 \equiv 5 - 1 + 4 = 8$.

The t_s as computed in (a) above work in this case as well. Since $t_1 = t_2 = t_8 \equiv 8 \equiv p \pmod q$, there are three possible hits. Observe that $T[234] = 151 \neq P$, $T[91011] = 635 \neq P$ but $T[345] = 514 = P$. Therefore, out of the three possible hits, $s = 2$ is a valid shift while the other two $s = 1, 8$ are spurious hits.

(b) Here $p = 29 \equiv 7 \pmod q$. Also we have to rework the t_s in this case. *Modulo q,*

$t_0 = 31 \equiv -3 + 1 \equiv 9$	$t_5 = 29 \equiv -2 + 9 = 7$	$t_{10} = 58 \equiv -5 + 8 = 3$
$t_1 = 15 \equiv -1 + 5 \equiv 4$	$t_6 = 95 \equiv -9 + 5 \equiv 7$	$t_{11} = 89 \equiv -8 + 9 = 1$
$t_2 = 51 \equiv -5 + 1 \equiv 7$	$t_7 = 56 \equiv -5 + 6 = 1$	$t_{12} = 97 \equiv -9 + 7 \equiv 9$
$t_3 = 14 \equiv -1 + 4 \equiv 3$	$t_8 = 63 \equiv -6 + 3 \equiv 8$	$t_{13} = 79 \equiv -7 + 9 = 2$
$t_4 = 42 \equiv -4 + 2 = 9$	$t_9 = 35 \equiv -3 + 5 = 2$	$t_{14} = 93 \equiv -9 + 3 \equiv 5$

Observe that $t_2 \equiv t_5 \equiv t_6 \equiv p \bmod q$ and, so, there are three possible hits. We have $T[34] = 51 \neq P$, $T[78] = 95 \neq P$ while $T[67] = 29 = P$. Thus, out of the three possible hits only one namely $s = 5$ is a valid shift while the other two namely $s = 2, 6$ are spurious hits.

Exercise 13.1. Working *modulo q* $= 13, 17$, how many spurious hits does the Rabin–Karp string-matcher encounter in the text T when looking for the pattern P, where P and T are as in Example 13.4.

Remark. When $P = a^m$, i.e., a repeated m times and $T = a^n$, then for every s, $0 \leq s \leq n - m$, $t_s \equiv p$ and, so, every $T[s + 1 \cdots s + m]$ needs to be compared with P. Comparison of $T[s + 1 \cdots s + m]$ with P gives m matching pairs and, therefore, the running time of the algorithm is $\theta((n - m + 1)m)$. This is the worst case running time of the Rabin–Karp matcher.

The Rabin–Karp method can be extended to the problem of matching a text string for the occurrence of any one of a given set of k patterns. We consider here the case $k = 2$. The method can then be extended to the case of any k quite easily. In fact the case $k = 2$ is quite indicative of the general case.

Let T be a text string of length n and P, P' be two patterns of length m, m', respectively. The string and the patterns are taken over radix-d digits and q is the chosen modulus. Let p, p' be the numbers determined by the strings P, P', respectively. Also, as before, t_s and t'_s are the numbers determined by the sequences $T[s+1 \ldots s+m]$ and $T[s+1 \ldots s+m']$, respectively.

Case (i). $m' = m$, i.e., the two patterns P, P' are of the same length. Then $t'_s = t_s$ for all $s \geq 0$. Find all $s \geq 0$ such that

$$t_s \equiv p \bmod q \quad \text{or} \quad t_s \equiv p' \bmod q.$$

If $t_s \equiv p \bmod q$, check if $T[s+1 \ldots s+m] = P$ and if $t_s \equiv p' \bmod q$, check if $T[s+1 \ldots s+m] = P'$. Observe that $T[s+1 \ldots s+m]$ cannot match P and P' both. If $T[s+1 \ldots s+m] = P$ or $T[s+1 \ldots s+m] = P'$, then s is a valid shift for the patterns P, P'. Otherwise even though $t_s \equiv p \bmod q$ or $t_s \equiv p' \bmod q$ may be true, s is a spurious hit for the patterns P, P'. This is equivalent to saying that

The set of valid shifts for P, P' is $\{s \geq 0 \mid s$ is a valid shift for P or s is a valid shift for $P'\}$.

Case (ii). $m' \neq m$. We may suppose that $m' > m$. Let $m' - m = k \geq 1$. Let $P'' = P'_m$. Then P, P'' are patterns of equal length. Any valid shift for P' will surely be a valid shift for P''. Let p'' be the number associated with the sequence P''. Then p' can be computed from p'' by

$$p' = d^k p'' + d^{k-1} P'[s+m+1] + \cdots + d P'[s+m'-1] + P'[s+m']$$

using only a finite number of arithmetic operations. For an s, $0 \leq s \leq n - m'$, the numbers t'_s determined by the sequence $T[s+1 \ldots s+m']$ can be computed from t_s similarly by

$$t'_s = d^k t_s + d^{k-1} P'[s+m+1] + \cdots + d P'[s+m'-1] + P'[s+m'].$$

Let s be a valid shift for P'' with $s \leq n - m'$. Then of course $t_s \equiv p'' \bmod q$. If $t'_s \equiv p' \bmod q$ is not true, ignore s. Otherwise, if $t'_s \equiv p' \bmod q$, check if $T[s+1 \ldots s+m'] = P'[m+1 \ldots m']$. If yes, s is a valid shift for P''. Hence, the set of valid shifts for P, P' is $\{s \mid 0 \leq s \leq n - m$ and s is a valid shift for P or $0 \leq s \leq n - m'$ and s is a valid shift for $P'\}$.

13.4. Pattern-Matching with Finite Automata

Definition. A deterministic finite-state machine without output also called a **deterministic finite-state automaton** $M = (S, s_0, \Sigma, A, \delta)$ consists of

1. a finite set S of states,
2. $s_0 \in S$ an initial or starting state,
3. Σ a finite set of input alphabet,
4. a function $\delta : S \times \Sigma \to S$ called a state-transition function,
5. a subset A of S consisting of accepting or final states.

In case the state-transition function is replaced by a transition function δ from $S \times \Sigma$ such that $\delta(s, a)$ for $s \in S$, $a \in A$, is a subset of S, the resulting machine is called a **non-deterministic finite-state automaton**. Since in this section we are concerned with only deterministic finite-state automaton, we drop the word deterministic and refer to it simply as a finite-state automaton. Initial state s_0 means the state in which the machine is before receiving any input. If at any instant the machine is in state s, then $\delta(s, a)$ is the state in which the machine is on receiving an input $a \in \Sigma$.

We can extend the state-transition function $\delta : S \times \Sigma \to S$ to a transition function $\delta : S \times \Sigma^* \to S$ recursively as follows:

$$\delta(s, \varepsilon) = s, \quad s \in S \quad \text{and} \quad \text{if } x = wa, \quad w \in \Sigma^*, a \in \Sigma, \text{ then}$$
$$\delta(s, x) = (\delta(s, w), a).$$

Finite-state automaton always starts functioning from the initial state s_0 and if w is a string over Σ such that $\delta(s_0, w)$ is in the subset A of S, we say that w is **accepted** by the machine. If $w \in \Sigma^*$ is such that $\delta(s_0, w)$ is not in A, we say that w is **rejected** by the machine. The transition function $\delta : S \times \Sigma^* \to S$ induces a function $\varphi : S \times \Sigma^* \to S$ by

$$\varphi(w) = \delta(s_0, w), \quad w \in \Sigma^*$$

and it is called the **final state function**. The name final-state function for φ is justified by the fact $\varphi(w)$ **is the final state reached by the machine starting from the initial state s_0 on processing the characters of w from left to right one by one**. It is also clear that $\varphi(wa) = \delta(\varphi(w), a)$ for $w \in \Sigma^*, a \in \Sigma$.

The state-transition function $\delta : S \times \Sigma \to S$ of a finite-state automaton may be represented in tabular form by writing the input characters $a_1, \ldots, a_k$ (say) in the top row, the states $s_0, s_1, \ldots, s_m$ as the leftmost column with the entry in the row of $s \in S$ and the column of $a \in \Sigma$ being the value $\delta(s, a)$. This function may also be given by a directed graph in which the vertices are the states $s_0, s_1, \ldots, s_m$ and states s, s' are connected by an edge from s to s' if there exists an $a \in \Sigma$ such that $\delta(s, a) = s'$. Also a is written over this edge indicating that it is a weighted graph. A loop over an $s \in S$ means that there is an $a \in \Sigma$ such that $\delta(s, a) = s$. The states which are in the accepting subset A of the machine are indicated by a pair of small concentric circles over each such state. As some examples, we construct state diagrams for the finite-state automaton $M = (S, s_0, \Sigma, A, \delta)$ where

(a) $S = \{s_0, s_1, s_2, s_3\}$, $\Sigma = \{0, 1\}$, $A = \{s_1, s_3\}$ and δ is given in Table 13.2.
(b) $S = \{s_0, s_1, s_2, s_3\}$, $\Sigma = \{0, 1\}$, $A = \{s_2\}$ and δ is given in Table 13.3.
(c) $S = \{s_0, s_1, s_2, s_3\}$, $\Sigma = \{0, 1\}$, $A = \{s_1, s_3\}$ and δ is given in Table 13.4.

The state diagrams are as given in Figs. 13.3(a)–(c), respectively.

String-matching automata
For every pattern P, there is a string-matching finite automaton. Construction of such automata is described by Aho, Hopcroft and Ullman [1] which has been simplified by Cormen *et al.* [5]. We follow the simplified version as given in [5].

Let $P[1 \ldots m]$ be a given pattern. The finite string-matching automaton that corresponds to P is defined as follows:

1. The state set $S = \{0, 1, 2, \ldots, m\}$.

Table 13.2		
	Input	
State	0	1
s_0	s_0	s_1
s_1	s_0	s_2
s_2	s_0	s_0
s_3	s_2	s_2

Table 13.3		
	Input	
State	0	1
s_0	s_1	s_2
s_1	s_3	s_2
s_2	s_3	s_3
s_3	s_3	s_3

Table 13.4		
	Input	
State	0	1
s_0	s_0	s_1
s_1	s_3	s_2
s_2	s_2	s_2
s_3	s_3	s_3

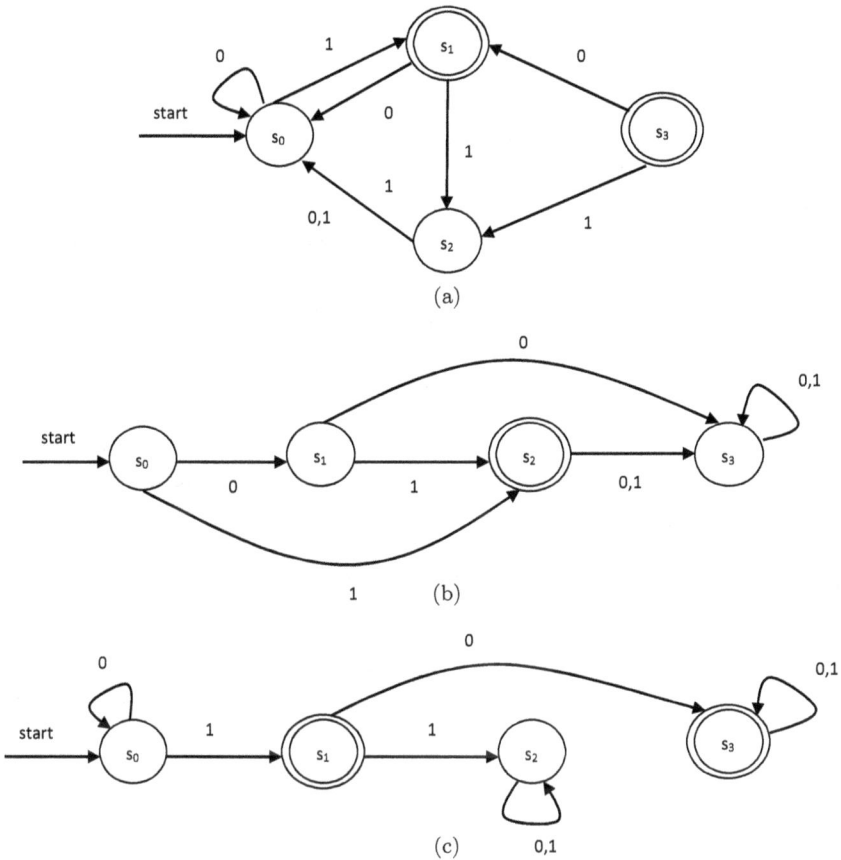

(a)

(b)

(c)

Fig. 13.3

2. The start state s_0 is 0.
3. The state m is the only accepting state.
4. The state-transition function $\delta : S \times \Sigma \rightarrow S$ is defined for any $s \in S$ and $a \in \Sigma$ (Σ the set of input alphabet) by $\delta(s, a) = \max\{k \mid P_k]P_s a\}$.

Since we have to use the definition of δ a number of times, we introduce an auxiliary function $\sigma : \Sigma^* \rightarrow S$ defined by

$$\sigma(w) = \max\{k \mid P_k]w\}$$

and it is called the **suffix function**.

Observe that $w = w\varepsilon = wP_0$ and $0 \in \{i \mid P_i]w\} = L$ (say). Since $|w|$ is finite, L is a finite set of non-negative integers. Therefore L has a unique maximal element, so, suffix function $\sigma(w) = \max\{k \mid P_k]w\}$ and the suffix function σ is well defined.

Since $P_0 = \varepsilon$, $P_0]w$, for every $w \in \Sigma^*$, $\sigma(w) \geq 0$. If $x, y \in \Sigma^*$ and $x]y$, let $y = zx$ for some $z \in \Sigma^*$. Suppose that $\sigma(x) = k$. Then there exists a $u \in \Sigma^*$ such that $x = uP_k$. Then $y = zuP_k$ and, therefore, $\sigma(x) = k \leq \sigma(y)$. Also since $\max\{k \mid P_k]\varepsilon\} = 0$, $\sigma(\varepsilon) = 0$.

With this notation, we have for $s \in S, a \in \Sigma$

$$\delta(s, a) = \sigma(P_s a).$$

Let φ be the final state function defined by finite-state string-matching automaton defined above. We now prove that σ itself is the final state function. To prove this we first need the following result.

Lemma 13.3. *For any string w and character a, if $k = \sigma(w)$, then*
  (i) $\sigma(wa) \leq k + 1$ and (ii) $\sigma(wa) = \sigma(P_k a)$.

Proof. Now $k = \sigma(w)$ implies that there exists $z \in \Sigma^*$ such that $w = zP_k$. Then $wa = zP_k a$ and if $r = \sigma(wa)$, there exists $u \in \Sigma^*$ such that

$$uP_r = wa = zP_k a. \tag{13.4}$$

This relation then implies that $uP_{r-1} = zP_k = w$ and the definition of k implies that $r - 1 \leq k$ or $r \leq k + 1$, i.e., $\sigma(wa) \leq \sigma(w) + 1$.

(ii) If $\sigma(P_k a) = s$, there exists $v \in \Sigma^*$ such that $P_k a = vP_s$. Therefore

$$uP_r = wa = zvP_s \tag{13.5}$$

and the definition of $\sigma(wa)$ implies that

$$\sigma(P_k a) \leq \sigma(wa). \tag{13.6}$$

By (i) $|P_r| = r = \sigma(wa) \leq k + 1 = |P_k| + 1 = |P_k a|$ and relation (13.4) then implies that $P_r]P_k a$. Therefore $\sigma(wa) = r \leq \sigma(P_k a)$. Combining this with the relation (13.6), we get $\sigma(wa) = \sigma(P_k a)$. $\square$

Theorem 13.4. *The suffix function σ is the final state function for this finite-state automaton.*

Proof. We have $\sigma(\varepsilon) = 0 = \varphi(\varepsilon)$. Suppose that $w \in \Sigma^*$, $a \in \Sigma$ and that $\sigma(w) = \varphi(w) = k$ (say). Then

$$
\begin{aligned}
\varphi(wa) &= \delta(\varphi(w), a) \quad &&\text{(by definition of } \varphi) \\
&= \delta(\sigma(w), a) \quad &&\text{(by induction hypothesis)} \\
&= \delta(k, a) \\
&= \sigma(P_k a) \quad &&\text{(by definition of } \sigma) \\
&= \sigma(wa) \quad &&\text{(by Lemma 13.3).}
\end{aligned}
$$

This completes induction. □

We now describe how the string-matching finite machine constructed yields all possible valid shifts. Let $T = T[1..n]$ be a given text of length n and P be the pattern of length m we are already working with. We examine the string T from left to right examining one character at a time. Then $\varphi(T_i)$ denotes the final state reached by the machine when the string T_i has been examined. It is clear because of $\varphi(T_i) = \sigma(T_i)$ that $\sigma(T_i) \leq m$. Moreover, $\sigma(T_i) = k$ if and only if $P_k]T_i$. Since m is the only accepting state, $\sigma(T_i) = m$ if and only if $P = P_m]T_i$. This is so if and only if $i - m$ is a valid shift. Hence the set of valid shifts is $\{i - m \mid i \leq n \text{ and } \sigma(T_i) = m\}$ and P_{k+1} is not a suffix of T_i.

In order to apply the finite-state string-matching automaton, we need to pre-compute the state-transition function $\delta : S \times \Sigma \rightarrow S$ which needs $(m+1)|\Sigma|$ computations $\delta(s, a) = \sigma(P_s a)$. Each $\sigma(P_s a)$ can be computed in a straightforward manner in a constant amount of time. By definition $\sigma(P_s a) = k$ where k is largest with $P_k]P_s a$. Therefore, the length k of P_k is $\leq |P_s a|$. Also P_k is a prefix of P_m which is of length m. Therefore $k \leq m$. Hence $k \leq \min\{m, s + 1\}$. We can now give the algorithm for computation of state-transition function.

The action at 2 has $(m + 1)|\Sigma|$ choices. The action at 4 and 5 has at most $m + 1$ choices. Also as s varies from 0 to m, the action at 3 has $m + 1$ choices. Hence the running time of the algorithm State-Transition Function Computation is $O(m^3|\Sigma|)$.

Next we describe the algorithm for string matching through finite automaton. Since we are given finite automaton, the state-transition function δ is given. We start searching the prefixes T_i for $0 \leq i \leq n$ of the

---

Algorithm 13.3. State-Transition Function Computation (P, Σ)

---

1. $m = $ length of pattern P
2. for $s = 0$ to m and for each $a \in \Sigma$, consider $P_s a$
3. choose $k = \min\{m, s + 1\}$
4. if $P_k] P_s a$, return k
5. if P_k is not a suffix of $P_s a$, put $k = k - 1$
6. repeat action at 4 and 5 till we arrive at k such that $P_k] P_s a$
7. return k

---

text $T[1 \ldots n]$. For any $s \in S$, $\delta(s, T_i)$ denotes the state arrived at after the string T_i has been scanned, the scanning beginning with the initial state s_0, i.e., $\delta(s, T_i) = (\delta(s, T_{i-1}), T[i])$.

Algorithm 13.4. Finite-Automaton-Matcher Algorithm (T, δ, m)

---

1. $n = \text{length}(T)$, $m = \text{length}(P)$, $\delta = $ the state-transition function
2. $s = 0$ is the initial state
3. for $i = 1$ to n
4. compute $\delta(s, T[i[)$
5. if $\delta(s, T[i]) = m$, return m and $i - m$ is a valid state

---

Since only one character $T[i]$ is being examined at every step and i varies from 1 to n as at 4 and 3 and every $\delta(s, T[i])$ takes a constant amount of time, the running time of the algorithm is $\theta(n)$. Observe that $\delta(s, T[i])$ as at step 4 is actually $\delta(\delta(\ldots \delta((\delta(s_0, T[1]), T[2])\ldots, T[i-1]), T[i])$.

We end this section with a couple of examples on the application of finite automata to the string matching problem.

Example 13.5. Using the state-transition function computation algorithm, compute the transition function δ when $P = a\ a\ b\ a\ b$ and $\Sigma = \{a, b\}$.

Solution. Here $|P| = m = 5$ and, so, we need to find $\delta(s, a)$ and $\delta(s, b)$ for $s = 0, 1, 2, 3, 4, 5$. Since we need to examine P_i, $0 \le i \le 5$, for every s, we record these as a first step.

$$P_1 = a, P_2 = a\ a, P_3 = a\ a\ b, P_4 = a\ a\ b\ a, P_5 = a\ a\ b\ a\ b = P.$$

(i) For $s = 0$, $k = \min\{|P_0x|, m\} = 1$. Since $P_1]P_0a$ but P_1 is not a suffix of P_0b, therefore

$$\delta(0, a) = 1 \quad \text{and} \quad \delta(0, b) < 1 \text{ and, so, } \delta(0, b) = 0.$$

(ii) For $s = 1$, $k = \min\{1 + 1, 5\} = 2$.
As $P_2 = aa]P_1a$ and neither P_2 nor P_1 is a suffix of P_1b, $\delta(1, a) = 2$ and $\delta(1, b) = 0$.

(iii) For $s = 2$, $k = \min\{2 + 1, 5\} = 3$.
Since P_3 is not a suffix of P_2a but P_2 is a suffix of P_2a, $\delta(2, a) = 2$. However $P_3 = P_2b$, and, so, $\delta(2, b) = 3$.

(iv) For $s = 3$, $k = \min\{3 + 1, 5\} = 4$.
Here $P_3a = P_4$ and, so, $\delta(3, a) = 4$.
However, none of P_4, P_3, P_2, P_1 is a suffix of $P_3b = a\,a\,b\,b$ and, so, $\delta(3, b) = 0$.

(v) For $s = 4$, $k = \min\{4 + 1, 5\} = 5$.
None of P_5, P_4, P_3 is a suffix of $P_4a = a\,a\,b\,a\,a$ but $P_2]P_4a$, $\delta(4, a) = 2$.
Since $P_4b = a\,a\,b\,a\,b = P_5$, $\delta(4, b) = 5$.

(vi) $s = 5$, $k = \min\{5 + 1, 5\} = 5$.
Since $P_5a = a\,a\,b\,a\,b\,a$, none of P_5, P_4, P_3, P_2 is a suffix of P_5a but $P_1]P_5a$. Therefore $\delta(5, a) = 1$.
Again $P_5b = a\,a\,b\,a\,b\,b$ and none of P_5, P_4, P_3, P_2, P_1 is a suffix of P_5b. Therefore $\delta(5, b) = 0$.

Example 13.6. Construct the string-matching automaton for the pattern $P = a\,a\,b\,a\,b$ and illustrate its operation on the text string

$$T = b\,a\,b\,a\,a\,b\,a\,a\,b\,a\,b\,a\,a\,b\,a\,b\,a\,a\,b\,a\,a.$$

Solution. Since P is of length 5, $m = 5$ and, therefore, $S = \{0, 1, 2, 3, 4, 5\}$, $s_0 = 0$, $A = \{5\}$, i.e., 5 is the only accepting state and $\delta : S \times \Sigma \rightarrow S$ is defined by $\delta(s, x) = \sigma(P_sx)$, $x \in \Sigma$. The state-transition function δ has been computed in Example 13.5. Now

$$\varphi(1) = \delta(0, T[1]) = \delta(0, b) = 0,$$

$$\varphi(2) = \delta(\varphi(1), T[2]) = \delta(0, a) = 1$$

$$\varphi(3) = \delta(\varphi(2), T[3]) = \delta(1, b) = 0$$

$$\varphi(4) = \delta(\varphi(3), T[4]) = \delta(0, a) = 1$$

$$\varphi(5) = \delta(\varphi(4), T[5]) = \delta(1, a) = 2$$

$$\varphi(6) = \delta(\varphi(5), T[6]) = \delta(2, b) = 3$$

$$\varphi(7) = \delta(\varphi(6), T[7]) = \delta(3, a) = 4$$

$$\varphi(8) = \delta(\varphi(7), T[8]) = \delta(4, a) = 2$$

$$\varphi(9) = \delta(\varphi(8), T[9]) = \delta(2, b) = 3$$

$$\varphi(10) = \delta(\varphi(9), T[10]) = \delta(3, a) = 4$$

$$\varphi(11) = \delta(\varphi(10), T[11]) = \delta(4, b) = 5 \rightarrow$$

$$\varphi(12) = \delta(\varphi(11), T[12]) = \delta(5, a) = 1$$

$$\varphi(13) = \delta(\varphi(12), T[13]) = \delta(1, a) = 2$$

$$\varphi(14) = \delta(\varphi(13), T[14]) = \delta(2, b) = 3$$

$$\varphi(15) = \delta(\varphi(14), T[15]) = \delta(3, a) = 4$$

$$\varphi(16) = \delta(\varphi(15), T[16]) = \delta(4, b) = 5 \rightarrow$$

$$\varphi(17) = \delta(\varphi(16), T[17]) = \delta(5, a) = 1$$

$$\varphi(18) = \delta(\varphi(17), T[18]) = \delta(1, a) = 2$$

$$\varphi(19) = \delta(\varphi(18), T[19]) = \delta(2, b) = 3$$

$$\varphi(20) = \delta(\varphi(19), T[20]) = \delta(3, a) = 4$$

$$\varphi(21) = \delta(\varphi(20), T[21]) = \delta(4, a) = 2.$$

Since $\varphi(11) = 5 = \varphi(16)$, the set of valid shifts is $\{11, 16\}$.

Exercise 13.2.

1. Find the state-transition function for the string matching automaton for the pattern $P = a\,b\,a\,b\,a\,a$ over the alphabet $\Sigma = \{a, b\}$.
2. Construct the string-matching automaton for the pattern $P = a\,b\,a\,a\,b\,a$ and illustrate its operation on the text string $T = a\,a\,b\,a\,a\,b\,a\,b\,a\,a\,b\,a\,b\,a\,a\,b\,a\,b\,a\,b\,a\,b\,a\,a\,b\,a\,a$ over the alphabet $\Sigma = \{a, b\}$.

3. We call a pattern P **non-overlapping** if $P_k]P_s$ implies $k = 0$ or $k = s$. Construct some non-overlapping patterns of length 4 and 5 over $\Sigma = \{a, b\}$. Describe the state-transition diagrams for the non-overlapping patterns constructed by you.

(Note that $a\ b\ b, a\ a\ b$ are two non-overlapping patterns of length 3, $a\ a\ b$ $b, a\ b\ b\ b$ are two non-overlapping patterns of length 4 and $a\ a\ a\ a\ b, a\ a$ $a\ b\ b, a\ a\ b\ b\ b, a\ b\ b\ b\ b$ are four non-overlapping patterns of length 5. Are there any other non-overlapping patterns of length 3, 4, 5?)

13.5. The Knuth–Morris–Pratt Algorithm

We now describe a string-matching algorithm which finds all occurrences of a pattern $P[1 \ldots m]$ in a text $T[1 \ldots n]$ of length n with linear running time. This algorithm is due to Knuth, Morris and Pratt. We follow the original approach of Knuth, Morris and Pratt [18] as also its simplified approach adopted by Cormen *et al.* [5]. The algorithm needs computing some auxiliary functions the processing time of which is $\theta(m)$.

13.5.1. Some auxiliary functions

We introduce three functions f, *next* as defined by Knuth, Morris and Pratt and the prefix function π from $\{1, 2, \ldots, m\}$ to $\{1, 2, \ldots, m - 1\}$.

For $j \geq 2$,

$f[j] =$ the largest integer $i < j$ such that $P_{i-1}]P_{j-1}$,

$next[j] =$ the largest integer $i < j$ such that $P_{i-1}]P_{j-1}$ and $P[i] \neq P[j]$

and for $j \geq 1$,

$\pi[j] =$ the largest integer $i < j$ such that $P_i]P_j$.

It is immediate from the definition that $\pi[j + 1] = 1 + \pi[j]$ for all j, $1 \leq j \leq m - 1$. Since $\pi[i] \geq 0$, we find that $f[2] = 1$ and $f[j] \geq 1$ for all $j \geq 2$. We cannot say the same thing about $next[j]$ because it depends on $P[j]$ being equal to $P[i]$ or not. We also define $f[1] = next[1] = 0$.

Once we have computed $f[k]$ for $k = 1, 2, \ldots, j$, it is fairly easy to compute $next[j]$. Also, computation of the function f is fairly easy in view of the following lemma.

Lemma 13.5. *Let $f[j] = i$ and $next[i] = l$. Then*

$$f[j+1] = \begin{cases} 1 + f[j] & \text{if } P[i] = P[j] \\ 1 + l & \text{if } P[i] \neq P[j] \text{ and } P[l] = P[j], \end{cases}$$

and $f[j+1] \leq 1 + next[l]$ if $P[i] \neq P[j]$ and $P[l] \neq P[j]$.

Proof. By definition of the function f, i is the largest integer $< j$ such that

$$P_{i-1}]P_{j-1}. \tag{13.7}$$

If $P[i] = P[j]$, it follows from (13.7) that $P_i]P_j$. If $i + 1$ is not the largest integer $< j + 1$, with $P_i]P_j$, then there exists a k, $i < k < j$, with $P_k]P_j$. Cancelling the rightmost characters $P[k]$ and $P[j]$ from this relation, we get $P_{k-1}]P_{j-1}$ with $k > i$ which contradicts the fact that $f[j] = i$. Hence $f[j+1] = 1 + i = 1 + f[j]$.

Now suppose that $P[i] \neq P[j]$.

The relation (13.7) together with $P[i] \neq P[j]$ imply that P_i is not a suffix of P_j. Therefore $f[j+1] \neq i+1$.

Let $f[j+1] = k+1$. Then k is the largest integer $< j$ such that $P_k]P_j$ and, so, k is an integer $< j$ such that

$$P_{k-1}]P_{j-1}. \tag{13.8}$$

If $f[j+1] \geq i+2$, then (13.8) implies that $i+2 \leq k+1 \leq i+1$ which is an absurdity. Hence $f[j+1] = k+1 \leq i(< j)$. Since $k < i$, it follows from (13.7) and (13.8) that $P_{k-1}]P_{i-1}$. Also $P[k] = P[j] \neq P[i]$. Therefore

$$l = next[i] \geq k. \tag{13.9}$$

Again, $next[i] = l$ implies that l is the largest integer $< i$ with

$$P_{l-1}]P_{i-1} \quad \text{and} \quad P[l] \neq P[i]. \tag{13.10}$$

Also $P[i] \neq P[j]$. Therefore $P[l]$ and $P[j]$ may or may not be equal. Suppose that

$$P[j] = P[l]. \tag{13.11}$$

Now (13.7) and (13.10) imply that $P_{l-1}]P_{j-1}$ which together with (13.11) implies that $P_l]P_j$. But k is the largest integer $< j$ such that $P_k]P_j$. Therefore $l \leq k$ and this together with (13.9) implies that $k = l$. Hence $f[j+1] = 1 + k = 1 + l = 1 + next[i]$.

We are now left to consider the case

$$P[j] \neq P[l]. \tag{13.12}$$

Equations (13.7), (13.8), (13.10) and $k \leq l < i$ imply that

$$P_{k-1}]P_{l-1}. \tag{13.13}$$

As $P[k] = P[j]$, it follows from (13.12) that $P[k] \neq P[l]$. Therefore $k \leq next[l]$. Hence $f[j+1] \leq 1 + next[l]$. This completes the proof of the lemma. □

Remark. If we let $next[l] = t$, then $k \leq t$ and t is the largest integer $< l$ such that $P_{t-1}]P_{l-1}$ and $P[t] \neq P[l]$. Again, as $P[l] \neq P[j]$, $P[t]$ and $P[j]$ may or may not be equal. If $P[t] = P[j]$, the argument used above shows that $k = t$ while if $P[t] \neq P[j]$ then $k \leq next[t]$. We can continue this argument and stop only when we arrive at a u with $k \leq next[u] = 0$.

In view of Lemma 13.5, we need to compute $f[j]$ and $next[j]$ simultaneously in the same table.

Lemma 13.6. *If* $f[j] = i$, *then*

$$next[j] = \begin{cases} i & \text{if } P[j] \neq P[i], \\ next[i] & \text{if } P[j] = P[i]. \end{cases}$$

Proof. Let $f[j] = i$ and $next[j] = k$.

If $P[j] \neq P[i]$, it follows from the definitions of $f[j]$ and $next[j]$ that $next[j] = i = f[j]$. Now, suppose that $P[j] = P[i]$. Then $k < i$. Also we have $P_{i-1}]P_{j-1}$ and $P_{k-1}]P_{j-1}$. Therefore $P_{k-1}]P_{i-1}$ and $P[k] \neq P[j] = P[i]$. Hence $next[i] \geq k$.

If $next[i] = l > k$, then $P_{l-1}]P_{i-1}$ and $P[l] \neq P[i]$. As $P_{i-1}]P_{j-1}$ and $l - 1 < i - 1$, $P_{l-1}]P_{j-1}$ and $P[l] \neq P[i] = P[j]$. This contradicts the choice of k, i.e., the definition of $next[j]$. Hence $k = next[j]$. □

We now consider some examples.

Example 13.7. Compute the $f, next$ and π functions for the pattern P when

(a) $P = a\,b\,c\,a\,b\,c\,a\,c\,b\,a\,b\,c$ over $\Sigma = \{a, b, c\}$.

(b) $P = a\,b\,a\,c\,a\,b\,a$ over the alphabet $\Sigma = \{a, b, c\}$.

(c) $P = a\,b\,a\,b\,a\,a\,b\,a\,b\,b\,a\,a\,b\,a\,b\,b\,a\,a\,b$ over the alphabet $\Sigma = \{a, b\}$.

Solution. (a) Tables 13.5–13.7 give the computations.

Since $f[2] = 1$ and $P[2] \neq P[1]$, $next\{2\} = 1$. As P_1 is not a suffix of P_2, $f[3] = 1$ and as $P[3] \neq P[1]$, $next[3] = 1$. Neither of P_2, P_1 is a suffix of P_3 so that $f[4] = 1$. However $P[4] = P[1]$ and, so, $next[4] = 0$ and $f[5] = 1 + 1 = 2$. Again $P[5] = P[2]$ implies that $next[5] = next[2] = 1$ and $f[6] = 2 + 1 = 3$. As $P[6] = P[3]$, $next[6] = next[3] = 1$ and $f[7] = 3 + 1 = 4$. Now $P[7] = P[4]$ and, so, $next[7] = next[4] = 0$ whereas $f[8] = 4 + 1 = 5$. As $P[8] \neq P[5]$, $next[8] = 5$ and $f[9] \leq 1 + next[5] = 2$. But P_1 is not a suffix of P_8 and, so, $f[9] = 1$. As $P[9] \neq P[1]$, $next[9] = 1$ and $f[10] \leq 1 + next[1] = 1$. Thus $f[10] = 1$. As $P[10] = P[1]$, $next[10] = 0$. Again $P[10] = P[1]$ implies that $f[11] = 1 + f[10] = 1 + 1 = 2$ and $next[11] = next[2] = 1$. As $P[11] = P[2]$, $f[12] = 2 + 1 = 3$ and $next[12] = next[3] = 1$.

For $1 \leq j \leq 11$, the computations for $\pi[j]$ follow from the relation $f[j + 1] = 1 + \pi[j]$. As P_8, P_6 are not suffixes of P_{12} but P_3 is a suffix of $P_{12} = P$, $\pi[12] = 3$.

We next give the tables (Tables 13.6 and 13.7) for the computation of these functions for the other two patterns but do not give the detailed reasoning for the same. Reasoning only for some typical entries will be given.

(b) $P = abacaba$ (see Table 13.6).

Recall Lemmas 13.5 and 13.6. If $f[j] = i$, then

$$f[j + 1] = 1 + f[j] \text{ and } next[j] = next[i] \text{ if } P[j] = P[i]$$
$$f[j + 1] \leq 1 + next[i] \text{ and } next[j] = i \text{ if } P[j] \neq P[i].$$

Table 13.5

j	1	2	3	4	5	6	7	8	9	10	11	12
$P[j]$	a	b	c	a	b	c	a	c	b	a	b	c
$f[j]$	0	1	1	1	2	3	4	5	1	1	2	3
$next[j]$	0	1	1	0	1	1	0	5	1	0	1	1
$\pi[j]$	0	0	0	1	2	3	4	0	0	1	2	3

Table 13.6

j	1	2	3	4	5	6	7
$P[j]$	a	b	a	c	a	b	a
$f[j]$	0	1	1	2	1	2	3
$next[j]$	0	1	0	1	0	1	0
$\pi[j]$	0	0	1	0	1	2	3

Table 13.7

j	1	2	3	4	5	6	7	8	9	10	11	12	13	14	15	16	17	18	19
$P[j]$	a	b	a	b	a	a	b	a	b	b	a	a	b	a	b	b	a	a	b
$f[j]$	0	1	1	2	3	4	2	3	4	5	1	2	2	3	4	5	1	2	2
$next[j]$	0	1	0	1	0	4	1	0	1	5	0	2	1	0	1	5	0	2	1
$\pi[j]$	0	0	1	2	3	1	2	3	4	0	1	1	2	3	4	0	1	1	2

(i) Now $f[2] = 1$, $P[2] \neq P[1]$ implies $f[3] \leq 1 + next[1] = 1$ and $next[3] = next[1] = 0$.

(ii) $f[3] = 1$, $P[3] = P[1]$ and, so, $f[4] = 1 + 1 = 2$ and $P[4] \neq P[2]$ so that $next[4] = next[2] = 1$.

(iii) $f[4] = 2$ and $P[4] \neq P[2]$ imply that $f[5] \leq 1 + next[2] = 2$. But P_1 is not a suffix of P_4 and, therefore, $f[5] = 1$. As $P[5] = P[1]$, $next[5] = next[1] = 0$.

(iv) $f[5] = 1$, $P[5] = P[1]$ imply $f[6] = 1 + f[5] = 2$. $P[6] = P[2]$ and, so, $next[6] = next[2] = 1$.

(v) $P[6] = P[2]$ and, so, $f[7] = 3$. As $P[7] = P[3]$, $next[7] = next[3] = 0$.

Now using the relation $f[j + 1] = 1 + \pi[j]$, the last row of the table is completed except for the entry $\pi[7]$. As 3 is the largest integer ≤ 3 such that P_3 is a suffix of P_7, $\pi[7] = 3$.

(c) $P = a\,b\,a\,b\,a\,a\,b\,a\,b\,b\,a\,a\,b\,a\,b\,b\,a\,a\,b$ (see Table 13.7).

(i) $f[2] = 1$ and $P[2] \neq P[1]$ imply $next[2] = 1$.

(ii) $f[2] = 1, P[2] \neq P[1]$ imply $f[3] \leq 1 + 1 = 2$. But P_1 is not a suffix of P_2. Therefore $f[3] = 1$. Also $P[3] = P[1]$ then shows that $next[3] = next[1] = 0$.

(iii) $f[3] = 1$ and $P[3] = P[1]$ imply that $f[4] = 2$. Also $P[4] = P[2]$ and, so, $next[4] = next[2] = 1$.

(iv) $f[4] = 2$, $P[4] = P[2]$ imply that $f[5] = 3$. Also $P[5] = P[3]$. Therefore $f[6] = 4$ and $next[5] = next[3] = 0$.

(v) $f[6] = 4$, $P[6] \neq P[4]$ imply $next[6] = 4$ and $f[7] \leq 1 + f[4] = 3$. As P_1 is a suffix of P_6 but P_2 is not a suffix of P_6, $f[7] = 2$.

(vi) $f[7] = 2$, $P[7] = P[2]$ together imply that $f[8] = 3$ whereas $next[7] = next[2] = 1$.

(vii) $f[8] = 3$, $P[8] = P[3]$ imply that $f[9] = 4$ and $next[8] = next[3] = 0$.

(viii) $f[9] = 4$, $P[9] = P[4]$ imply that $f[10] = 5$. Also $next[9] = next[4] = 1$.

(ix) $f[10] = 5$, $P[10] \neq P[5]$. Therefore $f[11] \leq 1 + next[5] = 1$. So $f[11] = 1$ and $next[10] = 5$.

(x) $f[11] = 1$, $P[11] = P[1]$ and, so, $f[12] = 2$ and $next[11] = next[1] = 0$.

(xi) $f[12] = 2$, $P[12] \neq P[2]$ so that $next[12] = 2$ and $f[13] \leq 1 + next[2] = 2$ and, as $P_1]P_{12}$, $f[13] = 2$.

(xii) $f[13] = 2$, and $next[13] = next[2] = 1$, $P[13] = P[2]$ and, so, $f[14] = 3$.

(xiii) $f[14] = 3$, $P[14] = P[3]$ so that $next[14] = next[3] = 0$ and $f[15] = 4$.

(xiv) $f[15] = 4$, $P[15] = P[4]$ so that $next[15] = next[4] = 1$ and $f[16] = 5$.

(xv) $f[16] = 5$, $P[16] \neq P[5]$ so that $next[16] = 5$ and $f[17] \leq 1 + next[5] = 1$. Therefore $f[17] = 1$.

(xvi) $f[17] = 1$, $P[17] = P[1]$ so that $next[17] = next[1] = 0$ and $f[18] = 2$.

(xvii) $f[18] = 2$, $P[18] \neq P[2]$ so that $next[18] = 2$ and $f[19] \leq 1 + next[2] = 2$. As P_1 is a suffix of P_{18}, $f[19] = 2$.

(xviii) $f[19] = 2$, $P[19] = P[2]$ and, so, $next[19] = next[2] = 1$.

Lastly $\pi[19] = 2$ as $P_2]P_{19}$ and none of the other P_i is a proper suffix of P_{19}.

Table 13.8

j	1	2	3	4	5	6	7	8	9
P[j]	A	B	C	D	A	B	D	C	A
f[j]	0	1	1	1	1	2	3	1	1
next[j]	0	1	1	1	0	1	3	1	0
π[j]	0	0	0	0	1	2	0	0	1

Table 13.9

j	1	2	3	4	5	6	7	8	9	10	11	12	13	14	15	16	17	18	19	20	21	22	23	24
P[j]	P	A	R	T	I	C	I	P	A	T	E	X	I	N	X	P	A	R	A	C	H	U	T	E
f[j]	0	1	1	1	1	1	1	2	3	1	1	1	1	1	1	2	3	4	1	1	1	1	1
next[j]	0	1	1	1	1	1	1	0	1	3	1	1	1	1	1	0	1	1	4	1	1	1	1	1
π[j]	0	0	0	0	0	0	0	1	2	0	0	0	0	0	0	1	2	3	0	0	0	0	0

Example 13.8. Compute the functions f, *next* and π for the patterns

(a) $P = A\ B\ C\ D\ A\ B\ D\ C\ A$
(b) $P = P\ A\ R\ T\ I\ C\ I\ P\ A\ T\ E\ X\ I\ N\ X\ P\ A\ R\ A\ C\ H\ U\ T\ E$

Solution. Tables 13.8 and 13.9 give the computations for the patterns P.

    (a) $P = A\ B\ C\ D\ A\ B\ D\ C\ A$ (see Table 13.8).

(i) $f[2] = 1$.

(ii) $f[2] = 1$, $P]2] \neq P[1]$ so that $next[2] = 1$ and $f[3] \leq 1 + next[1] = 1$. Thus $f[3] = 1$.

(iii) $f[3] = 1$, $P]3] \neq P[1]$ so that $next[3] = 1$ and $f[4] \leq 1 + next[1] = 1$. Thus $f[4] = 1$.

(iv) $f[4] = 1$, $P]4] \neq P[1]$ so that $next[4] = 1$ and $f[5] \leq 1 + next[1] = 1$. Thus $f[5] = 1$.

(v) $f[5] = 1$, $P[5] = P[1]$ so that $next[5] = next[1] = 0$ and $f[6] = 2$.

(vi) $f[6] = 2$, $P[6] = P[2]$ so that $next[6] = next[2] = 1$ and $f[7] = 3$.

(vii) $f[7] = 3$, $P]7] \neq P[3]$ so that $next[7] = 3$ and $f[8] \leq 1 + next[3] = 1 + 1 = 2$. But P_1 is not a suffix of P_7 and $f[8] = 1$.

(viii) $f[8] = 1$, $P]8] \neq P[1]$ so that $next[8] = 1$ and $f[9] \leq 1 + next[1] = 1$. Thus $f[9] = 1$.

(ix) $f[9] = 1$, $P[9] = P[1]$ so that $next[9] = next[1] = 0$.

Since P_1 is a suffix of $P_9 = P$, and none of the other P_i is a suffix of P_9, $\pi[9] = 1$.

(b) $P = P\ A\ R\ T\ I\ C\ I\ P\ A\ T\ E\ X\ I\ N\ X\ P\ A\ R\ A\ C\ H\ U\ T\ E$ (see Table 13.9).

(i) $f[2] = 1$, $P[2] \neq P[1]$ so $next[2] = 1$, $f[3] \leq 1 + next[1] = 1$ and $f[3] = 1$.

(ii) $f[3] = 1$, $P[3] \neq P[1]$ so $next[3] = 1$, $f[4] \leq 1 + next[1] = 1$ and $f[4] = 1$.

(iii) $f[4] = 1$, $P[4] \neq P[1]$ so $next[4] = 1$, $f[5] \leq 1 + next[1] = 1$ and $f[5] = 1$.

(iv) $f[5] = 1$, $P[5] \neq P[1]$ so $next[5] = 1$, $f[6] \leq 1 + next[1] = 1$ and $f[6] = 1$.

(v) $f[6] = 1$, $P[6] \neq P[1]$ so $next[6] = 1$, $f[7] \leq 1 + next[1] = 1$ and $f[7] = 1$.

(vi) $f[7] = 1$, $P[7] \neq P[1]$ so $next[7] = 1$, $f[8] \leq 1 + next[1] = 1$ and $f[8] = 1$.

(vii) $f[8] = 1$, $P[8\} = P[1]$ so $next[8] = next[1] = 0$, $f[9] = 1 + 1 = 2$.

(viii) $f[9] = 2$, $P[9\} = P[2]$ so $next[9] = next[2] = 1$, $f[10] = 1 + 2 = 3$.

(ix) $f[10] = 3$, $P[10] \neq P[3]$ so $next[10] = 3$, $f[11] \leq 1 + next[3] = 2$. As P_1 is not a suffix of P_{10}, $f[11] = 1$.

(x) $f[11] = 1$, $P[11] \neq P[1]$ so $next[11] = 1$, $f[12] \leq 1 + next[1] = 1$ and $f[12] = 1$.

(xi) $f[12] = 1$, $P[12] \neq P[1]$ so $next[12] = 1$, $f[13] \leq 1 + next[1] = 1$ and $f[13] = 1$.

(xii) By the argument as in the previous two cases
$next[13] = 1$, $f[14] = 1$, $next[14] = 1$, $f[15] = 1$,
$next[15] = 1$, $f[16] = 1$.

(xiii) $f[16] = 1$, $P[16\} = P[1]$ so $next[16] = next[1] = 0$ and $f[17] = 2$.

(xiv) $f[17] = 2$, $P[17] = P[2]$ so $next[17] = next[2] = 1$ and $f[18] = 3$.

(xv) $f[18] = 3$, $P[18] = P[3]$ so $next[18] = next[3] = 1$ and $f[19] = 4$.

(xvi) $f[19] = 4$, $P[19] \neq P[4]$ so $next[19] = 4$, and $f[20] \leq 1 + next[4] = 2$ but P_1 is not a suffix of P_{19} and, so, $f[20] = 1$.

(xvii) $f[20] = 1$, $P[20] \neq P[1]$ so $next[20] = 1$ and $f[21] \leq 1 + next[1] = 1$ or $f[21] = 1$.

(xviii) $f[21] = 1$, $P[21] \neq P[1]$ so $next[21] = 1$, and $f[22] = 1$.

(xix) $f[22] = 1$, $P[22] \neq P[1]$ so $next[22] = 1$, and $f[23] = 1$.

(xx) $f[23] = 1$, $P[23] \neq P[1]$ so $next[23] = 1$ and $f[24] = 1$.

(xxi) $f[24] = 1$, $P[24] \neq P[1]$ so $next[24] = 1$.

So far as the function π is concerned, we only need to compute $\pi[24]$ and the only possible suffix of P_{24} could be P_{12} but actually P_{12} is not a suffix of $P_{24} = P$. Hence $\pi[24] = 0$.

Example 13.9. Compute the functions f, $next$ and the prefix function π for the pattern $P = a\,b\,a\,b\,b\,a\,b\,b\,a\,b\,b\,a\,b\,a\,b\,b\,a\,b\,b$ when the alphabet is $\Sigma = \{a, b\}$.

Solution. The computations are given in Table 13.10.

(i) $f[2] = 1$, $P[2] \neq P[1]$, so $next[2] = 1$, $f[3] \leq 1 + next[1] = 1$ or $f[3] = 1$.

(ii) $f[3] = 1$, $P[3] = P[1]$, so $next[3] = next[1] = 0$, $f[4] = 1 + 1 = 2$.

(iii) $f[4] = 2$, $P[4] = P[2]$, so $next[4] = next[2] = 1$, $f[5] = 3$.

(iv) $f[5] = 3$, $P[5] \neq P[3]$, so $next[5] = 3$, $f[6] \leq 1 + next[3] = 1$ or $f[6] = 1$.

(v) $f[6] = 1$, $P[6] = P[1]$, so $next[6] = next[1] = 0$, $f[7] = 2$.

(vi) $f[7] = 2$, $P[7] = P[2]$, so $next[7] = next[2] = 1$, $f[8] = 3$.

Table 13.10

j	1	2	3	4	5	6	7	8	9	10	11	12	13	14	15	16	17	18	19
$P[j]$	a	b	a	b	b	a	b	b	a	b	b	a	b	a	b	b	a	b	b
$f[j]$	0	1	1	2	3	1	2	3	1	2	3	1	2	3	4	5	6	7	8
$next[j]$	0	1	0	1	3	0	1	3	0	1	3	0	1	0	1	3	0	1	3
$\pi[j]$	0	0	1	0	0	1	2	0	1	2	0	1	2	3	4	5	6	7	8

(vii) $f[8] = 3$, $P[8] \neq P[3]$, so $next[8] = 3$, $f[9] \leq 1 + next[3] = 1$ or $f[9] = 1$.

(viii) $f[9] = 1$, $P[9] = P[1]$, so $next[9] = next[1] = 0$, $f[10] = 2$.

(ix) $f[10] = 2$, $P[10] = P[2]$, so $next[10] = next[2] = 1$, $f[11] = 3$.

(x) $f[11] = 3$, $P[11] \neq P[3]$, so $next[11] = 3$, $f[12] \leq 1 + next[3] = 1$ or $f[12] = 1$.

(xi) $f[12] = 1$, $P[12] = P[1]$, so $next[12] = next[1] = 0$, $f[13] = 2$.

(xii) $f[13] = 2$, $P[13] = P[2]$, so $next[13] = next[2] = 1$, $f[14] = 3$.

(xiii) $f[14] = 3$, $P[14] = P[3]$, so $next[14] = next[3] = 0$, $f[15] = 4$.

(xiv) $f[15] = 4$, $P[15] = P[4]$, so $next[15] = next[4] = 1$, $f[16] = 5$.

(xv) $f[16] = 5$, $P[16] = P[5]$, so $next[16] = next[5] = 3$, $f[17] = 6$.

(xvi) $f[17] = 6$, $P[17] = P[6]$, so $next[17] = next[6] = 0$, $f[18] = 7$.

(xvii) $f[18] = 7$, $P[18] = P[7]$, so $next[18] = next[7] = 1$, $f[19] = 8$.

(xviii) $f[19] = 8$, $P[19] = P[8]$, so $next[19] = next[8] = 3$.

The only computation that we need is $\pi[19] = 8$ as P_8 is a prefix of $P_{19} = P$ and none of the P_i for $i > 8$ is a prefix of P.

13.5.2. The KMP procedure

Suppose that while matching the pattern P against the text T, a mismatch occurs at $T[j]$, i.e., $P[1 \ldots j-1] = T[s+1 \ldots s+j-1]$ but $P[j] \neq T[s+j]$. Here s is the shift for which the mismatch occurs. The aim is to find the least shift $s' > s$ such that $T[s'+1 \ldots s'+k-1] = P[1 \ldots k-1]$ where $s' + k = s + j$ but $P[k]$ and $T[s'+k]$ may be equal. Then k is the largest integer $< s + j - s' \leq j - 1$ with P and T satisfying the desired properties above. If $next[j] = t$, then t is the largest integer $\leq j - 1$ such that $P[1 \ldots t-1]$ is a suffix of $P[1 \ldots j-1] = T[s+1 \ldots s+j-1]$ and $P[t] \neq P[j]$. Since $P[j] \neq T[s+j]$, $P[t]$ and $T[s+j]$ may be equal. In particular, both k and t are largest positive integers $< j$ such that $P_{k-1}P_{j-1}$ and $P_{t-1}P_{j-1}$. Therefore $k = t = next[j]$ and the next least shift after s is $s + (j - t)$, i.e., we have to slide P to the right by $j - t$ steps.

The above argument clearly shows the process of pre-computing the next function. Since in this procedure we slide P to the right over $j - next[j] \geq 1$ symbols, it is a clear generalization of the naïve string matcher except when $next[j] = j - 1$ for all $j \geq 1$. But this is ruled out as $next[j] = t$ is the largest integer $< j$ such that $P_{t-1}]P_{j-1}$. Moreover while moving from shift s to shift $s + j - t$, we do not have to compare the characters of $P[1 \ldots t-1]$ with the characters of $T[s+j-t+1 \ldots s+j-1]$ as the same have already been matched.

We now illustrate the above procedure through a couple of examples.

Example 13.10. Using KMP procedure find all possible shifts of the pattern $P = A\ B\ C\ D\ A\ B\ D\ C\ A$ in the text

$$T = A\ D\ B\ C\ A\ B\ D\ A\ C\ B\ A\ B\ C\ D\ A\ B\ D\ C\ A\ D\ B\ C\ D\ A\ B.$$

Solution. We write the pattern P above the text T initially aligning the characters of P with those of T starting with $T[1]$. An upward arrow below an entry indicates a point of mismatch:

```
A B C D A B D C A
A D B C A B D A C B A B C D A B D C A D B C D A B
  ↑
```

Comparing the respective characters of P and T, we find that there is a mismatch at $P[2]$. We refer to Example 13.8(a) where we have computed the next function for this pattern P. Since $next[2] = 1$, we need to slide the pattern P to the right by $j - next[j] = 2 - 1 = 1$ step. The new alignment of P with T is given by

```
  A B C D A B D C A
A D B C A B D A C B A B C D A B D C A D B C D A B
  ↑
```

Now there is a mismatch at the point $P[1]$. As $next[1] = 0$, we need to slide P to the right by $1 - 0 = 1$ step. The new alignment of P and T is indicated by

```
    A B C D A B D C A
A D B C A B D A C B A B C D A B D C A D B C D A B
    ↑
```

The current mismatch is at $P[1]$. As $next[1] = 0$, we need to slide P to the right by one step. The new alignment of P and T is indicated by

```
        A  B  C  D  A  B  D  C  A
A  D  B  C  A  B  D  A  C  B  A  B  C  D  A  B  D  C  A  D  B  C  D  A  B
        ↑
```

Again the current mismatch is at $P[1]$ and we need to slide P to the right by one step. The new alignment of P and T is given below:

```
           A  B  C  D  A  B  D  C  A
A  D  B  C  A  B  D  A  C  B  A  B  C  D  A  B  D  C  A  D  B  C  D  A  B
           ↑
```

The current mismatch is at $P[3]$ and $next[3] = 1$. Therefore, we need to slide P to the right by $3 - 1 = 2$ steps and the resulting alignment of P and T is given by

```
              A  B  C  D  A  B  D  C  A
A  D  B  C  A  B  D  A  C  B  A  B  C  D  A  B  D  C  A  D  B  C  D  A  B
              ↑
```

Now the mismatch is at $P[1]$ so that we need to slide P to the right by only $1 - 0 = 1$ step and the new alignment of P and T is as below:

```
                 A  B  C  D  A  B  D  C  A
A  D  B  C  A  B  D  A  C  B  A  B  C  D  A  B  D  C  A  D  B  C  D  A  B
                 ↑
```

The current mismatch occurs at $P[2]$ so that we have to slide P to the right by $2 - 1 = 1$ step again. The new alignment of P and T is as given below:

```
                    A  B  C  D  A  B  D  C  A
A  D  B  C  A  B  D  A  C  B  A  B  C  D  A  B  D  C  A  D  B  C  D  A  B
                    ↑
```

Now again the mismatch is at $P[1]$ and we slide P to the right by 1 step. The resulting alignment is as below:

```
                       A  B  C  D  A  B  D  C  A
A  D  B  C  A  B  D  A  C  B  A  B  C  D  A  B  D  C  A  D  B  C  D  A  B
                       ↑
```

Once again we need to slide P by only one step and the resulting alignment is

```
                          A  B  C  D  A  B  D  C  A
A  D  B  C  A  B  D  A  C  B  A  B  C  D  A  B  D  C  A  D  B  C  D  A  B
```

and there is a perfect match. Thus the first valid shift is at $s = 10$. To look for other valid shifts, we need to slide P by one step. The alignment of P and T is then given by

```
                A B C D A B D C A
A D B C A B D A C B A B C D A B D C A D B C D A B
                    ↑
```

Again we need to slide P by one step and the resulting alignment is

```
                A B C D A B D C A
A D B C A B D A C B A B C D A B D C A D B C D A B
                  ↑
```

This time also we need to slide P only by one step to get the new alignment of P and T as below:

```
                  A B C D A B D C A
A D B C A B D A C B A B C D A B D C A D B C D A B
                    ↑
```

Sliding P to the right by one step, the new alignment of P and T is given by

```
                  A B C D A B D C A
A D B C A B D A C B A B C D A B D C A D B C D A B
                    ↑
```

The present mismatch is at $P[3]$ and we need to slide P to the right by $3 - 1 = 2$ steps. The new alignment of P and T is as given below:

```
                      A B C D A B D C A
A D B C A B D A C B A B C D A B D C A D B C D A B
                    ↑
```

There is a mismatch at $P[1]$ and the text T has been exhausted. Hence there is no other valid shift in the case. Thus, the only valid shift is $s = 10$.

Example 13.11. Using KMP procedure find all possible valid shifts of the pattern $P = a\,b\,c\,a\,b\,c\,a\,c\,b\,a\,b\,c$ in the text

$$T = b\,a\,b\,c\,b\,a\,b\,c\,a\,b\,c\,a\,a\,b\,c\,a\,b\,c\,a\,b\,c\,a\,c\,b\,a\,b\,c\,a\,b\,c\,b\,a.$$

Solution. For this we refer to Example 13.7(a) for the next function computation for this pattern P.

We write the pattern P above the text T initially aligning the characters of P with those of T starting with $T[1]$. An upward arrow below an entry indicates a point of mismatch:

```
a b c a b c a c b a b c
b a b c b a b c a b c a a b c a b c a b c a c b a b c a b c b a
↑
```

Comparing the corresponding characters of P and T, we find that there is a mismatch at $P[1]$. Since $next[1] = 0$, we need to slide P to the right by $j - next[j] = 1 - 0 = 1$ step. The new alignment of P and T is as given below:'

```
  a b c a b c a c b a b c
b a b c b a b c a b c a a b c a b c a b c a c b a b c a b c b a
  ↑
```

Now there is a mismatch at the point $P[4]$. Since $next[4] = 0$, we need to slide P to the right by $4 - 0 = 4$ steps. The new alignment of P and T is as indicated below:

```
      a b c a b c a c b a b c
b a b c b a b c a b c a a b c a b c a b c a c b a b c a b c b a
      ↑
```

This time mismatch occurs at $P[8]$ and we need to slide P to the right by $8 - next[8] = 8 - 5 = 3$ steps. Then the new alignment of P and T is

```
      a b c a b c a c b a b c
b a b c b a b c a b c a a b c a b c a b c a c b a b c a b c b a
      ↑
```

Now a mismatch is found at $P[5]$ so that we have to slide P to the right by $5 - next[5] = 4$ steps. The alignment of P and T after sliding becomes

```
      a b c a b c a c b a b c
b a b c b a b c a b c a a b c a b c a b c a c b a b c a b c b a
      ↑
```

This time a mismatch is observed at the point $P[8]$ and we have to slide P to the right by $8 - next[8] = 8 - 5 = 3$ steps. The resulting alignment of P and T is as below:

```
      a b c a b c a c b a b c
b a b c b a b c a b c a a b c a b c a b c a c b a b c a b c b a
```

There is a perfect match this time and the valid shift is $s = 15$. Then we slide P to the right by 1 step to get the new alignment

```
      a b c a b c a c b a b c
b a b c b a b c a b c a a b c a b c a b c a c b a b c a b c b a
      ↑
```

Now a mismatch is observed at the point $P[1]$ and we have to slide P to the right by just 1 step. The resulting alignment of P and T is as given below:

$$a\ b\ c\ a\ b\ c\ a\ c\ b\ a\ b\ c$$
$$b\ a\ b\ c\ b\ a\ b\ c\ a\ b\ c\ a\ a\ b\ c\ a\ b\ c\ a\ b\ c\ a\ c\ b\ a\ b\ c\ a\ b\ c\ b\ a$$
$$\uparrow$$

There is a mismatch at the point $P[1]$ and we need to slide P to the right by 1 step. The resulting alignment of P and T is as given below:

$$a\ b\ c\ a\ b\ c\ a\ c\ b\ a\ b\ c$$
$$b\ a\ b\ c\ b\ a\ b\ c\ a\ b\ c\ a\ a\ b\ c\ a\ b\ c\ a\ b\ c\ a\ c\ b\ a\ b\ c\ a\ b\ c\ b\ a$$
$$\uparrow$$

Now we find a mismatch at $P[5]$ so that we have to slide P to the right by $5 - next[5] = 4$ steps finding that T is exhausted and P after sliding goes beyond T. Thus we find no new perfect match. The process stops at this point. Hence there is only one valid shift at $s = 15$.

Observe that Example 13.10 is not a good example for an application of KMP procedure.

Lemmas 13.5 and 13.6 immediately lead to the following algorithms for pre-computing $f[j]$ and $next[j], j \geq 1$. We also give the KMP algorithm.

Running time of KMP string matcher

As a first step when there is a mismatch between $P[j]$ and $T[j]$, the pattern P is shifted by $j - next[j] \geq 1$ steps and j character comparisons are made. After the pattern shift no character comparison is needed up to $T[j - 1]$. After the shift character comparisons are made

---

Algorithm 13.5. Algorithm Pre-Computing $f[j+1]$

Given a pattern P of length m and that $f[j]$ and $next[1], \ldots, next[j-1]$ have already been computed

1. let $f[j] = i$ and $j > i_1 > \cdots > i_k$ be integers such that $i_1 = i$ and $i_r = next[i_{r-1}], r \geq 2$
2. if $P[j] = P[i]$, then $f[j+1] = i+1$
3. if $P[j] \neq P[i]$, $k \geq 2$ be the first integer with $P[j] \neq P[i_2], \ldots, P[i_{k-1}]$ but $P[j] = P[i_k]$ or $i_k = 0$. Then
4. $f[j+1] = 1 + i_k = 1 + next[i_{k-1}]$

---

from $T[j]$ onwards. If the next mismatch occurs at $T[k]$ the number of new comparisons made is $k - j + 1$ and the total number of character comparisons is $j + k - j + 1 = k + 1$. Let a mismatch occurring at $T[l]$ correspond to the position $P[l']$ of P after the pattern P has been moved to the right from the previous mismatch position. Let $T[k]$, $T[l]$ with $k < l$ be two consecutive mismatches. The number of character comparisons to be made is l'. However $P_{next[k']}$ being a suffix of $P_{k'}$, the entries just before $P[k']$ need not be compared. Hence the number of character comparisons needed to come to the mismatch $T[l]$ is $l' - k' + 1 = l - k + 1$. Thus the total number of character comparisons to come to the mismatch $T[l]$ is $l - k + 1 +$*the number of comparisons needed to come to the mismatch*

Algorithm 13.6. Algorithm Pre-Computing *next[j]*, $j \geq 1$

Given a pattern P of length m

1. $j = 1, next[j] = 0$
2. for $j < m$, set $t = f[j]$ while $t > 0$ and $P[t] \neq P[j]$, $next[j] = t$
3. if $P[t] = P[j]$, then $next[j] = next[t]$

Algorithm 13.7. The *KMP* String Matching Algorithm

---

1. $P = P[1 \ldots m]$ is the pattern of length m, $m > 0$
2. $T = T[1 \ldots n]$ is the text of length n, $n \geq m$
3. initially place P above T so that P is in alignment with $T[1 \ldots m]$
4. if $P[j] \neq T[j]$, $j > 0$, slide P to the right through $j - next[j]$ steps
5. P_{i-1} matches with $T[j - i + 1 \ldots j - 1]$ and $P[i]$ may or may not equal $T[j]$
6. at an intermediate step, P is in alignment with $T[k-j+1 \ldots k-j+m]$ with $P[j] \neq T[k]$
7. repeat step 4 till $j > m$ in which case a complete match has been obtained or T has been exhausted
8. in case a complete match has been obtained but T has not been exhausted, slide P right by one step
9. repeat step 4 till a complete match is obtained or T has been exhausted
10. repeat steps 8 and 9 till finally T is exhausted

$T[k]$. Let $T[j]$, $T[k]$, $j < k$ be the first two mismatches. As already observed above, the number of character comparisons needed to come to the mismatch $T[j]$ is j so that the number of character comparisons needed to come to the mismatch $T[k]$ is $k - j + 1 + j = k + 1$. If $T[l]$ is the third mismatch in succession, then the number of character comparisons to come to $T[l]$ is $l - k + 1 + (k + 1) = l + 2$. Continuing this way, if $T[s]$ is the last mismatch when the text T is exhausted and it is the rth mismatch, the total number of character comparisons is $s + r - 1 \leq 2n - 1 < 2n$. Hence the running time of this algorithm is $\theta(n)$.

Running time of pre-computing the functions f and *next*

Observe that computation of the functions f and *next* are inter-dependent. Given $f[j]$, to compute $f[j + 1]$ we need to know $next[1], next[2], \ldots, next[j - 1]$ and for computing $next[j]$ we need to know $f[j]$. Let $f[j] = i$ be given and let $i_1 = i\langle = f[j]\rangle, i_2 = next[i_1], i_3 = next[i_2], \ldots, i_k = next[i_{k-1}]$. In view of the definition of f and *next*, $f[t] < t$ and $next[t] < t$. Therefore $j > i_1 > i_2 > \cdots > i_k$. In particular $k < j$. For finding $f[j + 1]$, we have to compare $P[j]$ with $P[i_1], \ldots, P[i_k]$ till k is the first index for which $P[j] = P[i_k]$. Thus for the computation of $f[j + 1]$, at most $k \leq j - 1$ comparisons are needed. Also at most l comparisons, where l is the smallest index with $P[j] = P[i_l]$, are needed for the computation of $next[j]$. Thus, given $f[j]$ and $next[1], next[2], \ldots, next[j - 1]$, the number of comparisons needed to obtain both $f[j + 1]$ and $next[j]$ is at most $j - 1$. Since $f[1] = next[1] = 0$ and $f[2] = 1$, running time for the computation of $f[j]$ and $next[j]$ for all $j \leq m$ is $1 + 2 + \cdots + m - 1 = \frac{m(m-1)}{2} = O(m^2)$.

For KMP procedure/algorithm, we follow the original approach of Knuth, Morris and Pratt [18] as simplified by Cormen *et al.* [5].

Chapter 14

Sorting Networks

We have earlier considered sorting algorithms in which only one operation could be executed at a time. Now we consider sorting algorithms in which many operations can be performed simultaneously. These algorithms are based on a comparison-network model in which more than one comparison operations may occur simultaneously. Such operations are said to be **in parallel**. The sorting networks we study have a fast sorting or ordering capability.

We begin by introducing comparison elements or comparators which make up a comparison network, give (i) a network for merging two sorted sequences, (ii) a bitonic network both of which as also other networks make use of the 0–1 principle. We also prove the 0–1 principle.

14.1. Comparison Networks

The basic element of sorting networks is a **comparison element** or a **comparator** or a **compare-exchange operation** which consists of two inputs, say x, y along two parallel wires and two outputs x', y' with x' on the upper wire and y' on the lower wire where $x' = \min\{x, y\}$ and $y' = \max\{x, y\}$. Pictorially such a comparator may be represented as in Fig. 14.1.

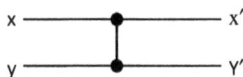

Fig. 14.1

439

The numbers appearing on the left are the inputs while the outputs are on the right.

We assume that each compare-exchange operator takes a constant amount of time.

Given a set of n wires each with an input value, the set of comparators which can operate in parallel, i.e., which can all be applied simultaneously is called a **comparator stage**. A composition of certain comparator stages is called **comparison network**. The number of distinct comparator stages is called the **depth** of the comparison network. Observe that two comparators can be applied simultaneously if and only if they do not have a common wire.

Definition. Given a comparison network, the depth of a wire is zero at the initial input. For a gate or a comparator with two inputs of depth d_1 and d_2, the depth on the output wire is $1 + \max\{d_1, d_2\}$. We may also call this the depth of the gate.

Lemma 14.1. *The depth of a comparison network is the maximum depth of an output wire.*

Proof. If a comparison network has only one comparator stage, consider any one comparator. The depth of both the input wires of this comparator is 0. Therefore, the depth of the output wire of this comparator is $1 + \max\{0, 0\} = 1$. Thus the lemma holds in this case. Suppose that the result holds when a comparison network has k comparator stages. Now consider a comparison network with $k + 1$ comparator stages. Ignoring the last comparator stage which is applied last of all, we get a network with k comparator stages. Therefore, by induction hypothesis, the depth of the truncated network is k, i.e., k is the maximum depth of an output wire. For any gate in the $(k + 1)$th comparator stage which involves a wire of depth k of the truncated network, depth of an output wire is $1 + \max\{k, j\}$ for some $j \leq k$ and, so, it is $k + 1$. Clearly $k + 1$ is the maximum of the depths of output wires of the network. This completes induction and, so, the lemma. (Recall that if no gate in the $(k + 1)$th comparator stage has input wire of depth k, then we do not get $(k + 1)$th comparator stage). $\square$

Example 14.1. Consider the comparison networks as given in Fig. 14.2.

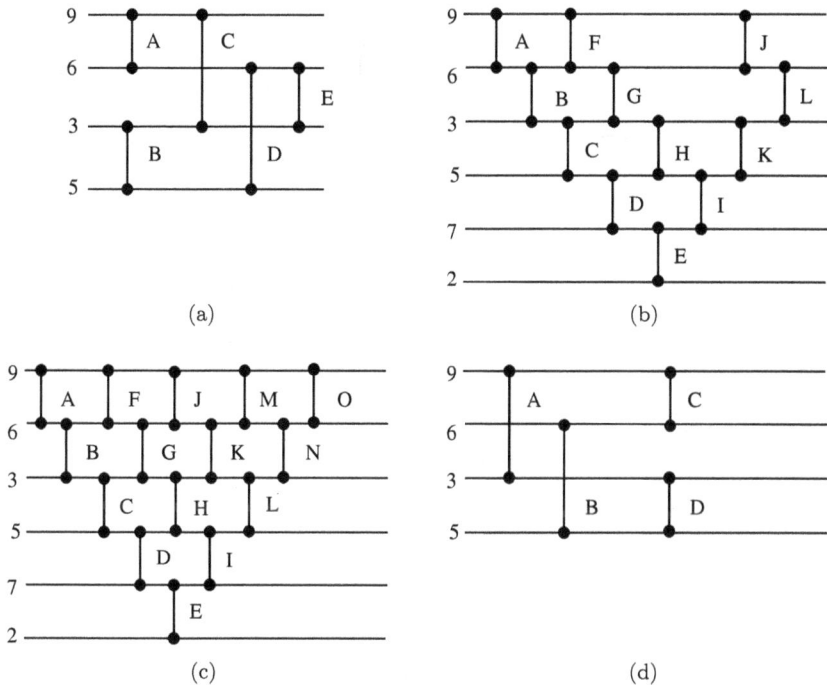

(a)

(b)

(c)

(d)

Fig. 14.2

In Fig. 14.2(a), gates A and B can be applied together. Since the comparators C, D, E have one wire in common with A and one wire in common with B, the gates A, B give one comparator stage. The comparators C, D can be applied together and since E has one wire in common with D, the C, D give another comparator stage and finally E constitutes the third comparator stage. Therefore, the network has depth 3. Also the sequence $9, 6, 3, 5$ gets transformed by the network of Fig. 14.2(a) to the sequence $3, 5, 6, 9$ and the steps involved are given by Fig. 14.3(a).

Consider now the network of Fig. 14.2(b). Initially, only the comparator A has two input wires at depth 0 and B can be applied only after A has been applied. Once both A and B have been applied, both F, C can be applied together. When the comparators A, B, C and F have been applied, comparators D and G can be applied simultaneously. Now, when the comparators A, B, C, D, F and G have been applied, the comparators E, H, J can be applied

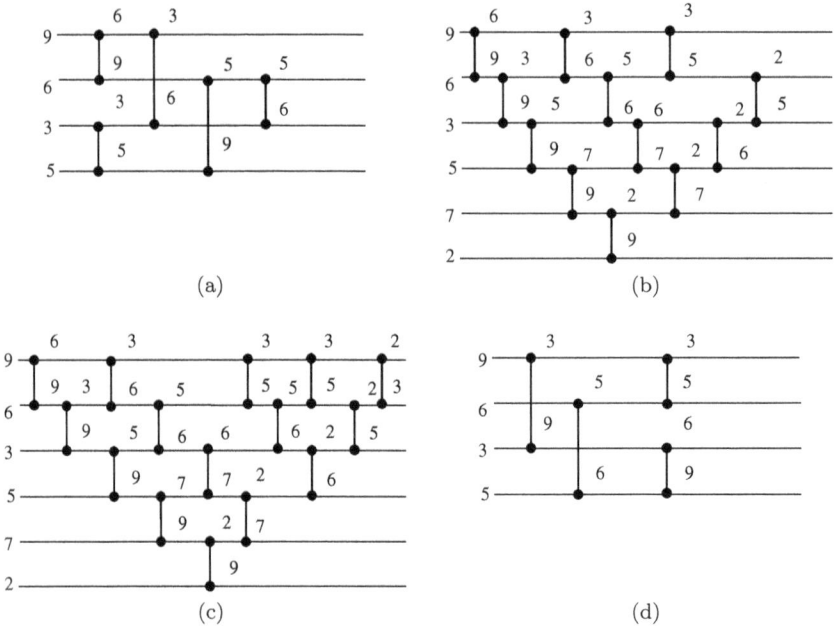

(a)

(b)

(c)

(d)

Fig. 14.3

simultaneously. After the comparators A, B, C, D, F, G, H and J have been applied, the comparator I can be applied and then comparators K and L can be applied in that order. Thus, the comparator stages are $\{A\}, \{B\}, \{C, F\}, \{D, G\}, \{E, H, J\}, \{I\}, \{K\}, \{L\}$ and the depth of the network is 8. Observe that depths of the various comparators are $A(1), B(2), C(3), F(3), D(4), G(4), E(5), H(5), J(5), I(6), K(7)$ and $L(8)$.

After the execution of this network, the given sequence of numbers gets transformed as under to the sequence 3, 2, 5, 6, 7, 9.

Omitting the details, we find that comparator stages for the network of Fig. 14.2(c) are $\{A\}, \{B\}, \{C, F\}, \{D, G\}, \{E, H, J\}, \{I, K\},$ $\{L, M\} \cdot \{N\}, \{O\}$ and the depth of the network is 9. After the execution of this network the sequence 9, 6, 3, 5, 7, 2 gets transformed to the sequence 2, 3, 5, 6, 7, 9.

Again omitting the details we find that the comparator stages of the network of Fig. 14.2(d) are $\{A, B\}, \{C, D\}$ and the depth of the network is 2. Also the sequence 9, 6, 3, 5 gets transformed by this network to the sequence 3, 5, 6, 9.

Observe that the networks of Figs. 14.2(a) and 14.2(d) transform the sequence 9, 6, 3, 5 to the same sequence although one of the networks is of depth 2 while the other is of depth 3.

The execution of all the comparators in one comparator stage takes a constant time or, so to say, a unit time. Therefore, the running time of a comparison network is $O(d)$, where d is the depth of the network. The final result is actually obtained in exactly d steps.

Definition. A **sorting network** is a comparison network such that for any sequence $a_1, \ldots, a_n$, the output sequence $b_1, \ldots, b_n$ is monotonically increasing, i.e., $b_1 \leq b_2 \leq \cdots \leq b_n$. The **size** of a sorting network is the number of gates or comparators in the network.

Every comparison network need not be a sorting network. For example, the comparison network of Fig. 14.2(b) is not a sorting network.

Example 14.2. If n is a power of 2, it is possible to construct an n-input, n-output comparison network of depth $\lg n$ in which the top output wire always carries the minimum input value and the bottom output wire always carries the maximum input value.

Solution. Let $n = 2^k$, the minimum input value be a, and the maximum input value be b. Suppose that a occurs on the wire r and b occurs on the wire s. Applying the parallel comparators $(1, r)$ and (s, n) we shift a on the top wire and b on the bottom wire. Ignore for the moment the top and bottom wires and the resulting $n - 2$ wires be denoted by $1, 2, \ldots, n - 2$. Any comparator network on these $n - 2$ wires will always leave the positions of the given minimum input and the given maximum input values unchanged. We thus need to construct a comparator network of depth $k - 1$ on the $n - 2$ wires numbered $1, 2, \ldots, n - 2$.

We first consider some special cases.

The case $n = 4$ is trivial as we only need to have one more comparator, namely $(1, 2)$. There are a couple of exceptions. One in which b occurs on the top wire and a occurs on the second or third wire. We first apply the comparator $(1, 4)$ and then the comparator $(1, 2)$ or $(1, 3)$ as the case may be. We have already got a network (desired) of depth 2. The case when a

occurs on the bottom wire and b occurs on the second or third wire may be dealt with similarly. When a occurs on the bottom wire and b occurs on the top wire, one comparator needed is $(1, 4)$ and the other is $(2, 3)$ both of which can be applied simultaneously. But in this case we may apply the comparators $(1, 3), (2, 4)$ in parallel and then apply $(1, 2), (3, 4)$ which can be applied simultaneously. Thus in this case also we get a desired network of depth 2.

The case $n = 8 = 2^3$. When a is on the bottom wire and b is on the top wire, we may either apply the interchanger a, b or apply $(1, 7), (2, 8)$ in parallel and then apply $(1, 2), (7, 8)$ again in parallel (which are of depth 2). In either case we ignore the top and bottom wires and name the remaining 6 wires 1 to 6. In the first case we apply the comparators $(1, 3), (2, 4)$ which do not increase the depth of the network and then apply $(3, 5), (4, 6)$ which are at depth 2 and finally apply the comparator $(5, 6)$ which is at depth 3 and we get a network of depth 3. In the second case, we have already obtained a network of depth 2 and apply the comparators $(3, 5), (4, 6)$ simultaneously without increasing the depth of the network obtained. Finally, we apply the comparators $(1, 5), (2, 6)$ in parallel at depth 3 and, so, we get a network of depth 3.

In case a is on the ith wire and b on the jth wire, we apply the comparators $(1, i), (j, 8)$ and $(r, s), r, s$ both different from $i, j, 2 \leq r \leq 7, 2 \leq s \leq 7$ and such that these are all in parallel. Thus, by applying this comparator stage of depth 1 the minimum input value a has been brought on the top wire and the maximum input value b on the bottom wire. Then apply the comparators $(2, 7)$ and $(2, 6)$ in turn to get a network of depth 3.

We now consider the general case $n = 2^k, k \geq 3$. First observe that if we have a comparator network of depth d, let (i, j) be one of the comparators at depth d. Then for any $l, l \neq i, j, (i, l)$ is a comparator of depth $d + 1$ and introducing this comparator we obtain a network of depth $d + 1$.

Suppose that a is on the ith wire and b is on the jth wire. Apply the comparators $(1, i), (j, n)$ and $(r, s), r, s$ different from $i, j, 2 \leq r, s \leq n - 1$ and which are all in parallel to get a network of depth 1 with a on the top wire and b on the bottom wire. Then apply the above procedure which adds 1 to the depth of the network at every step but avoiding comparators of the form $(1, r)$ and (s, n) till we get a network of depth k.

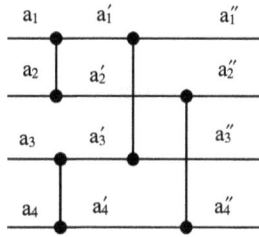

Fig. 14.4

We now consider the general case $n = 2^k$, $k \geq 1$. Suppose that the given sequence of numbers is $a_1, a_2, \ldots, a_n$ with a_i occurring on the ith wire. When $k = 1$, a network consisting of just one comparator $(1, 2)$ does the job. When $k = 2$, let us consider a pair of parallel comparators $(1, 2)$, $(3, 4)$. Let the output sequence on application of this pair of comparators be a_1', a_2', a_3', a_4' so that $a_1' < a_2'$ and $a_3' < a_4'$. We next apply the parallel comparators $(1, 3)$, $(2, 4)$ and suppose that the output sequence is $a_1'', a_2'', a_3'', a_4''$ where $a_1'' < a_3''$ and $a_2'' < a_4''$. Thus a_1' is the smaller of a_1, a_2 and a_3' is the smaller of a_3, a_4. Then the comparator $(1, 3)$ makes a_1'' as the smaller of a_1', a_3' and, therefore, the smallest of a_1, a_2, a_3, a_4. A similar reasoning shows that a_4'' is the largest of a_1, a_2, a_3, a_4. Thus a network of depth 2 applied to the input sequence results in the top wire carrying the smallest number and the bottom wire carrying the largest number (see Figure 14.4).

Now suppose that $k \geq 2$ and that the result holds for all $m = 2^l$ numbers with $l < k$. Consider the given sequence $a_1, a_2, \ldots, a_n$ as consisting of two sequences $a_1, \ldots, a_{\frac{n}{2}}$ on the wires $1, 2, \ldots, \frac{n}{2}$ and $a_{\frac{n}{2}+1}, \ldots, a_n$ on the wires $\frac{n}{2} + 1, \ldots, n$. By the induction hypothesis, we can construct two networks which can be applied in parallel and each is of depth $k - 1$. On applying the two networks in parallel, we find that the top wire carries the smallest of the numbers $a_1, \ldots, a_{\frac{n}{2}}$, the $\frac{n}{2}$th wire carries the largest of the numbers $a_1, \ldots, a_{\frac{n}{2}}$, the $(\frac{n}{2}+1)$th wire carries the smallest of the numbers $a_{\frac{n}{2}+1}, \ldots, a_n$ and the nth wire carries the largest of the numbers $a_{\frac{n}{2}+1}, \ldots, a_n$. Next we apply two comparators $(1, \frac{n}{2} + 1)$, $(\frac{n}{2}, n)$ in parallel and there results a network of depth $k - 1 + 1 = k$ in which the top wire carries the smallest and the bottom wire carries the largest of the numbers $a_1, a_2, \ldots, a_n$. This completes induction and we can construct a network of depth k for the numbers $a_1, a_2, \ldots, a_n$, with $n = 2^k$, which results in the top wire carrying the smallest and the bottom wire carrying the largest of the numbers $a_1, a_2, \ldots, a_n$.

Example 14.3. Prove that any sorting network on n inputs has depth at least $\lg n$.

Solution. We prove the result by contradiction. Suppose that there exists a sorting network on n inputs which has length $< \lg n$. Then the network has less than $\lg n$ comparator stages for any input sequence of length n. Let $n = 3$ so that $\lg n < 2$. Therefore the network sorts every sequence of length 3 in ≤ 1 comparator stages. Consider a sequence 3, 1, 2. All the comparators for a sequence of length 3 are $(1, 2)$, $(1, 3)$, $(2, 3)$ no two of which can be applied in parallel. Thus there are three comparator stages. Application of any one of the three comparators cannot sort the sequence under consideration. Indeed exactly two comparators are needed to sort the given sequence. This is a contradiction. Hence the depth of any sorting network on n inputs is $\geq \lg n$.

14.2. Batcher's Odd–Even Mergsort

Our interest here is to construct a systematic sorting network, i.e., a network that sorts a given set of numbers in ascending (i.e., increasing) order. A sorting network is constructed using an elegant mergsort algorithm developed by Batcher (cf.[3]). We first prove this rule of odd–even merge.

Let $A = a_1, a_2, \ldots, B = b_1, b_2, \ldots$ be two ascendingly ordered sequences of numbers and $A_o = a_1, a_3, \ldots, B_o = b_1, b_3, \ldots$ be the subsequences of A, B, respectively, consisting of odd indexed terms of the sequences and A_e, B_e be the subsequences $a_2, a_4, \ldots$ and $b_2, b_4, \ldots$ of even indexed terms of A and B, respectively. Let $D = d_1, d_2, d_3, \ldots$ be the ordered merge of the sequences A_o, B_o and $E = e_1, e_2, e_3, \ldots$ be the ordered merge of the sequences A_e, B_e. Let $c_1, c_2, c_3, \ldots$ be the ordered merge of the sequences A, B.

Theorem 14.2. *With the notations as above, for every* $i \geq 1$, $c_{2i} = \min\{d_{i+1}, e_i\}$, $c_{2i+1} = \max\{d_{i+1}, e_i\}$, *where* $c_1 = \max\{d_1\} = d_1$.

Proof. Let $i \geq 1$ and in the $i + 1$ terms $d_1, d_2, \ldots,$ let k of these be from A_o and $i + 1 - k$ of these be from B_o. Then $d_{i+1} \geq a_{2k-1}$ and $d_{i+1} \geq b_{2i+1-2k}$. Since A, B are increasingly ordered, $d_{i+1} \geq 2k - 1$ terms of A and $d_{i+1} \geq 2i + 1 - 2k$ terms of B. Therefore $d_{i+1} \geq 2i$ terms of the merge of A, B and, so of $2i$ terms of the ordered merge of A, B. Thus

$$d_{i+1} \geq c_{2i}. \tag{14.1}$$

Now consider i terms $e_1, e_2, \ldots, e_i$ of the ordered merge of A_e, B_e and suppose that k of these come from A_e and $i - k$ of these come from B_e. Then $e_i \geq a_{2k}$ and $e_i \geq b_{2i-2k}$. Again, the sequences A, B being ordered increasingly, $e_i \geq 2k$ elements of A, $e_i \geq 2i - 2k$ elements of B, i.e., $e_i \geq 2i$ elements of the merge of A, B and, hence, of the ordered merge of A, B. Thus

$$e_i \geq c_{2i}. \tag{14.2}$$

Now consider $2i + 1$ terms of $c_1, c_2, \ldots$ and suppose that k of these come from A and $2i + 1 - k$ of these come from B. Suppose k is even, say $k = 2l$. Then

$$c_{2i+1} \geq a_1, a_3, \ldots, a_{2l-1}, \text{ i.e., } c_{2i+1} \geq l \text{ terms of } A_o.$$

Similarly $c_{2i+1} \geq i + 1 - l$ terms of B_o. Therefore, $c_{2i+1} \geq i + 1$ terms of the merge of A_o, B_o and, so of ordered merge of A_o, B_o. Thus

$$c_{2i+1} \geq d_{i+1}. \tag{14.3}$$

Again, $c_{2i+1} \geq a_2, a_4, \ldots, a_{2l}$ and $c_{2i+1} \geq b_2, b_4, \ldots, b_{2i-2l}$, so that, $c_{2i+1} \geq i$ terms of the merge of A_e, B_e and, so, of ordered merge of A_e, B_e. Thus

$$c_{2i+1} \geq e_i. \tag{14.4}$$

Next, suppose that k is odd, say $k = 2l + 1$. Then $c_{2i+1} \geq a_1, a_3, \ldots, a_{2l+1}$ (i.e., $l + 1$ terms of A_o) and $c_{2i+1} \geq b_1, b_3, \ldots, b_{2i-1-2l}$ (i.e., $i - l$ terms of B_o), i.e., $c_{2i+1} \geq i + 1$ terms of the merge of A_o, B_o and, so, of ordered merge of A_o, B_o. Thus $c_{2i+1} \geq d_{i+1}$. A similar argument also shows that $c_{2i+1} \geq e_i$. Thus (14.3) and (14.4) hold whether k is odd or even. Now (14.1) – (14.4) imply that $c_{2i} \leq \min\{d_{i+1}, e_i\}$ and $c_{2i+1} \geq \max\{d_{i+1}, e_i\}$. Therefore $c_{2i} \leq \min\{d_{i+1}, e_i\} \leq \max\{d_{i+1}, e_i\} \leq c_{2i+1}$.

Since every term of $d_1, d_2, d_3, \ldots$ and every term of $e_1, e_2, e_3, \ldots$ appears in the sequence $c_1, c_2, c_3, \ldots$, we must have $c_{2i} = \min\{d_{i+1}, e_i\}$ and $c_{2i+1} = \max\{d_{i+1}, e_i\}$. $\qquad\square$

As a consequence of the above procedure, we get the following method of finding ordered merge of the sequences A, B provided the number of elements in A is equal to the number of elements in B.

Once we have obtained $d_1, d_2, \ldots, d_n$ and $e_1, e_2, \ldots, e_n$, splash the second sequence with the first, i.e., form the sequence $d_1, e_1, d_2, e_2, \ldots, d_n, e_n$. Then the ordered merge of A and B is obtained from this sequence, by leaving d_1 and e_n untouched in their positions and applying the comparators (e_i, d_{i+1}) for every i, $1 \leq i \leq n-1$.

Example 14.4. As an application of the above procedure, find ordered merge of the sequences $A = 3, 6, 9, 12, 13, 15$ and $B = 1, 4, 5, 7, 8, 14$.

Solution. The subsequences

A_o, B_o, A_e, B_e are $A_o = 3, 9, 13$; $B_o = 1, 5, 8$; $A_e = 6, 12, 15$; $B_e = 4, 7, 14$. Since the lengths of the sequences are small, we find that

$$d_1 = 1, d_2 = 3, d_3 = 5, d_4 = 8, d_5 = 9, d_6 = 13,$$

$$e_1 = 4, e_2 = 6, e_3 = 7, e_4 = 12, e_5 = 14, e_6 = 15.$$

Splashing the two sequences, we get

$$1, 4, 3, 6, 5, 7, 8, 12, 9, 14, 13, 15.$$

Now applying compare exchange operators for pairs of consecutive terms starting from the second term, we get the ordered merge of A and B as

$$1, 3, 4, 5, 6, 7, 8, 9, 12, 13, 14, 15.$$

We can also apply the procedure as given in Theorem 14.2 for finding ordered merge of sequences which are not necessarily of equal lengths.

When n is even, the time complexity of the odd–even mergsort = the running time of the mergsort of the sequences A_o, B_o + the running time of the mergsort of the sequences A_e, B_e + time taken for splashing E with

D and then applying the compare exchange operators (e_i, d_{i+1}), $1 \le i \le n-1$. There are $n-1$ insertions of e_i between d_i and d_{i+1}, insertion of e_n at the end, i.e., after d_n and the $n - 1$ comparators (e_i, d_{i+1}) all of which can be applied in parallel. Each of these operations takes a unit time. Therefore, if we let $Mergsort(A, B) = Mergsort(n, n)$ denote the running time or time complexity of finding ordered merge of A, B, then

$$Mergsort(n, n) = 2\, Mergsort\left(\frac{n}{2}, \frac{n}{2}\right) + n. \tag{14.5}$$

This is a fairly simple recurrence relation and if n is a power of 2, say 2^k, then we have the following theorem.

Theorem 14.3. $Mergsort(n, n) = n \lg n = k2^k$.

Proof. For $n = 1$, we have to merge a_1 with b_1 and for this we only need to apply one comparator (a_1, b_1) and, so, only one unit of time is taken. For $n = 2$, let the sequences be a_1, a_2; b_1, b_2. Then comparators (a_1, b_1), (a_2, b_2) in parallel give the sequences (d_1, d_2); (e_1, e_2). Again only one comparator (d_2, e_1) finally gives the sorted merge sequence. Thus, two units of time give the result. Thus, the result holds for $n = 2$. Now suppose that $n = 2^k$, $k \ge 2$ and that the result holds for all $m < k$. By (14.5) and then by induction hypothesis,

$$Mergsort(2^k, 2^k) = 2Mergsort(2^{k-1}, 2^{k-1}) + 2^k$$
$$= 2\{(k - 1)2^{k-1}\} + 2^k$$
$$= (k - 1)2^k + 2^k = k2^k = n \lg n.$$

$\square$

Example 14.5. Find the ordered merge of the sequences $A = 3, 6, 9, 12, 13, 15, 19, 26$ and $B = 1, 4, 5, 7, 8, 14, 20, 27$.

Solution. The subsequences A_o, B_o, A_e, B_e in this case are

$$A_o = 3, 9, 13, 19; \ B_o = 1, 5, 8, 20;$$

$$A_e = 6, 12, 15, 26; \ B_e = 4, 7, 14, 27.$$

We have now to find the mergsort $D = d_1, d_2, \ldots$ of A_o and B_o and the mergsort $E = e_1, e_2, \ldots$ of the sequences A_e, B_e. Although the sequences A_o, B_o, A_e, B_e are of small lengths and we can find D and E straightaway, we break the four sequences again into sequences of odd and even terms. We consider the sequences

$$A_{oo} : 3, 13; \quad B_{oo} : 1, 8; \quad A_{oe} : 9, 19; \quad B_{oe} : 5, 20;$$

$$A_{eo} : 6, 15; \quad B_{eo} : 4, 14; \quad A_{ee} : 12, 26; \quad B_{ee} : 7, 27.$$

Ordered merge of

$$A_{oo} \text{ and } B_{oo} \text{ is } 1, 3, 8, 13; A_{oe} \text{ and } B_{oe} \text{ is } 5, 9, 19, 20;$$

$$A_{eo} \text{ and } B_{eo} \text{ is } 4, 6, 14, 15; A_{ee} \text{ and } B_{ee} \text{ is } 7, 12, 26, 27.$$

Splashing the ordered merge of A_{oo} and B_{oo} with the ordered merge of A_{oe} and B_{oe} and also the ordered merge of A_{eo} and B_{eo} with the ordered merge of A_{ee} and B_{ee}, respectively, we get the sequences

$$1, 5, 3, 9, 8, 19, 13, 20 \text{ and } 4, 7, 6, 12, 14, 26, 15, 27.$$

Now applying the comparators $(i, i+1)$ to these sequences leaving the first and last entries untouched, we get the ordered merges D and E as

$$D : 1, 3, 5, 8, 9, 13, 19, 20 \ E : 4, 6, 7, 12, 14, 15, 26, 27.$$

Now splashing D and E, we get

$$1, 4, 3, 6, 5, 7, 8, 12, 9, 14, 13, 15, 19, 26, 20, 27.$$

Applying the comparators to every pair $(i, i+1)$ of consecutive terms, leaving the first and last terms untouched, we get the desired ordered merge of the given sequences A, B as below:

$$1, 3, 4, 5, 6, 7, 8, 9, 12.13, 14, 15, 19, 20, 26, 27.$$

In order to apply the Batcher's odd–even merge, we need to sort the splash of A_o, B_o and of A_e, B_e to obtain D and E. But the odd–even merge itself can be applied to systematically construct a sorting network.

Given a sequence $(a[1], a[2], \ldots, a[n])$ of length n, let m be the smallest integer $\geq \frac{n}{2}$ and divide the given sequence into subsequences $(a[1], \ldots, a[m])$ and $(a[m+1], \ldots, a[n])$ of lengths m and $n-m$, respectively. Both these sequences being of smaller length than n, suppose that sorting networks have been constructed for sorting these subsequences. Then $(a[1], \ldots, a[m])$ and $(a[m+1], \ldots, a[n])$ are sorted sequences and we apply Batcher's odd–even merge to obtain a sorted merge of these sequences which is thus the ordered form of the given sequence.

The building blocks of constructing this sorting network are the construction of such networks of length $n = 2, 3, 4, 5$. However, to make the procedure clearer, we construct these networks for $n = 2, 3, 4, 5, 6, 7$.

Example 14.6. (a) The case $n = 2$ is trivial. To sort the sequence $(a[1], a[2])$, we need to merge the sequences $(a[1])$ and $(a[2])$ which amounts to just applying $compex(a, 1, 2)$. (We write $compex(a, 1, 2)$ for the comparator $(a[1], a[2])$ or just $(1, 2)$.)

(b) Now consider a sequence $(a[1], a[2], a[3])$ of length 3. Break this sequence into subsequences $(a[1], a[2])$ and $(a[3])$. Sort $(a[1], a[2])$ by using $compex(a, 1, 2)$ to get sorted sequences $(a[1], a[2])$ and $(a[3])$. The corresponding odd and even sequences are $(a[1]), (a[3])$ and $(a[2])$. Apply $compex(a, 1, 3)$ to the sequence $(a[1], a[3])$ and we get sorted sequences $(a[1]), (a[3])$ and $(a[2])$. Shuffle these to get the sequence $(a[1]), (a[2], a[3])$. Now correct this sequence, i.e., apply $compex(a, 2, 3)$ to get the sorted sequence. Translating the above steps, we get the network of wires as given in Fig. 14.5 which is a network of depth 3.

(c) Consider a sequence $(a[1], a[2], a[3], a[4])$ of length 4.

Divide the sequence into subsequences $(a[1], a[2])$ and $(a[3], a[4])$. Sort these two sequences by merging $(a[1)], (a[2])$ and $(a[3]), (a[4])$, i.e., apply $compex(a, 1, 2)$ and $compex(a, 3, 4)$ so that $(a[1], a[2])$ and $(a[3], a[4])$ are sorted. Merge the odd term sequences $(a[1]), (a[3])$ and even term sequences $(a[2]), (a[4])$, i.e., apply $compex(a, 1, 3)$ and $compex(a, 2, 4)$. Thus the sequences $(a[1], a[3])$ and $(a[2], a[4])$ are sorted. Shuffle these two sequences to get $(a[1], a[2], a[3], a[4])$. Correct this sequence by applying $compex(a, 2, 3)$ and we get a sorted sequence $(a[1], a[2], a[3], a[4])$. Taking care of the various steps including compex

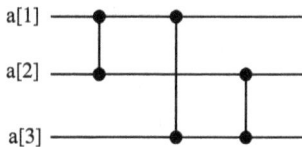

Fig. 14.5

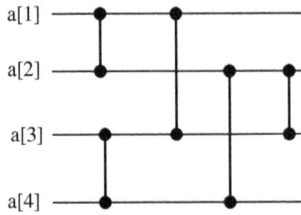

Fig. 14.6

operations, we get the sorting network of Fig. 14.6 which is again a sorting network of depth 3.

(d) Consider a sequence $(a[1], a[2], a[3], a[4], a[5])$ of length 5.

Divide the sequence into subsequences $(a[1], a[2], a[3])$ and $(a[4], a[5])$. Sort these sequences by applying the networks obtained in (b) and (a). Thus we obtain sorted sequences $(a[1], a[2], a[3])$ and $(a[4], a[5])$. To find the merge sort of these sorted sequences, consider the odd sequences $(a[1], a[3])$ and $(a[4])$ and the even sequences $(a[2])$ and $(a[5])$. We need to merge the odd sequences $(a[1])$, $(a[4])$ which is done by applying $compex(a, 1, 4)$ so that $(a[1], a[4])$ is a sorted sequence. Now shuffle this sequence with $(a[3])$ to get the sequence $(a[1], a[3], a[4])$. Correct this sequence, i.e., apply $compex(a, 3, 4)$ so that $(a[1], a[3], a[4])$ is a sorted sequence.

Merge $(a[2])$ and $(a[5])$, i.e., apply $compex(a, 2, 5)$ to get a sorted sequence $(a[2], a[5])$. We thus have the sorted sequences $(a[1], a[3], a[4])$ and $(a[2], a[5])$. Shuffle these sequences to get the sequence $(a[1], a[2], a[3], a[5], a[4])$. Correct this sequence by applying $compex(a, 2, 3)$ and $compex(a, 4, 5)$ and we get the sorted sequence

Keeping track of the various comparators used, the sorting network is as given in Fig. 14.7 and is of depth 6.

(e) Consider a sequence $(a[1], a[2], a[3], a[4], a[5], a[6])$ of length 6.

Divide the sequence into subsequences $(a[1], a[2], a[3])$ and $(a[4], a[5], a[6])$. Apply the sorting network as in (b) to get sorted sequences $(a[1], a[2], a[3])$ and $(a[4], a[5], a[6])$. We need to find mergsort of these sequences. Consider the odd and even sequences $(a[1], a[3])$, $(a[4], a[6])$ and $(a[2])$, $(a[5])$. Merge the even sequences, i.e., apply $compex(a, 2, 5)$ to get a sorted sequence $(a[2], a[5])$.

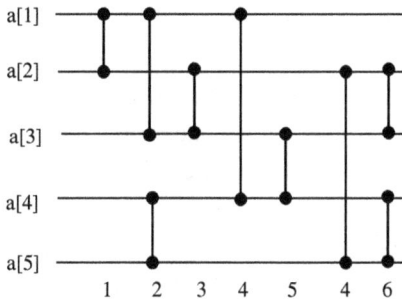

Fig. 14.7

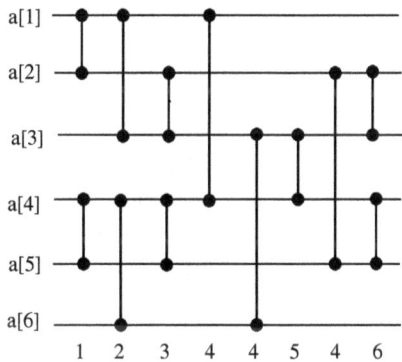

Fig. 14.8

For the odd sequences, we again get the odd and even sequences $(a[1])$, $(a[4])$, and $(a[3])$, $(a[6])$. Merge these odd and even sequences, i.e., apply $compex(a, 1, 4)$ and $compex(a, 3, 6)$ to get sorted sequences $(a[1], a[4])$ and $(a[3], a[6])$. Shuffle these sequences to get the sequence $(a[1], a[3], a[4], a[6])$. Correct this sequence, i.e., apply $compex(a, 3, 4)$ to get a sorted sequence $(a[1], a[3], a[4], a[6])$.

Now shuffle this sequence with the even sorted sequence $(a[2], a[5])$ to get the sequence $(a[1], a[2], a[3], a[5], a[4], a[6])$. Correct this sequence, i.e., apply $compex(a, 2, 3)$ and $compex(a, 4, 5)$ to get the sorted sequence $(a[1], a[2], a[3], a[4], a[5], a[6])$.

The sorting network as above is thus given by Fig. 14.8 which is of depth 6.

(f) Consider a sequence $(a[1], a[2], a[3], a[4], a[5], a[6], a[7])$ of length 7.

Divide the sequence into subsequences $(a[1], a[2], a[3], a[4])$ and $(a[5], a[6], a[7])$. Applying sorting networks of examples (c) and (b), respectively, to these sequences, we obtain sorted sequences $(a[1], a[2], a[3], a[4])$ and $(a[5], a[6], a[7])$. To find the mergsort of these sequences, we consider the odd and even sequences $(a[1], a[3])$, $(a[5], a[7])$ and $(a[2], a[4])$, $(a[6])$. For the odd sequences, we again get the odd and even sequences as $(a[1])$, $(a[5])$ and $(a[3])$, $(a[7])$. Merge these pairs of odd and even sequences. We only need apply $compex(a, 1, 5)$ and $compex(a, 3, 7)$ to get sorted odd–even sequences $(a[1], a[5])$, and $(a[3], a[7])$. Shuffle these sequences to get the sequence $(a[1], a[3], a[5], a[7])$. Correct this sequence: apply $compex(a, 3, 5)$ to get the sorted odd sequence $(a[1], a[3], a[5], a[7])$ of the given sequence.

Next, consider the even sequences $(a[2], a[4])$, $(a[6])$. Odd and even sequences of these sequences are $(a[2])$, $(a[6])$ and $(a[4])$. Merge the odd sequences: apply $compex(a, 2, 6)$ to get the sorted sequence $(a[2], a[6])$. Shuffle this sequence with the sorted even sequence $(a[4])$ to get the sequence $(a[2], a[4], a[6])$. Correct this sequence: apply $compex(a, 4, 6)$ to get the sorted even sequence $(a[2], a[4], a[6])$ of the given sequence.

Shuffle the sorted odd and even sequences of the given sequence to get $(a[1], a[2], a[3], a[4], a[5], a[6], a[7])$. Correct this sequence: apply $compex(a, 2, 3)$, $compex(a, 4, 5)$, $compex(a, 6, 7)$ to get the given sequence in sorted form.

This sorting network is given in Fig. 14.9 and is of depth 6.

Examples 14.6(a)–(f) are fairly indicative of the application of Batcher's odd–even merge sort for the construction of sorting network of length n. Also observe that the depth of sorting network of length n is 1 for $n = 2$, 3 for $n = 3$, 4 and 6 for $n = 5, 6, 7$.

Exercise 14.1. Apply Batcher's odd–even mergsort and sort the sequences:

(a) 11,5,3,2,6,9,7,4,13,10,
(b) 10,6,3,7,5,9,8,1,11,
(c) 14,11,6,9,12,15,10,11,1,2,3.

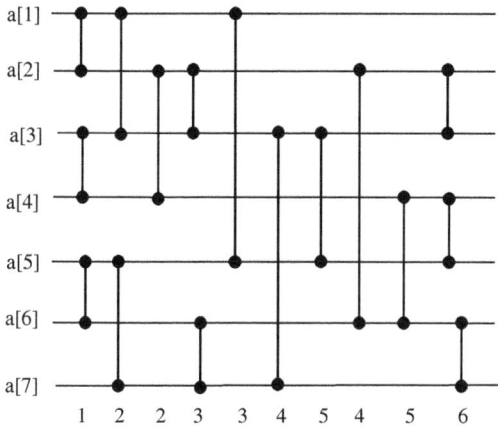

Fig. 14.9

14.3. The Zero-One Principle

To test the validity of an n-wire sorting network we need to check $n!$ (factorial n) permutations of n numbers, and this number of permutations increases significantly if n is large. The number of test cases is significantly reduced by using the so-called zero-one principle. The **zero-one principle** states that if a comparison network correctly sorts all binary sequences of length n, then it sorts correctly any input sequence of length n. To prove the zero-one principle, we first prove the following lemma.

Lemma 14.4. *If a comparison network transforms an input sequence* $a = (a_1, a_2, \ldots, a_n)$ *into the output sequence* $b = (b_1, b_2, \ldots, b_n)$, *then for any monotonically increasing function* f, *the network transforms the input sequence* $f(a) = (f(a_1), f(a_2), \ldots, f(a_n))$ *into the output sequence* $f(b) = (f(b_1), f(b_2), \ldots, f(b_n))$.

Proof. Consider a single comparator with input sequence (x, y) and output sequence (x', y') where $x' = \min\{x, y\}$, $y' = \max\{x, y\}$. If $x \leq y$, then $x' = x$, $y' = y$ and the function f being monotonically increasing, $f(x) \leq f(y)$ and $\min\{f(x), f(y)\} = f(x) = f(x') = f(\min\{x, y\})$, $\max\{f(x), f(y)\} = f(y) = f(y') = f(\max\{f(x), f(y)\})$.

If, on the other hand, $x > y$, then $x' = y$, $y' = x$ and $f(x) > f(y)$ so that $\min\{f(x), f(y)\} = f(y) = f(x') = f(\min\{x, y\})$, $\max\{f(x), f(y)\} = f(x) = f(y') = f(\max\{f(x), f(y)\})$. Thus in either

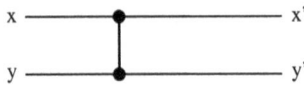

Fig. 14.10

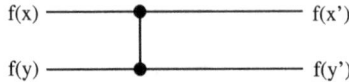

Fig. 14.11

case, the comparator transforms the sequence $(f(x), f(y))$ onto the output sequence $(f(x'), f(y'))$ (see Figs. 14.10 and 14.11).

We now claim: If a wire assumes the value a_i when the input sequence $a = (a_1, a_2, \ldots, a_n)$ is applied to the comparison network, then the wire assumes the value $f(a_i)$ when the input sequence $f(a) = (f(a_1), f(a_2), \ldots, f(a_n))$ is applied to the network.

We prove this claim by induction on the depth d of the wire. Suppose that depth of the wire is 0. Then a_i is just the input value on the wire when $a = (a_1, a_2, \ldots, a_n)$ is applied to the network. Thus, when the input sequence $f(a) = (f(a_1), f(a_2), \ldots, f(a_n))$ is applied to the network, then the value on the wire under consideration is the initial value $f(a_i)$ itself. Thus, the claim holds for depth $d = 0$. Suppose that a_i is the value on the wire at point p at depth $d \geq 1$ when the input sequence a is applied. Then a_i is one output of a comparator G the input values of which are, say, b_i and b_j. Let the other output value of G be a_j. The b_i, b_j are the values of wires of depth $d - 1$ and depth $\leq d - 1$. Therefore, by induction hypothesis, the values assumed by these wires when the input sequence $f(a) = (f(a_1), f(a_2), \ldots, f(a_n))$ is applied to the network are $f(b_i), f(b_j)$. Then the output values of the comparator G, i.e., the values assumed by the wires on the right side of G, are $f(a_i), f(a_j)$ (by the observation made in the beginning of the proof). This proves that the value assumed by the wire under consideration at depth d which assumed the value a_i when the sequence a is applied is $f(a_i)$ when the input sequence $f(a) = (f(a_1), f(a_2), \ldots, f(a_n))$ is applied. This completes induction. This proves, in particular, that if the comparison network yields an output sequence $b = (b_1, b_2, \ldots, b_n)$ when an input sequence $a = (a_1, a_2, \ldots, a_n)$ is applied, then the network transforms the input sequence $f(a) = (f(a_1), f(a_2), \ldots, f(a_n))$ into the output sequence $f(b) = (f(b_1), f(b_2), \ldots, f(b_n))$. □

Theorem 14.5 (The 0–1 principle). A comparison network that sorts all 2^n sequences of zeros and ones of length n is a sorting network and, so, it sorts all sequences of length n of arbitrary numbers.

Proof. We prove the result by contradiction. Let N be a network of n wires that correctly sorts all binary sequences of length n but there is a sequence, say $(a_1, a_2, \ldots, a_n)$ of arbitrary numbers that is not correctly sorted by N. Let $a_i < a_j$ be two numbers that are not correctly sorted, i.e., in the output the number a_j appears before a_i. Define a function f by

$$f(x) = \begin{cases} 0 & \text{for } x \leq a_i, \\ 1 & \text{for } x > a_i, \end{cases}$$

which is clearly a monotonically increasing function. Let the output sequence for the input sequence $(a_1, a_2, \ldots, a_n)$ be $(b_1, b_2, \ldots, b_n)$. By our assumption, there exist k, l, $1 \leq k < l \leq n$ such that $b_k = a_j$ and $b_l = a_i$. By Lemma 14.4, the output sequence for the input sequence $(f(a_1), f(a_2), \ldots, f(a_n))$ is $(f(b_1), f(b_2), \ldots, f(b_n))$. $= (\ldots, 1, \ldots, 0, \ldots)$ which shows that the binary sequence $(f(a_1), \ldots, f(a_i), \ldots, f(a_j), \ldots, f(a_n)) = (\ldots, 0, \ldots, 1, \ldots)$ is not sorted. This contradicts our assumption that all binary sequences are sorted by the network. Hence the zero-one principle holds. □

We consider a couple of examples first of which is a nice application of Lemma 14.4.

Example 14.7. A comparison network with n inputs correctly sorts the input sequence $(n, n-1, \ldots, 2, 1)$ if and only if it correctly sorts the $n-1$ zero-one sequences $(1, 0, 0, \ldots, 0), (1, 1, 0, \ldots, 0), \ldots, (1, 1, \ldots, 1, 0)$ of length n.

Proof. Suppose that a comparison network correctly sorts the input sequence $(n, n-1, \ldots, 2, 1)$ so that the output sequence is $(1, 2, \ldots, n)$.

For integer r, $1 \leq r \leq n - 1$ define a function f_r by

$$f_r(x) = \begin{cases} 0 & \text{for } 1 \leq x \leq r, \\ 1 & \text{for } x > r. \end{cases}$$

By Lemma 14.4 the sequence $(f_r(n), f_r(n-1), \ldots, f_r(2), f_r(1))$ gets transformed by the network into the output sequence $(f_r(1), \ldots, f_r(r), f_r(r+1) \ldots, f_r(n))$, i.e., the sequence $(a_n, a_{n-1}, \ldots, a_1)$ with $a_1 = \cdots = a_r = 0$ and $a_{r+1} = \cdots = a_n = 1$ gets transformed into the sequence $(b_1, b_2, \ldots, b_n)$ with $b_1 = \cdots = b_r = 0$, $b_{r+1} = \cdots = b_n = 1$. Since this is true for all r, $1 \leq r \leq n - 1$, the sequences $(1, 0, 0, \ldots, 0)$, $(1, 1, 0, \ldots, 0), \ldots, (1, 1, \ldots, 1, 0)$ are all correctly sorted.

Conversely, suppose that a comparison network sorts the $n - 1$ binary sequences $(1, 0, 0, \ldots, 0)$, $(1, 1, 0, \ldots, 0), \ldots, (1, 1, \ldots, 1, 0)$ of length n. Since the network sorts the sequence $(1, 0, 0, \ldots, 0)$, it must contain the comparator $(1, n)$. This comparator then transforms the other $n - 2$ sequences into the sequences $(0, 1, 0, \ldots, 0, 1)$, $(0, 1, 1, 0, \ldots, 0, 1), \ldots, (0, 1, 1, \ldots, 1)$. Again, since the sequence $(0, 1, 0, \ldots, 0, 1)$ is corrected, the network must contain the comparator $(2, n - 1)$. Applying this comparator to the $n - 3$ sequences $(0, 1, 0, \ldots, 0, 1)$, $(0, 1, 1, 0, \ldots, 0, 1), \ldots, (0, 1, 1, \ldots, 1, 0, 1)$ the sequences get transformed to $(0, 0, \ldots, 1, 1)$, $(0, 0, 1.0, \ldots, 0, 1, 1), \ldots, (0, 0, 1, \ldots, 1)$ the first and last of which are already sorted. Continuing this argument, we find that the comparison network contains the comparators $(1, n)$, $(2, n - 1)$, $\ldots, (r, n - r + 1)$ where $r = \frac{n}{2}$ if n is even and $r = \frac{n-1}{2}$ if n is odd. It is straightforward to check that application of these comparators (in parallel) transforms the sequence $(n, n - 1, \ldots, 2, 1)$ into the sequence $(1, 2, \ldots, n)$, i.e., the network sorts the sequence $(n, n - 1, \ldots, 2, 1)$.

Example 14.8. Prove that an n-input sorting network must always contain at least one comparator between the ith and $(i + 1)$th lines for all i, $1 \leq i \leq n - 1$.

Proof. Let $i \geq 1$ and choose an input sequence $a = (a_1, a_2, \ldots, a_n)$ with $a_i = 1$ and $a_j = 0$ for $1 \leq j \leq i - 1$, $a_{i+1} = 0$ and $a_j = 1$ for $j > i + 1$. Since the given network N sorts all sequences of length n, it must sort the sequence a. But to sort this sequence we need a comparator $(i, i+1)$. Thus N contains at least one comparator $(i, i + 1)$ for every $i \geq 1$. $\qquad\square$

14.4. A Bitonic Sorting Network

Definition. A sequence which is first monotonically increasing and then monotonically decreasing or can be circularly shifted to become so is called a **bitonic sequence**. A sequence $a_1 \ldots a_n$ with $a_i = 0$ for every i or $a_i = 1$ for every i is called a **clean sequence**.

Sequences $(1, 5, 7, 9, 4, 3, 2)$, $(7, 9, 4, 1, 3, 5)$, $(8, 6, 4, 2, 3, 5)$ are bitonic but the sequences $(1, 3, 1, 3)$, $(8, 6, 2, 5, 3)$ are not bitonic.

Lemma 14.6. *A binary bitonic sequence (i.e., a bitonic sequence every entry of which is either 0 or 1) is always of the form $0^i 1^j 0^k$ or $1^i 0^j 1^k$ for some $i, j, k \geq 0$, where by a^i we mean a sequence of length i in which every entry is a.*

Proof. A binary monotonically increasing sequence is of the form $0^i 1^j$ while a binary monotonically decreasing sequence is of the form $1^i 0^j$ for some $i, j \geq 0$. Concatenation of a binary increasing sequence followed by a decreasing binary sequence is then of the form $0^i 1^j 0^k$ for some $i, j, k \geq 0$. If $j \geq 1$, giving $k + 1$ or more but less than $k + j$ cyclic shifts to this sequence results in a sequence of the form $1^r 0^s 1^t$. Observe that giving k cyclic shifts to this sequence with $0 < k < s + t$ results in a sequence of the form $0^i 1^r 0^l$ for some $i, r, l \geq 0$. Hence every binary bitonic sequence is of the form $0^i 1^j 0^k$ or $1^i 0^j 1^k$ for some $i, j, k \geq 0$. $\square$

Observe that every monotonically increasing or decreasing sequence is always bitonic. Then every sequence of length 1 or 2 is bitonic. It is also fairly easy to see that every sequence of length 3 is bitonic. However, every sequence of length 4 need not be bitonic as already observed that the sequence $(1, 3, 1, 3)$ is not bitonic.

Definition. Consider a network of n wires where n is even. A **half-cleaner** is a comparison network with comparators $(i, i + \frac{n}{2})$ for $1 \leq i \leq \frac{n}{2}$.

A half-cleaner with n inputs is denoted by **Half-cleaner**$[n]$. Observe that all the $\frac{n}{2}$ comparators $(i, i + \frac{n}{2})$, $1 \leq i \leq \frac{n}{2}$, can be performed in parallel and, as such, the **depth** of a Half-cleaner$[n]$ is **one**.

To see the effect of a Half-cleaner$[n]$ on a bitonic sequence, consider a bitonic sequence $A = a_1 a_2 \ldots a_{2n}$ of length $2n$. For $1 \leq i \leq n$, let

$$d_i = \min\{a_i, a_{i+n}\} \text{ and } e_i = \max\{a_i, a_{i+n}\} \tag{14.6}$$

Theorem 14.7. *The sequences $d_1 d_2 \ldots d_n$ and $e_1 e_2 \ldots e_n$ are both bitonic and $\max\{d_1, d_2, \ldots, d_n\} \leq \min\{e_1, e_2, \ldots, e_n\}$.*

Proof. Let j be an integer, $1 \leq j \leq 2n$, such that

$$a_1 \leq a_2 \leq \cdots \leq a_j \geq a_{j+1} \geq \cdots \geq a_{2n} \qquad (14.7)$$

If $j \leq n$, we can consider the bitonic sequence $b_1 b_2 \ldots b_{2n}$ with $b_i = a_{2n+1-i}$ and for this sequence we find that $b_1 \leq b_2 \leq \cdots \leq b_{2n+1-j} \geq b_{2n-j} \geq \cdots \geq b_{2n}$ with $n < 2n+1-j < 2n$ and $\min\{b_i, b_{i+n}\} = d_{n+1-i}$ and $\max\{b_i, b_{i+n}\} = e_{n+1-i}$. Thus, without loss of generality, we consider the bitonic sequence A satisfying (14.7) with j, $n < j < 2n$.

Suppose that $a_n \leq a_{2n}$. Then $a_i \leq a_{i+n}$ for every i, $1 \leq i \leq n$ and, so, $d_i = a_i$, $e_i = a_{i+n}$. Therefore, $d_1 \leq d_2 \leq \cdots \leq d_n$ and $e_1 \leq \cdots \leq e_{j-n} \geq e_{j-n+1} \geq \cdots \geq e_n$ so that $d_1 d_2 \ldots d_n$ and $e_1 e_2 \ldots e_n$ are both bitonic. It is clear that $\max\{d_1, d_2, \ldots, d_n\} = a_n$ and $\min\{e_1, e_2, \ldots, e_n\} = \min\{a_{n+1}, \ldots, a_j, a_{j+1}, \ldots, a_{2n}\} = \min\{a_{n+1}, a_{2n}\}$. Therefore $\max\{d_1, d_2, \ldots, d_n\} \leq \min\{e_1, e_2, \ldots, e_n\}$.

Now suppose that $a_n > a_{2n}$. This together with $a_{j-n} \leq a_j$ implies that there exists a k with $j \leq k < 2n$ such that $a_{k-n} \leq a_k$ but $a_{k-n+1} > a_{k+1}$. Then $a_{k-n+r} > a_{k+r}$ for all r, $1 \leq r < 2n - k$. Therefore

$$d_i = a_i, e_i = a_{i+n} \text{ for } 1 \leq i \leq k - n, \qquad (14.8)$$

$$d_i = a_{i+n}, e_i = a_i \text{ for } k - n < i \leq n. \qquad (14.9)$$

Since $k - n < n$, it follows from (14.8) and (14.9) that

$$d_i \leq d_{i+1} \text{ for } 1 \leq i \leq k - n \text{ and } d_i \geq d_{i+1} \text{ for } k - n \leq i < n.$$

Thus $d_1, d_2, \ldots, d_n$ is a bitonic sequence. Also

$$\max\{d_1, d_2, \ldots, d_n\} = \max\{a_1, \ldots, a_{k-n}, a_{k+1}, a_{k+2}, \ldots, a_{2n}\}$$

$$= \max\{a_{k-n}, a_{k+1}\}.$$

For e_i, again using (14.8) and (14.9), we get

$$e_i = a_{i+n} \text{ for } 1 \leq i \leq j - n \text{ and, so, } e_1 \leq e_2 \leq \cdots \leq e_{j-n},$$

$$e_i = a_{i+n} \text{ for } j - n \leq i \leq k - n \text{ and, so,}$$

$$e_{j-n} \geq e_{j-n+1} \geq e_{j-n+2} \geq \cdots \geq e_{k-n},$$

$$e_i = a_i \text{ for } k - n \leq i \leq n \text{ and, so, } e_{k-n+1} \leq e_{k-n+2} \leq \cdots \leq e_n.$$

Also $e_n = a_n \leq a_{1+n} = e_1$. Thus, we have

$$e_1 \leq e_2 \leq \cdots \leq e_{j-n} \geq e_{j-n+1} \geq \cdots \geq e_{k-n}, \tag{14.10}$$

$$e_{k-n+1} \leq e_{k-n+2} \leq \cdots \leq e_n \leq e_1. \tag{14.11}$$

The two sequences (14.10) and (14.11) lead to a bitonic sequence

$$e_{k-n+1} \leq \cdots \leq e_n \leq e_1 \leq e_2 \leq \cdots \leq e_{j-n} \geq e_{j-n+1} \geq \cdots \geq e_{k-n}.$$

The sequence

$$e_{k-n+1}, e_{k-n+2}, \ldots, e_n, e_1, \ldots, e_{k-n} \tag{14.12}$$

is obtained from the sequence $e_1, e_2, \ldots, e_n$ by giving $2n - k$ cyclic shifts. Hence the sequence $e_1, e_2, \ldots, e_n$ is bitonic. Also

$$\min\{e_1, e_2, \ldots, e_n\} = \min\{e_{k-n+1}, e_k\} = \min(a_{k-n+1}, a_k).$$

Since $k - n < n$, $k - n + 1 \leq n$ and, therefore, $a_{k-n} \leq a_{k-n+1}$. Also $a_{k+1} < a_{k-n+1}$, $a_{k-n} \leq a_k$ and $a_{k+1} \leq a_k$. Therefore

$$\max\{a_{k-n}, a_{k+1}\} \leq \min\{a_{k-n+1}, a_k\} \text{ or}$$

$$\max\{d_1, d_2, \ldots, d_n\} \leq \min\{e_1, e_2, \ldots, e_n\}.$$

$\square$

Corollary 14.8. *A Half-cleaner $[2n]$ applied to a binary bitonic sequence transforms it to a sequence*

(i) *elements in the top half of which are smaller than the elements in the bottom half;*

(ii) *one of the halves is clean, and the other is bitonic.*

This is equivalent to saying that the output sequence is of the form $0^k 1^{2n-k}$ with $k \geq 0$.

Theorem 14.2 of Batcher suggests a simple recursive construction of Bitonic-Sorter[n] for $n = 2^k$ for some $k \geq 1$.

Applying Half-cleaner[n] to the given bitonic sequence we find that the output is two bitonic sequences in which every element in the upper half is less than or equal to the smallest of the elements in the lower half. The number of gates used is $\frac{n}{2} = 2^{k-1}$ and the depth of the network is 1. Now repeat the above procedure, i.e., apply Half-Cleaner [$\frac{n}{2}$] to each of the two bitonic sequences. The number of gates used will be $2 \times 2^{k-2} = 2^{k-1}$ and the depth of the comparison network is 1 again. There result four bitonic sequences B_1, B_2, B_3, B_4. Each of these sequences is of length 2^{k-2} and the largest element of B_i is less than or equal to the smallest element of $B_{i+1}, 1 \leq i \leq 3$. Continue repeating the argument till we arrive at 2^{k-1} bitonic sequences of length 2 each and having the property that the larger element in the ith sequence is less than or equal to the smaller of the two elements in the $(i + 1)$th sequence. Finally, applying the Half-cleaner[2] to the bitonic sequences of length 2 each, we get a sorted sequence. The total number of gates used is $1 \times 2^{k-1} + 2 \times 2^{k-2} + 2^2 \times 2^{k-3} + \cdots + 2^{k-1} \times 1 = k \times 2^{k-1} = (\frac{n}{2}) \lg n$ and depth of the network is $1 + 1 + \cdots + 1$ (k terms) $= k = \lg n$.

Theorem 14.9. *Bitonic-Sorter[n] sorts bitonic sequences of length $n = 2^k$ using $(\frac{n}{2})k = (\frac{n}{2}) \lg n$ gates and it is of depth $k = \lg n$.*

The number of comparators used in Bitonic-Sorter[n] is equal to the number of gates in the sorter. Therefore, the running time of Bitonic-Sorter[n] is $O(n \lg n)$.

Let $D(n)$ and $G(n)$ denote the depth and the number of gates used in a Bitonic-Sorter[n]. Observe that $D(n) = 0$ and $G(n) = 0$ when $n = 1$. If $n = 2^k, k \geq 1$ and a Half-Cleaner[n] is applied to the bitonic sequence, it adds 1 to the depth and $\frac{n}{2}$ to $G(n)$ and there result two bitonic sequences of length $\frac{n}{2}$. Therefore

$$D(n) = 1 + D\left(\frac{n}{2}\right) \text{ and } G(n) = \frac{n}{2} + 2G\left(\frac{n}{2}\right).$$

Proof of Theorem 14.9 can also be given by easy induction on $k \geq 1$.

We now construct Bitonic-Sorter[n] for $n = 4, 8, 16$. We give these as networks of wires and omit details.

Example 14.9. Construct bitonic sorters to sort the bitonic sequences.

(a) 3, 2, 4, 5;
(b) 9, 8, 7, 5, 3, 2, 4, 6;
(c) 7, 9, 5, 4, 3, 1, 2, 6;
(d) 9, 7, 6, 2, 1, 3, 4, 5, 8, 10, 13, 14, 16, 15, 12, 11.

Solution. (a)

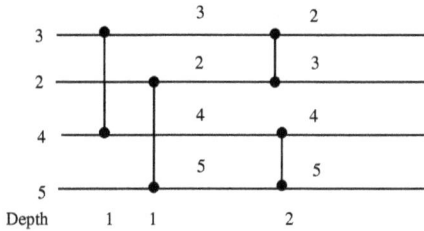

Fig. 14.12

The depth is 2, the number of gates is 4 and the sorted sequence is 2, 3, 4, 5.

(b)

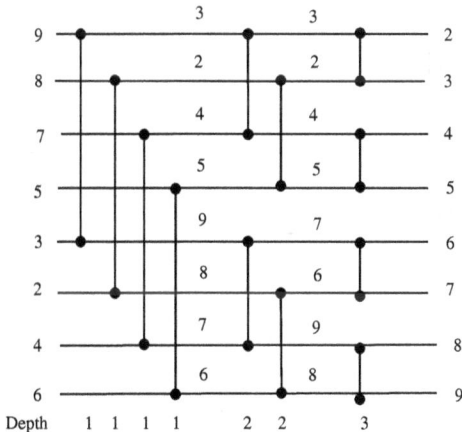

Fig. 14.13

The depth is 3, the number of gates is 12 and the sequence is finally sorted to 2, 3, 4, 5, 6, 7, 8, 9.

(c)

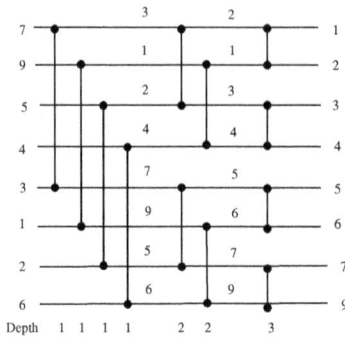

Fig. 14.14

The depth of the network is 3, the number of gates is 12 and the sequence is sorted to 1, 2, 3, 4, 5, 6, 7, 9.

(d)

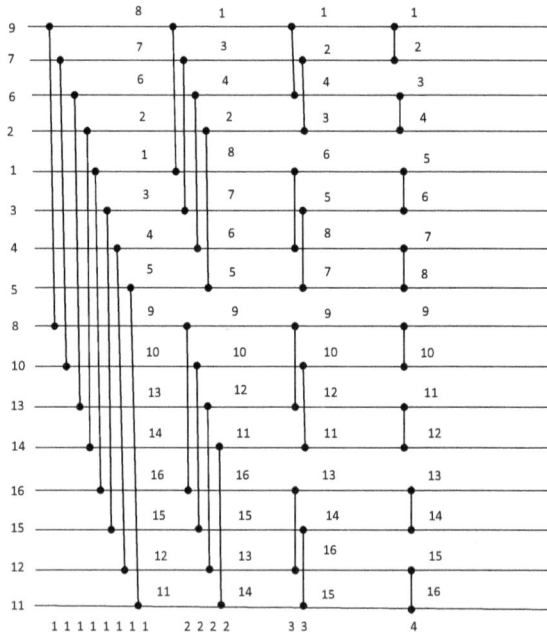

Fig. 14.15

The depth of the network is 4 and the number of gates is $8 + 2 \times 4 + 4 \times 2 + 8 \times 1 = 32$ (Figs. 14.12–14.15).

Since Bitonic-Sorter $[n]$ sorts every bitonic sequence of length $n = 2^k$, if there is a sequence of length n which is not sorted by Bitonic-Sorter$[n]$, then the sequence is not bitonic.

Example 14.10. Decide if the sequence 9, 8, 7, 5, 2, 3, 6, 4 is bitonic.

Solution. The given sequence is of length 8 and we apply Bitonic-Sorter $[8]$ to this sequence. As a network of wires, we have the following figure (Fig. 14.16):

We observe that the sequence is not sorted as 6 appears on the fourth wire while 5 appears on the fifth wire. Thus the given sequence is not bitonic.

After an application of the gates $(i, i + \frac{n}{2})$, for $i = 1, 2, 3, 4$ and $n = 8$, there result two subsequences 2, 3, 6, 4 and 9, 8, 7, 5. We find that every element in the top subsequence is $\nleq$ every element of the bottom subsequence. This violates Theorem 14.9 and, so, right after this step we could conclude that the given sequence is not bitonic.

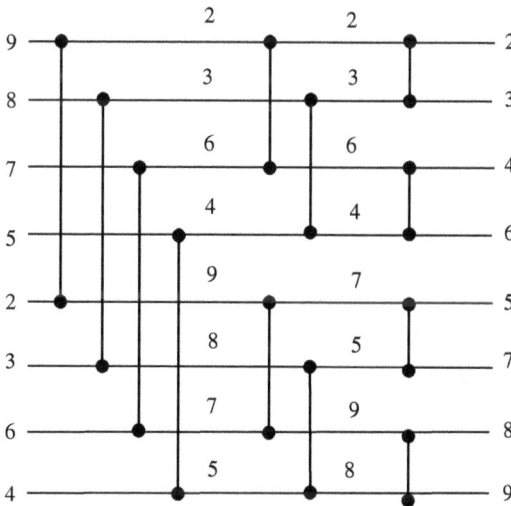

Fig. 14.16

Exercise 14.2.

1. Prove that the sequence 2, 3, 2, 3 is not bitonic.
2. Decide if the sequence 9, 7, 8, 6, 2, 4, 3, 1 is bitonic or not.
3. Prove that the sequence 8, 6, 3, 5, 9, 7, 2, 4 is not bitonic.

So far we have considered sorting of bitonic sequences of length n when n is a power of 2. Now suppose that n is not a power of 2. There exists an integer k such that $2^k < n < 2^{k+1}$. Then $k < \lg n < k + 1$. Given a bitonic sequence, by giving some cyclic shifts, if necessary, we can always bring it to the form of a sequence which is an increasing sequence followed by a decreasing sequence. Therefore, without loss of generality, we may suppose the given bitonic sequence is

$$a_1 \leq a_2 \leq \cdots \leq a_i \geq a_{i+1} \geq \cdots \geq a_n.$$

Let $a < \min\{a_1, a_n\}$. Introducing $2^{k+1} - n$ terms each equal to a at the end of this sequence we get a bitonic sequence of length 2^{k+1}. Bitonic-Sort$[2^{k+1}]$ is a sorting network of depth $k + 1$ which sorts the transformed sequence of length 2^{k+1}. Then ignore the first $2^{k+1} - n$ terms from this sorted sequence each of which is equal to a. There then results the sorted form of the given bitonic sequence. The depth of this sorting network is $k + 1 \leq 2k$ and, so, $k + 1 \leq 2k < 2 \lg n$, i.e., the depth of the sorting network is $O(\lg n)$.

Observe that to make the given bitonic sequence of length n into a bitonic sequence of length 2^{k+1}, we may not introduce all the $2^{k+1} - n$ terms less than a at the end or beginning of the given sequence but may introduce some of these at the beginning and the others at the end so that the resulting sequence is an increasing sequence of length 2^k followed by a decreasing sequence of length 2^k.

We have given an example to see that if a bitonic sorter does not sort a sequence, then the given sequence is not bitonic.

Exercise 14.3. Does there exist a non-bitonic sequence which is sorted by the Bitonic= Sorter$[n]$?

14.5. Another Merging Network

Earlier we have discussed Batcher's Odd–Even Merging Network. Having already discussed sorting of bitonic sequences, we can consider a merging network which seems much simpler than Batcher's odd–even merging network. Given two sorted sequences $a = a_1, a_2, \ldots, a_m$ and $b = b_1, b_2, \ldots, b_n$, we can form a bitonic sequence of length $m + n$ by concatenating a with b reversed, i.e., we form the sequence $a_1, a_2, \ldots, a_m, b_n, \ldots, b_1$. We can then sort this sequence using Bitonic-Sorter$[m + n]$ the depth of which is $\lg(m + n)$.

Suppose that $m = n$. Let the result of Half-cleaner$[2n]$ applied to the bitonic sequence $a_1, a_2, \ldots, a_n, b_n, \ldots, b_1$ be $A_1, A_2, \ldots, A_n, B_n, \ldots, B_1$. Then $A_i = \min\{a_i, b_{n-i+1}\}$, $B_i = \max\{a_i, b_{n-i+1}\}$ for $1 \leq i \leq n$.

Now consider the sequence $a_1, a_2, \ldots, a_n, b_1, \ldots, b_n$ which is concatenation of the two sequences. Apply the comparators $(i, 2n - i + 1)$ to this sequence for $1 \leq i \leq n$. Let the result of applying the n comparators in parallel be $C_1, C_2, \ldots, C_n, D_1, D_2, \ldots, D_n$. Then $C_i = \min\{a_i, b_{n-i+1}\}$, $D_i = \max\{a_i, b_{n-i+1}\}$ for all i, $1 \leq i \leq n$. Thus $C_i = A_i$ and $D_i = B_i$ for all i, $1 \leq i \leq n$. We thus have the following result.

Lemma 14.10. *Given two sorted sequences $a = a_1, a_2, \ldots, a_n$ and $b = b_1, b_2, \ldots, b_n$, an application of the comparison network $[(i, 2n - i + 1) \mid 1 \leq i \leq n]$ to the concatenation $ab = a_1, a_2, \ldots, a_n, b_1, \ldots, b_n$ leads to a sequence $A_1, \ldots, A_n, B_1, \ldots, B_n$ in which the sequences $A_1, \ldots, A_n$ and $B_1, \ldots, B_n$ are bitonic and $\max\{A_1, \ldots, A_n\} \leq \min\{B_1, \ldots, B_n\}$.*

*It then follows from this lemma that applying Bitonic-Sorter$[n]$ to the sequences $A_1, \ldots, A_n$ and $B_1, \ldots, B_n$ we get the sorted merge of the sequences $a_1, a_2, \ldots, a_n$ and $b_1, b_2, \ldots, b_n$. The network which is comparison network $[(i, 2n - i + 1) \mid 1 \leq i \leq n]$ followed by the network $\{$Bitonic-Sorter$[n]$, Bitonic-Sorte$[n]\}$ is called **Merging-Network**$[2n]$ or simply **Merger**-$[2n]$. In case we merge two sorted sequences of length $\frac{n}{2}$ each, then Merger$[n]$ and Bitonic-Sorter$[n]$ differ in the first step only which is of depth 1 and, therefore,*

$$\text{Depth of Merger}[n] = \text{depth of Bitonic-Sorter}[n] = \lg n.$$

Also the number of comparators of Merge[n] = the number of comparators of Bitonic-Sorter[n] = $\frac{n}{2} \lg n$.

Next suppose that n is not a power of 2. Let k be the smallest positive integer such that $n < 2^k$. Let α be an arbitrarily large integer. In particular, it is larger than every element of the sequence $a = a_1, a_2, \ldots, a_n$ and we are required to merge the monotonic sequences $a_1 \leq \cdots \leq a_{\frac{n}{2}}$ and $a_{\frac{n}{2}+1} \leq \cdots \leq a_n$. Let $m = \frac{(2^k - n)}{2}$ and extend the sequences by introducing m terms each equal to α at the end of these two sequences. We thus get a sequence $b_1 b_2 \ldots b_t$ of length $t = n + 2m = 2^k$, where $b_i = a_i$ for $1 \leq i \leq \frac{n}{2}$, $b_i = \alpha$ for $\frac{n}{2} < i \leq \frac{n}{2} + m$ and $n + m < i \leq n + 2m = 2^k$ and $b_{\frac{n}{2}+m+i} = a_{\frac{n}{2}+i}$, for $1 \leq i \leq \frac{n}{2}$. Then Merge $[2^k]$ applied to the sequence $b_1 b_2 \ldots b_{n+2m}$ gives a sorted sequence in which the first n terms are the a_i in some order and the last $2m$ terms are each equal to α. Ignoring the last $2m$ terms of this sequence we get the sequence a in sorted form.

Depth of Merger[2^k] $= k$ and number of comparators used in Merger[2^k] $= \frac{1}{2} 2^k k = k2^{k-1}$.

Also depth of Merger-[n] $=$ depth of Merger-[2^k] and number of comparators used in Merger[n] $=$ number of comparators used in Merger [2^k]. Therefore depth of Merger[n] $= k$ and number of comparators used in Merger-[n] $= k2^{k-1}$. Since $2^{k-1} < n < 2^k$, $k - 1 < \lg n < k$ and $(k - 1)n < n \lg n < kn$ so that $\frac{1}{2} n \lg n < \frac{1}{2} kn = k2^{k-1}$. Therefore depth of Merger[n] $= \Omega(\lg n)$ and number of gates in Merger[n] $= \Omega(n \lg n)$.

Example 14.11. Merge together and sort the monotonic sequences

(a) 5, 7, 8, 9 and 2, 3, 4, 6;
(b) 3, 6, 8, 9 and 1, 2, 4, 5.

Solution. We concatenate the given sorted sequences and then apply Merger[8] to get the final sorted sequences (Figs. 14.17 and 14.18).

The depth of each of the networks is $3 = \lg 8$.

Example 14.12. Merge together and sort the monotonically increasing sequences

(a) 3, 4, 6, 7, 9, 10, 12, 16 and 1, 2, 5, 8, 11, 13, 14, 15;
(b) 3, 5, 6, 8, 9, 11, 12, 13 and 1, 2, 4, 7, 10, 14, 15, 16.

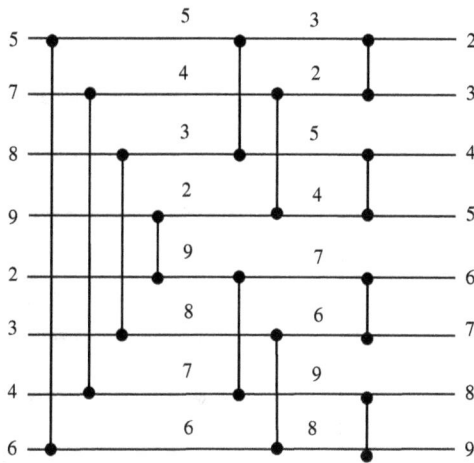

Fig. 14.17

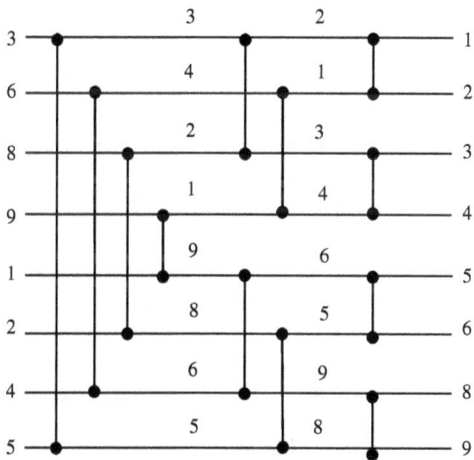

Fig. 14.18

Solution. We concatenate the two sorted sequences and then apply Merger[16]. After the initial step, we get two bitonic sequences of length 8 each and apply Bitonic-Sorter[8] to these two bitonic sequences and get the final sorted sequences (Figs. 14.19 and 14.20).

The depth of each of the network is $4 = \lg 16$ and the number of comparators is $8 \times 4 = 32 = \left(\frac{1}{2}\right)16\lg 16$.

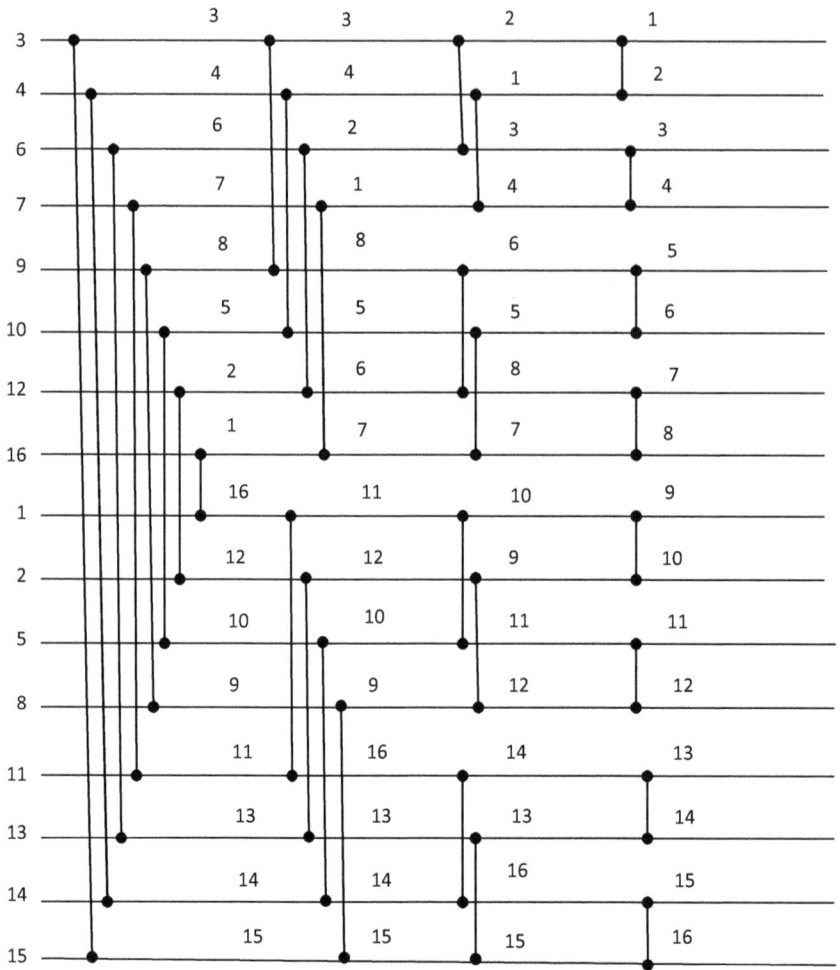

Fig. 14.19

Lemma 14.10 says that the comparison network $[(i, n - i + 1)$ $|1 \leq i \leq \frac{n}{2}]$ applied to the concatenation of two sorted sequences $a_1 \leq \cdots \leq a_{\frac{n}{2}}$ and $a_{\frac{n}{2}+1} \leq \cdots \leq a_n$ leads to concatenation of two bitonic sequences. However, there do exist non-sorted sequences such that the above comparison network applied to concatenation of these sequences leads to concatenation of two bitonic sequences and Bitonic-Sorter$[\frac{n}{2}]$

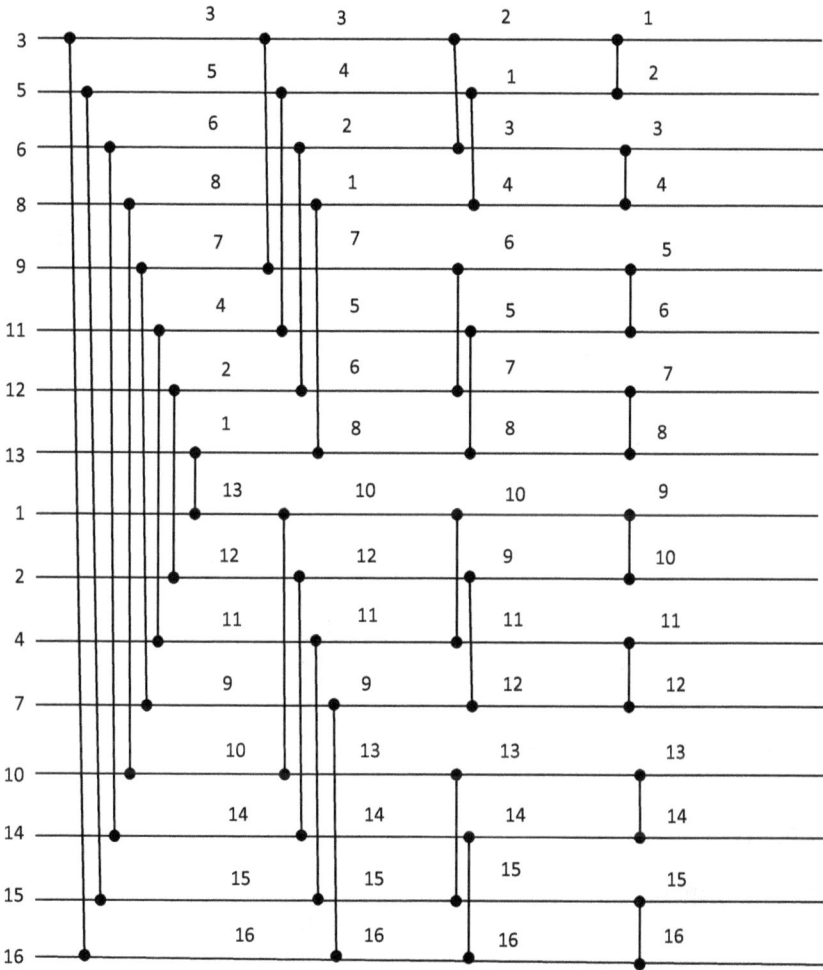

Fig. 14.20

applied to the two bitonic subsequences then leads to sorting of the given sequence $a_1, a_2, \ldots, a_n$.

Example 14.13. Consider the sequence $8, 6, 3, 5, 9, 7, 2, 4$ of length 8 which is neither bitonic nor concatenation of two sorted sequences. Considered as network of wires, we have the following figure (Fig. 14.21).

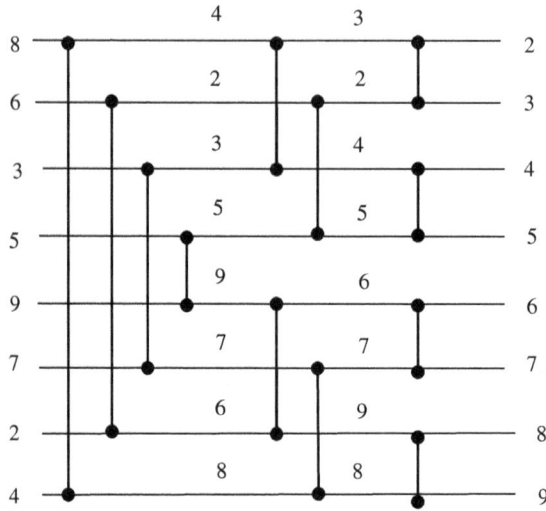

Fig. 14.21

Observe that application of the comparison network $(i, 9 - i)$, $1 \leq i \leq 4$ leads to the sequence 4, 2, 3, 5, 9, 7, 6, 8 which is concatenation of two bitonic sequences 4, 2, 3, 5 and 9, 7, 6, 8. Bitonic-Sorter then naturally leads to sorting of the given sequence.

14.6. A Sorting Network

We have now all the tools and machinery for constructing a sorting network that sorts all sequences of length n. We first consider the case when $n = 2^k$ a power of 2. Any single number or integer may be regarded as a sorted sequence of length 1. Given a sequence of length 2, we may regard it as concatenation of two sorted sequences of length 1. Using Merge-[2] we get a sorted form of the given sequence. Next, if we are given a sequence of length 4 which is concatenation of two sorted sequences of length 2 each, we use the network Merge-[4] to get a sorted form of the given sequence of length 4. Continuing this process we may get a sorted form of the sequence of length $n = 2^k$ which is concatenation of two sorted sequences of length 2^{k-1} each. The actual process of constructing a sorted network is taking the above steps in a certain systematic way. Given a sequence $a = a_1, a_2, \ldots, a_n$ of length n, we may regard it as concatenation of

$\frac{n}{2}$ sequences of length 2 each as $(a_1a_2), (a_3a_4), \ldots, (a_{n-1}a_n)$. Applying Merge-[2] to each of the pair of subsequences $a_{2i-1}a_{2i}, a_{2i+1}a_{2i+2}$ for $1 \le i \le \frac{n}{2} - 1$, we get a concatenation of $\frac{n}{4}$ sorted sequences of length 4 each, say $(b_1b_2b_3b_4) \ldots (b_{n-3}b_{n-2}b_{n-1}b_n)$.

Now apply Merge-[4] to each of the pair of sequences $b_{4i-3}b_{4i-2}$ $b_{4i-1}b_{4i}$ and $b_{4i+1}b_{4i+2}b_{4i+3}b_{4i+4}$ for $1 \le i \le \frac{n}{4} - 1$, we get concatenation of $\frac{n}{8}$ sorted sequences of length 8 each. We continue the process till we get sorted form of the sequence we started with. Let $D(n)$ denote the depth of the network constructed in this fashion. Since the depth of Merge-[n] is $\lg n$, $D(n)$ is related to $D(\frac{n}{2})$ by the rule

$$D(n) = D\left(\frac{n}{2}\right) + \lg n. \tag{14.13}$$

This is so because in $k = \lg n$ steps we are able to arrive at concatenation of two sorted sequences of length $\frac{n}{2} = 2^{k-1}$. Also observe that going from 4 sorted sequences of length $\frac{n}{4}$ each to two sorted sequences of length $\frac{n}{2}$, we use Merge-$[\frac{n}{2}]$ to two pairs of sequences in parallel. We can solve the recurrence relation (14.13) as follows:

$$D(n) = D(\frac{n}{2}) + \lg n = D(2^{k-1}) + \lg 2^k = D(2^{k-1}) + k$$

$$= D(2^{k-2}) + (k-1) + k$$

$$\vdots$$

$$= D(2) + 2 + 3 + \cdots + (k-1) + k$$

$$= 1 + 2 + 3 + \cdots (k-1) + k = \frac{k(k+1)}{2}$$

$$= \frac{\lg n + \lg^2 n}{2} = \theta(\lg^2 n),$$

where we write $\lg^2 n$ for $(\lg n)^2$. In particular, we can sort a given sequence of length n in $O(\lg^2 n)$ time.

Now, suppose that n is not a power of 2. Let k be the smallest positive integer with $n < 2^k$. Let α be an arbitrarily large number and enlarge the given sequence $a_1, a_2, \ldots, a_n$ to a sequence of length 2^k by taking $a_{n+i} = \alpha$ for $1 \le i \le 2^k - n$. We can then sort the enlarged sequence and its depth is $\frac{k(k+1)}{2}$. Now $2^{k-1} < n < 2^k$ and, so, $k - 1 < \lg n < k$.

Thus, $\frac{k(k-1)}{2} < \frac{(\lg n)(1+\lg n)}{2} < \frac{k(k+1)}{2}$. Therefore, the depth of the network is $\Omega((1 + \lg n) \lg n)$.

We now consider the number of comparators for the case $n = 2^k$. Since the number of comparators at every step of Merge-$[n]$ is $\frac{n}{2}$, the number of gates in Sorter-$[n] = \frac{n}{2}D(n) = (\frac{n}{4})(\lg n)(1 + \lg n)$.

Example 14.14. Construct sorting networks for the sequences:

(a) $9, 8, 7, 5, 2, 3, 6, 4$;
(b) $8, 9, 6, 3, 1, 2, 5, 4$;
(c) $4, 16, 9, 6, 10, 12, 3, 7, 14, 2, 11, 1, 5, 15, 8, 13$;
(d) $11, 3, 8, 13, 6, 9, 12, 5, 4, 10, 14, 2, 1, 7, 16, 15$.

Solution. We construct Sorter-$[n]$ in each case without elaborating the steps involved (Figs. 14.22 and 14.23).

In the case of examples (a), (b) the depth of the network is $6 = \frac{(1+\lg 8)\lg 8}{2}$ and the number of comparators is $1 \times 4 + 2 \times 2 + 1 \times 4 + 4 \times 1 + 2 \times 2 = 6 \times 4 = \frac{4(1+\lg 8)\lg 8}{2} = 4 \times$ depth of Sorter-$[8]$.

In the case of example (c) the depth of the network so far is 6 and the number of comparators is $6 \times 8 = 48$. The result is concatenation of

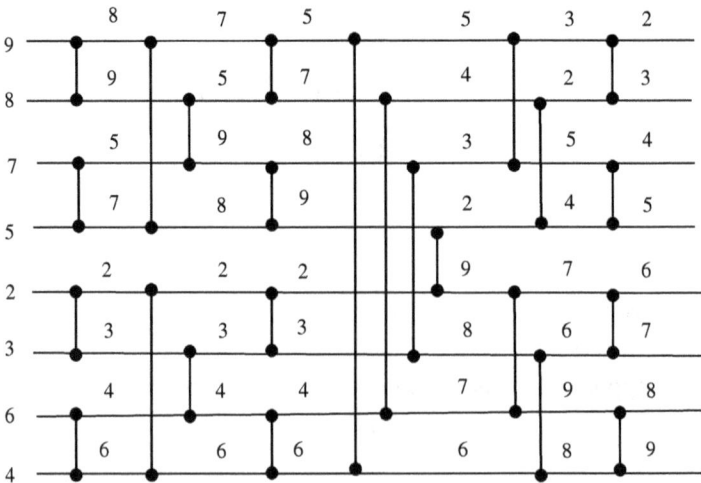

Fig. 14.22

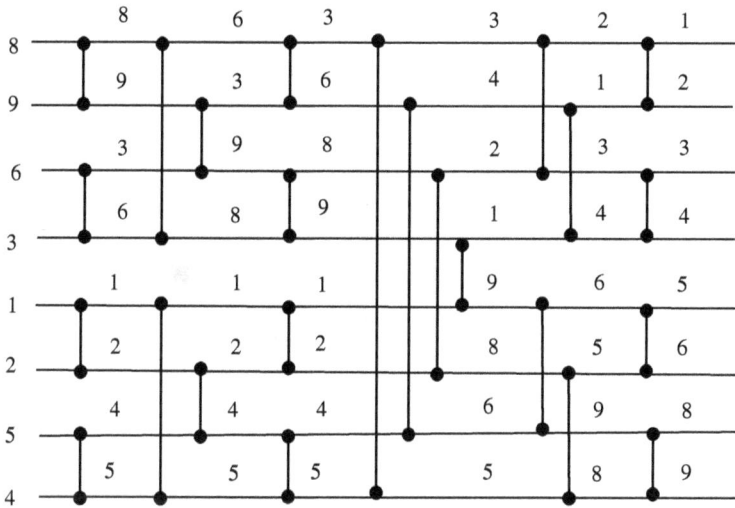

Fig. 14.23

two sorted sequences of length 8 each and we now need to apply Merge-[16] to get the final sorted sequence. The sorted sequences of length 8 arrived at are the same that we merged in Example 14.12(a). So, at this stage we need to take the same steps as taken in that example to get the final sorted sequence. The depth of the Merge-[16] is 4 and the number of comparators is 32. Therefore, for the Sorter-[16], the depth is $6 + 4 = 10 = \frac{(1+\lg 16)\lg 16}{2}$ and the number of comparators is $32 + 48 = 80 = \frac{(\lg 16)(1+\lg 16)16}{4}$ (Fig. 14.24).

In the case of example (d) the depth of the network so far is 6 and the number of comparators is $6 \times 8 = 48$. The result is concatenation of two sorted sequences of length 8 each and we now need to apply Merge-[16] to get the final sorted sequence. The sorted sequences of length 8 arrived at are the same that we merged in Example 14.12(b). So, at this point we need to take the same steps as taken in that example to get the final sorted sequence. The depth of the Merge-[16] is 4 and the number of comparators is 32. Therefore, for the Sorter-[16], the depth is $6 + 4 = 10 = \frac{(1+\lg 16)\lg 16}{2}$ and the number of comparators is $32 + 48 = 80 = \frac{(\lg 16)(1+\lg 16)16}{4}$ (Fig. 14.25).

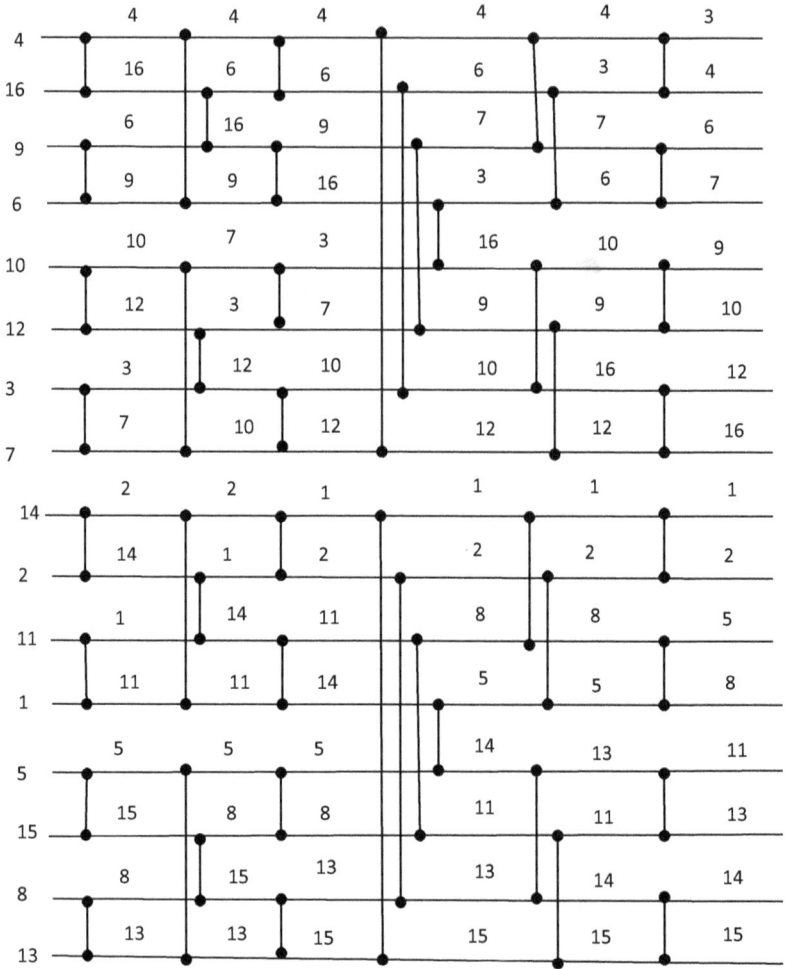

Fig. 14.24

Exercise 14.4.

1. Using Batcher's odd–even merge, find the merge sort of the sequences

   (a) 2, 5, 8, 11, 13, 16, 21, 26 and 4, 9, 12, 18, 23, 27, 31, 35;
   (b) 2, 5, 8, 11, 13, 16 and 4, 12, 18, 27, 31;
   (c) 2, 8, 11, 16, 21, 27 and 4, 9, 17, 22, 30, 32, 35.

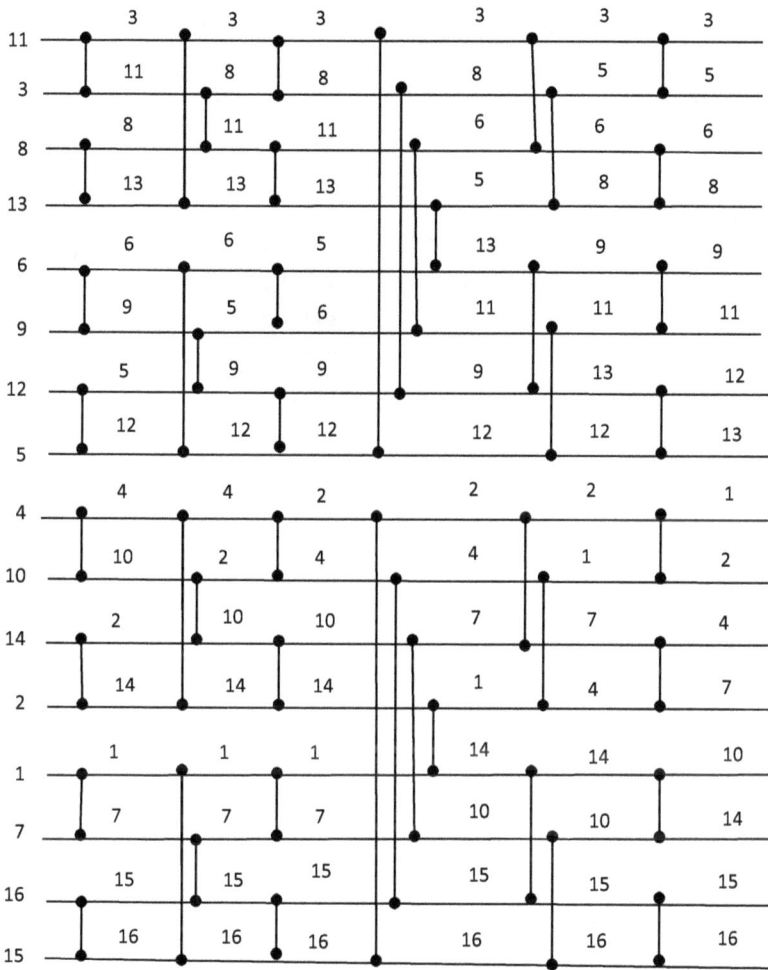

Fig. 14.25

2. Find the sorted merge of the sequences in Exercise 1(a)–(c) above using Merge-[n].

3. Using Sorter-[8], sort the sequence 2, 4, 22, 7, 5, 2, 13, 11, 15, 23, 32, 10, 4, 17, 12, 28, 35.

4. Merge the two sequences as in Exercise 1(a)–(c) above using bitonic merging.

5. Prove that applying a monotonically increasing function to a sorted sequence produces a sorted sequence.
6. Find the number of all possible binary bitonic sequences of length n.
7. If every element in one binary sequence is atleast as small as every element in another binary sequence, prove that one of the two binary sequences is clean.
8. Prove that a comparison network that can sort any bitonic sequence of 0's and 1's, can sort any bitonic sequence of arbitrary numbers.
9. Prove that a comparison network that can merge any two monotonically increasing sequences of 0's and 1's can merge any two monotonically increasing sequences of arbitrary numbers.

Chapter 15

NP-Complete Problems

Most of the algorithms we have studied so far were polynomial-time algorithms, i.e., algorithms having time complexity $O(n^k)$ where n is the size of the input instance and k is a constant positive integer. However, not all problems can be solved in polynomial time — there do indeed exist such problems as cannot be solved in polynomial time. Problems that are solvable by polynomial-time algorithms are called easy or **tractable**. Those problems that cannot be solved by polynomial-time algorithms are called **hard** or **intractable**.

There do exist several problems which are not solvable by polynomial-time algorithms. For example, the Traveling Salesman Problem involving n cities to be visited is one such. There is no algorithm which solves this problem in general and the only way to solve it is by brute force. There are $(n - 1)!$ possible tours, we need to find the cost for each tour and then to look for a tour with minimum cost. Therefore solving this problem by brute force requires $(n - 1)!$ computations. Observe that $(n - 1)! > 2^n$ for all $n \geq 6$ and, so, the time complexity for such a solution is even more that exponential.

In general, there are three types of problems — P, NP and $NP\text{-}complete$. There is another class of problems, i.e., $Co\text{-}NP$ but we do not consider these here (although we may define this class). Every problem in class P is in class NP but "is every problem in class NP in class P" is still an interesting open problem. Problems in the class $NP\text{-}complete$ are in the class NP and are the most difficult ones in the class NP. The problems in the class $NP\text{-}complete$ have an interesting property that if one of these is solvable in polynomial time, then every other problem in the class is solvable in polynomial time (cf. [5]).

Most problems that have been considered so far in the earlier chapters are in P. In this chapter, our effort is to study and identify *NP-complete* problems.

15.1. Classes P, NP and NP-Complete

Problems that are solvable in polynomial time form the class P. Thus

$P = \{X \mid X$ is a problem that can be solved in time $O(n^k)$ for some

constant integer k, where n is the size of the input to the problem$\}$.

For example, given a weighted graph G and two vertices u, v of G, finding the shortest path from u to v (i.e., a path from u to v such that the sum of the weights of edges on the path is shortest among weights of all paths from u to v) is a problem in P. Finding product of two square matrices of order n is another problem in P.

The class *NP* consists of those problems that are **verifiable** in polynomial time. So

$NP = \{X \mid X$ is a problem such that if we are given or somehow are

able to guess/find a solution to the problem, then we can verify

that the solution is correct in time which is polynomial in the

size of input to the problem$\}$.

If X is a problem that can be solved in polynomial time, then any certificate/proposed solution for this problem can also be verified in polynomial time. Thus P is a subclass of *NP* (i.e., $P \subseteq NP$).

We next consider two other examples of problems in *NP*.

Problem 15.1 (Hamiltonian Cycle Problem or HC). *Let* $G = (V, E)$ *be a given connected graph. A simple cycle in* G *is called a **Hamiltonian cycle** if it traverses each vertex of* G *exactly once. "Does* G *have a Hamiltonian cycle?" is called the **Hamiltonian Cycle Problem** or just **HC**. A certificate for this problem shall be a sequence* $(v_1, v_2, \ldots, v_n)$ *of vertices of* G *with* $|V| = n$. *It can be verified in polynomial time if* *the edges (v_i, v_{i+1}) for $1 \leq i \leq n - 1$ and (v_n, v_1) are in E. If the graph is represented by the adjacency lists, all that we need to check is if $v_{i+1} \in \text{adj}(v_i)$ for $1 \leq i \leq n - 1$ and $v_1 \in \text{adj}(v_n)$, i.e., a total number of almost $n\,|E|$ comparisons. This proves that HC is an NP-problem.*

Problem 15.2 (Satisfiability or just SAT). *Let $x_1, x_2, \ldots, x_n$ be Boolean variables so that every x_i takes the value "true" (i.e., 1) or "false" (i.e., 0). A Boolean formula is an expression that is formed from the variables $x_1, x_2, \ldots, x_n$ using Boolean expressions "And" ($\wedge$), "Or" ($\vee$), "negation" ($-$) and paranthesis. For simplicity of notation, we write for variables x, y or Boolean expressions e, f*

$$x \wedge y = xy, x \vee y = x + y, e \wedge f = ef, e \vee f = e + f.$$

A Boolean formula or expression e is in **Conjunctive Normal Form** (CNF) if it is of the form $C_1 \cdot C_2 \cdot \ldots \cdot C_m$, where for every i, $1 \leq i \leq m$, $C_i = x'_j + x'_k + \cdots$ with $x'_j = x_j$ or $\overline{x_j}$ etc., i.e., C_i is the disjunction of a finite number of x'_j with x'_j equal to the Boolean variable x_j or its negation $\overline{x_j}$. Each one of $C_1, C_2, \ldots, C_m$ is called **clause** and $x'_j, x'_k, \ldots$ are called **literals**. The problem **Satisfiability (SAT)** or **CNF Satisfiability** (**CNF SAT**) consists in finding a combination of values of the variables $x_1, x_2, \ldots, x_n$ which when used in the conjunctive normal form expression e or $C_1 \cdot C_2 \cdot \ldots \cdot C_m$ make its value 1. Equivalently, it consists in deciding if a combination of values of $x_1, \ldots, x_n$ results in giving the expression e, or $C_1 \cdot C_2 \cdot \ldots \cdot C_m$, the value 1. Observe that the expression $C_1 \cdot C_2 \cdot \ldots \cdot C_m$ gets the value 1 if and only if every clause C_1 gets the value 1.

If each clause in the expression $C_1 \cdot C_2 \cdot \ldots \cdot C_m$ is the disjunction of the same number k of literals, the corresponding CNF SAT is called **k-CNF SAT** or just **k-SAT**. A certificate for solution of k-SAT is a combination of values $a_1, a_2, \ldots, a_n$ of the n variables $x_1, x_2, \ldots, x_n$. In n operations we can find the corresponding values of $\overline{x_1}, \overline{x_2}, \ldots, \overline{x_n}$. Consider any clause $C = x'_{i_1} + x'_{i_2} + \cdots + x'_{i_k}$. If even one literal appearing in C has the value 1, then putting the given values of x'_{i_j}, we find that C takes the value 1. Thus in a (linear) polynomial time it can be verified if the given certificate solves C. Since an expression e is a product of a finite number of clauses with k literals and a product of polynomials is again a polynomial it can be verified in polynomial time if the given certificate or combination of values of $x_1, x_2, \ldots, x_n$ produces a solution to the expression e. Hence k-SAT is in NP and, in particular, the problem 3-SAT is in the class NP.

In order to define the class of *NP-complete* problems, we need to recall the concept of reduction transforms. For the purpose of *NP-complete* problems, we need to consider only decision problems.

15.2. Optimization Problems as Decision Problems

Most problems of interest are optimization problems. These are problems in which each feasible solution has associated with it a value and we are looking for a feasible solution with best value. Optimization problems include the TSP and given a weighted graph G and vertices u, v of G, finding a shortest path from u to v and or also finding the shortest path length from u to v. A problem in which a solution is "yes" (or true or 1) or "no" (or false or 0) is called a decision problem. For example, Hamiltonian Cycle (HC) problem in a simple connected graph is a decision problem.

To every optimization problem, there corresponds a related decision problem. We consider some examples.

Example 15.1 (Traveling Salesman Problem or TSP). A salesman is required to visit a number of cities in a trip for the sales of a particular product. Distances (or costs of travel) between every pair of cities are given. In what order should he travel so that (i) he visits every city precisely once and returns to the home city (i.e., the city he started his trip from) and (ii) the distance traveled (or cost of travel) is the least?

Equivalently, if the given n cities are numbered $1, 2, \ldots, n$ and distance between cities i and j (or the cost of travel from i to j) is $w(i, j)$, we are required to find a cyclic permutation π of the numbers $1, 2, \ldots, n$ such that $w(\pi) = \sum_{1 \leq i \leq n} w(i, \pi(i))$ is minimum.

As a decision problem TSP may be proposed as follows.

As above, the n cities that the salesman has to visit be numbered $1, 2, \ldots, n$ and $w(i, j)$ be the distance between cities i and j. Let also a positive number W be given. Does there exist a cyclic permutation $(1, \pi(1), \pi^2(1), \ldots, \pi^{n-1}(1))$ of the numbers $1, 2, \ldots, n$ such that $\sum_{1 \leq i \leq n} w(i, \pi(i)) \leq W$?

Definition. Given a graph $G = (V, E)$, a subset V' of V is called a **vertex cover** of G if every edge of G is incident with at least one vertex in V'.

Example 15.2 (Vertex Cover or VC). Given a graph $G = (V, E)$, find a smallest vertex cover of G.

As a decision problem Vertex Cover (or VC) may be stated as follows.

Given a graph $G = (V, E)$ and a positive integer k, does there exist a vertex cover V' of G with $|V'| \leq k$?

Example 15.3. Let S be a finite set and F a family of non-empty subsets S_i of S such that $S = \bigcup S_i$. A subfamily F' of F is called a **(set) cover** of S if $S = \bigcup_{S_i \in F'} S_i$.

Set Cover Problem (or **SC** in short) is to find a smallest subfamily F' of F which covers S.

The corresponding (or equivalent) decision problem states:

Given a finite set S, a family F of non-empty subsets S_i of S with $S = \bigcup S_i$ and a positive integer k, does there exist a subfamily F' of F with $|F'| \leq k$ which covers S?

Example 15.4. Let a graph $G = (V, E)$ be given. A subset S of V is called an **independent set** of G if whenever $u, v \in S$, then there is no edge in G between u and v.

Independent Set Problem is then to find the largest independent subset of G.

Equivalent decision version of this problem says:

Given a graph G and a positive integer k, does there exist an independent set in G with at least k elements?

Example 15.5. Let $G = (V, E)$ be a graph. A subset C of V is called a **clique** in G if every two vertices in C are adjacent in G. Alternatively a clique in G is a complete subgraph of G.

The k-**clique problem** in G is to find a clique in G with k vertices. Sometimes the clique problem may also be stated as to find a largest clique in G. As a decision problem it says the following:

Given a graph $G = (V, E)$ and a positive integer k, does there exist a clique in G with at least k vertices?

15.3. Reducibility or Reductions

For studying *NP-completeness* of problems, we need an important concept of reducibility of problems.

Definition. Let X and Y be two decision problems with Q_X and Q_Y the questions of the problems X and Y, respectively. Let A be a polynomial time algorithm that maps an instance I_X of X onto an instance I_Y of Y. If the answer of Q_X for I_X equals the answer of Q_Y for I_Y then the algorithm A is called a **transform** that transforms X to Y.

   If there exists a transform that transforms problem X to problem Y, then we say that **problem X reduces to problem** Y and we indicate it by $X <_P Y$.

   If X and Y are two problems such that $X <_P Y$ and $Y <_P X$ the two problems X, Y are said to be **polynomially equivalent**. We express it by writing $X \simeq_P Y$.

Definition. A decision problem X is **NP-complete** if

(a) X is in *NP* (i.e., any given or guessed solution of X can be verified in polynomial time);
(b) every problem in *NP* is reducible to X in polynomial time.

Remark. Let X, Y and M be decision problems with $X <_P Y$ and $Y <_P M$. Then there exist polynomial time algorithms A, B such that A transforms an instance I_X of X to an instance I_Y of Y and then B transforms instance I_Y to I_M. Then the algorithm A followed by algorithm B transforms I_X to I_M. Also the answer of question Q_X for $I_X =$ answer for question Q_Y for $I_Y =$ answer for question Q_M for I_M. This proves that $X <_P M$ and the relation $<_P$ is transitive.

Theorem 15.1. *A problem X is* NP-complete *if*

(1) X is in NP; and
(2) there exists an NP-complete *problem M that reduces polynomially to X.*

Proof. Let Y be an arbitrary problem in *NP*. The problem M being *NP-complete*, $Y <_P M$. Since M is reducible to X, $M <_P X$. The relation $<_P$ being transitive, $Y <_P X$. Therefore X is *NP-complete*.   □

We now study some examples of reducibility.

Theorem 15.2. *Let $G = (V, E)$ be a graph and S be a subset of V. Then S is an independent set if and only if $V - S$ is a vertex cover of G.*

Proof. Suppose that S is an independent set. Consider an arbitrary edge $e = (u \cdot v)$ of G. Then, by definition of independent set, u, v cannot both be in S. Thus, at most one of u, v is in S. Hence at least one of u, v is in $V - S$. Therefore $V - S$ is a vertex cover of G.

Conversely, suppose that $V - S$ is a vertex cover of G. Let $u, v \in S$. Then neither $u \in V - S$ nor $v \in V - S$. The subset $V - S$ being a vertex cover of G, it follows that there is no edge in G between u and v. Hence S is an independent set. $\square$

As an immediate consequence of the above theorem, we have the following corollary.

Corollary 15.3. *Independent set decision problem and vertex cover decision problem are polynomially equivalent problems.*

Proof. Let a graph $G = (V, E)$ and an integer k be an instance of the independent set problem and let S be an independent set with $|S| \geq k$. Since $V - S$ is a vertex cover of G and $|V - S| \leq |V| - k$, then $G = (V, E), n - k$ (with $n = |V|$) is the corresponding instance of the vertex cover problem. It follows from Theorem 15.2 that answer to the independent set problem is yes if and only if the answer to the vertex cover problem is yes. $\square$

Theorem 15.4. *CNF-SAT $<_P$ clique decision problem.*

Proof. Let $F = C_1 C_2 \ldots C_k$ be a propositional formula in CNF. Let $x_1, x_2, \ldots, x_n$ be the variables in F. Using F we construct a graph $G = (V, E)$ which will have the properties we desire G to have for the clique problem. Take

$$V = \{(\sigma, i) \mid \sigma \text{ is a literal in } C_i\} \text{ and}$$

$$E = \{((\sigma, i), (\delta, j)) \mid i \neq j \text{ and } \delta \neq \overline{\sigma}\}.$$

Suppose that F is satisfiable. Then there is a set of truth values for x_i, $1 \leq i \leq n$, such that each clause C_i in F is true with the given assignment of the variables. Then in each clause C_i there is at least one literal σ (say) which is true. Choosing exactly one true literal from each clause, we form a subset $S = \{(\sigma, i) \mid \sigma$ is true in the clause $C_i\}$ of V. Clearly $|S| = k$. For $i \neq j$, choose vertices (σ, i), (δ, j) of G in S. If $\delta = \overline{\sigma}$, since σ is true, δ is false which is a contradiction. Therefore $\delta \neq \overline{\sigma}$. By the construction of G, it follows that there is an edge in G connecting these two vertices. Therefore S is a k-clique in G.

Conversely, suppose that G has a clique $K = (V', E')$ with $|V'| \geq k$. If $|V'| > k$, there exists an i and literals σ, τ in C_i such that (σ, i), (τ, i) are in V'. But (σ, i), (τ, i) are not connected in G which contradicts the hypothesis that K is a clique. Hence $|V'| = k$ and for every i, $1 \leq i \leq k$, there exists a unique literal σ in C_i such that (σ, i) is in V'. Let $S = \{\sigma \mid (\sigma, i) \in V'\}$. If $\sigma, \overline{\sigma}$ both are in S for some σ, then there exist i, j such that (σ, i), $(\overline{\sigma}, j)$ are in V'. Then $i \neq j$ and (σ, i), $(\overline{\sigma}, j)$ are not adjacent in G which is a contradiction, Hence only one of $\sigma, \overline{\sigma}$ can be in S. To the variables x_i assign the values $x_i = $ true if $x_i \in S$, $x_i = $ false if $\overline{x_i} \in S$ and assign the values true/false arbitrarily to the other variables. Then every clause C_i has at least one literal which takes the value true so that every C_i is true and hence F is true.

Thus F is satisfiable if and only if G has a clique of size $\geq k$. $\qquad\square$

Corollary 15.5. *3-SAT$<_P$ CDP.*
Before considering another result on reduction, we consider an example on the construction of a graph G for a given propositional formula.

Example 15.6. Consider a propositional formula $F = (x_1 + \overline{x_2} + x_3)(\overline{x_1} + x_2 + \overline{x_3})$. The graph G that corresponds to F has six vertices $V = \{(x_1, 1), (\overline{x_2}, 1), (x_3, 1), (\overline{x_1}, 2), (x_2, 2), (\overline{x_3}, 2)\}$ and six edges with $E = \{((\sigma, 1), (\delta, 2)) \mid \delta \neq \overline{\sigma}\}$. The graph G is then given in Fig. 15.1.

The graph G is clearly bipartite and this is always the case if the propositional formula F from which G arises has only two clauses.

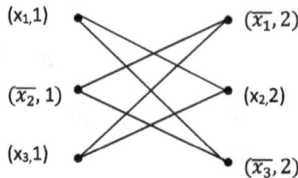

Fig. 15.1

However, if F has three or more clauses, the graph G is never bipartite. In the graph of Fig. 15.1 there are six 2-cliques with vertex sets

$$\{(x_1, 1), (x_2, 2)\}, \{(x_1, 1), (\overline{x_3}, 2)\}, \{(\overline{x_2}, 1), (\overline{x_1}, 2)\},$$

$$\{(\overline{x_2}, 1), (\overline{x_3}, 2)\}, \{(x_3, 1), (\overline{x_1}, 2)\}, \{(x_3, 1), (x_2, 2)\}.$$

There are no cliques with three or more vertices. Indeed in any bipartite graph there are no cliques of size three or more.

Theorem 15.6. *The Clique Decision Problem (CDP)* $<_P$ *Vertex Cover Decision Problem (VCDP).*

Proof. Let $G = (V, E)$ and an integer k define an instance of CDP. Suppose that $|V| = n$. To construct an instance of VCDP, we need a graph G' which has a vertex cover of size $\leq n - k$ if and only if G has a clique of size $\geq k$. Let G' be the complement graph of G, i.e., $G' = (V, \overline{E})$, where $\overline{E} = \{(u, v) \mid u, v \in V$ and $(u, v) \notin E\}$.

Let $K = (V', E')$ be any clique in G. Consider an edge (u, v) in $\overline{E}$. Then (u, v) is not an edge in E and, in particular, it is not in K. Therefore u, v cannot both be in V'. Therefore, at least one of u, v is in $V - V'$ and it follows that $V - V'$ is a vertex cover in G'. Also $|V - V'| = n - |V'|$ so that if $|V'| \geq k$, then $|V - V'| \leq n - k$.

Next, suppose that S is a vertex cover of G'. Take $V' = V - S$. Let $u, v \in V'$ so that neither of u, v is in S. Therefore u, v are not adjacent in G' which then implies that (u, v) is in E. Thus every pair of vertices in V' determines an edge in E and, hence, V' determines a clique in G.

Since there are $\frac{n(n-1)}{2}$ ordered pairs of elements of V and the number of edges m in E is $\leq \frac{n(n-1)}{2}$, G' can be obtained in polynomial time. $\quad\square$

Remark. The above proof can also be rephrased and we can have VCDP $<_P$CDP. Therefore **the vertex cover decision problem and the clique decision problem are polynomially equivalent.**

Since the relation $<_P$ of reducibility is transitive, combining Theorem 15.6 and Corollary 15.5, we get the following theorem.

Theorem 15.7. *3-SAT* $<_P$ *VCDP.*

Set cover decision problem (SCDP). Given a finite set S of elements and a family $F = \{S_1, S_2, \ldots, S_m\}$ of non-empty subsets of S and k a positive integer, does there exist a collection of at most k of these subsets whose union equals S?

Theorem 15.8. $VCDP <_P SCDP.$

Proof. Let a graph $G = (V, E)$ and a positive integer k be an instance of Vertex Cover Problem. Take $S = E$ and for every $u \in V$, take $S_u =$ the set of edges in E incident with u. Since every edge in E adjoins two vertices in V, $S = \bigcup_{u \in V} S_u$.

Claim. S can be covered by k subsets S_u if and only if G has a vertex cover of size $\leq k$.

Suppose that S is covered by k subsets $S_{u_1}, S_{u_2}, \ldots, S_{u_k}$. Then every edge e in $E(= S)$ is adjacent to at least one of the vertices $u_1, \ldots, u_k$ showing that $\{u_1, \ldots, u_k\}$ is a vertex cover of G of size k.

On the other hand if $u_1, \ldots, u_k$ is a vertex cover of G, every edge of G is incident with at least one of these vertices and so every edge is in S_u for some u in the given vertex cover of G. Hence S is covered by the k subsets $S_{u_1}, S_{u_2}, \ldots, S_{u_k}$. $\square$

The next result gives an interesting example of reduction of a problem to a special case of the problem.

Theorem 15.9. $CNF\text{-}SAT <_P 3\text{-}SAT.$

Proof. Let I be an instance of CNF-SAT so that I is a product of a finite number of clauses and each clause is conjunction (sum) of a finite number of literals. There may be some clauses having only one literal each, some having two literals each, some having three literals each and the rest having ≥ 4 literals each.

Claim. If C is a clause having

(i) one literal, i.e., $C = (a)$, then C is true if and only if $S = (a, x, y)(a, \bar{x}, y)(a, x, \bar{y})(a, \bar{x}, \bar{y})$ is true for literals x, y both different from a;

(ii) only two literals, say $C = (a, b)$, then C is true if and only if $S = (a, b, y)(a, b, \bar{y})$ is true for all literals y different from a, b;

(iii) $k \geq 4$ literals, say, $C = (a_1, a_2, \ldots, a_k)$, then C is true if and only if $S = (a_1, a_2, y_1)(\bar{y_1}, a_3, y_2) \ldots (\bar{y_{k-4}}, a_{k-2}, y_{k-3})(\bar{y_{k-3}}, a_{k-1}, a_k)\}$ is true for all literals $y_1, y_2, \ldots, y_{k-3}$ not occurring in C.

Proof. Claim (ii). For literals a, b, x,
$$S = (a + b + x)(a + b + \overline{x}) = (a + b)(a + b) + (a + b)(x + \overline{x}) + x\overline{x} = (a + b) + (a + b) = (a + b)$$ and (a, b) is true if and only if S is true.

Claim (i) For literals a, x, y, $S = (a + x + y)(a + x + \overline{y})(a + \overline{x} + y)(a + \overline{x} + \overline{y}) = (a + x)(a + \overline{x}) = aa + a(x + \overline{x}) + x\overline{x} = a$. Thus (a) is true if and only if S is true.

Claim (iii). Let $k \geq 4$, $C = (a_1, \ldots, a_k)$ and $S = (a_1, a_2, y_1)$ $(\overline{y_1}, a_3, y_2) \ldots (\overline{y_{k-4}}, a_{k-2}, y_{k-3})(\overline{y_{k-3}}, a_{k-1}, a_k)$.

Suppose that S is true. Then all the clauses (a_1, a_2, y_1), $(\overline{y_i}, a_{i+2}, y_{i+1})$, $1 \leq i \leq k - 4$, $(\overline{y_{k-3}}, a_{k-1}, a_k)$ are true. Suppose that none of the a_i is true. Then y_1 is true and a simple induction on i shows that y_i is true for all i, $1 \leq i \leq k - 3$. But then $(\overline{y_{k-3}}, a_{k-1}, a_k)$ is false, which is a contradiction. Hence if S is true, at least one of the literals a_i is true and, so, C is true.

Now suppose that C is true. Then at least one of the literals, say a_i, is true. Assign the value true to each y_j, $1 \leq j \leq i - 2$ and the value false to the rest of the y_j. Then the clauses $a_1 + a_2 + y_1, \overline{y_1} + a_3 + y_2, \ldots, \overline{y_{i-2}} + a_i + y_{i-1}$ are all true and the clauses $\overline{y_{i-1}} + a_{i+1} + y_i, \ldots, \overline{y_{k-4}} + a_{k-2} + y_{k-3}, \overline{y_{k-3}} + a_{k-1} + a_k$ are all true because $\overline{y_{i-1}}, \overline{y_i}, \ldots, \overline{y_{k-3}}$ are all true. Hence the conjunctive normal form S is true.

In the instance I replace each clause with only one literal by a product of four clauses each having three literals as in claim (i), each clause with two literals by a product of two clauses each having three literals as in claim (ii), each clause with $k \geq 4$ literals by a product of $k - 2$ clauses each having three literals as in claim (iii) and keep all the clauses in I with three literals each unchanged. Let the result of these replacements be I'. Then I' is an instance of 3-SAT. Clearly I' is obtained from I in polynomial time and, also, I is true if and only if I' is true. $\square$

Our aim next is to prove that Vertex Cover Problem is reducible to the Hamiltonian Circuit Problem.

Theorem 15.10. *VCP $<_P$ HCP, i.e., the vertex cover problem is polynomially reducible to the Hamiltonian circuit problem.*

Proof. Let a graph $G = (V, E)$ and a positive integer k be given as an instance of *VCP*. We need to construct a graph H and then prove that G

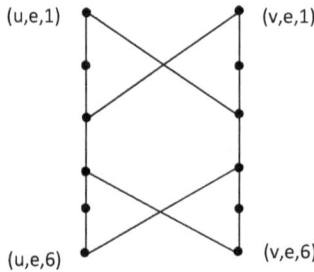

Fig. 15.2

has a vertex cover with cardinality $\leq k$ if and only if H has a Hamiltonian circuit.

Let X be a set of arbitrarily chosen elements $a_1, a_2, \ldots, a_k$ which are called **cover-choosing elements**. For each vertex $e = (u, v)$ in G, choose a subgraph H_e with 12 vertices and 14 edges as given in Fig. 15.2. The vertices of H_e are denoted by $(u, e, 1), (u, e, 2), \ldots, (u, e, 6)$ and $(v, e, 1), (v, e, 2), \ldots, (v, e, 6)$. The vertices $(u, e, 1), (v, e, 1)$ are called and denoted by u_e-top, v_e-top, respectively, and the vertices $(u, e, 6), (v, e, 6)$ are respectively denoted by and called u_e-bot and v_e-bot (bot for bottom). Each subgraph H_e corresponding to edge e of G is called a **cover-enforcer** (or **cover-tester**) **component**. So, for the construction of the graph H we have constructed $m \,(= |E|)$ subgraphs no two of which are joined in any way and with the k other elements. To introduce edges for joining the cover-enforcing subgraphs, we consider for every vertex u of G the edges of G incident with u and order these edges arbitrarily, say $e_{i1}, e_{i2}, \ldots, e_{id}$, where $d = \deg(u)$. Introduce edges by joining the vertices $(u, e_{ij}, 6)$ and $(u, e_{ij+1}, 1)$ for $j = 1, 2, \ldots, d - 1$. The cover-enforcing subgraphs joined in this way are said to form a covering-thread for u. This **covering-thread** is shown in Fig. 15.3.

Each covering thread introduces $d - 1$, where $\deg(u) = d$, new edges. Also, we introduce edges joining the initial vertex $(u, e_{i1}, 1)$ and the terminal vertex $(u, e_{id}, 6)$ (both shown encircled in Fig. 15.3) to every vertex in X giving $2k$ new edges. The construction of H is now complete. The graph H has $12|E| + k$ vertices and $14|E| + \sum_{u \in V}(d(u) - 1) + 2k|V|$ edges, since every vertex u of G gives rise to one covering thread.

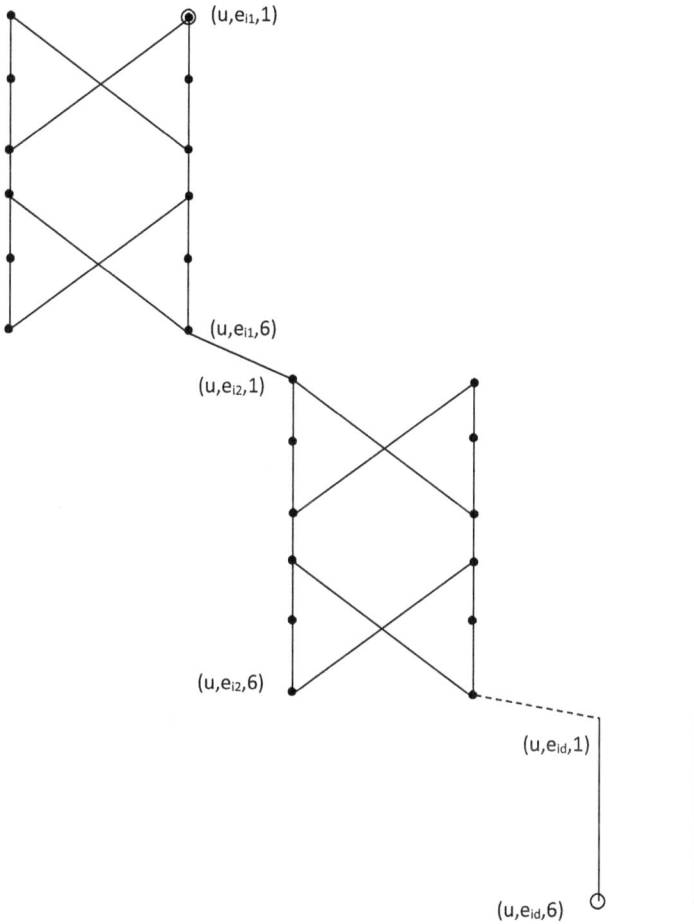

Fig. 15.3

It is clear that the graph H is obtained in polynomial time given the graph G and k.

Since only (possibly) the four corner vertices $(u, e, 1)$, $(u, e, 6)$, $(v, e, 1)$, $(v, e, 6)$ of the cover-enforcer subgraph H_e are adjacent to some vertices outside of H_e and none of the other eight vertices of H_e is adjacent to any vertex outside of H_e, a Hamiltonian circuit of H (if one exists) visits the vertices of H_e (all of which have to be visited by such a circuit) in only one of the three possible ways as shown in Figs. 15.4(a)–(c).

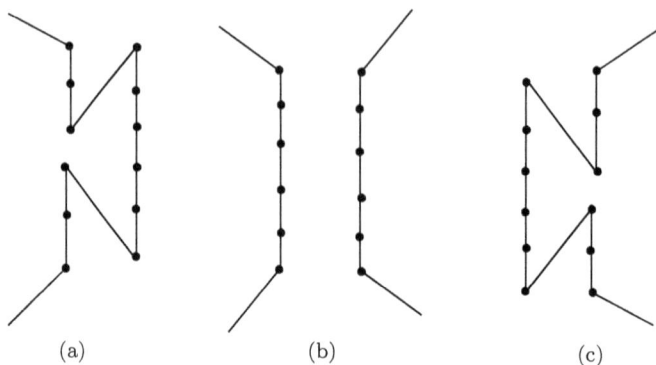

(a)   (b)   (c)

Fig. 15.4

In this figure only the edges actually appearing in the Hamiltonian circuit are shown except for Figs. 15.4(a) and 15.4(c) where one extra edge appears which has been shown as $\neq$ to indicate that the circuit does not include this edge.

We now prove that the graph G has a vertex cover of order $\leq k$ if and only if H has a Hamiltonian circuit.

Suppose that G has a vertex cover W of order $\leq k$. By adjoining a few vertices to W, if necessary, we may assume that order of W is k and let $\{v_{i1}, v_{i2}, \ldots, v_{ik}\} = W$. For every $j, 1 \leq j \leq k - 1$, we construct a Hamiltonian path P_j between a_j and a_{j+1} and a path P_k between a_k and a_1. For constructing P_j, we start with a_j, go to the initial vertex $(v_{ij}, e, 1)$ of H_e where e is the first edge in the ordering of the edges of G incident with v_{ij} (and not already visited by any earlier path) and visit the covering thread for v_{ij} or part thereof. The various cover testing components on this thread are to be visited in one of the three ways as indicated in Figs. 15.4(a)–(c). If on this thread appears a component H_f and $f = (v_{ij}, w)$ with w again in W, then only the right or left half of H_f is visited as Fig. 15.4(b) and the tracing of the path stops here. If w is not in W, then the whole of H_f is visited according to Fig. 15.4(a) or 15.4(c) and the path proceeds to the next component on this thread. When the whole thread is visited or it stops earlier, it stops on the vertex $(v_{ij}, f, 6)$ which happens to be a terminal vertex of a thread and, so, we can go along an edge to a_{j+1}. Since the cover-enforcing component or part thereof already visited in a path P_j is not considered again in a path, no edges of H are visited more than once. Also every edge of G is incident with at least one vertex in W, no component H_e is left out. Now combining the paths $P_1, P_2, \ldots, P_k$ in order, we get a Hamiltonian path in H.

Conversely, suppose that we are given a Hamiltonian circuit C in H. Divide C into k parts $P_1, P_2, \ldots, P_k$, each part starting with an a_j and ending with a_{j+1} for $j = 1$ to k with $a_{k+1} = a_1$. In view of the structure of cover enforcer components and the way these are joined together and the way the elements of X are joined to vertices in various H_e, if $(v_{ij}, e, 1)$ is the vertex of H appearing next to a_j in P_j the path P_j traverses the whole of the thread for v_{ij} or part thereof. Thus every edge e for which H_e or part thereof is traversed in P_j is adjacent to v_{ij}. Since C is the composition of $P_1, P_2, \ldots, P_k$ and every cover enforcing component is traversed by C, every edge e of G is adjacent to at least one of the vertices $v_{i1}, v_{i2}, \ldots, v_{ik}$ and $\{v_{i1}, v_{i2}, \ldots, v_{ik}\}$ is a vertex cover of G.

This completes the proof of our claim and, hence, VCP $<_P$ HCP. $\quad\square$

As an application of the procedure as given in the proof of the theorem, we consider a couple of examples.

Example 15.7. Consider a graph $G = (V, E)$, where $V = \{u, v, w\}$ and $E = \{e, f\}$ with $e = (u, v)$, $f = (v, w)$ and $k = 1$. The graph G is as given in Fig. 15.5.

There being two edges, each edge is represented by a subgraph of 12 vertices and 14 edges. The vertices of the subgraph corresponding to the edge e are denoted by $(u, e, 1), (u, e, 2), \ldots, (u, e, 6)$ and $(v, e, 1), \ldots, (v, e, 6)$ as in Fig. 15.6. The vertices of the subgraph corresponding to the edge f are represented similarly. The vertices $(u, e, 1), (v, e, 1)$ are also called and denoted by u_e-top, v_e-top, respectively, and the vertices $(u, e, 6), (v, e, 6)$ are, respectively, denoted by and

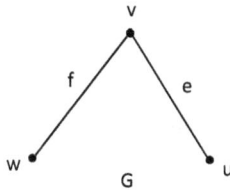

Fig. 15.5

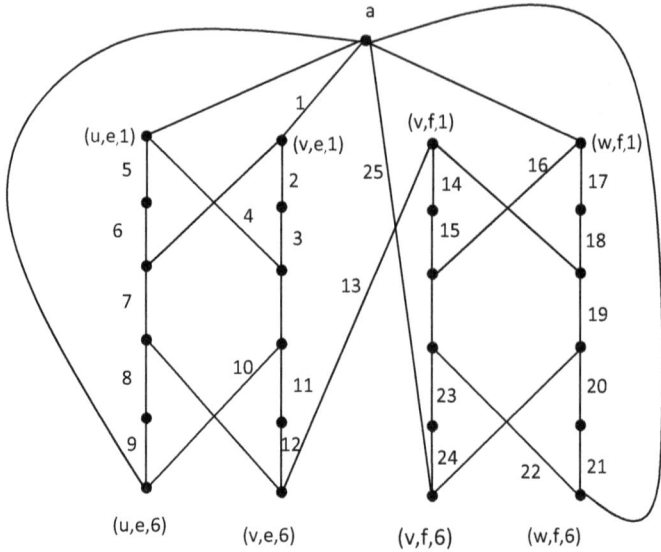

Fig. 15.6

called u_e-bot and v_e-bot (bot for bottom). Each subgraph corresponding to edge e of G is denoted by H_e and is called (as mentioned in the proof of the last theorem) **a cover-enforcer** (or **cover-tester**). Corresponding to the k vertices in a vertex cover of G, we choose k arbitrary elements $a_1, a_2, \ldots, a_k$ and call these **cover-choosing elements**. In the present case $k = 1$ and, so, there is only one cover-choosing element a (say). The instance H for the Hamiltonian-Circuit problem has as described, $12|E| + k$ vertices which in the present case is $12 \times 2 + 1 = 25$. The graph H already has $14|E| = 28$ edges. We now describe how the other edges are created in H. For each vertex u (say) of G, we consider the edges of G incident with u and order them arbitrarily as, say, $e_1, e_2, \ldots, e_d$ where $d = \deg(u)$. We connect the vertices $(u, e_1, 1)$ and $(u, e_d, 6)$ to the k vertex choosing elements — thus giving $2k$ edges for every vertex of G. This gives $2kn, n = |V|$, new edges in H. So far the cover-testing components H_e are not connected to each other. Again consider the edges of G incident with u as above. For $i, 1 \le i \le d - 1$, introduce the edges $((u, e_i, 6), (u, e_{i+1}, 1))$. This gives $d - 1$ new edges. The construction of the graph H is now complete. Observe that the total number of edges in H

is $14|E|+2k|V|+\sum_{u\in V}(\deg(u)-1)$. (Refer also to the proof of Theorem 15.8.)

In the present case under consideration, let the edges incident with the vertices be: $u : e; v : e, f; w : f$. Then the edges introduced in H are obtained by connecting

$$a \text{ to } (u, e, 1), (u, e, 6), (v, e, 1), (v, f, 6), (w, f, 1), (w, f, 6),$$

together with an edge connecting $(v, e, 6)$ to $(v, f, 1)$.

A vertex cover being $\{v\}$, for a Hamiltonian cycle we start with the edge joining a to $(v, e, 1)$. The corresponding Hamiltonian cycle for the present example is given by the following edges numbered 1 to 25 in order as in Fig. 15.6 i.e.,

$$a \to (v, e, 1) \to (v, e, 2) \to (v, e, 3) \to (u, e, 1) \to (u, e, 2) \to \cdots$$
$$\to (u, e, 6), \to (v, e, 4), \to (v, e, 5), \to (v, e, 6), \to (v, f, 1),$$
$$\to (v, f, 2), \to (v, f, 3), \to (w, f, 1), \to \cdots \to (w, f, 6),$$
$$\to (v, f, 4), \to (v, f, 5), \to (v, f, 6), \to a.$$

On the other hand, if we are given, say, the Hamiltonian circuit as above, then the cover-choosing element is joined to the vertices $(v, f, 6)$ and $(v, e, 1)$, i.e., it is connected to a vertex in H_e and a vertex in H_f. In G, v is the common vertex of the edges e and f and, so, $\{v\}$ is a vertex cover of G.

Example 15.8. Consider a graph $G = (V, E)$ as given in Fig.15.7 and $k = 2$. Let the edges e, f, g be, respectively, $(u, v), (v, w), (w, u)$. Let the edges incident with the vertices be ordered as: $u : e, g; v : e, f; w : f, g$.

The corresponding instance of the Hamiltonian circuit problem consists of three cover-enforcing components each having 12 vertices and 14 edges and two cover-choosing elements a, b (say). The components H_e and H_g are connected by joining the vertices $(u, e, 6)$ and $(u, g, 1)$, the components H_e and H_f are joined by connecting $(v, e, 6)$ to $(v, f, 1)$, and the components H_f , H_g are connected by joining $(w, f, 6)$ and $(w, g, 1)$. Then every one of the vertices $(u, e, 1), (u, g, 6), (v, e, 1), (v, f, 6), (w, f, 1), (w, g, 6)$ is joined to both

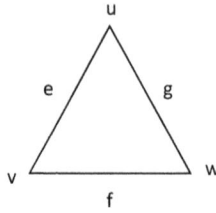

Fig. 15.7

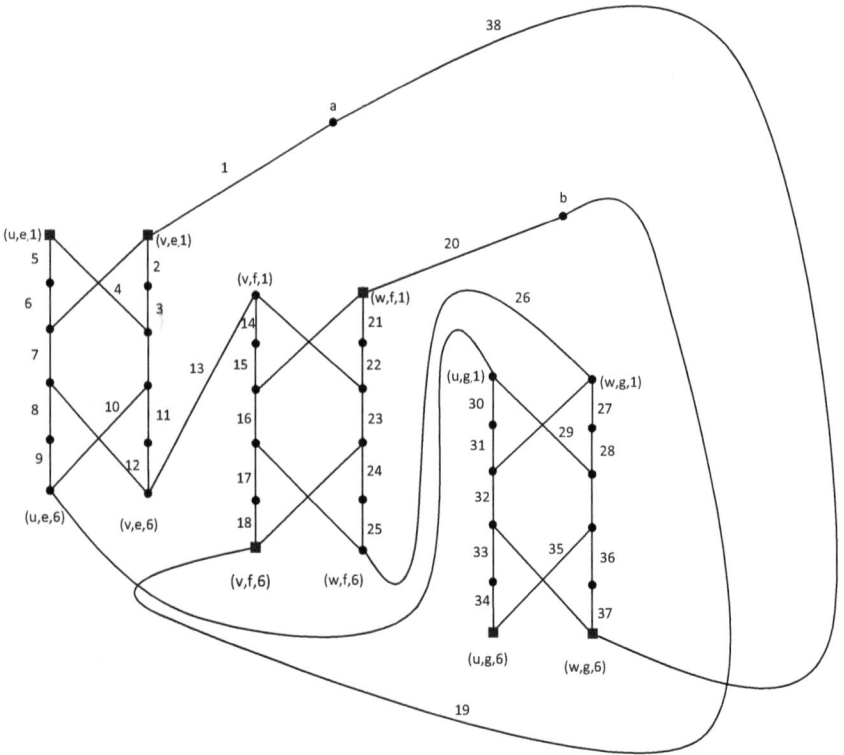

Fig. 15.8

the cover-choosing vertices a, b giving 12 edges. The construction of H is then as given in Fig. 15.8.

The six vertices to be joined to the cover choosing vertices a, b are shown by thick dots. Consider a vertex cover $\{v, w\}$. We need to join a to $(v, f, 1)$. Then a Hamiltonian circuit starting at a is given by following the edges 1 to 38 in order as indicated in Fig. 15.8. Explicitly the Hamiltonian circuit is

$$a \to (v, e, 1) \to (v, e, 2) \to (v, e, 3) \to (u, e, 1) \to \cdots \to (u, e, 6)$$
$$\to (v, e, 4) \to (v, e, 5) \to (v, e, 6) \to (v, f, 1) \to \cdots$$
$$\to (v, f, 6) \to b \to (w, f, 1) \to \cdots \to (w, f, 6)$$
$$\to (w, g, 1) \to (w, g, 2) \to (w, g, 3) \to (u, g, 1) \to \cdots$$
$$\to (u, g, 6) \to (w, g, 4)$$
$$\to (w, g, 5) \to (w, g, 6) \to a.$$

Observe that only four edges connecting other vertices of H which actually occur in the Hamiltonian circuit are shown.

Given the above Hamiltonian circuit, since a is connected to a vertex of H_e and a vertex of H_g and the only vertex in G common to e and g being u, the circuit forces a to be u. The vertex b of H is connected to vertices $(v, f, 6)$ and $(w, f, 1)$ both in H_f. Since f is an edge connecting the vertices v and w, b may be taken as either of these two vertices. Therefore, the circuit yields two vertex covers $\{u, v\}$ and $\{u, w\}$ each having two vertices.

On the other hand, given a vertex cover $\{v, w\}$, we choose two vertices a, b for H and construct H as above. One of the Hamiltonian circuits of H is then given as above.

Let the edges incident with vertices be now ordered as: u : e, g; v : e, f; w : g, f. Then in the instance H of the Hamiltonian circuit problem the components H_e, H_g are connected by joining the vertices $(u, e, 6)$, $(u, g, 1)$, the components H_e, H_f are joined by connecting the vertices $(v, e, 6)$, $(v, f, 1)$ and the components H_f, H_g are connected by joining vertices $(w, g, 6)$, $(w, f, 1)$. Then the ends $(u, e, 1)$, $(u, g, 6)$, $(v, e, 1)$, $(v, f, 6)$, $(w, g, 1)$ and $(w, f, 6)$ of the three covering threads every one of which is to be connected to the cover selecting vertices are shown encircled and only those edges adjacent to a, b will be shown which actually appear in a Hamiltonian circuit corresponding to a given vertex cover.

Consider a vertex cover $\{v, w\}$ of G. We construct a Hamiltonian circuit in H by first joining a to $(v, e, 1)$. Then we traverse the cover-enforcing component H_e as in Fig. 15.4(c). Having arrived at vertex $(v, e, 6)$, we go to the vertex-tester component H_f through the edge $((v, e, 6), (v, f, 1))$. If we traverse the component H_f as in Fig. 15.4(a), we will arrive at the vertex $(v, f, 6)$ which then has to be joined to cover-choosing element b. The only way to go to the remaining component H_g is by going from b to $(w, g, 1)$ or $(u, g, 6)$. There is then no way to traverse H_g and go to a through a Hamiltonian path. Therefore, the only way that we can traverse H_f is as in Fig. 15.4(b). (Observe that $f = (v, w)$ where v, w both are in the vertex cover.) By doing that only the left half of H_f is traversed and we arrive at $(v, f, 6)$ which needs to be joined to the cover choosing element b. We connect b to $(w, g, 1)$ and traverse H_g as in Fig. 15.4(c) and arrive at the vertex $(w, g, 6)$. We now go along the edge from $(w, g, 6)$ to $(w, f, 1)$ and then traverse the right-hand arm of H_f. Traversal of all the three cover-enforcing components H_e, H_f, H_g is complete. To get the corresponding Hamiltonian circuit, we join the vertex $(w, f, 6)$ to a. The Hamiltonian circuit is thus given by going along the edges 1 to 38 as indicated in Fig. 15.9. Explicitly the circuit is

$$a \rightarrow (v, e, 1) \rightarrow (v, e, 2) \rightarrow (v, e, 3) \rightarrow (u, e, 1) \rightarrow \cdots \rightarrow (u, e, 6)$$
$$\rightarrow (v, e, 4) \rightarrow (v, e, 5) \rightarrow (v, e, 6) \rightarrow (v, f, 1) \rightarrow \cdots \rightarrow (v, f, 6)$$
$$\rightarrow b \rightarrow (w, g, 1) \rightarrow (w, g, 2) \rightarrow (w, g, 3) \rightarrow (u, g, 1) \rightarrow \cdots$$
$$\rightarrow (u, g, 6) \rightarrow (w, g, 4) \rightarrow (w, g, 5) \rightarrow (w, g, 6) \rightarrow (w, f, 1)$$
$$\rightarrow \cdots \rightarrow (w, f, 6) \rightarrow a.$$

Conversely, suppose that the Hamiltonian circuit is given as above. We break this circuit into two paths by following the edges numbered 1 to 19 and 20 to 38, the two paths being from a to b and from b to a, respectively. The edge numbered 1 joins a to $(v, e, 1)$ so that we may take v as a vertex in a vertex cover, In the second path, the vertex choosing element b is joined to $(w, g, 1)$ and as such w is the next element in the vertex cover. Thus the corresponding vertex cover is $\{v, w\}$.

Exercise 15.1. Given a graph G, a vertex cover W of G, and an edge $e = (u, v)$ with u, v both in W, let H_e be a cover-enforcing component determined by e. Decide if the vertices $(u, e, 6)$ and $(v, e, 6)$ of H_e are the terminal vertices of a covering thread for some vertex w in W.

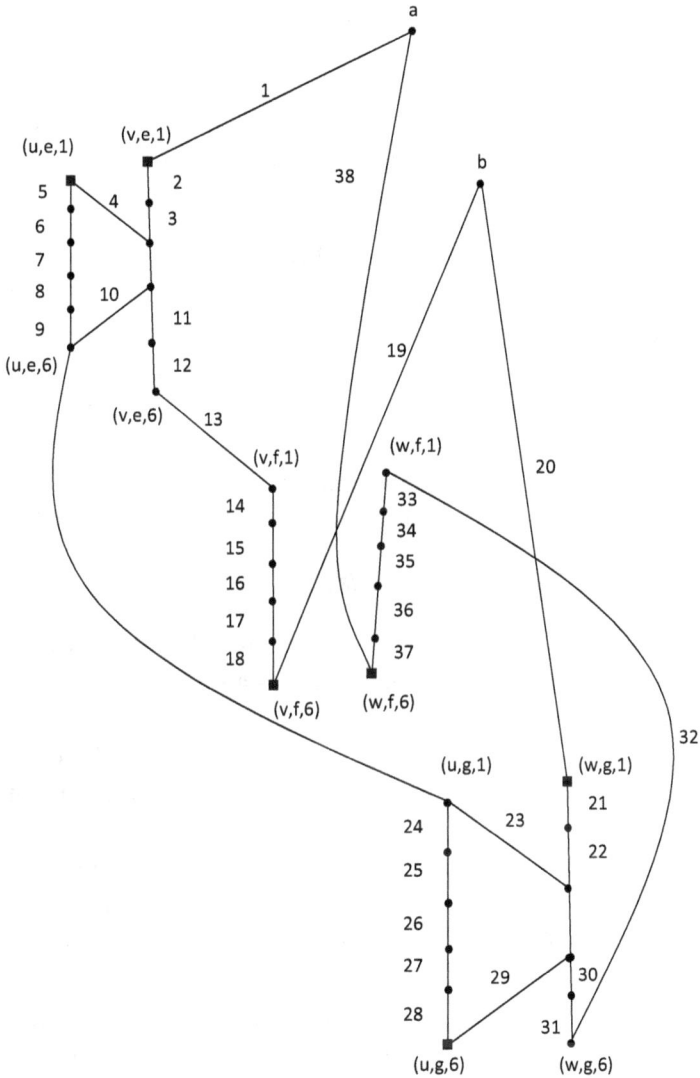

Fig. 15.9

15.4. Some NP-Complete Problems

As already proved (Theorem 15.1) a problem *X* in *NP* is *NP-complete* if there exists an *NP-complete* problem *M* such that $M \leq_P X$. Thus *NP-completeness* of a problem depends on the existence of some other

NP-complete problem. This then gives rise to the question: Does there exist some *NP-complete* problem? This question is settled in the affirmative by the following theorem of Cook a proof of which is available in [5]. Proof of the theorem of Cook that SAT is in P if and only if $P = NP$ is given in [14]. Formula Satisfiability or CNF-Satisfiability or just SAT was in fact the first problem proved to be *NP-complete*.

Theorem 15.11 (Cook's theorem). *SAT is NP-complete.*
We use this theorem as a base and prove *NP-completeness* of some other problems. Although hundreds of problems are known to be *NP-complete*, we have chosen only very few of these that we could present nicely using the reducibility results already proved in the previous section.

Although we accept Cook's theorem on *NP-completeness* of the SAT problem, we prove the easier part of this result in the following theorem.

Theorem 15.12. *The problem SAT is in NP.*

Proof. Let F be a binomial formula in CNF. Then F is a product of clauses $C_1, C_2, \ldots, C_m$ where each clause is a sum of k literals for some k. We may assume that $k \geq 3$. Let the variables under consideration be $x_1, x_2, \ldots, x_n$. Given a set of values $a_1, a_2, \ldots, a_n$ (say) of the variables, each clause C_i can be obtained using $k - 1$ (addition) computations. Therefore, the m clauses can be evaluated using $m(k - 1)$ computations. Then F can be computed from the C_i in at most $m - 1$ (multiplicative) computations. Thus F is evaluated by $m(m - 1)(k - 1)$ computations. The maximum number of conjunctive clauses in n variables is $n(2n - 1)$. Therefore, $m \leq n(2n - 1)$. Therefore, running time for the evaluation of F is $\leq (k - 1)n(2n - 1)(2n^2 - n - 1)$, i.e., polynomial time in the number of variables involved. Once F is evaluated, we just need to see if it is 1 or not. This shows that in polynomial time it can be decided if the given set of values of the variables provides a solution to the satisfiability of F. □

As a particular case of this we have the following corollary.

Corollary 15.13. *3-SAT Problem is in NP.*
Combining Theorems 15.1, 15.9, 15.11 and Corollary 15.13, we get the following particular case of Cook's theorem.

Theorem 15.14 (Cook). *3-SAT problem is NP-complete.*

Theorem 15.15. *Clique problem (CP in short) is NP-complete.*

Proof. Let $G = (V, E)$ be a graph with $|V| = n$. Since any clique in G, if it exists, has $\leq |V|$ vertices, let k be a positive integer $\leq n$. A certificate for a possible solution of the clique problem is a subgraph H of G with k vertices. For a vertex u of H, let $\text{adj}_H(u)$, $\text{adj}_G(u)$ denote the adjacency class of u as a vertex of H and as a vertex of G, respectively. Then $\text{adj}_H(u) = H \cap \text{adj}_G(u)$.

We need to verify whether H is a complete subgraph on its k vertices or not. It is clear that H is a complete graph on its k vertices if and only if $\text{adj}_H(u) = k - 1$ for every u in H. Since $|\text{adj}_G(u)| \leq n$, a computation of $(a) \cap \text{adj}_G(u)$ for an $a \in H$ requires at most n comparisons. Hence the computation of $\text{adj}_H(u)$ for all the k vertices of H requires at most nk comparisons. Once all the $\text{adj}_H(u)$ have been obtained, all that needs to be checked is if for any vertex u of H, we have $\text{adj}_H(u) < k - 1$. If there is no such u, then H is a complete subgraph on k vertices. Hence a polynomial time verification leads to whether H is a k-clique or not. Therefore clique problem is in *NP*. That the clique problem (CP) is *NP-complete* then follows Theorems 15.1, 15.14 and Corollary 15.5. □

Theorem 15.16. *Vertex cover problem (VCP in short) is NP-complete.*

Proof. Let $G = (V, E)$ be a graph with $|V| = n$ and k be a positive integer. A certificate for a possible solution of VCP is a subset V' of V. If $|V'| > k$, then V' is not a solution to the problem and, so, we assume that $|V'| \leq k$. In order to check if V' is a vertex cover of G or not, we need to check if every edge of G is incident with at least one vertex in V' or not. Consider an arbitrary edge e of G. Then e is incident with vertices u, v (say). We just look for u, v in the set V'. If at least one of u, v is in V', then e is incident with at least one vertex in V'. This checking for u and v in V' can be done in a finite number of steps — the number of steps being $\leq 2n$. Since G has only a finite number of $|E| = m$ edges, the number of steps required to check whether every edge of G is incident with at least one vertex in V' is $\leq 2mn \leq n^3$, as the number of edges in G is at most n^2 (we can assume without loss of generality that G is a simple graph without any loops). Hence VC is a problem in *NP*. It then follows from Theorems 15.1, 15.6 and 15.15 that VCP is *NP-complete*.   □

Theorem 15.17. *Independent set problem (ISP) is NP-complete.*

Proof. Let $G = (V, E)$ be a graph with $|V| = n$ and k be a positive integer. A certificate for a possible solution of the independent set problem is a subset V' of V. Recall that V' is an independent set (of vertices) if whenever $u, v \in V'$, then there is no edge in G incident with u, v. If $|V'| < k$, then V' is not a solution of the independent set problem. So, we may assume that $|V'| \geq k$. Observe that V' is an independent set if and only if $V' \cap \text{adj}(v) = \varphi$ for every $v \in V'$. Since $|V'| \leq n$, we have to look up at most n adjacency lists (one each for $v \in V'$) and see if the at most n^2 sets $V' \cap \text{adj}(v)$, $v \in V'$ are empty sets or not. If these are empty, then V' is indeed a solution of the problem and it has been verified in $\leq n^4$ steps, i.e., in polynomial time. Of course we can assume the graph G to have been given through adjacency lists. Hence ISP is in *NP*. It then follows from Theorems 15.1, 15.16 and Corollary 15.3 that ISP is *NP-complete*.   □

Theorem 15.18. *Set cover problem (or SCP) is NP-complete.*

Proof. Let S be a finite set, F a family of non-empty subsets S_i of S and k a positive integer. A certificate for a possible solution of the set cover problem is a subfamily F' of F. If $|F'| > k$, then F' is not a solution of the set cover problem. So, we can suppose that $|F'| \leq k$. Renaming the elements of F, if necessary, suppose that $S_1, S_2, \ldots, S_r, r \leq k$, are the members of F'. Let $x \in S$. A finite number of comparisons shows if $x \in S_1$ or not. If not, we check if x belongs to S_2 or not. We continue this process till we find an i such that $x \in S_i$, $1 \leq i \leq r$. If there is no such i, then we find that F' is not a cover of S. Then we adopt the above procedure for another element y of S and check if y is in some S_j, $1 \leq j \leq r$ or not. We continue this process of checking till we arrive at an $s \in S$ which is not in any S_i, $1 \leq i \leq r$. In this case F' is not a cover of S. If we find that every element of S is in some S_i, $1 \leq i \leq r$, F' is a cover of S. The number of comparison checks made is $\leq kn^2 < n^3$, where $n = |S|$. Hence in a polynomial number of steps we can verify if F' is a solution of the set cover problem and, therefore, the set cover problem is in *NP*. That set cover problem is *NP-complete* then follows from Theorems 15.1, 15.8 and 15.16. □

Theorem 15.19. *Hamiltonian cycle problem is NP-complete.*

Proof. Let $G = (V, E)$ be a graph with $|V| = n$. A certificate for a possible solution of the Hamiltonian cycle problem is a sequence $s = (x_1, \ldots, x_n)$ of vertices of G. If $x_i = x_j$ for some $i, j, 1 \leq i, j \leq n$, then s is not a Hamiltonian path. So, we may suppose that $x_i \neq x_j$ for any $i \neq j$. We may suppose that G is represented by adjacency lists. We check if x_{i+1} belongs to $\text{adj}(x_i)$ or not for $i = 1$ to n with $x_{n+1} = x_1$. If there exists an $i, 1 \leq i \leq n$, such that x_{i+1} is not in $\text{adj}(x_i)$, then s is not a Hamiltonian cycle. On the other hand, if x_{i+1} is in $\text{adj}(x_i)$ for every $i, 1 \leq i \leq n$, then s is a Hamiltonian cycle and answer to the Hamiltonian cycle problem is yes. The number of comparison checks required to check whether x_{i+1} is in $\text{adj}(x_i)$ or not is $\deg(x_i)$. Hence the total number of comparison checks required is $\leq \sum_{1 \leq i \leq n} \deg(x_i) \leq n(n-1)$. Hence a polynomial number of steps and, so, a polynomial time is required to check whether s is a Hamiltonian cycle or not. This proves that HC problem is in *NP*; the result then follows from Theorems 15.1, 15.10 and 15.16. □

Theorem 15.20. *Traveling sales person problem is NP-complete.*

Proof. Let the n cities to be visited by the sales person be numbered (and denoted by) $1, 2, \ldots, n$ and let $w(i, j)$ denote the cost (or distance) for going from city i to city j. Let W be a given positive number.

We first prove that TSP is in *NP*.

A certificate for a possible solution of TSP is a permutation π of the set $S = \{1, 2, \ldots, n\}$. The sales person is able to visit each city on a tour and return to the starting city if and only if π is a cyclic permutation of length n. Observe that π is a cyclic permutation if and only if $S = \{1, \pi(1), \pi^2(1), \ldots, \pi^{n-1}(1)\}$. By using i comparison checks we can see if $\pi^i(1) = 1$ or not. If $\pi^i(1) = 1$ for some $i < n$, then π is not a cyclic permutation of length n and, so, π is not a solution of TSP. Now suppose that π is cyclic permutation of length n (it will need $n - 1$ comparison checks). The tour taken by the sales person can then be taken as

$$1 \to \pi(1) \to \pi^2(1) \to \cdots \to \pi^i(1) \to \pi^{i+1}(1) \to \cdots$$
$$\to \pi^{n-1}(1) \to \pi^n(1) = 1.$$

The total cost of tour is $C = \sum_{0 \le i < n} w(\pi^i(1), \pi^{i+1}(1))$ which can be calculated by n additive compositions. If $C > W$, then π is not a solution while if $C \le W$, then π is a solution to the TSP. Thus it is verified in polynomial time if π is a solution of the TSP or not. Hence TSP is in *NP*.

We next prove that $HC <_P TSP$.

Let a connected graph $G = (V, E)$ with $|V| = n$ be an instance of the Hamiltonian cycle problem. For the sake of simplicity we may take $V = \{1, 2, \ldots, n\}$. Form a complete graph K_n on the n vertices in V. Then E is a subset of the edge set of K_n. For the vertices i, j in V, define the weight $w(i, j)$ (for the edge (i, j) in K_n) by

$$w(i, j) = \begin{cases} 1 & \text{if the edge } (i, j) \text{ is in } E, \\ 2 & \text{if the edge } (i, j) \text{ is not in } E. \end{cases}$$

Then the corresponding instance of TSP is (K_n, w, n). Since K_n is obtained by joining i, j if these are not already adjacent in G, thereby constructing only $\frac{n(n-1)}{2} - |E|$ edges or that adding $< n^2$ vertices in the adjacency lists of G and the weight function w is defined for all the $\frac{n(n-1)}{2}$ edges of K_n, the instance (K_n, w, n) is obtained from G in polynomial time.

Suppose that G has a Hamiltonian cycle h (say). Then h is a cyclic permutation of length n of V so that h leads to a tour of the n cities in V. Each edge on this tour being an edge in G, sum of the weights of the edges on the tour is exactly n. Conversely, suppose that there is a tour of cost $\le n$ in K_n. Thus there is a permutation π of V of length n with $\sum_{0 \le i < n} w(\pi^i(1), \pi^{i+1}(1)) \le n$. Since $w(i, j) \ge 1$ for all i, j in V, we must have $\sum_{0 \le i < n} w(\pi^i(1), \pi^{i+1}(1)) = n$ so that $w(\pi^i(1), \pi^{i+1}(1)) = 1$ for all i. It follows from the definition of the weight function w that every edge $(\pi^i(1), \pi^{i+1}(1))$ say (k, j) is an edge in E. Hence π provides, in fact is, a Hamiltonian cycle in G. This completes the proof that $HC <_P TSP$.

The result then follows from Theorems 15.1 and 15.19. □

Exercise 15.2.

1. Prove that the independent set problem for a connected graph every vertex of which is of degree 2 is not *NP-complete*. Give an algorithm for finding an independent set in such a graph.
2. Prove that the problem 2-SAT is not *NP-complete*.

The proof that vertex cover problem is reducible to Hamiltonian cycle problem is based on the proof of the problem given in [5,15]. The proof of Theorem 15.9 is based on the proof of [2, Theorem 10.4].

Bibliography

[1] Aho, A. V., Hopcroft, J. E. and Ullman, J. D., *Data Structures and Algorithms*, Pearson Education (Singapore), Delhi, 2003.

[2] Aho, A. V., Hopcraft, J. E. and Ullman, J. D., *The Design and Analysis of Computer Algorithms*, Pearson Education (Singapore), Delhi, 2003.

[3] Batcher, K. E., Sorting networks and their applications, in *Proc. AFIPS*, Vol. **32**, pp 307–314, 1968.

[4] Bellman, R., On a routing problem, *J. Quart. Appl. Math.* **16**(1), 87–90, 1958.

[5] Cormen, T. H., Leiserson, C. E., Rivest, R. L. and Stein, C., *Introduction to Algorithms*, 2nd edn., Prentice-Hall of India, New Delhi, 2006.

[6] Dasgupta, S., Papadimitrion, C. H. and Vazirani, U. V., *Algorithms*, McGraw-Hill Higher Education, 2007.

[7] Deo, N., *Graph Theory with Applications to Engineering and Computer Science*, Prentice-Hall of India, New Delhi, 2007.

[8] Dijkstra, E. W., A note on two problems in connection with graphs, *Numer. Math.*, **1**, 269–271, 1959.

[9] Floyd, R. W., Algorithm 97 (shortest path), *Commun. ACM*, **5**, 345, 1962.

[10] Garey, M. R., Johnson, D. S. and Stockmeyer, L., Some simplified NP-complete graph problems, *Theoret. Comput. Sci.* **1**, 237–267, 1976.

[11] Goodrich, M. T. and Tamasia, R., *Algorithm Design: Foundations, Analysis & Internet Examples*, Wiley, 2001/2004.

[12] Harary, F., *Graph Theory*, Addison-Wesley, Reading, MA, 1960.

[13] Har-Peled, S., *Graduate Algorithms*, Lecture Notes in Computer Science, Vol. **373**, Lecture Notes, Sorting Networks Springer, 2014.

[14] Horowitz, E., Sahni, S. and Rajasekaran, S., *Fundamentals of Computer Algorithms*, Galgotia Publications, New Delhi, 2004.

[15] Jungnickel, D., *Graphs, Networks and Algorithms*, Springer-Verlag, Berlin, 1999.

[16] Kleinberg, J. and Tardos, E., *Algorithm Design*, Pearson Education, 2005.

[17]   Knuth, D. E., Optimum binary search trees, *Acta Inform.* **1**, 14–25, 1971.

[18]   Knuth, D. E., Morris, Jr. J. H. and Pratt, V. R., Fast pattern matching in strings, *SIAM J. Comput.* **6**(2), 323–350, 1977.

[19]   Levintin, A., *Introduction to the Design and Analysis of Algorithms*, Pearson Education, 2011/2012.

[20]   Liu, C. L., *Elements of Discrete Mathematics*, 2nd edn., Tata McGraw-Hill Publishing Co. Ltd., 2002.

[21]   Miller, B. N., *Problem Solving with Algorithms and Data Structures Using Python*, Brad Miller, David Ranum, 2014.

[22]   Puntambekar, A. A., *Design and Analysis of Algorithms*, Technical Publications, 2011.

[23]   Rosen, K. H., *Discrete Mathematics and its Applications*, Tata McGraw-Hill, New Delhi, 2003/2005.

[24]   Sedgewick, R. and Wayne, K., *Algorithms*, 4th edn., Addison-Wesley Professional, Pearson Education, 2011.

[25]   Sridhar, S., *Design and Analysis of Algorithms*, Oxford University Press, 2014.

[26]   Strassen, V., Gaussian elimination is not optimal. *Numerische Mathematik*, **14**(3), 354–356, 1969.

[27]   Vazirani, V., *Approximation Algorithms*, Springer-Verlag, Berlin, 2001.

[28]   Vermani, L. R. and Vermani, S., *A Course in Discrete Mathematical Structures*, Imperial College Press, London, 2012.

[29]   Burfield, C., Floyd–Warshall algorithm, available at http://math.mit.edu/rothvoss/18.304.1PM/Presentations/1-Chandler-18.304lecture1.pdf, (accessed on 20-2-2013).

[30]   CMPS 102, Introduction to analysis of algorithms, 2003, asymptotic growth of functions, available at http://www.academia.edu/5330713/ CMPS_102_Introduction_to_Analysis_of_Algorithms_Fall_2003_Asymptotic_Growth_of_Functions (accessed on 8-6-2015).

[31]   Gross, J. L., Course notes on growth of functions (for use with Discrete Mathematics and its Applications by Rosen), available at https://users.encs.concordia.ca/~bui/pdf/slides0.pdf (accessed on 8-6-2015).

[32]   Zaguia, Dr- CSI 2101-W08 Growth of function, available at http://www.site.uottawa.ca/~lucia/courses/2101-09/classnotes/GrowthFunctions08.pdf (accessed on 8-6-2015).

[33]   Growth of functions, available at https://www.math.fsu.edu/~pkirby/mad2104/SlideShow/s4_4.pdf (accessed on 8-6-2015).

[34]   Growth of functions, available at http://www.cs.bme.hu/~sali/thalg/bigoandrecurrences.pdf (accessed on 8-6-2015).

[35] Computer science 373, lecture 16, NP-hard problems, 2002, available at http://www.dcs.fmph.uniba.sk/~fduris/VKTI/3color_HamCycle.pdf (accessed on 11-6-2015).

[36] ICS 161: Design and analysis of algorithms, Lecture Notes, NP completeness, 1996, available at https://www.ics.uci.edu/~eppstein/161/960312. html (accessed on 11-6-2015).

[37] Geeks for Geeks, NP-completeness, set I (introduction), available at https:// www.geeksforgeeks.org/np-completeness-set-1/ (accessed on 11-6-2015).

[38] CMSC 451: Reductions and NP-completeness, Deptt of Computer Science, University of Maryland, College Park (Based on Section 8.1 of Algorithm Design by Kleinberg and Tardos), available at https://www.cs.cmu.edu/ -ckigdt/broinfm-lectures/npcomplete.pdf (accessed on 11-6-2015).

[39] Heap sort tutorials & problems, available at https://www.hackerearth.com/ practice/algorithms/sorting/heap-sort/tutorial/ (accessed on 7-7-2017).

[40] Programiz-heap sort algorithm, available at https://www.programiz.com/ dsa/heap-sort (accessed on 7-7-2017).

[41] Kleinberg, J. and Tardos, E., Algorithm design-greedy algorithms, available at https://www.cs.princeton.edu/~wayne/kleinberg-tardos/pearson/ 04GreedyAlgorithms.pdf (accessed on 18-06-2015).

[42] Greedy introduction, available at http://www.personal.kent.edu/ ~rmuhamma/Algorithms/MyAlgorithms/Greedy/greedyIntro.htm (accessed on 18-06-2015).

[43] Sorting networks, available at http://www.iti.fh-flensburg.de/lang/ algorithmen/sortieren/networks/sortieren.htm (accessed on 9-5-2014).

[44] Har-Peled, S., Sorting networks, December 2012, available at https:// courses.engr.illinois.edu/cs573/fa2012/lec/lec/19_sortnet.pdf (accessed on 9-5-2014).

[45] Montanaro, A., All pairs shortest paths, COMS21103, November 2013, available at https://people.maths.bris.ac.uk/~csxam/teaching/dsa/allpairs. pdf (accessed on 18-6-2015).

[46] Geeks for Geeks, Johnson's algorithm for all-pairs shortest paths, available at https://www.geeksforgeeks.org/johnsons-algorithm/ (accessed on 18-6-2015).

[47] CSE341T/CSE549T, More all pairs shortest paths, Lecture 13, 2014, available at https://www.cse.wustl.edu/~angelee/archive/cse341/ fall14/handouts/lecture13.pdf (accessed on 15-10-2014).

[48] All pairs shortest paths problem: Floyd's algorithm, available at http://lcm. csa.iisc.ernet.in/dsa/node164.html (accessed on 18-6-2015).

[49] All-pairs shortest path with Floyd–Warshall, CPSC 490 graph theory: finding all-pairs shortest paths and counting walks, available at https://www.ugrad.cs.ubc.ca/~cs490/Spring05/notes/graph_part3.pdf (accessed on 18-6-2015).

[50] Floyd–Warshall all–pairs shortest path, available at https://www.cs.usfca.edu/~galles/visualization/Floyd.html (accessed on 18-6-2015).

[51] CS 360 Lecture Notes, Lecture 23: All pairs shortest paths-Floyd–Warshall algorithm, available at http://faculty.ycp.edu/~dbabcock/PastCourses/cs360/lectures/lecture23.html (accessed on 18-6-2015).

[52] CS 360, Lecture Notes, Lecture 21: Single source shortest paths-Bellman–Ford algorithm, available at http://faculty.ycp.edu/~dbabcock/PastCourses/cs360/lectures/lecture21.html (accessed on 18-6-2015).

[53] Umut Acar, Shortest weighted paths II, parallel and sequential data structures and algorithms-lecture 13, 15-210 (Spring 2013), available at https://www.cs.cmu.edu/afs/cs/academic/class/15210-s13/www/lectures/lecture13.pdf (accessed on 18-06-2015).

[54] More all pairs shortest paths, available at https://www.cse.wustl.edu/~angelee/archive/cse341/fall14/handouts/lecture13.pdf (accessed on 18-06-2015).

[55] Lecture 10 Sorting networks, available at http://fileadmin.cs.lth.se/cs/Personal/Rolf_Karlsson/lect10.pdf (accessed on 5-9-2014).

[56] Chapter 19 Sorting networks, available at https://courses.engr.illinois.edu/cs573/fa2012/lec/lec/19_sortnet.pdf (accessed on 5-9-2014).

[57] Matuszek, D. (2002), Backtracking, available at https://www.cis.upenn.edu/~matuszek/cit594-2012/Pages/backtracking.html (accessed on 5-9-2014).

Index

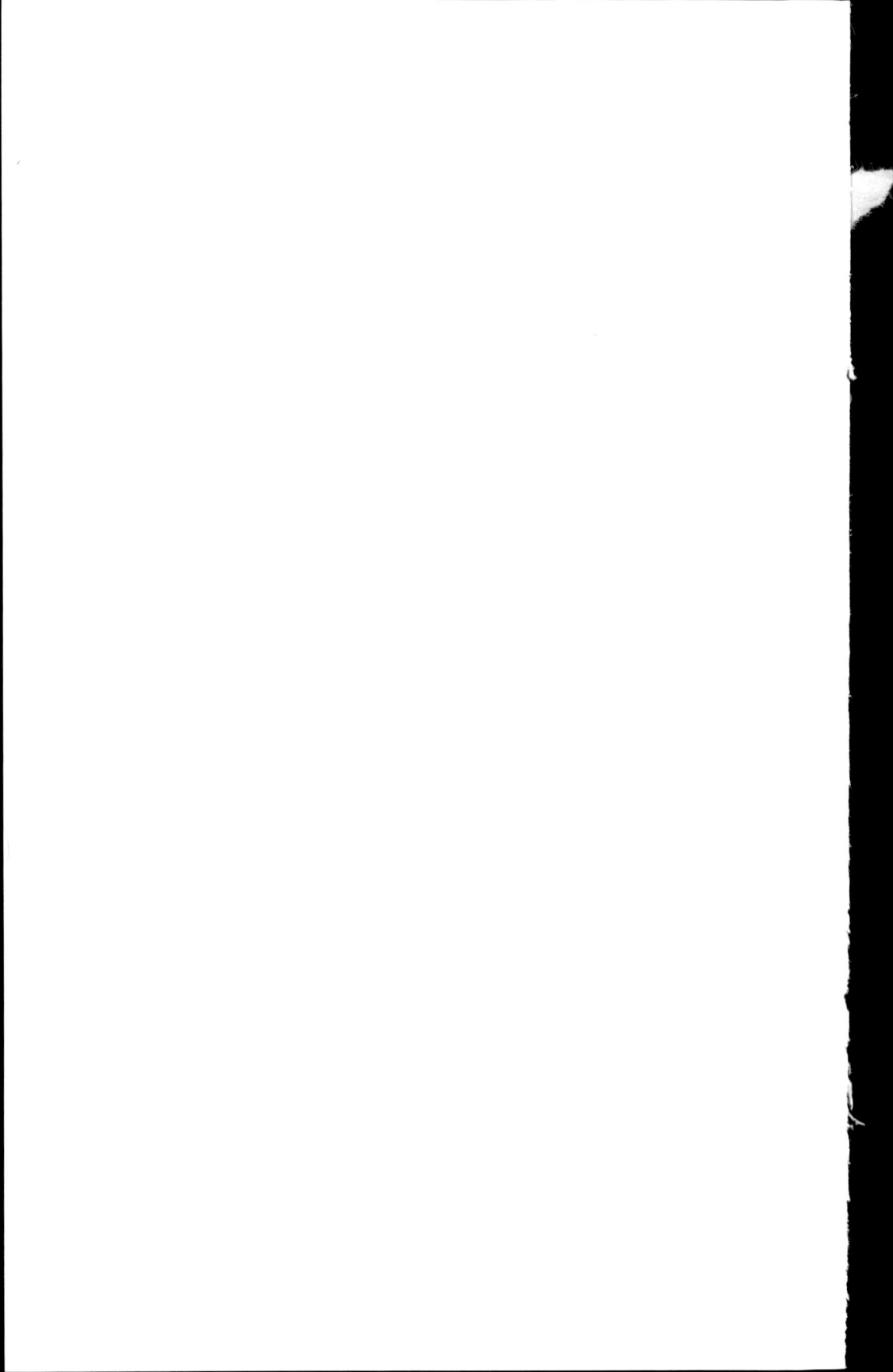